Foreword

Tribology is the science that deals with the design, friction, wear and lubrication of interacting surfaces in relative motion (as in bearings or gears). Composite materials, one of the most rapidly growing classes of materials, are being used increasingly for such tribological applications. Yet, by now, much of the knowledge on the tribological behaviour of composite materials is empirical and very limited predictive capability currently exists. Nevertheless, it has been attempted in recent years to determine to what degree phenomena governing the tribological performance of composites can be generalized and to consolidate interdisciplinary information for polymer-, metal- and ceramic matrix composites [1]. The importance of promoting better knowledge in the mundane areas of friction, lubrication and wear, in general, has been further demonstrated by quite a number of important articles and books published on these topics in recent years (see, e.g., refs. [2–8]). Along with a comprehensive "Encyclopedia of Composite Materials" [9] and a variety of other tribology handbooks cited in a previous volume on "Friction and Wear of Polymer Composites" [10], these references should build up a powerful basis for finally understanding the complexity of composites' tribology in the same way as it is the case for isotropic materials.

One step in this direction shall be made by this volume, which covers a wide range of subjects extending from fundamental research on the tribological characteristics of various multi-phase materials up to final applications of composites in wear loaded, technical components. An attempt was made to exemplify this in the design of the book cover. First and central to all efforts in tribology is basic research work. Therefore, the background of the cover shows the worn surface of a carbon fibre/glass matrix composite, with some evidence of fibre/matrix thinning and transfer film formation as a result of sliding wear contact with a metallic counterpart. The fundamental aspects are further illustrated in the lower left of the cover with a schematic drawing of the wear process which may occur if sharp counterpart asperities scratch on the surface of a composite material consisting of hard particles in a softer matrix. Such a situation may be encountered, e.g., in tribology of composites for magnetic recording systems. The figure in the upper right of the cover shows schematically the build up of an automotive brake system, indicating the ultimate technology application, e.g., of composites used as frictional materials.

Although the general topic of this volume is "advances in composite tribology", it was chosen to emphasize the "messier" materials aspects in preference to the more classical fluid dynamics and hydrodynamic lubrication. The rationale for this is that most authors are, after all, writing for a materials audience where there is not a widespread appreciation of the field of tribology. An additional objective was to emphasize, besides the science of composite tribology, the great practical aspect of the field in many industrial applications; therefore, authors were selected who are engaged in applied research as well as those in more academic activities.

After a preface by B.J. Briscoe, in which the fundamental principles governing the friction and wear mechanisms of composite materials are outlined, the book is structured into four main sections. The first one is dedicated to the tribology of polymer composites. It can be considered as an up-date and extension of the first volume in this Composite Materials Series [10]; the chapters included are dealing with new polymer composite developments for applications in which low friction and high wear resistance at different external loading situations are required. G.W. Ehrenstein and J. Song report on an interesting method of self-reinforcing various kinds of thermoplastic polymers in order to achieve excellent sliding wear resistance and low coefficients of friction at room temperature conditions and below. The performance of short-fiber reinforced, high-temperature resistant thermoplastics under reciprocating dry friction, as it occurs in compressors or beverage filling machines, is illustrated by H.H. Kausch and A. Schelling. From a more general point of view, A.M. Häger, and M. Davies discuss the problems (a) which one encounters in sliding wear testing of short-fibre reinforced polymer composites and (b) how to interpret the results with the help of new analytical methods. Further, a number of sliding wear data of internally lubricated, high-temperature resistant polymers against metallic counterparts, as generated in a wide temperature range (up to 220°C), are presented. A wide survey of the existing literature on friction and wear of discontinuous and continuous fibre reinforced composites with thermoplastic and thermosetting matrices is presented by U.S. Tewari and J. Bijwe. In conclusion of this section, the editor of this book, K. Friedrich, describes various models which have been developed by internationally known tribologists for predicting the frictional coefficient and wear rate of multi-phase materials under abrasive of sliding wear loading. In addition, experimental results and synergistic effects on the wear of polymeric hybrid composites are presented, which lead to a hypothetical model for hybrid composites with optimal wear resistance.

The second main section of this volume deals with the tribological aspects of ceramic-, glass- and metal matrix composites. K. Fukuda and M. Ueki illustrate the usefulness of SiC-whisker reinforced ceramic matrix composites for dry sliding applications against metallic counterparts (e.g. in relation to potential applications in the automotive sector). In the spirit of the now-classic treatment of mechanical data in the form of fracture or failure maps, P.K. Rohatgi, Y. Liu and S.C. Lim explored various domains of mechanisms for friction and wear of sliding couples made of metallic and ceramic matrix composites, thus producing another "map", namely wear-mechanisms maps. In a more practically related sense, i.e. with regard to the usefulness for various high-performance, energy efficient applications (engine components, drive shafts, or connecting rods) S. Wilson and A. Ball investigated the sliding, abrasive, cavitation and erosive wear performance of particulate reinforced aluminium matrix composites. A more detailed study on the subject of erosion of metal matrix composites by I.G. Greenfield contributes to a more fundamental understanding of the wear mechanisms involed under this type of wear loading. Two additional chapters in this section demonstrate further

ADVANCES IN
COMPOSITE TRIBOLOGY

COMPOSITE MATERIALS SERIES

Series Editor: R. Byron Pipes, Center for Composite Materials, University of Delaware, Newark, Delaware, USA

Cover illustration – Background photo: Surface of carbon fibre/glass matrix composite worn against smooth steel in antiparallel direction. Schematics: asperity scratch on the surface of a particulate composite (lower left), and automotive brake system (upper right).

Composite Materials Series, 8

ADVANCES IN COMPOSITE TRIBOLOGY

edited by

Klaus Friedrich

Universität Kaiserslautern, Kaiserslautern, Germany

ELSEVIER
Amsterdam—London—New York—Tokyo   1993

ELSEVIER SCIENCE PUBLISHERS B.V.
Sara Burgerhartstraat 25
P.O. Box 211, 1000 AE Amsterdam, The Netherlands

Library of Congress Cataloging-in-Publication Data

Advances in composite tribology/edited by Klaus Friedrich.
    p.   cm. -- (Composite materials series; 8)
   Includes bibliographical references and indexes.
   ISBN 0-444-89079-3 (alk. paper)
   1. Composite materials--Surfaces. 2. Tribology. I. Friedrich,
Klaus, 1945-. II. Series.
TA418.9.C6A292 1993
620. 1'1892--dc20

93-10612
CIP

ISBN 0-444-89079-3 (vol. 8)
ISBN 0-444-42525-X (Series)

friction and wear characteristics of advanced ceramic composites (B. Prakash) on the one hand, and of unidirectionally oriented carbon fibre reinforced glass matrix composites on the other (Z. Lu).

Although the previous chapters are somehow related to certain applications, it is mainly the performance of the materials under particular friction and wear conditions which is discussed. In the following section, chapters are found which especially focus on the tribological use of composite materials for special applications. H. Jacobi and U. Nowak give a comprehensive review of the tribological problems with particulate composites used for magnetic recording systems. G.W. Stachowiak has reviewed an enormous amount of data (from the literature and from own studies) on the friction and wear of polymers, ceramics and composites for biomedical applications. A special topic, namely the high-speed tribology of polymer composites, as it may be encountered in clutches and brakes, is described by I. Narisawa. In addition, G. Crosa and I.J.R. Baumvol demonstrate in their chapter how asbestos-free composites must be built up, in order to function as frictional materials, mainly for automotive brake applications.

The fourth main section of this volume relates to the problem how the friction and wear characteristics of polymeric composites can affect other requirements of major concern for their practical application. T.A. Stolarski discusses the subject of rolling contact fatigue of polymers and polymer composites. Another contact problem, fretting, very well-known for metals to reduce the fatigue life in a rather unpredictable manner, is discussed by K. Schulte, K. Friedrich and O. Jacobs in their chapter on fretting and fretting fatigue of advanced composite laminates. And finally, there also exist tribological facets of polymer composites fabrication, which will be further outlined by D.R. Williams.

When considering the content as a whole, it becomes clear that much research has been done and a lot of progress has been made towards the use of composite materials in a wide field of tribological applications. But still much more research remains to be done before the mechanical designer will have a set of predictive, materials engineering-related models for a more reliable use of composites as tribo-materials.

Nevertheless, a little step forward has been made with this book. Therefore, the editor is very thankful to the authors of the chapters. They helped to approach this final objective by their valuable contributions. Further thanks are due to Professor R.B. Pipes, editor-in-chief of this Elsevier Composite Materials Series, for the invitation to organize this particular volume.

August, 1992 K. Friedrich

References

[1] P.K. Rohatgi, P.J. Blau and C.S. Yust, eds, Tribology of Composite Materials (ASM International, Materials Park, OH, 1990).
[2] K.-H. Zum Gahr, Microstructure and Wear of Materials (Elsevier, Amsterdam, 1987).
[3] Y. Yamaguchi, Tribology of Plastic Materials (Elsevier, Amsterdam, 1990).
[4] Z. Rymuza, Tribology of Miniature Systems (Elsevier, Amsterdam, 1989).

[5] W.A. Glaeser, Materials for Tribology (Elsevier, Amsterdam, 1992).

[6] B. Bushan and B.K. Gupta, Handbook of Tribology, Materials, Coatings, and Surface Treatments (McGraw-Hill, New York, 1991).

[7] D. Dowson, C.M. Taylor and M. Godet, eds, Mechanics of Coatings, Tribology Series, Vol. 17 (Elsevier, Amsterdam, 1990).

[8] D. Dowson, C.M. Taylor, M. Godet and D. Berthe, eds, Tribology Design of Machine Elements, Tribology Series, Vol. 14 (Elsevier, Amsterdam, 1989).

[9] S.M. Lee, ed., International Encyclopedia of Composites, Vol. 1–6 (VCH Publishers, New York, 1990).

[10] K. Friedrich, ed., Friction and Wear of Polymer Composites, Composite Materials Series, Vol. 1 (Elsevier, Amsterdam, 1986).

CONTENTS

A more detailed contents list is given at the beginning of each chapter.

Chapter 6 (K. Friedrich)

Wear models for multiphase materials and synergistic effects in polymeric hybrid composites   209

PART III. TRIBOLOGY OF CERAMIC, GLASS AND METAL MATRIX COMPOSITES

Chapter 7 (K. Fukuda and M. Ueki)

Tribology of ceramic matrix composites against metals   277

Chapter 8 (P.K. Rohatgi, Y. Liu and S.C. Lim)

Wear mapping for metal and ceramic matrix composites   291

Chapter 9 (S. Wilson and A. Ball)

**Performance of metal matrix composites under various tribological conditions
311**

Chapter 10 (Z. Lu)

**Tribological properties of unidirectionally oriented carbon fibre reinforced glass
matrix composites   367**

Chapter 11 (B. Prakash)

Friction and wear characteristics of advanced ceramic composite materials   405

Chapter 12 (I.G. Greenfield)

Erosion of metal matrix composites   451

PART IV. TRIBOLOGY OF COMPOSITE MATERIALS FOR SPECIAL APPLICATIONS

Chapter 13 (H. Jacobi and U. Nowak)

Tribology of composites for magnetic tape recording   469

Chapter 14 (G.W. Stachowiak)

**Friction and wear of polymers, ceramics and composites in biomedical applications
509**

PART I

The Tribology of Composite Materials:
A Preface

Advances in Composite Tribology
edited by K. Friedrich

Chapter 1

The Tribology of Composite Materials: A Preface

B.J. BRISCOE

Department of Chemical Engineering, Imperial College, London, UK

Contents

Abstract

An effective and economical means of rationalising tribological behaviour is available through the application of the two-term non-interactive model. It defines two regimes, with different sizes, of friction and hence energy dissipation in contacts. By the same token it defines two regimes of damage and hence wear. The value, or not, of this approximation is considered in the context of composite systems. The important additional theorem is the incorporation and comparison of the relative scales of the phases, in the composite component, with the scales of the dimensions of the primary energy-dissipation zones initially introduced into the two-term non-interacting model. Sometimes the model is directly applicable and on other occasions, it is not. In the latter case special considerations are required and the model is not directly applicable. It is this circumstance which sets aside composite tribology from the tribology of nominally homogeneous systems.

1. Introduction

The word "composite" is coined from the Latin adjective "compositus" to mean a material "made up of various parts...". The definition naturally begs the question as to whether there is such a common thing as an homogeneous solid. The truth is that such animals are quite uncommon and anyway heterogeneities will be found at some length scale. This is a fine point; we all know what we mean by a "composite". It is a multiphase system where the components are identifiable and as such have quite distinct properties; normally we are preoccupied with the mechanical properties and, in this review, the tribological properties. The essence is naturally the idea that somehow we may provide a synergy of action with the components. However, the formulation of materials for a particular purpose is an inevitable compromise.

In the context of tribology, there is a useful theorem [1,2] which presupposes some knowledge of the length scale associated with the frictional dissipation process and the resulting damage and, ultimately, wear which accrues as a result. The point may be put differently; in a tribological context: "when is a composite a composite?". For example, if the frictional work is dissipated over a large volume, compared with the scale of the secondary-phase domain size, then a continuum argument is probably sufficient and principles developed for homogeneous phases will arguably suffice. If, however, the frictional work and damage are localised to volumes comparable with the domain sizes, the new ideas, which specifically address composite systems, must be introduced.

It is a shame to add a complication here but the potentially high anisotropic nature of certain composite systems, fibre-reinforced matrices for example, do require some modification to this general proposition. The aspect ratio of a fibre may exceed 10^3 units. Thus, in one orientation it may provide an "homogeneous" matrix, whilst in another not. Simply, in the former the fibres may be oriented in a plane parallel to the interface and in the latter they are normal to this plane. This idea is reconsidered later in the chapter.

These rather nebulous ideas are clarified by a consideration of our best ways of addressing the tribological characteristics of nominally homogeneous solids. The approach here has been to tacitly ascribe homogeneous mechanical or chemical properties to the energy dissipating components. It has, however, been often supposed that a component may evolve its character as a natural consequence of the sliding proces: dislocations may multiply, thermal transformations occur, strain softening or hardening could develop, oxidation may result and so on. These things apart, the essence of the established wisdom in tribology has been to adopt continuum principles; obviously this principle has an intrinsic potential weakness for composite systems, which is addressed again later. For the moment we may usefully review, in generic terms, the accepted ideas which have provided a useful framework to interpret and rationalise tribological observations for homogeneous bodies.

1.1. Friction dissipation and damage processes

Friction, or rather the magnitude of the frictional force, is just a description of the level of energy dissipated when contacting bodies move over each other; they may either roll or slide or have a combination of both. Friction models seek to do two things: prescribe the extent of the dissipation regime and the conditions which prevail there and also quantify the intrinsic energy-dissipation characteristics of this region. The latter is tantamount to predicting the magnitude of the frictional force. Those interested in wear would also wish to define and quantify the damage and, ultimately, material mass removed as a consequence of the frictional work.

These ideas embody the elementary principles incorporated within the "two-term non-interacting model of friction", popularised, in particular, by Phillip Bowden and David Tabor [3–6]. Careful consideration of the model highlights its weaknesses but also emphasises its merits. It is a useful basis for reviewing the domain-scale effects in composite tribology which are the obvious features which would distinguish the treatment of "composite tribology" from the description of the tribology of nominally homogeneous materials.

1.2. The two-term model

The original idea was that the frictional work dissipated may be separately and independently ascribed to two specific regimes: an interfacial zone and a subsurface domain. Figure 1 is a now familiar pictorial description of this proposition. The interface zone is usually considered to be quite narrow in its extent and arises because of asperity interactions, which induce adhesion and traction at discrete contact zones [7]. It is common now to attribute an interface shear stress to these junctions [8,9], although there is inevitably some debate as to the nature and extent of the primary energy-dissipation zone; the French school with Godet would detail this process in terms of "third body" formation and velocity accommodation locii [10]. Classically, this interface zone has been regarded as the "adhesion component of friction". The other component is usually referred to as the ploughing or deformation component, and includes subsurface dissipation processes, which encompass a much larger volume of material. The virtue of the two term model is that it provides a relatively simple means of aportioning the frictional work in the material and also offers an indication of the stresses, strains and temperatures which prevail in these regions. This vehicle then gives a means of prescribing the work done and the type of damage which may result as a consequence of the frictional work dissipated. It is primarily from this basis that wear models are developed and, in the present review, the tribology of composite systems may be interpreted and even optimised. Clearly, since a subjective length scale is adopted in this approach, the treatment of multiphase systems produces a special complica-

TWO TERM MODEL OF FRICTION
<hr />

FOR THE TWO CASES WHERE ONE
OF THE MATERIAL IS NOT RIGID

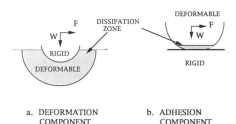

a. DEFORMATION   b. ADHESION
   COMPONENT   COMPONENT

Fig. 1. The pictorial description of the two term non interacting model of friction, indicating the classical means of undertaking experiments which emphasise each with respect to the other. The deformation component (a) involves significant surface deformations in volumes whose dimensions are comparable with the apparent contact dimensions or perhaps the asperity contact dimensions. In the adhesion component, the work is supposed to be dissipated initially at an interface zone, or at a region near an interface plane (b). It is assumed that the two components act independently, which is a gross, but nevertheless often effective, approximation.

tion. With this complication is, of course, the opportunity to formulate systems which optimise tribological responses.

1.3. Generic types of damage processes

In general, damage induced by friction will be of two extreme forms: chemical damage and mechanical damage. For most purposes it is sufficient to assume that chemical damage is entirely thermally activated, although there are suggestions that the prevailing strain may enhance such processes [11]; these processes would be classified as chemical wear or corrosive wear. Mechanical damage may be usefully divided into two extreme forms: a unit process, where material cohesion is lost by a single stress cycle, and a fatigue process, where repeated stress cycles are necessarily involved. Unit damage processes would occur in extreme forms of abrasion or in chip-forming machining processes. A fatigue process could involve crack initiation and propagation. In other cases, dislocation development or the evolution of an oriented surface structure may occur where the material progressively changes its properties as a result of the sliding process.

Finally, we may naively ascribe the cohesive failure processes as being intrinsically ductile or brittle fractures and comment that special chemical factors may be involved as a result of environmental stress induced cracking or environmental plasticisation. Additionally, there may be stress, but more likely thermally induced phase changes which are involved.

Figure 2 is a scheme of the types of damage processes envisaged in a very crude first approximation where no distinction is made between unit or fatigue processes.

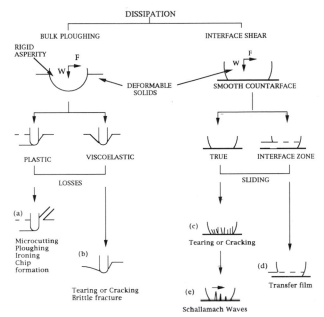

Fig. 2. An outline of nominal damage modes for "bulk" and interface deformations which supposes two broad regimes of material response: ductile and brittle. The bulk category is considered further in fig. 3. The interface mode considers two types of sliding mechanism: true interface sliding and interface zone sliding. True interface sliding (which could involve Schallamach waves; see ref. [1]) is an intrinsically brittle (or elastic) response. Damage involves tearing or cracking since the necessary interfacial intrinsic ductility is absent. Interface zone sliding (or shear) presumes that an interface zone acts as the primary region of mechanical energy dissipation. The shear plane (or zone), sometimes termed the velocity accommodation locus, is within the deformed body. This emphasises a ductile character and also the formation of transformed surface layers and transferred films. These are extreme distinctions.

Briefly, two dissipation regimes, termed "bulk" (ploughing) and "interface" (adhesion component), are shown. In the bulk zone, plastic, brittle and viscoelastic responses are shown. The former may produce topographical distortions or machined-out chips. In the viscoelastic regimes cracking due to surface traction and subsurface deformations are indicated.

Similar types of processes are indicated within the interface zone, but an important distinction is made which involves the dimension and localisation of the sliding plane. "True" sliding conveys the idea of an interfacial fracture located at the original interface. "Interface-zone" shear suggests a failure zone or plane within the softer (more ductile) contacting body.

1.4. Brief comments on wear processes

Wear arises as a consequence of the damage induced by the frictional work. In the primary dissipation zones mechanical disruption may take place, but ultimately virtually all of the frictional work will appear as heat. Thermally induced degradation (chemical wear) will result. Simple wear models will presuppose a mechanical damage mechanism, say unit fracture, and then correlate the observed wear with a separately established material property, say the fracture toughness. The Ratner–Lancaster correlation is an example of such a model for unit fracture damage processes [12]. There are several difficulties here. It will be assumed that the chosen intrinsic material property is not changed by the sliding process and also that the contact conditions, which define the stress levels, are not changed as well. Debris inclusion or counterface topographical modification will naturally result and change the contact conditions. This is inevitable. It is also not immediately obvious that there will be a direct relationship between the rate of damage and the rate of wear. Debris must be expelled from the contact to facilitate wear, and this process will be a function of the contact and sliding geometry; this is well-recognised in fretting wear [13].

2. Examples of tribological response for composite systems

In the remainder of this review I have selected examples which fit within the framework outlined above. The prime subdivision is the distinction between *interface* and *subsurface* friction damage and wear modes.

2.1. Subsurface modes

The point about length scales was made earlier. The extent of the subsurface deformation zone will be typically of the order of the contact dimensions; micro-contacts will normally occur with the apparent contact area, but it is the latter which is the controlling variable here. If the scale of the secondary phase is very much smaller than the characteristic contact length, then a continuum argument is likely to be sufficient. The composite will have a response which may be adequately described by a bulk property, and so no special ideas are needed. A good example would be an elastomer filled with fine particulate fillers, such as carbon black being deformed by a large, well-lubricated indentor. The friction of the elastomer will be proportional to $E^{-1/3} \tan \delta$ where E and $\tan \delta$ are the Young's modulus and loss properties of the composite [4,14]. If tearing of the elastomer occurs the bulk tear strength is apparently a good indicator of the damage sensitivity and, indeed, wear characteristics [15,16]. Actually, fracture toughness models seem more appropriate for interpreting abrasion by "coarse" counterfaces and fatigue models seem more viable for smoother substrates [17]. The latter is demonstrated by the effective performance of certain fatigue inhibitors in suppressing certain types of elastomeric wear [18]. The same types of approximation are also valuable for ductile flows. Figure 3 shows an example of the way in which frictional work may be aportioned within a contact [16]. The example is for a gamma-damaged, filled PTFE, over which is slid a sharp rigid cone of semi angle $\pi - \theta$. The various regimes are cited in fig. 2. Hysteresis loss, which involves no damage and also no permanent changes in surface topography, occurs in viscoelas-

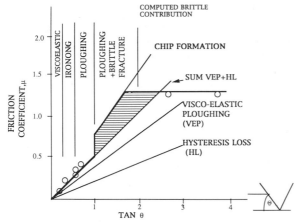

Fig. 3. A schematic interpretation of the various potential deformation contributions when a cone (whose semi-complement angle is θ) is slid over an unlubricated sample of a gamma-damaged PTFE under a constant load. The friction coefficients are indicated [12]. The regimes indicated are: (i) viscoelastic ploughing with nascent and impermanent groove formation, (ii) ironing; where modest plastic displacements are induced in a direction orthogonal to the sliding direction, (iii) ploughing; significant plastic flow of a more pronounced form occurs, including the generation of a prow ahead to the indentor, (iv) ploughing accompanied by brittle cracking; the cracks are of two types – in the plastically deformed groove and in regions adjacent to the groove. The cracks within the groove appear to be induced by surface tractions whilst those outside the groove result from the gross distortions of the material. In both cases the cracks are roughly transverse to the sliding direction, (v) chip forming; with sharp cones chip development occurs. The predictions of two friction models are shown: viscoelastic hysteresis loss (HL) and viscoelastic (plastic) ploughing (VEP). The sum HL + VEP is shown and accounts for the friction in the ironing and ploughing regions. More work is dissipated when brittle cracking occurs. The shaded region shows the friction contribution computed from the stick–slip component of the friction trace.

tic ploughing. Viscoelastic–plastic ploughing involves viscoelastic ploughing combined with irreversible plastic distortion. The terms "ironing" and "ploughing" denote the severity of the plastic deformation. The cracking or brittle failures may occur within the plastically deforming groove or in the material adjacent to the groove. The experimental points shown are estimates of the various components of the friction such as hysteresis loss (HL) and viscoelastic ploughing (VEP). The shaded region above the line shown as "sum VEP + HL" is an estimation of the cracking or tearing work done derived from the stick–slip nature of the friction trace.

The essential feature of this group is that bulk-response descriptions are sufficient but must naturally take into account the influence of the separate phases upon the dissipation and damage sensivity of the material. The scale of the deformation is large compared with the scale of the structures which exist within the solids. When this is not the case, special considerations are necessary, although it is not entirely clear how such types of systems could be treated without resort to numerical modelling.

As a general comment, in conclusion to this section, we may note that composite formulations will often not be greatly advantageous in this mode of energy dissipation. Perhaps, the inclusion of a second phase may broaden a viscoelastic peak in an elastomer and, hence, prevent pronounced viscoelastic subsurface heating in a particular range of contact frequencies [19]. However, additonal loss mechanisms, and indeed damage mechanics, may be introduced. The interface zones between the various phases will provide intrinsic sources of weakness and, hence, new modes of damage and wear [20].

The secondary phases will, however, often provide two rather important secondary attributes for the contact: an increase in creep resistance and, hence, load-bearing capacity and also, for certain thermally sensitive matrices, an improvement in thermal conductivity. The latter, for the case of organic matrices, will often be of major advantage since here the ability of the contact to dissipate heat is a major practical limitation. Other factors, such as the prospect of a reduction in thermal expansion, may also be important for systems, such as journal bearings, where the operating clearances must be maintained [21]. The dissipation of static electricity is commonly achieved by inclusion of electrically conducting secondary phases.

Thus, in final conclusion, we may say that composites may not offer a special tribological synergy for the tribological response in the bulk deformation modes. Nevertheless, the composite may provide a number of special secondary attributes for the supporting structure.

2.2. Interface modes

Whilst the composite may not offer remarkable synergistic effects in subsurface modes of friction, damage and wear, this is certainly not the case for the interface modes. It is here that the true value of multiphase systems emerges. It is convenient to invoke two intrinsic types of composite character; see fig. 4. In some instances we consider a "soft" phase dispersed in a "hard" phase (fig. 4a) and the converse (fig. 4b). Examples of both are encountered in practice and have been studied at some length in fundamental investigations.

2.2.1. "Soft" phases disposed in "hard" phases
There are two extremes of behaviour [2,22]. The "soft" phase may be essentially immobile in the hard phase (PTFE or high molecular weight silicone fluids, graphite, MoS_2, carbon fibres, in high-temperature thermoplastics and thermosetting resins, inorganic glasses in ceramics, lead in "white metal"). Alternatively, the secondary phase may be able to migrate within the host matrix and appear at the sliding contact (amides in low-density polyethylene, low molecular weight silicone fluids in polystyrene).

The general idea is that the soft phase will segregate at the sliding interface to provide a self-regenerating "lubricant layer" [22]. The mobile additives migrate and the immobile ones are produced as a consequence of the wear of the host matrix [23]. The composite, whose host matrix is essentially non-self-lubricating,

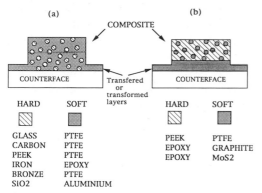

Fig. 4. Two extreme categories of composite formulations, invoking the idea of "soft" (self-lubricating) and "hard" (high-friction) components. A transformed and transfer layer is indicated, and some practical examples are listed.

becomes self-lubricating by the generation of a self-lubricating interface film which is variously termed a transformed layer, a third body or a segregated layer. Often, a layer of similar composition is produced upon the counterface.

Much has been written about the requirements and characteristics of effective transfer layers [24]. The belief is that they should be very thin; thicker films may provide unacceptable thermal barriers or be mechanically unstable. They should "smooth out" surface roguosities and adhere strongly to the counterface; if the films detach, transfer wear processes may become a dominant wear mechanism as the depleted film is repeatedly replenished. The literature contains much discussion on the ways in which the adhesion between transfer films and counterfaces may be improved [21,25,26].

The basic requirement in this type of process is to produce a stable interface layer by the action of mechanical work, within which the frictional work may be dissipated with a minimum of subsurface damage. The secondary phase plays a critical role in the provision of this new phase.

There is some reference in the literature to the environmental sensitivity of systems which operate by this mechanism [27]. The classical example, although not entirely appropriate in this context, is the case of the lead oxide–copper oxide-PTFE system (DU, Glacier Metals), where modest increases in ambient water concentration were found to produce prolific increases in wear [28]. The consensus is that water somehow induces the abhesion of the transfer film and, as a result, accelerates the wear process. Stable, strongly adherent and coherent transfer films seem to produce low friction and low wear. This is a case of the extreme adverse influence of the environment upon the behaviour of a sliding contact. There are cases however, where the environment, fortuitously or deliberately provided, may improve the behaviour of sliding systems. There are numerous examples observed in all facets of material selection. For organic polymers the most often cited example is that of nylon 6; it is often referred to as the Rubenstein effect [29]. In this case ambient water plasticises the nylon surface producing an effective boundary lubricant or, more accurately, modified surface layer. For metals, oxide layers produce comparable effects [30]. With alumina, surface hydration may create beneficial lubricating gels [31]. Clearly, there are prospects to provide environmentally induced beneficial surface layers by the inclusion of appropriate phases within composite systems.

The case of carbon, or graphite, filled polymer is a good example. The lubricating action of the carbon depends upon the availability of water and perhaps oxygen [32]. If these ingredients are removed from the ambient the quality of the transferred lubricant layer is undermined. Metals scuff in a gross way in the absence of oxygen [33]. Many ceramic contacts benefit from the presence of relatively high ambient water contents [31].

2.2.2. "Hard" phases in "soft" matrices

The basic premise is that the "soft" matrix is itself an intrinsically good lubricant but does not have the additional merits of dimensional stability. The most frequently encountered case is that of PTFE reinforced with hard phases such as glass fibres or metallic particles [34,35]. The "hard" phase will provide dimensional stability for the whole assembly, but also for the contact. The idea is that the hard phase "rides upon", and hence supports, the load on the soft phase, which acts as its own lubricant. The hard phase may also optimise the surface chemistry (by "cleaning") and topography (by polishing) of the counterface.

2.2.3. Orientation and anisotropy

A feature of many commercial composites is their highly anisotropic character. The secondary phase may have one length scale comparable in size with that of the primary interface dissipation zone, but another which corresponds, in the extreme, to the dimensions of the body. For a long and oriented fibre-reinforced system this will be seen where the fibres are oriented either in the plane of the interface or are normal to it, respectively. Clearly such circumstances will undermine the value of the length-scale theorem introduced earlier. The implications are not entirely obvious but will certainly depend upon the system in question. For "soft" oriented phases in "hard" matrices it may be of no consequence. The soft phase, say PTFE or aramid fibres in a phenolic resin, will be transferred and redistributed in a way which is largely unaffected by its orientations [36].

The case of highly oriented hard phases in "soft" matrices is of more interest since the integrity of the structure is more likely to be maintained. Glass and perhaps carbon fibre systems, in, say, a resin matrix, are obvious examples which have been studied. Where the fibres are oriented normal to the interface there would be an attractive prospect of an enhanced load-bearing capacity, but at the same time a likelihood of propagating fibre–matrix interface rupture deep into the structure of the body [2]. Where the fibre orientation lies in the plane of contact, damage propagation would be minimised, but its action will be localised near the matrix–fibre interfaces. Delamination processes will be encouraged [1,37].

3. Final remarks

The essential idea is that energy is dissipated in sliding contacts in two rather different interface domains: a large subsurface domain of the order of the contact dimensions and a very thin interface zone of the order of asperity dimensions.

The composite systems will, or should, provide a viable support structure to support the load and dissipate the frictional heat. When the mechanical energy is primarily dissipated in the subsurface, composite systems offer no very special advantage. Indeed, the multiplicity of their interfaces offers additional sources of weakness and has extra prospects for matrix damage and wear. The composite entity does, however, provide mechanical stability and also the attractions of means of dissipating the frictional work.

When the friction is dissipated near the interface the composite systems have many potential advantages. They may, by their own action, produce transformed counterfaces or interfaces which are intrinsically self lubricating. The intervention of the environment may facilitate this transformation. The subjectivity of these statements is entirely within the ethic of the the current wisdom of our understanding of friction, damage and wear processes for homogeneous systems. The difficulty in the interpretation of the behaviour of composite systems is compounded by several factors not specifically addressed in the treatment of homogeneous systems. To mention some: the scale length of the deformation zone as compared with the scale of the dispersion; the environmental and mechanical transformation of the secondary phase; and the perceived properties of the transformed interface.

4. A conclusion

By admission all systems have composite properties at some length scale; this fact is often forgotten. The study and examination of the tribological behaviour of fabricated composite systems focusses upon the intrinsic limitations of the basic understanding of the processes involved in tribological behaviour generally. Through the study of, and observation of, the tribology of sensibly formulated composite systems may emerge a clear and more profound understanding of the subject of tribology. In this sense, this book offers a major and critical evaluation of the state of understanding of the principles of tribology and its ability to serve the practical and commercial needs of this technology generally, and particularly in the context of composite systems.

References

[1] K. Friedrich, ed., Friction and Wear of Polymer Composites (Elsevier, Amsterdam, 1986). See in particular ch. 8.
[2] B.J. Briscoe and P.J. Tweedale, Proc. Am. Soc. Mater. Conf. on Tribology of Composite Materials, Oak Ridge, TN, May 1990, eds P.K. Rohatgi, P.J. Blau and C.S. Yust (A.S.M. International, Ohio, 1990).
[3] F.P. Bowden and D. Tabor, Friction and Lubrication of Solids, Vol. 1 (Clarendon Press, Oxford, 1950).
   F.P. Bowden and D. Tabor, Friction and Lubrication of Solids, Vol. 2 (Clarendon Press, Oxford, 1964).

[4] B.J. Briscoe and D. Tabor, in: Polymer Surfaces, eds D. Clark and J. Feast (Wiley, Chichester, 1978).

[5] H. Czichos, in: Friction and Wear of Polymer Composites, ed. K. Friedrich (Elsevier, Amsterdam, 1986) ch. 1.

[6] K.L. Johnson, in: Friction and Traction, eds D. Dowson, C.M. Taylor, M. Godet and D. Berthe (Westbury House, IPC Press, Guildford, 1981).

[7] B.J. Briscoe, in: Friction and Wear of Polymer Composites, ed. K. Friedrich (Elsevier, Amsterdam, 1986) ch. 2.

[8] I. Singer and H. Pollock, eds, Fundamentals of Friction: Macroscopic and Microscopic Processes, Proc. NATO Symposium on Friction (Kluwer Academic Publishers, Dordrecht, 1992).

[9] B.J. Briscoe and D.C.B. Evans, Proc. R. Soc. London A 380 (1982) 389.

[10] M. Godet, Wear 100 (1984) 437–452.

[11] G. Heinicke, ed., Tribochemistry (Carl Hansen Verlag, Berlin, 1984).

[12] B.J. Briscoe and P.D. Evans, Wear 133 (1989) 47.

[13] K.-H. Zum Gahr, Microstructure and Wear of Materials (Elsevier, Amsterdam, 1987).
L. Vincent, Y. Berthier, M.C. Dubourg and M. Godet, Wear 153 (1992) 135.

[14] D. Tabor, Proc. R. Soc. London A 229 (1955) 198.
K.C. Ludema and D. Tabor, Wear 9 (1966) 329.

[15] J.A. Greenwood and D. Tabor, Proc. Phys. Soc. 71 (1958) 989.

[16] B.J. Briscoe, P.D. Evans and J.K. Lancaster, J. Phys. D 20 (1987) 346.

[17] Y. Uchiyama, in: Proc. J.S.L.E. Int. Conf., Tokyo (Elsevier, Amsterdam, 1985) p. 48-4.

[18] G.M. Bartenev and V.V. Laurentev, Friction and Wear of Polymers (Elsevier, Amsterdam, 1981).

[19] G.L. Wannop and J.F. Archard, Proc. Inst. Mech. Eng. London 187 (1973) 615.
B.J. Briscoe, in: Friction and Wear of Polymer Composites, ed. K. Friedrich (Elsevier, Amsterdam, 1986) ch. 2.

[20] H. Ishida, ed., Controlled Interphases in Composite Materials (Elsevier, Amsterdam, 1990).

[21] Anonymous, ESDU 87007 Update of 76029, NCT (Risley, Washington, 1987).

[22] B.J. Briscoe and P.J. Tweedale, in: Proc. 6th Int. Conf. on Composite Materials and 2nd European Conf. on Composite Materials, 5/282 Imperial College, London, eds F.L. Matthews, N. Buskell, J.M. Hodgkinson and J. Morton (Elsevier, London, 1987) p. 187.

[23] B.J. Briscoe, V. Mustafaev and D. Tabor, Wear 19 (1972) 399.
B.J. Briscoe, M. Steward and A. Groszek, Wear 42 (1977) 99.

[24] B.J. Briscoe, Acta Metall. 24(5) (1990) 839. See also various chapters in ref. [1].

[25] N.P. Suh and N. Saka, eds, Fundamentals of Tribology (MIT Press, Cambridge, MA, 1980) several chapters.

[26] D. Dowson, C.M. Taylor, M. Godet and D. Berthe, eds, Interface Dynamics (Elsevier, Amsterdam, 1988) various chapters.

[27] J.K. Lancaster, ed., Speciality Bearings, Special Issue, Tribol. Int. 15(5) (1982).
J.K. Lancaster, in: Tribology in the 80's, NASA Conference Publication 2300 (NASA, Cleveland, OH, 1984).
J.K. Lancaster, Tribol. Int. 23 (1990) 371.

[28] G.C. Pratt, Trans. J. Plastics Inst. 32 (1964) 255.

[29] C. Rubenstein, J. Appl. Phys. 32(8) (1961) 1445.
B.J. Briscoe, B.H. Stuart, S. Sebastian and P.J. Tweedale, Wear (1993) , San Francisco.

[30] T.F.J. Quinn, Wear 153 (1992) 179.
W.A. Glaeser, ed., Materials for Tribology (Elsevier, Amsterdam, 1992) various parts.

[31] M.G. Gee, Wear 153 (1992) 201 and references therein.

[32] Z. Eliezer, in: Friction and Wear of Polymer Composites, ed. K. Friedrich (Elsevier, Amsterdam, 1986) ch. 6.
J.K. Lancaster, in: Wear of Non-Metallic Materials, eds D. Dowson, C.M. Taylor and M. Godet (Mechanical Engineering Publications Ltd., 1978).

[33] D.H. Buckley, Surface Effects in Adhesion, Friction, Wear and Lubrication (Elsevier, Amsterdam, 1981).
N. Gane, P. Pfeltzer and D. Tabor, Proc. R. Soc. London A 340 (1974) 495.

[34] T.A. Blanchet and F.E. Kennedy, Wear 153 (1992) 229.

[35] B.J. Briscoe and P.J. Tweedale, Inst. Phys. Conf. Ser. No. III, New Materials Applications (I.O.P. Publications, London, 1990).

[36] E. Hornbogen, in: Friction and Wear of Polymer Composites, ed. K. Friedrich (Elsevier, Amsterdam, 1986) ch. 3.

T. Tsukizoe and N. Ohmae, in: Friction and Wear of Polymer Composites, ed. K. Friedrich (Elsevier, Amsterdam, 1986) ch. 7.

PART II

Tribology of Polymer Composites

Advances in Composite Tribology
edited by K. Friedrich

Chapter 2

Friction and Wear of Self-Reinforced Thermoplastics

J. SONG and G.W. EHRENSTEIN

Lehrstuhl für Kunststofftechnik, Universität Erlangen-Nürnberg, W-8520 Erlangen, Germany

Contents

Abstract

Semicrystalline thermoplastics can be self-reinforced by oriented crystallisation, whereby improved tribological and mechanical properties are achieved. The effects of self-reinforcement on the tribological properties of polymeric materials are discussed with particular emphasis on influences of tribological systems and operating parameters. Self-reinforced polymers are interesting subjects for some fundamental experiments on tribology. The property of self-reinforcement, reinforcing materials without changing their adhesive property, enables the investigation of the correlation between the mechanical and tribological properties. In addition, the effects of self-reinforcement on the tribological properties are compared with those of fillers. Finally, with help of SEM graphs of worn surfaces, the change of wear mechanisms caused by the self-reinforcement and its influence on the wear resistance of polymeric materials are studied. The discussion of the effects of time–temperature equivalence gives some new aspects of wear calculation. Based on the results of the investigation some suggestions are made for the development of high-performance materials for tribological application.

1. Introduction

The tribological properties of polymeric materials can be largely varied by modification. Polymeric materials can be modified basically by means of
– variation of chemical structures,
– addition of fillers,
– variation of crystalline structures and their orientation.
Since there is no additive needed for the third possibility, it is generally called self-reinforcement.

The objective of this chapter is to discuss the influences of self-reinforcement on the tribological properties of polymeric materials. The basic idea of improving tribological properties by self-reinforcement is based on the adhesion, and deformation theories of tribology and the recent achievements of self-reinforcement of polymeric materials. One of the most investigated polymeric materials for self-reinforcement is PE-HD. It shows a weak adhesion, which leads to low friction. On the other hand, its wear resistance is very low because of its low strength and modulus [1].

The mechanical properties of PE-HD can be largely improved by flow-induced crystallisation, which merely influences the adhesion property [2]. This method of improving the tribological properties has apparently the following advantages:

There is no filler used for self-reinforcement. The addition of fillers influences several properties of polymeric materials coincidentally, which may lead to some disadvantages for the tribological properties. Hard fillers can increase the abrasiveness of the composite materials [3] and polar fillers the adhesion, which increases the friction [4].

Beside the potential of improving the tribological properties, self-reinforcement enables some interesting fundamental tribological experiments. The influence of the mechanical properties, e.g., can be directly studied by varying the mechanical properties with different self-reinforcement without changing the adhesion of the materials, which is hardly avoidable when using fillers. The tribological tests were carried out on pin-on-disk and journal bearing test systems with varying test parameters in order to show the influences of the tribological parameters and systems on the tribological properties. The investigation was carried out with PE-HD, PP and LCP. PA 66 and several thermoplastic-based composites were used for the comparison purpose.

2. Theory

2.1. Friction

Friction between sliding surfaces results in complex molecular-mechanical inter-actions. These interactions are the deformation of the roughness, ploughing by debris and roughnesses of the harder counterpart and the adhesion between the contact surfaces. Since ploughing is a kind of deformation, the friction force can be devided into a deformative component and an adhesive component [1,5–7]:

$$F_f = F_a + F_d,\tag{1}$$

where F_f is the friction force, F_a is the adhesive friction force and F_d is the deformative friction force.

The friction coefficient can be expressed analogously by:

$$f = f_a + f_d.\tag{2}$$

The value of each component depends on the conditions of the sliding surfaces, which can be influenced by the characteristics of the involved materials, the topography of the surfaces and the environmental conditions [1,7,8]. These can be summarised as three basic aspects which influence the dry friction process (friction without liquid lubricants) [6]:
– the real contact surfaces,
– the shear strength of the contacts,
– the way in which the material is sheared and fractured in and around the contact zone in the sliding process.

2.2. The adhesive friction

Examining the separating process of the contacts of the sliding partner, we get the following equation for the adhesive friction force F_a [1,9]:

$$F_a = \tau_s A_r.\tag{3}$$

The shear strength of the contacts, τ_s, depends on the adhesive properties of the contacting surfaces. A_r is the real contact surface. Analogously, there is the following expression for the adhesive friction coefficient:

$$f_a = \frac{\tau_s A_r}{F_N}.\tag{4}$$

The determining factors for the adhesive friction coefficient are the adhesive interaction and the real contact surface.

2.2.1. Adhesive interaction

Adhesion is defined as the intermolecular interaction between solid-state bodies. In order to separate contacting surfaces, the least energy W_{ab} needed is

$$W_{ab} = \gamma_a + \gamma_b - \gamma_{ab},\tag{5}$$

where γ_a and γ_b are the surface energies of the separated surfaces and γ_{ab} is the boundary surface energy of the two contacting surfaces.

2.2.2. The real contact area

The acting radius of the different adhesive forces is small [10]. Therefore the effective zone of adhesion includes only the real contact areas. The real contact area is the result of penetration of roughness under normal load. For an elastic contact the real contact area, A_r, is approximately proportional to the normal force, F_N, and inversely proportional to Young's modulus E:

$$A_r = c(E/F_N)^n, \tag{6}$$

where $n < 1$ [10].

The relation can be physically more precisely expressed by the following equation:

$$\Phi = \frac{A_r}{A_0} = 1 - \exp\left(-\beta\frac{p}{E}\right), \tag{7}$$

where Φ is the relative contact area, which is defined as the quotient of the real contact area and the nominal geometric contact area, β is the roughness coefficient and p is the normal pressure. The above equation shows that the influence of Young's modulus and the normal pressure is very small for a large relative contact area. The equation for the adhesive component of the friction coefficient can be then derived from eqs. (4) and (6):

$$f_a = \frac{\tau_s}{p} \cdot \left[1 - \exp\left(-\beta\frac{p}{E}\right)\right]. \tag{8}$$

The adhesive component of the friction coefficient can be reduced by increasing Young's modulus and by decreasing the shear strength of the contacts.

2.3. The surface topography and the depth of penetration

The depth of penetration is an important influencing factor for both the adhesive and the deformative component of the friction and for the wear.

2.3.1. The surface topography

The surface topography can be characterised with different methods [1,10]. One of these methods is the so-called profile diagram, which is especially suitable to show the relation between the surface topography, the depth of penetration and the real contact area (fig. 1). The contact length can be considered as a simplified, one-dimensional description of the real contact area for a given depth of penetration. The real contact area increases obviously with increasing depth of penetration. Since the deformative resistance in the tangential direction is proportional to the depth of penetration and the adhesive resistance is proportional to the real contact area, the correlation between the adhesive and deformative components of friction is visible. In the engineering praxis instead of h_{max} the average roughness

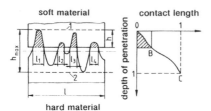

Fig. 1. Profile diagram (left), 1: top line of the profile; 2: bottom line of the profile. Contact-length diagram (right).

R_a and the average peak-to-valley height R_z are often used to characterise the surface roughness.

2.3.2. The depth of penetration

For elastic deformation there is the following relation between material property, normal load and depth of penetration [11]:

$$h = c h_{max}^{n_1} \left(\frac{p}{E} \right)^{n_2},$$

(9)

where h is the depth of penetration, h_{max} is the maximal roughness, c is the proportionality factor, p is the normal pressure, E is Young's modulus and n_1, n_2 are exponential factors, $0 < n_1$, $n_2 < 1$.

The equation for plastic deformation is [11]:

$$h = c h_{max} \left(\frac{p}{\sigma_{dF}} \right)^{n_2},$$

(10)

where σ_{dF} is the compression yield point.

The materials properties which influence the depth of penetration are Young's modulus, the compression yield point or the hardness. The maximal roughness of the harder counterpart has a stronger influence on the depth of penetration under a higher normal pressure than under a lower normal pressure.

2.4. The deformative component of friction

The above discussion leads to another important aspect of friction, namely the deformative component of friction, which results in the form-locking characteristic in the tangential direction. This resistance can be only overcome by deformation or fracture of the material. Considering the fact that the deformation resistance is proportional to the depth of penetration, eq. (9), and that the friction represents a kind of dissipation of mechanical energy which is proportional to the mechanical loss factor, tan δ, the following relationship is suggested for the deformation component of fricion [10,12,13]:

$$f_d = \xi \left(\frac{p}{E} \right) \tan \delta,$$

(11)

where ξ depends on the characteristics of the surface.

2.5. Wear

In this chapter wear of materials is quantitatively expressed by the wear factor k, which is defined as

$$k = \frac{w_v}{sF_N}, \tag{12}$$

where w_v is the volume of the worn material, s is the sliding distance and F_N is the normal load. Measurement using the weight of the worn material is not very suitable for polymeric materials, since the worn polymeric material is not always separated from the specimens. This is especially true for tough polymeric materials. Therefore we prefer the measurement using loss volume to that using the weight.

By the nature of the wear mechanisms, polymer wear can be tentatively divided into abrasive and fatigue. Adhesive wear is only important for the first contact of two virgin surfaces. By the fatigue wear multiple deformations of the polymer occur during the friction process at separate points of real contact, and this leads to fracture and subsequent tearing of the material [6].

The fatigue wear resistance depends on the fatigue strength of a material and the magnitude and the frequency of deformation, depending upon the sliding velocity, pressure and temperature.

An increase in roughness of the hard surface during the wear process leads to an increase in local stress. When the local stress exceeds a certain value the wear will change to abrasive wear, whereby the tearing of material occurs in a single traveling of the hard surface over the soft counterpart. Rigid materials wear easier abrasively than tough materials [6,10].

From the nature of the wear mechanisms it can be seen that materials wear faster by an abrasive mechanism than by a fatigue mechanism. The following criterion for the transition from fatigue wear to abrasive wear is suggested in refs. [2,10]:

$$\frac{h_{max}}{R} p \geqslant \tfrac{1}{2}(\sigma_0 - 2\tau_s), \tag{13}$$

where h_{max} is the maximal roughness, R is the radius of curvature of the asperities, p is the normal pressure, σ_0 is the strength of the material and τ_s is the shear strength of the material.

The tribological parameters on the left side are usually given conditions. The material properties on the right side are the main concern of a materials scientist. In order to avoid abrasive wear, the detrimental type of wear, a high strength of the material and a weak adhesion between the contact areas are preferable.

Wear particles are found only in the real contact area, which is inversely proportional to the hardness of the material for plastic contact and Young's modulus of the material for elastic contact. The resisting force of the motion is the friction force fF_N and the separating work is the product of the strength σ_0 and the rupture strain or elongation yield point for materials with a yield point, ε. According to this consideration, wear can be expressed by the following relationship for plastic contact [5,6]:

$$w_v = c \frac{fF_N}{H\sigma_0\varepsilon},$$ (14)

and for elastic contacts:

$$w_v = c \frac{fF_N}{E\sigma_0\varepsilon}.$$ (15)

The wear decreases with increasing Young's modulus and strength.

In the wear calculation the tribological and environmental parameters have to be considered. In refs. [14–17] this is expressed by

$$w = w_p w_h w_T,$$ (16)

where w is the total wear, w_p is the pressure dependent part, w_h is the surface roughness dependent part and w_T is the temperature dependent part of wear.

2.6. Effects of self-reinforcement and fillers on the tribological properties

Improved tribological properties of polymeric materials are mostly achieved with the help of reinforcing and/or lubricating fillers [3,18]. Graphite carbon or other modifications of carbon, molybdenum disulphide (MoS_2), PTFE, PE and synthetic oils are often used as lubricating fillers which eliminate or reduce the direct adhesive interaction. The lubricating effect of polymeric fillers is partly achieved by a decrease in surface energy. There is the following expression for the surface energy of a copolymer [19]:

$$\gamma = y_a\gamma_a + y_b\gamma_b,$$ (17)

where γ is the surface energy and y is the mole fraction. A degressive relation between the fraction of fillers and their contribution to the total surface energy was observed for composite materials. A possible accompanying effect of the lubricating fillers is that they lead to a weak bond in the material, which can reduce the strength of the material.

Glass fibres, carbon fibres and aramide fibres, sometimes glass spheres, metal pulver and mineral pulver are used as reinforcing fillers to increase the strength and the wear resistance of polymeric materials. Diverse variants of reinforced polymeric materials, which are made using various combinations of fillers, different shapes of fillers and different processing processes, are compared in refs. [20–24].

A possible accompanying effect of the reinforcing fillers is, especially if they are polar, that the friction coefficient increases, because of the high surface energy and the high shear strength [2,4]. In addition, the abrasiveness, namely the wear of the counterpart, can be increased by wear resistant fillers [3,4].

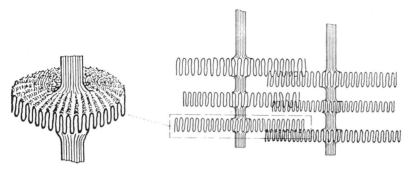

Fig. 2. Model of shish-kebab structures.

Another less known way to improve the tribological properties of polymeric materials is the self-reinforcement of polymeric materials. Self-reinforcement can be achieved in different ways [25]:
– self-reinforcement by means of oriented crystallisation,
– self-reinforcing liquid crystal polymers.

The self-reinforcement of semicrystalline polymeric materials with flexible chain molecules, such as PE-HD and PP, is based on flow-induced crystallisation, whereby so-called shish-kebab structures are induced [26]. Shish-kebab structures are highly oriented crystalline structures and are shown in fig. 2. It is well known that polymer molecules have covalent bonds with higher strength (bonding energy 200 to 400 kJ/mol) along the longitudinal axis and intermolecular bonds (mostly van der Waals bonds) with lower strength (bonding energy less than 1/50 of covalent bonds) in the perpendicular direction [20]. Therefore, improved mechanical properties of polymeric materials can be achieved by increasing the number of oriented chain molecules and assuring their orientation in the loading direction [27,28]. The results of investigations on PE-HD show that by varying the parameters of injection molding, such as the geometrical boundary conditions for melt flow, melt temperature, velocity of injection and holding pressure, it is possible to induce the highly oriented shish-kebab structures, which markedly improve the mechanical properties of semicrystalline thermoplastics. From analysis of the relation between tribological properties and mechanical properties we know that improved mechanical properties, especially strength and modulus, lead to improved tribological properties. This improvement is in addition to the improvement of other mechanical properties. Since the abrasiveness and the polarity of materials are not largely changed, the disadvantages of fillers can be avoided.

Thermotropic liquid-crystalline polyesters are a new class of polymeric materials, which have unique molecular and solid-state structures, flow characteristics and mechanical properties. Liquid-crystalline substances have been known for almost a century. The materials are characterised as having structures intermediate between isotropic fluids and three-dimensional ordered crystals. Due to molecular order in solution or in the melt, liquid-crystalline main-chain polymers can be highly chain-extended by proper processing, and as a result very high moduli can be obtained. Therefore, liquid-crystalline polymeric materials are also called self-reinforcing polymeric materials. Injection-moulded plaques were found to consist mainly of three, highly anisotropic, flow-induced macrolayers: two skins with a core in between. The skin macrolayer has a distinct structural gradient comprised of three subdivisions, from the surface inward: a highly oriented top layer, several oriented sublayers and a less oriented inner zone. The top layer is fibrillar in nature and the sublayers consist of stacks of interconnected microlayers. In the core, no well-defined substructure was observed, yet molecular orientation perpendicular to the injection direction represented the localised flow patterns [29].

2.7. Summary and valuation of theory

Friction and wear are very complex processes, in which the characteristics of the counterpart material play a very important role. The theoretical models above can only give a qualitative analysis, since not all factors are taken into account. Such as the topographic and structural changes of surfaces where the changes in material properties because of the high local temperature are often not avoidable. These processes are mathematically difficult to describe. In the above study, only the influences of the materials properties have been analysed, since this is the main concern of a materials scientist. In summary, the relationships between tribological properties and material properties show that the tribological properties can be improved by enhanced mechanical properties of polymers and reduced adhesion in the contact area. Nevertheless, as presumption of the theory is taken constant tribological conditions of the system. In case that the surface properties of the counterpart greatly change, there is often weak correlation between the tribological and mechanical properties. Nevertheless, some recommendations can be given on the basis of the theoretical analysis of tribology research to date. Below are general recommendations to guide concepts of improving the tribological properties of polymeric materials:
- the adhesion component of friction can be reduced by using polymeric materials and fillers with low surface energies and by lubricants,
- the adhesion component of friction can also be reduced by decreasing the real contact surface,
- a high stiffness or a high yield point of polymeric materials is preferable in order to reduce both the adhesion component and the deformation component of friction. This can be achieved by self-reinforcement or reinforcing fillers,
- a low friction coefficient, a high modulus and a high strength lead to low wear.
One of the most important advantages of improving the materials properties by means of fillers is the versatility, which enables to tailor a material to fit the needs. However, some secondary effects of fillers can be observed in composite materials. The mechanical properties can be greatly improved by using the technology of self-reinforcement, whereby the deformation component of friction decreases and the wear resistance increases without worsening the adhesion property.

3. Experimental materials

3.1. Self-reinforced specimens

3.1.1. Semicrystalline thermoplastic materials

A PE-HD was used for the fundamental investigation on effects of self-reinforcement on tribological properties. PE-HD is especially suitable for self-reinforcement with its C–C covalent bonds and its lanky and linear molecules [27].

In order to investigate the transferability of the technology of self-reinforcement a PP was tested as well. The basis of self-reinforcement is oriented crystallisation. The basic condition for oriented crystallisation is a high molecular orientation. It is shown that molecules can be oriented much more effectively by extensional flow than by shear flow. Extensional flow is an accelerated flow with a positive velocity gradient, which can be achieved by means of a nozzle geometry (fig. 3). The overflow at the end of the specimen allows a virtually steady flow in the test region. The mold for a journal bearing specimen is designed analogously (fig. 4). The properties of self-reinforced materials depend strongly on the processing parameters, these are shown in tables 1–3.

3.1.2. Liquid-crystalline polymers

In contrast with self-reinforcement by flow-induced crystallisation of semicrystalline thermoplastic materials the geometry dependent flow conditions do not influence the crystallisation of liquid-crystalline polymers, but only the orientation of the rigid segments [30].

The processing parameters and the mechanical properties of the tested liquid-crystalline polymers are shown in table 4.

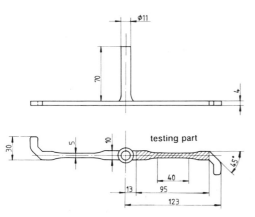

Fig. 3. Speciment geometry used for the mechanical and pin-on-disk test.

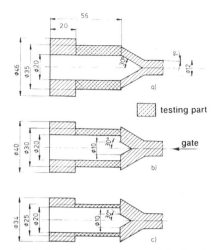

Fig. 4. Specimen geometry used for bearing test.

TABLE 1
Processing parameters and mechanical properties of the PE-HD specimens for the material tests.

	Normal	Self-reinforced				
		type 1	type 2	type 3	type 4	type 5
Processing parameters						
Mold temperature [°C]	30	45	35	45	55	65
Melt temperature [°C]	240	155	148	148	148	146
Holding pressure [bar]	40	1715	1715	1715	1715	1715
Injection rate [mm/ns]	7	180	180	180	180	180
Tensile properties						
Young's modulus [N/mm^2]	1100	5700	6000	7600	8100	8300
Strength [N/mm^2]	30	78	100	110	100	105
Compression properties						
Compression modulus [N/mm^2]	1200	–	3500	3100	2800	2700
Compression yield point [N/mm^2]	36	–	50	~ 50	~ 50	50

TABLE 2
Processing parameters of the PE-HD bearing.

	PE-HD	
	normal	self-reinforced
Mold temperature [°C]	40	70
Melt temperature [°C]	220	153
Holding pressure [bar]	600	1700
Injection rate [mm/s]	100	100

TABLE 3
Processing parameters of the PP specimens for the material tests.

	PP	
	normal	self-reinforced
Processing parameters		
Mold temperature [°C]	60	30
Melt temperature [°C]	260	182
Holding pressure [bar]	600	1715
Injection rate [mm/s]	80	180
Tensile properties		
Young's modulus [N/mm²]	1300	3400
Strength [N/mm²]	30	80
Compression properties		
Compression modulus [N/mm²]	1400	2040
Compression yield point [N/mm²]	35	65

3.1.3. *Thermoplastic-based composite materials*

For comparison purposes some thermoplastics and thermoplastic-based composite materials were tested. The components of the used materials and the mechanical properties are shown in tables 5 and 6.

4. Experimental

The theoretical models can only give a qualitative analysis, since not all factors are taken into account. Quantitatively the tribological performance can be deter-

TABLE 4
Processing parameters and mechanical properties of LCP.

LCP	Type 1 Ultrax KR 4002	Type 2 Ultrax KR 4003	Type 3 KU1-9221
Processing parameters			
Mold temperature [°C]	100	100	–
Melt temperature [°C]	320	320	–
Holding pressure [bar]	500	500	–
Injection rate [mm/s]	110	180	–
Tensile properties			
Young's modulus [N/mm²]	7500 ± 500 *	20000 ± 1400 *	20000
Strength [N/mm²]	90 ± 10 *	120 ± 16 *	200
Compression properties			
Compression modulus [N/mm²]		9700 ± 1400 *	
Compression yield point [N/mm²]		75 ± 4 *	

* Standard deviation.

TABLE 5
Fillers and mechanical properties of PA and PA-based composites.

Fillers * [wt%]	PA 66 (water uptake: 1.6–2.3%, moist conditioned)						
	mod. 1	mod. 2	mod. 3	mod. 4	mod. 5	mod. 6	un-mod.
CF	20	–	–	20	–	–	–
GF	–	30	–	–	30	–	–
AF	–	–	–	–	–	40	–
PTFE	–	15	18	5	15	–	–
SiO_2	–	2	2	–	–	–	–
MoS_2	–	2	2	–	–	–	–
Tensile properties							
Strength [N/mm^2]	190	130	60	170	140	170	85
Young's modulus [10^3 N/mm^2]	19.5	10	2.5	18.5	9.3	15	3
Compression properties							
Compression yield point [N/mm^2]	165	150	55	140	130	–	–
Compression modulus [10^3 N/mm^2]	5	4	1.6	4.2	3.3	–	–

* CF: carbon fibre, GF: glass fibre, AF: aramide fibre.

mined only by means of tribological tests. The tribological properties must be evaluated in connection with a tribological system. Two categories of test systems were used for the investigations:
– a model test system with a pin-on-disk apparatus,
– a simulation test system with a journal bearing apparatus.
    The model test system was used for the fundamental investigations whereby the effects of the self-reinforcement and the fillers on the tribological properties are compared, whereas the simulation test system was mainly used for the examination of the transferability of the results to the machine parts.

4.1. Pin-on-disk apparatus

    The pin-on-disk test system consisted of a horizontal rotating disk fabricated from a hard high-alloy steel. The disk is changeable, so that the surface roughness

TABLE 6
Fillers and mechanical properties of a PEEK and a PEEK-based composite.

	PEEK mod.	PEEK
CF [wt%]	10	–
PTFE [wt%]	15	–
Strength [N/mm^2]	130	100
Young's modulus [N/mm^2]	10 000	3600

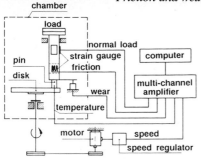

Fig. 5. Pin-on-disk apparatus.

of the counterpart can be varied by changing the disk. The sample was a pin of polymer, which was mounted on a sample holder as shown in fig. 5 and pressed against the rotating disk.

The relative sliding velocity between the pin and disk was smoothly variable over a wide range by adjusting the rotation speed of the disk. The friction force was calculated from the torque as measured by the strain gauge mounted in the vertical arm. The wear-induced position of the pin holder was measured by a displacement pick-up. The data obtained during the test was recorded and evaluated with a computer.

In order to investigate the influence of temperature on the tribological properties, the pin-on-disk apparatus is put in a temperature-regulated chamber which enables a temperature range of −40°C to 130°C.

The temperature of sliding surface was measured with a thermocouple embedded in the pin. The accuracy of the measurement depends on the depth of embedding. It was calculated that the temperature rise 0.5 mm above the sliding surface equals 90% of the temperature rise of the sliding surface for a pin with a diameter of 6 mm [2]. The temperature gradient in the contact zone and in the pin was measured with a thermoviewer. The thermograph (fig. 6) shows that the temperature within a thin layer of 1 mm in the contact zone does not vary much.

4.2. Journal bearing test apparatus

The journal bearing test apparatus has two test units which enables simultaneous testing of two journal bearings (fig. 7). The friction moment between the

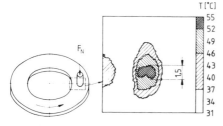

Fig. 6. Thermograph of the contact zone during a pin-on-disk test.

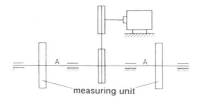

measuring unit

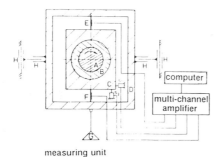

computer

multi-channel
amplifier

measuring unit

A: steel shaft, B: polymer bearing
C: bearing holder, D: frame
E: plate for friction measurement
F: plate for friction measurement and transfer of force
G: load, H: bearing
S_1, S_2: displacement pick up

Fig. 7. Journal bearing test apparatus.

rotating steel shaft (A) and the polymer bearing (B) is measured with the wire strain gauges on the plates (E) and (F). The wear of the bearing is measured with the displacement pick-ups (S_1 and S_2). Analogous to the pin-on-disk tests, the friction coefficient and the radial wear factor are used to describe the friction and wear properties. Figure 8 shows the shift of the bearing center due to wear.

Because of the shift of the bearing center the normal force F_Z is not the bearing load F_N, but its projection on the contact surface normal. Therefore, the following equation should be used to calculate the friction coefficient [31]:

$$f = \frac{F_f}{\sqrt{F_N^2 - F_f^2}},$$ (18)

where F_f is the friction force.

For the radial wear factor, k_R, the following relation is used [32]:

$$k_R = \frac{\Delta Z_R}{sp},$$ (19)

$$p = \frac{F_N}{bD},$$ (20)

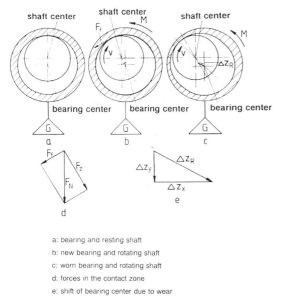

a: bearing and resting shaft
b: new bearing and rotating shaft
c: worn bearing and rotating shaft
d: forces in the contact zone
e: shift of bearing center due to wear

Fig. 8. Forces in the contact zone and shift of the bearing center due to wear.

where ΔZ_R is the shift of the bearing center due to wear, s the sliding distance, p the normal pressure, b the width of the bearing and D the inside diameter of the bearing. The outside diameter of the polymer bearing is first turned to fit into the bearing holder. The outside surface of the bearing is then grooved in the axial direction, in order to get the bearing well-glued to the holder. The grooves are about 0.2 mm deep and uniformly distributed over the surface.

The computer aided data acquisition system is similar to that of the pin-on-disk apparatus.

5. Results and discussion

5.1. Characterisation of specimens

The most common way to determine the degree of self-reinforcement is a mechanical tensile test. Nevertheless, it is not always possible to take a specimen from an injection molded part, such as a bearing, for the tensile tests. Therefore it is necessary to have other possibilities for the characterisation of specimens whereby only a small piece of specimen is needed.

Differential scanning calorimetry (DSC) and the measurement of retardation with a polarisation microscope were used to characterise the degree of self-reinforcement of semicrystalline thermoplastics.

Dynamic mechanical analysis was used to investigate the mechanical properties with changing temperature and the adhesion properties of polymeric materials were characterised by surface energy measurement.

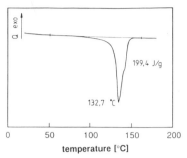

Fig 9. Melting peak in the DSC curve of a self-reinforced PE-HD specimen (type 1, see table 2), core area.

5.1.1. DSC-curves
5.1.1.1. PE-HD.
The DSC tests show that the DSC curve of a self-reinforced PE-HD fundamentally differs from that of a normal PE-HD (figs. 9 and 10). The following characteristics in the DSC curves were noticed:

The DSC curve of the self-reinforced PE-HD specimen has a shoulder above the melting peak temperature (about 133°C), which indicates a second melting peak at about 140°C (fig. 9). Figure 10 shows the DSC curve of a normal PE-HD specimen with only one melting peak (peak temperature: 132°C). A second melting peak at higher temperature is generally interpreted as an increase in the number of thermally stable crystalline structures.

The formation of thermally stable crystalline structures is not uniformly distributed over the cross section of the specimen. A specimen from the core area shows a marked shoulder, while the melting peak shoulder of a specimen from the margin is hard to find (compare figs. 9 and 11). The reason for this is the difference in flow conditions over the cross section, which influences the oriented crystallisation. Comparing the melting enthalpy, it is found that a self-reinforced PE-HD specimen has a higher melting enthalpy than a normal PE-HD specimen (figs. 9 and 10). A self-reinforced PE-HD specimen has a wider melting peak than a normal PE-HD specimen (fig. 12). This confirms a second crystalline phase, i.e.

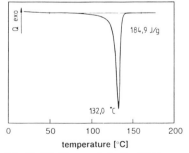

Fig 10. Melting peak in the DSC curve of a normal PE-HD specimen (see table 2).

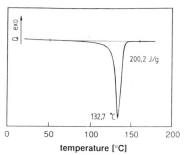

Fig 11. Melting peak in the DSC curve of a self-reinforced PE-HD specimen (type 1, see table 2), margin.

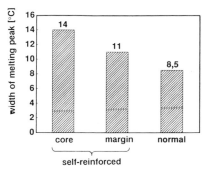

Fig 12. Comparison of the width of the melting peak of self-reinforced and normal PE-HD specimens at $\frac{1}{3}$ peak height.

the shish-kebab structures. The above-mentioned characteristics of DSC curves can appear in other variants. One example of this is shown in fig. 13, where shifting of the melting peak due to self-reinforcement to higher temperature is noticed.

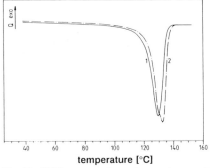

Fig 13. DSC curves of specimens from a PE-HD journal bearing (see table 3), 1 = normal, 2 = self-reinforced.

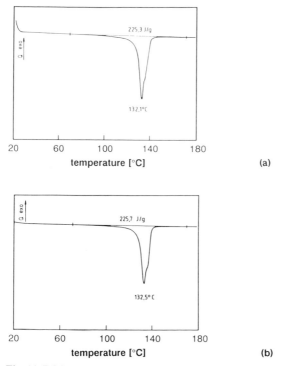

Fig. 14. DSC curves of PE-HD specimens with different degrees of self-reinforcement, a: type 2, b: type 4 (table 2).

The results of the DSC measurements can be summarised as follows:

The self-reinforcement in PE-HD can be recognised by means of DSC. The main characteristic of self-reinforcing crystalline structures is their high temperature resistance, which leads to a wider melting peak and/or a shoulder of the melting peak above the normal melting temperature and/or a shift of the melting peak to a higher temperature.

Nevertheless, a quantitative measurement of the degree of self-reinforcement is not possible. Figure 14 shows DSC curves of two specimens with different degrees of self-reinforcement (type 2 and type 4, see table 2). Even with a big difference in Young's modulus the difference in the DSC curves is not greater than the scatter of the measurements.

5.1.1.2. PP. A higher melting enthalpy and a wider melting peak due to self-reinforcement can be noticed by PP as well (fig. 15).

5.1.1.3. LCP. The difference between the DSC curves of a semicrystalline and an amorphous liquid-crystalline polymer specimen is shown in fig. 16. While both of them have a glass transition temperature of about 130°C, a small melting peak can only be noticed for the semicrystalline LCP.

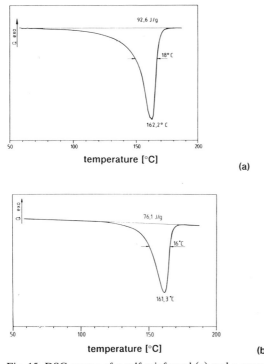

(a)

(b)

Fig. 15. DSC curves of a self-reinforced (a) and a normal (b) PP specimen.

5.1.2. Results of the measurement of retardation on PE-HD

The average retardation of a self-reinforced and a normal PE-HD specimen is compared in fig. 17.

PE-HD is self-reinforced by oriented crystalline structures (shish-kebab structures) and it is highly crystalline. Therefore, the orientation of the crystalline structures can approximately be found by means of measurement of retardation,

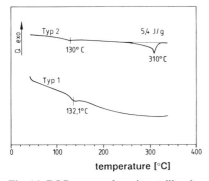

Fig. 16. DSC curves of semicrystalline (type 2) and amorphous (type 1, table 5) LCP.

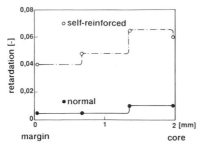

Fig. 17. Comparison of retardation across the cross section of a self-reinforced (type 4, table 2) and a normal PE-HD specimen.

which enables an estimation of the degree of self-reinforcement. Figure 18 shows the correlation between Young's modulus and the retardation.

Measurement of retardation was used for the classification of self-reinforced PE-HD bearings. The profiles of retardation across the cross section are shown in fig. 19. There is an orientation maximum at the inside margin, the tribological loaded area. In addition, it can be noticed that a stronger cross section reduction leads to a higher orientation, as expected (table 7).

5.1.3. Dynamic mechanical analysis (DMA) on PE-HD

The DMA curves which show the storage moduli and mechanical loss factors as function of temperature, are shown in fig. 20. Normal PE-HD has a lower storage modulus and a higher mechanical loss factor over the temperature range from room temperature to more than 100°C. The difference becomes greater with increasing temperature. From the deformation theory of friction we know that the deformation component of friction is proportional the mechanical loss factor (see eq. (11)). Therefore an improvement of the tribological properties by self-reinforcement, especially at higher temperature can be expected.

5.1.4. Contact angle – surface energy

The purpose of measurement of the contact angle is to examine the influence of self-reinforcement on surface energy. The contact angles of investigated polymeric materials with different test liquids are summarised table 8.

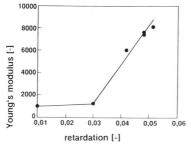

Fig. 18. Correlation between Young's modulus and the retardation of PE-HD.

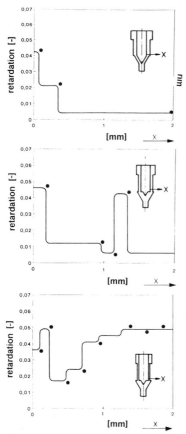

Fig. 19. Retardation profile across the cross section of a normal (a), a slightly self-reinforced (b) and a highly self-reinforced (c) PE-HD bearing (fig. 4, table 3).

The surface energies were then calculated from the contact angles by the geometric-mean method [19] and determined by conversion tables of contact angles to surface tensions [33]. While the geometric-mean method enables the determination of the polar component, γ^p, and the dispersion component, γ^d, the second method can only give the total surface energy (table 9).

TABLE 7
Correlation between retardation and die geometry.

Die geometry (fig. 4)	Cross section	Average retardation
a	increase	0.010
b	slight reduction	0.022
c	reduction	0.038

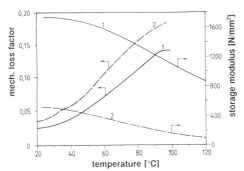

Fig. 20. DMA curves of a self-reinforced (type 2, table 2) and a normal PE-HD specimen, 1 = self-reinforced, 2 = normal.

TABLE 8
Contact angle, φ, of polymeric materials with different test liquids (with standard deviation).

	Contact angle φ [degree]		
	water	α-bromo-naphthalene	methylene iodide
Normal PE-HD	75 ± 3.5	28 ± 2.1	44 ± 1.6
Self-reinforced PE-HD (type 1, skin)	77 ± 5.5	29 ± 1.0	50 ± 2.3
Self-reinforced PE-HD (type 1, skin removed)	72 ± 5.1	34 ± 5.2	41 ± 4.4
Normal PP	84 ± 1.6	31 ± 4.5	61 ± 1.2
Self-reinforced PP	84 ± 1.6	35 ± 2.0	52 ± 2.3
LCP (type 3)	63 ± 7.1	21 ± 1.7	51 ± 6.7

TABLE 9
Surface energy, γ, of PE-HD, PP and LCP.

	Surface energy [mJ/m^2]					
	conversion tables [33]			geometric-mean method [19]		
	water	α-bromo-naphthalene	methylene iodide	γ	γ^d	γ^p
Normal PE-HD	39	39	39	38.5	33.5	5.0
Self-reinforced PE-HD (type 1, skin)	37	39	36	36.4	29.3	7.1
Self-reinforced PE-HD (type 1, skin removed)	40	37	41	41.6	33.3	8.3
Normal PP	33	39	31	30.0	24.0	6.0
Self-reinforced PP	33	38	35	34.0	29.9	4.1
LCP (type 3)	46	42	36	42.6	25.7	16.9

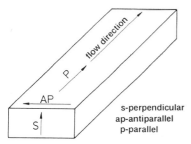

Fig. 21. Sliding directions.

s-perpendicular
ap-antiparallel
p-parallel

The surface energies of PE-HD and PP in the normal and self-reinforced state calculated from the contact angles with three different test liquids do not differ much from each other. The surface energies are obviously not influenced by self-reinforcement.

5.2. Tribological properties in the pin-on-disk system

5.2.1. Sliding directions
Due to the oriented crystalline structures, the self-reinforced specimens have a highly pronounced anisotropy. Therefore, it is necessary to test the materials in defined sliding directions. The reference direction is the flow direction of the melt in which the crystalline structures are oriented. The sliding direction parallel to the flow direction is defined as the p-direction. The s (perpendicular) – and ap (antiparallel) – directions are defined analogously (fig. 21).

5.2.2. The effects of self-reinforcement on the tribological properties
Figure 22 shows the friction coefficient and wear of normal and self-reinforced PE-HD as function of test time. The wear process is obviously slowed down by self-reinforcement and the friction coefficient is reduced. The investigation of PP shows a similar tendency. The evaluated results are summarised in table 10.

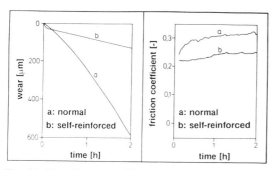

Fig. 22. The friction coefficient and wear of normal (a) and self-reinforced (b) PE-HD (type 1) as function of time. s-direction, steel disk, $R_z = 1.3$ μm, $T_U = 23°C$, $v = 0.5$ m/s, $p = 4$ N/mm².

TABLE 10
Effect of self-reinforcement on the tribological properties. s-direction, steel, $p = 4$ N/mm^2, $v = 0.5$ m/s, $T_U = 23°C$.

	Wear factor [10^{-6} mm^3/Nm]	Friction coefficient
$R_z = 1.3$ μm		
Normal PE-HD	39 ± 3 *	0.29 ± 0.03 *
Self-reinforced PE-HD (type 1)	10 ± 2 *	0.25 ± 0.02 *
$R_z = 0.4$ μm		
Normal PP	100	0.41
Self-reinforced PP	26	0.38
$R_z = 2.8$ μm		
Normal PP	210	0.46
Self-reinforced PP	100	0.32

* Standard deviation.

The results confirm the result of theoretical analysis that improved mechanical properties lead to improved tribological properties. The tribological mechanisms were investigated explicitly in the following tests.

Since the effect of self-reinforcement is much more conspicuous on PE-HD than on PP, the remaining tests are mainly carried out on PE-HD.

5.2.3. Correlation between Young's modulus and the tribological properties

The correlation between Young's modulus and the tribological properties were investigated with specimens with different degrees of self-reinforcement. An interesting phenomenon was observed. There is not a linear correlation between the tensile modulus and wear (fig. 23). The highest tensile modulus does not lead to the highest wear resistance.

A linear correlation between the wear factor and the compression modulus gives an explanation for this phenomenon (fig. 24):

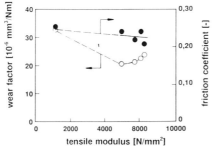

Fig. 23. Wear factor and friction coefficient as function of tensile modulus. Steel disk, $R_z = 2.8$ μm, $T_U = 23°C$, $v = 0.5$ m/s, $p = 8.8$ N/mm^2.

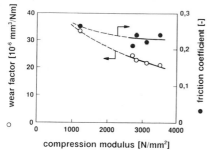

Fig. 24. Wear factor and friction coefficient as function of the compression modulus. Steel disk, $R_z = 2.8$ µm, $T_U = 23°C$, $v = 0.5$ m/s, $p = 8.8$ N/mm².

It is known that there is a big difference between tensile and compression mechanical properties of aramide fibres and aramide fibre reinforced materials, which have a similar structure to self-reinforced PE-HD. This phenomenon can be explained by microbuckling and a weak interface [22,34,35]. Since a specimen is loaded under compression in the pin-on-disk test, the tribological process is influenced mainly by the compression properties of the materials. The results reveal that the wear of the materials decreases markedly with increasing compression modulus.

The difference in the degree of self-reinforcement does not have a well-defined influence on the friction coefficient. The following mechanism could give a reasonable explanation for this:

While a higher stiffness reduces the real contact surface, which leads to a lower adhesion, it results in a higher resistance force in the sliding direction, which leads to higher friction.

The same phenomenon, i.e. that the wear factor is inversely proportional to the compression modulus, is also observed with LCP (table 11).

5.2.4. Effects of the sliding direction
5.2.4.1. Self-reinforced PE-HD.   In order to investigate the effects of orientation on the tribological properties, the specimens were tested in three sliding directions. The results are shown in fig. 25. There is no big difference between the

TABLE 11
Compression moduli and tribological properties of LCP (type 3). Steel disk, $R_z = 2.8$ µm, $v = 0.5$ m/s, $p = 3$ N/mm², $T_U = 23°C$, s-direction.

	Compression modulus * [N/mm²]	Wear factor [10^{-6} mm³/Nm]	Friction coefficient
Highly oriented	5500 ± 1100	0.6	0.4
Weakly oriented	4100 ± 1000	1.0	0.42

* With standard deviation.

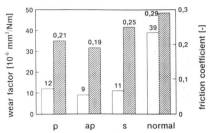

Fig. 25. Effects of the sliding direction on the tribological properties of self-reinforced PE-HD (type 1). Steel disk, $R_z = 1.3$ μm, $T_U = 23°C$, $v = 0.5$ m/s, $p = 4$ N/mm^2.

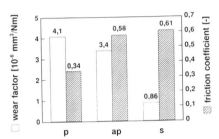

Fig. 26. Effects of the sliding direction on the tribological properties of a semicrystalline LCP (type 2). Steel disk, $R_z = 2.8$ μm, $T_U = 23°C$, $v = 0.5$ m/s, $p = 3.14$ N/mm^2.

self-reinforced PE-HD specimens. In comparison with the normal PE-HD specimens the wear factor and the friction coefficient decrease by the self-reinforcement in all the three sliding directions. This tendency was confirmed for other
tribological parameters as well [36].
5.2.4.2. LCP. LCP shows marked orientation dependent tribological properties
(fig. 26). The same effects of the sliding direction were also observed with an
amorphous LCP, which has a much higher wear factor than a semicrystalline LCP
(fig. 27). A possible explanation for the highly pronounced anisotropy seems to be
the weak interface between the fibrous segments, as shown in fig. 28.

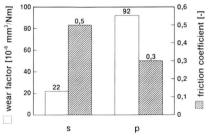

Fig. 27. Effects of the sliding direction on the tribological properties of an amorphous LCP (type 1). Steel disk, $R_z = 2.8$ μm, $T_U = 23°C$, $v = 0.5$ m/s, $p = 3.14$ N/mm^2.

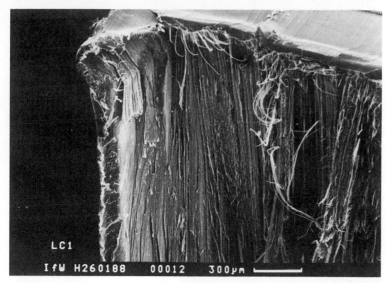

Fig. 28. A photograph of an LCP fractured surface.

5.2.5. Influence of the tribological parameters
5.2.5.1. Normal load.
The influence of normal load on the wear factor and friction coefficient of self-reinforced PE-HD is shown in fig. 29.

The wear factor increases slightly with increasing normal load. Therefore, the same normal load should be used for the tests for the comparison of materials with each other.

The normal load influences the wear property in a complicated way. The real contact surface, which determines the area of the deformed zone and the friction force, is influenced by the normal load. The higher the normal load, the higher the contact surface temperature. Since a slight decrease of the friction coefficient is observed, we know that it is the Amontons friction whereby the deformative component of the friction is larger than the adhesion component of the friction.

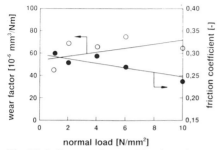

Fig. 29. Influence of normal load on the wear factor and friction coefficient of self-reinforced PE-HD (type 1). Steel disk, $R_z = 1.6$ μm, $T_U = 23°C$, $v = 0.5$ m/s.

TABLE 12
Influence of normal load on the tribological properties. Steel disk, $v = 0.5$ m/s, $T_U = 23°C$.

Normal load [N/mm²]	PE-HD (type 1)				PA66	
	$R_Z = 1.6$ μm		$R_Z = 1.1$ μm		$R_Z = 3$ μm	
	k	f	k	f	k	f
1	45	0.3	–	–	–	–
2	69	0.28	–	–	3.5	0.75
4	66 ± 8 *	0.29 ± 0.03 *	0.74	0.16	7.7	0.6
6	75	0.27	0.63	0.16	11	0.4
8	–	–	0.64	0.16	–	–
10	64	0.24	1.1	0.16	18	0.48

* Standard deviation, k: wear factor [10^{-6} mm³/Nm], f: friction coefficient.

The same tendency for the influence of the normal load is observed with PA 66 as well (table 12).

5.2.5.2. *Sliding velocity.* Two effects of the sliding velocity on the tribological properties were investigated:
- The influence of the sliding velocity on the stick–slip effect. According to ref. [37] a strongly marked falling friction coefficient with increasing sliding velocity leads to the stick–slip effect.
- Correlation between temperature rise or heat flow and sliding velocity.

5.2.5.2.1. *LCP.* LCP shows a marked tendency to stick–slip, especially with a smooth steel counterpart ($R_Z < 1$ μm). The tests with different sliding velocity show that the friction coefficient decreases with increasing sliding velocity (fig. 30). Furthermore, a high friction coefficient was observed, which was also considered as one of the reasons for stick–slip motion [38]. Another reason for the stick–slip motion should be the fracture of contact surface due to the weak interface between the fibrous segments of LCP, which can be observed on the SEM photograph (fig. 31). The fracture leads to irregular friction coefficient–sliding velocity characteristics.

5.2.5.2.2. *PA.* A falling friction coefficient with increasing sliding velocity was not observed (fig. 32), although PA showed a strongly marked stick–slip motion. A

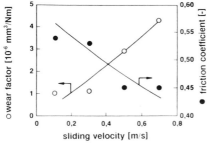

Fig. 30. Wear factor and friction coefficient of LCP (type 2) as function of the sliding velocity. Steel disk, $R_z = 0.4$ μm, $T_U = 23°C$, $p = 4$ N/mm².

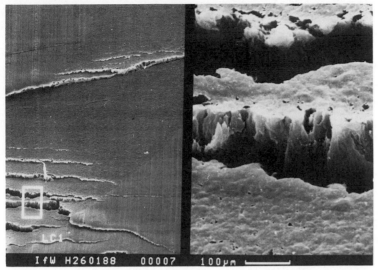

Fig. 31. Fracture of contact surface of LCP.

high friction coefficient and marked change in friction property should be, there-fore, the main reasons for stick–slip motion. A distinctively decreasing wear factor with increasing sliding velocity was noticed. This phenomenon will be discussed later.

5.2.5.2.3. PE-HD. The friction coefficient does not vary much with increasing sliding velocity and is low in value (fig. 33). These preferable characteristics lead to a quiet sliding.

5.2.5.3. Surface roughness of the steel counterpart. The wear factor and friction coefficient of self-reinforced and normal PE-HD as function of the surface roughness of the steel counterpart are shown in fig. 34. Sliding on a smooth steel surface, the adhesion component of friction is dominant. While the deformation component of friction increases with increasing surface roughness, an increase in friction coefficient and wear factor was noticed. This tendency is stronger marked

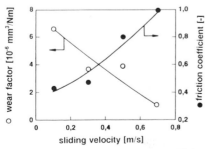

Fig. 32. Wear factor and friction coefficient of PA66 as function of sliding velocity. Steel disk, $R_z = 0.4$ μm, $T_U = 23°C$, $p = 4$ N/mm^2.

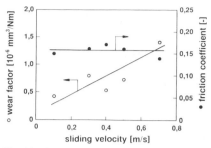

Fig. 33. Wear factor and friction coefficient of self-reinforced PE-HD as function of sliding velocity. Steel disk, $R_z = 1.1$ µm, $T_U = 23°C$, $p = 4$ N/mm².

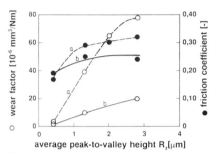

Fig. 34. Dependence of the wear factor and friction coefficient on the surface roughness of the steel counterpart. $T_U = 23°C$, $v = 0.5$ m/s, $p = 4$ N/mm², a: self-reinforced, b: normal.

for the normal specimens than for the self-reinforced specimens. This reveals that the deformation component of friction is influenced by the self-reinforcement.

5.2.6. Effects of fillers on the tribological properties

Since the effects of fillers are already investigated by many authors, we will not discuss them in detail. Nevertheless, our own tests are necessary to obtain the comparable results, because the data from the literature differ very often from each other and the test conditions are not given completely.

5.2.6.1. Carbon fibres. The wear factor is strongly decreased by addition of carbon fibres (table 13).

TABLE 13

Effect of carbon fibres on the tribological properties of PA66. $R_z = 2.8$ µm, $p = 3.14$ N/mm², $v = 0.5$ m/s, $T_U = 23°C$.

Material	Wear factor [10^{-6} mm³/Nm]	Friction coefficient
PA66	9.0	0.5
PA66 + 20% CF	0.9	0.33

TABLE 14
Effect of aramide fibres on the tribological properties of PA66. Steel disk, $R_z = 2.8$ μm, $p = 3.14$ N/mm², $v = 0.5$ m/s, $T_U = 23°C$.

Material	Wear factor [10^{-6} mm³/Nm]	Friction coefficient
PA66	9.0	0.5
PA66 + 40% AF	2.4	0.4

TABLE 15
Effect of glass fibres on the tribological properties of PA66. Steel disk, $R_z = 2.8$ μm, $p = 3.14$ N/mm², $v = 0.5$ m/s, $T_U = 23°C$.

Material	Wear factor [10^{-6} mm³/Nm]	Friction coefficient
PA66 mod. 3 (without GF)	1.2	0.23
PA66 mod. 2 (with 30% GF)	0.6	0.30

5.2.6.2. Aramide fibres. Aramide fibres show similar effects. However, they are not as marked as those of carbon fibres (table 14).

5.2.6.3. Glass fibres. In contrast to carbon fibres and aramide fibres the friction coefficient is increased by addition of glass fibres. The wear factor is decreased (table 15).

5.2.6.4. PTFE. PTFE is a typical lubricant filler. Both the friction coefficient and the wear factor are decreased by addition of PTFE. The same effects are noticed with pure and carbon fibre-reinforced polyamide (table 16).

5.2.7. Effects of temperature
5.2.7.1. Effects of temperature rise by friction heat on the tribological properties. The temperature rise by friction heat can be enhanced by both increasing the normal load and the sliding velocity. It is investigated whether there is a certain correlation between the temperature rise and the tribological properties which is independent of the way of heat production. The results show that it is not the case

TABLE 16
Effect of PTFE on the tribological properties of PA66. Steel disk, $R_z = 2.8$ μm, $p = 3.14$ N/mm², $v = 0.5$ m/s, $T_U = 23°C$.

Material	Wear factor [10^{-6} mm³/Nm]	Friction coefficient
PA66	9	0.50
PA66 + 18% PTFE	1.2	0.23
PA66 + 20% CF	0.9	0.33
PA66 + 20% CF +5% PTFE	0.7	0.27

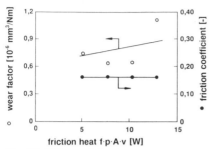

Fig. 35. Effects of the friction heat on the tribological properties of self-reinforced PE-HD (type 1). Steel disk, $R_z = 1.1$ μm, $T_U = 23°C$, $p = 4$–10 N/mm², $v = 0.5$ m/s.

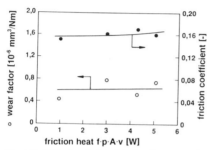

Fig. 36. Effects of the friction heat on the tribological properties of self-reinforced PE-HD (type 1). Steel disk, $R_z = 1.1$ μm, $T_U = 23°C$, $p = 4$ N/mm², $v = 0.1$–0.5 m/s.

(figs. 35–38). The tribological properties are influenced by the way of heat production.

5.2.7.2. Effects of the temperature dependence of the mechanical properties on the tribological properties. The mechanical properties of polymeric materials vary with environmental temperature. These in turn influence the tribological properties.

Figure 39 shows the dependence of the tribological properties of normal and self-reinforced PE-HD on the environmental temperature. It is observed that the

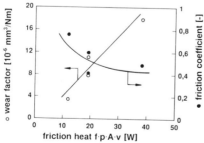

Fig. 37. Effects of the friction heat on the tribological properties of PA66. Steel disk, $R_z = 3$ μm, $T_U = 23°C$, $p = 2$–10 N/mm², $v = 0.5$ m/s.

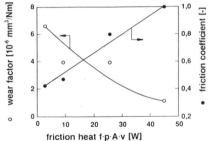

Fig. 38. Effects of the friction heat on the tribological properties of PA66. Steel disk, $R_z = 3$ μm, $T_U = 23°C$, $p = 4$ N/mm^2, $v = 0.1–0.7$ m/s.

wear factor and friction coefficient of normal PE-HD increase with increasing environmental temperature much faster than those of self-reinforced PE-HD.

5.3. Tribological properties for a journal bearing system

The tribological conditions in a journal bearing system are very different from those in a pin-on-disk system:
– While most of the debris is moved away from the contact area in a pin-on-disk system, much of it remains in the contact area in a journal bearing system. This fact makes the sliding condition in both systems different.
– In order to avoid a marked deflection of the shaft, which causes a nonuniform load in the axial direction of the bearing, a lower bearing load has to be used. Therefore a longer test time is needed to get measurable wear.
– Due to the difference in the tooling methods the same surface roughness is not a guarantee for the comparability of the surface topography.
– While the pressure on the contact surface in a pin-on-disk system is approximately constant, there is an elliptical pressure distribution on the contact surface in a journal bearing system.
The friction coefficients and radial wear factors of a PA 66, a normal and a self-reinforced PE-HD bearing are summarised in table 17.

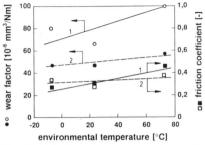

Fig. 39. Tribological properties of normal and self-reinforced PE-HD as function of the enviromental temperature. 1 = normal, 2 = self-reinforced, steel disk, $R_z = 2$ μm, $p = 4$ N/mm^2, $v = 0.5$ m/s.

TABLE 17
Tribological properties of PE-HD. Steel shaft, $p = 0.5$ N/mm^2 (eq. (20)), $v = 0.2$ m/s, $T_U = 23°C$.

R_z [μm]	Material	Radial wear factor [10^{-6} mm^3/Nm]	Friction coefficient
0.8	self-reinforced PE-HD	6.2	0.19 ± 0.04 *
	normal PE-HD	12 ± 1.5 *	0.18 ± 0.03 *
	PA66	5.2 ± 1.9 *	0.51 ± 0.09 *
3	self-reinforced PE-HD	49 ± 2 *	0.21 ± 0.06 *
	normal PE-HD	65 ± 12 *	0.21 ± 0.02 *
	PA66	7	0.55

* standard deviation.

Due to the larger contact area the friction heat in the journal bearing system is higher than in the pin-on-disk system. In order to avoid a high temperature rise of the contact surface, a lower sliding velocity was used.

The self-reinforced bearings have a lower wear factor than the normal bearings. However, no improvement in friction coefficient was observed. It has to be mentioned that the self-reinforcement of the bearing is not as high as that of the specimens used for the pin-on-disk tests.

5.4. Wear mechanism – SEM tests

SEM photographs of the worn surfaces of PE-HD are shown in fig. 40. The transverse marks on the worn surface of the self-reinforced PE-HD demonstrate a larger component of fatigue wear and the rough surface with clear grooves parallel to the sliding direction on the worn surface of the normal PE-HD demonstrate a larger component of abrasive wear. This change of wear mechanism is the result of

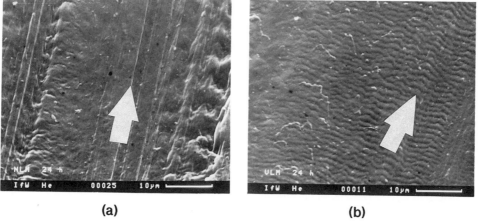

(a)   (b)

Fig. 40. SEM photographs of the worn surface of PE-HD. Steel disk, $R_z = 1.3$ μm, $p = 1$ N/mm^2, $v = 0.5$ m/s, $T_U = 23°C$, a: normal, b. self-reinforced. Arrows show the sliding direction.

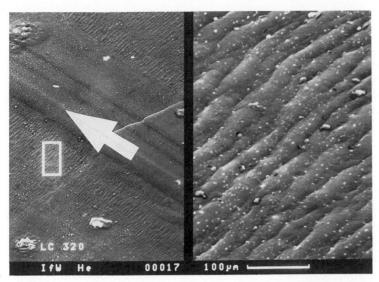

Fig. 41. SEM photograph of the worn surface of LCP (type 2). Steel disk, $R_z = 2.8$ μm, $p = 8.8$ N/mm^2, $v = 0.5$ m/s, $T_U = 23°$C. Arrows show the sliding direction. The scale is valid for the right side.

the improved mechanical properties. The favourable wear type of the self-reinforced PE-HD leads to low wear factor.

The same effects are observed on the worn surface of wear-resistant LCP with marks transverse to the sliding direction, which indicates a larger component of fatigue wear (fig. 41).

TABLE 18
Comparison of different ways for improvement of tribological properties.

Method	Efficient application	Positive effects	Negative effects
Lubricant fillers	adhesive materials	lower friction, possibly wear	worsening of mechanical properties, possibly wear resistance
Reinforcing fillers	low wear resistent materials	lower wear, possibly friction	anisotropy, possibly higher adhesion
Self-reinforcement	low wear resistent, self-reinforceable materials	lower wear, slight reduction of friction	anisotropy possible

6. Discussion

6.1. Self-reinforcement – a new way for improvement of tribological properties

The results of the investigations show that self-reinforcement represents a way to improve the tribological properties. From the viewpoint of materials technology the effects of different ways for improvement of tribological properties can be systematically compared with each other (table 18).

Table 18 is only a simplified summary. The boundary between lubricant and reinforcing fillers is fluent. The wear of a low wear resistent material can be reduced by lubricant fillers because of lower friction. However, worse mechanical properties is often an accompanying effect, which is always a negative effect on a function-integrated design. For the reinforcing fillers this is also the case. They could both decrease and increase friction. Self-reinforcement enables the improvement of tribological properties without the negative effects mentioned above.

6.2. New aspects of design of sliding elements

The product of sliding velocity and normal load has been used to calculate the working life of sliding elements. The fundamentals for this calculation are:
– a wear factor which is independent of the normal load, and
– a wear factor which is independent of the sliding velocity.
These are not definitely confirmed by the results of investigations.

6.2.1. Sliding velocity

Both increasing and approximately constant wear factors with increasing sliding velocity were observed (fig. 42).

The reason for this behaviour seems to be two compensating effects of the sliding velocity on the viscoelastic property of polymeric materials:
– time–temperature equivalence, WLF-equation [37,39]
– temperature dependance of mechanical properties, for example Young's modulus.

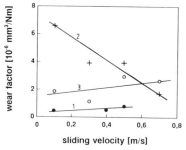

Fig. 42. Effects of the sliding velocity on the wear factor. Steel disk, $p = 4$ N/mm^2, $T_U = 23$°C. 1: self-reinforced PE-HD, $R_z = 1.1$ μm, 2: PA66, $R_z = 0.4$ μm, 3: LCP (type 2), $R_z = 0.4$ μm.

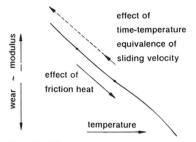

Fig. 43. Effects of the sliding velocity on wear.

The principle of time–temperature equivalence tells us that a change of loading velocity (sliding velocity for tribology) has an equivalent effect as a change of temperature. A high loading velocity leads to a high Young's modulus. On the other hand, a high sliding velocity leads to a high temperature rise. Young's modulus and strength of polymeric materials decrease with increasing temperature. The wear factor decreases with decreasing modulus, as our results show. The effect of the sliding velocity is an addition of both effects. Therefore a strongly marked time–temperature effect leads to a decreasing wear factor with increasing sliding velocity. This is shown in fig. 43 as a dotted line. A weaker time–temperature effect is shown as a full line. In this case the temperature dependence of the mechanical properties is dominant. Therefore a slightly increasing wear factor with increasing sliding velocity can be observed.

6.2.2. Normal load

Increasing wear factors with increasing normal load were observed (fig. 44). Based on the linear dependence of the wear factor of Young's modulus, as our results show, the nonlinear stress–strain curve of polymeric materials should be the explanation for the increasing wear factor with increasing normal load (fig. 45). Due to the nonlinearity of the curve the secant modulus decreases with increasing normal load.

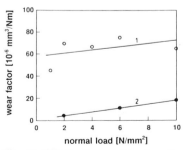

Fig. 44. Effects of normal load on the wear factor. Steel disk, $v = 0.5$ m/s, 1: self-reinforced PE-HD (type 1), $R_z = 1.6$ μm, 2: PA66, $R_z = 0.4$ μm.

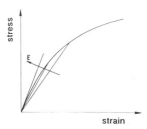

Fig. 45. Dependence of secant modulus of stress (normal load).

6.3. Tribological tests – comparison of pin-on-disk and journal bearing systems

Comparing the results from pin-on-disk and journal bearing systems under comparable conditions, we observed comparable wear factors (table 19). The friction coefficient of PE-HD in the journal bearing system is much lower than that in the pin-on-disk system. The reason for this phenomenon could be the large amount of powdery debris which remains in the contact area.

6.4. Findings in tribology

Although the adhesion and deformation theories can be used to explain the tribological behaviour of polymeric materials, a quantitative estimation of the tribological properties is not possible. One of the important reasons for this is that tribological properties have a highly pronounced dependance on the tribological conditions. However, the theories are important tools for understanding of the results.

6.4.1. Correlation between the wear property and the compression modulus
The compression modulus is one of the determining materials characteristics for the tribological properties. There is no linear correlation between the tribological properties and the tensile modulus if it differs from the compression modulus. The latter phenomenon can be often observed with self-reinforced materials and composite materials. In order to get preferable tribological properties, the compression modulus has to be improved.

TABLE 19
Comparison of the tribological properties for pin-on-disk and journal bearing systems.

	PE-HD		PA66	
	k [10^{-6} mm^3/Nm]	f	k [10^{-6} mm^3/Nm]	f
Pin-on-disk *	71	0.3	9	0.54
Journal bearing **	65	0.21	7	0.55

* $R_z = 3$ μm, $T_U = 23°C$, $v = 0.3$ m/s, $p = 4$ N/mm^2.
** $R_z = 3$ μm, $T_U = 23°C$, $v = 0.2$ m/s, $p = 0.5$ N/mm^2.

6.4.2. Wear mechanisms

The investigation on wear mechanisms shows the correlation between the mechanical properties, the wear mechanisms and the wear property. By the nature of its basic process, polymer wear can be tentatively divided into fatigue and abrasive. Material is removed after repeated loading by the fatigue mechanism while in one single load by the abrasive mechanism. Therefore a larger component of fatigue wear leads to a higher wear resistance and this can be achieved by means of reinforcement. This effect is shown by tribological and SEM investigation on normal and self-reinforced PE-HD.

6.4.3. Friction mechanisms

The investigation shows that many theories of tribology are only valid in a small range. According to eq. (8) the adhesion component of the friction coefficient decreases with increasing Young's modulus. According to eq. (11) the deformation component of the friction coefficient decreases with increasing Young's modulus as well. Therefore a strongly marked reduction of the friction coefficient is expected. The results show that the effect of self-reinforcement on the deformation component is obviously much more marked than that on the adhesion component. Two facts were observed (fig. 34):
- The decrease of the friction coefficient is only observed when the surface of steel disk is rough ($R_z > 1$ µm).
- Only a slight decrease of the friction coefficient is observed.

Since deep penetration of the roughness of a smooth disk is not possible, the adhesion friction is dominant in this case, and Young's modulus has only weak influence on the real contact area. This example shows the difficulty of theoretical estimation of tribological properties and the importance of tribological tests under defined conditions.

6.4.4. Temperature

Wear and friction increase with strongly increasing environmental temperature. This tendency can be slowed down by a higher heat stability of the mechanical properties, which can be achieved e.g. by self-reinforcement (fig. 39).

7. Conclusions

The subject of this chapter is the investigation of effects of self-reinforcement by means of flow-induced crystallisation and the use of LC-Polymers, on the tribological properties. Tribological properties are properties of a system which are influenced by many factors. These factors are theoretically analysed. The experiments are designed in such a way, that these influencing factors, such as temperature, mechanical load, anisotropy and difference in pin-on-disk system and journal bearing system, are taken into consideration.

The effects of self-reinforcement on the tribological properties are compared with those of fillers. The results of the investigation show that self-reinforcement represents a way to reinforce a material without worsening the adhesive property of the material. Therefore it is especially favourable to use self-reinforcement on a weak adhesive polymer, e.g. PE-HD, in order to get advantageous tribological properties.

Self-reinforced polymers are interesting subjects for some fundamental experiments of tribology as well. The special property of self-reinforcement, reinforcing material without changing its adhesive property, enables the investigation of the correlation between the mechanical and the tribological properties. The results show that the compressive modulus, which is often different from the tensile modulus for fibre-reinforced, anisotropic polymers, influences the tribological properties strongly. A high compressive modulus is desirable in order to develop a high-performance material for tribological application.

The investigation at elevated temperature reveals the effects of the temperature dependent mechanical properties of the material on the tribological properties. The tribological properties under elevated temperature can be improved by a higher stability of the mechanical properties at elevated temperature. This can be obtained e.g. by means of self-reinforcement, whereby a highly temperature-resistant crystalline structure is induced.

The velocity and the pressure influence the tribological properties in a complicated way because of the viscoelastic behaviour of polymers. The investigation shows that the effect of velocity on the temperature rise could be partly or totally compensated by the effects of time–temperature equivalence. On the other hand, the influence of pressure is enhanced because of the viscoelastic behaviour of polymers. Therefore, instead of the pressure–velocity product (pv), the influence of the velocity and the pressure should be evaluated seperately in the calculation of wear.

With help of SEM graphs of worn surfaces the change of wear mechanism caused by the self-reinforcement and its influence on the wear resistance of the polymer are studied. The wear resistance can be largely increased by minimising the abrasive wear.

In addition to the tribological properties in a pin-on-disk system, the tribological properties in a journal bearing system and the correlation of the results for different systems are investigated.

Acknowledgement

Our special thanks are addressed to the German Science Foundation (DFG) and the Industrial Research Foundation (AIF) for financial support for the investigation.

List of symbols

A_r	real contact surface
A_0	nominal geometric contact area
c	proportionality factor
E	Young's modulus

f	friction coefficient
f_a	adhesive friction coefficient
f_d	deformative friction coefficient
F_a	adhesive friction force
F_d	deformative friction force
F_f	friction force
F_N	normal force
h	depth of penetration
h_{max}	maximal roughness
H	hardness
k	wear factor
k_R	radial wear factor
n, n_1, n_2	exponential factors
p	normal pressure
R	radius of curvature of the asperities
R_a	average roughness
R_z	average peak-to-valley height
s	sliding distance
w	total wear
w_p	pressure dependent part of wear
w_T	temperature dependent part of wear
w_v	volume of the worn material
W_{ab}	least energy needed to separate contacting surfaces
y	mole fraction
β	roughness coefficient
γ	surface energy
γ_a	surface energy of surface a
γ_b	surface energy of surface b
γ_{ab}	boundary surface energy b
$\tan \delta$	mechanical loss factor
ε	rupture strain or elongation yield point
ξ	parameter depending on the surface
σ_0	strength of the material
σ_{dF}	compression yield point
τ_s	shear strength of the contacts
Φ	relative contact area
φ	contact angle

References

[1] G. Erhard, Zum Reibungs- und Verschleißverhalten von Polymerwerkstoffen, PhD Dissertation (Universität Karlsruhe (TH), Karlsruhe, 1980).
[2] J. Song, Reibung und Verschleiß eigenverstärkter Polymerwerkstoffe, VDI Fortschrittsber. Reihe 5, Nr. 220 (VDI-Verlag, Düsseldorf, 1991).
[3] K. Friedrich, Friction and Wear of Polymer Composites (Elsevier, Amsterdam, 1986).

[4] J. Song and G.W. Ehrenstein, Comparison of tribological properties of thermoplastic-based composites and self-reinforced thermoplastics, in: Proc. ASM Int. Conf. on Tribology of Composite Materials, eds P.K. Rohatgi, P.J. Blau and C.S. Yust (ASM Int. Math. Park, 1990).

[5] H. Uetz and J. Wiedemeyer, Tribologie der Polymere (Carl Hanser Verlag, München, 1985).

[6] G.M. Bartenev and V.V. Lavrentev, Friction and Wear of Polymers (Elsevier, Amsterdam, 1981).

[7] N.P. Suh and H.C. Sin, The genesis of friction, Wear 69 (1981) 91–114.

[8] D. Tabor, Friction – The present state of our understanding, J. Lubrication Technol. 103 (1981) 169–179.

[9] F.P. Bowden and D. Tabor, Reibung und Schmierung fester Körper (Springer, Berlin, 1959).

[10] I.V. Kragelski, M.N. Dobycin and V.S. Kombalov, Grundlagen der Berechnung von Reibung und Verschleiß (VEB Verlag Technik, Berlin, 1982).

[11] G. Polzer and F. Meißner, Grundlagen zu Reibung und Verschleiß (VEB Deutscher Verlag für Grundstoffindustrie, Leipzig, 1978).

[12] D.F. Moore, Principles and Applications of Tribology (Pergamon Press, Oxford, 1975).

[13] L.H. Lee, Effect of surface energetics on polymer friction and wear, in: Advances in Polymer Friction and Wear, ed. L.H. Lee (Plenum Press, New York, 1974) pp. 31–68.

[14] G. Erhard and E. Strickle, Berechnung der statischen Tragfähigkeit von Querlagern aus thermoplastischen Kunststoffen, Konstruktion 29(6) (1977) 237–241.

[15] G.W. Ehrenstein and H.-G. Walger, Berechnen von Kunststoff-Gleitlagern mit EDV, Kunststoffe 65(10) (1975) 702–709.

[16] D. Wimmer, Gleitelemente. Werkstattblatt, Gruppe B, Nr. 1054 (Hoppenstedt Technik Tabellen Verlag GmbH, 1988) pp. 1–11.

[17] G. Erhard and E. Strickle, Gleitelemente aus thermoplastischen Kunststoffen, Kunststoffe 62 (1972) 2–9, 232–234, 282–288.

[18] W.A. Bely, A.I. Swiridjonok and W.M. Kenjko, Einige Besonderheiten bei der Reibung von selbstschmierenden Polymerkompositionen, Schmierungstechnik 7(7) (1976) 203–206.

[19] S. Wu, Polymer Interface and Adhesion (Marcel Dekker, New York, 1982).

[20] G.W. Ehrenstein, Polymer-Werkstoffe, Struktur und mechanisches Verhalten (Hanser, Munich, 1978).

[21] D. Hull, An Introduction to Composite Materials (Cambridge University Press, London, 1981).

[22] W. Michaeli and M. Wegener, Einführung in die Technologie der Faserverbundwerkstoffe (Hanser, Munich, 1989).

[23] M.R. Piggot, Failure processes in fiber composites. In: Failure of Plastics, eds W. Brostow and R.D. Corneliussen (Hanser, Munich, 1986).

[24] W.W. Tsai and H.T. Hahn, Introduction to Composite Materials (Technomic Publishing Company, Westport, CT, 1980).

[25] J. Peterman, Eigenverstärkung von Kunststoffen. Verbundwerkstoffe und Werkstoffkunde in der Kunststofftechnik (VDI-Verlag, Düsseldorf, 1982) 83–99.

[26] A.J. Pennings and J.M.A.A. van der Mark, Hydrodynamic induced crystallization of polymers from solution, Rheol. Acta 10 (1971) 174–186.

[27] Cl.G. Maertin, Zur Eigenverstärkung von Polyethylen im Spritzguß, PhD Dissertation (Universität Kassel, 1988).

[28] G.W. Ehrenstein and J. Song, Strukturelle Beeinflussung von Gleitelementen aus Polymerwerkstoffen zur Erzielung eines günstigeren Reibungs- und Verschleißverhaltens, Research Report Eh60/28–2 (DFG, Kassel, 1988).

[29] T. Weng, A. Hiltner and E. Baer, Hierarchical structure in a thermotropic liquid-crystalline copolyester, J. Mater. Sci. 21 (1986) 744–750.

[30] Z. Ophir and Y. Ide, Injection molding of thermotropic liquid crystal polymers, Polymer Eng. Sci. 23(14) (1983) 792–796.

[31] G. Regnault, Prüfung feintechnischer Kunststoff-Gleitlager, PhD Dissertation (ETH Zürich, 1979).

[32] J. Song and G.W. Ehrenstein, Zum Reibungs- und Verschleißverhalten eigenverstärkter Kunststoffe, in: Reibung und Verschleiß bei metallischen und nichtmetallischen Werkstoffen, ed. K.-H. Zum Gahr (DGM Informationsgesellschaft Verlag, Oberursel, 1990).

[33] A.W. Neumann, D.R. Absolom, D.W. Francis and C.J. van Oss, Conversion tables of contact angles to surface tensions, Sep. Purif. Methods 9(1) (1980) 69–163.

[34] M.G. Dobb, The production, properties and structure of high-performance poly(p-phenylene terephthalamide) fibres, in: Strong fibers, eds W. Watt and B.V. Perov (Elsevier, Amsterdam, 1985).
[35] P.K. Mallick, Fiber-reinforced Composites (Marcel Dekker, New York, 1988).
[36] J. Song, Cl.G. Maertin and G.W. Ehrenstein, The effect of self-reinforcement on the tribological behaviour of thermoplastics, Proc. ANTEC 88, SPE, Atlanta (Society of Plastics Engineers, Brookfield, 1988).
[37] D.F. Moore, Principles and Applications of Tribology (Pergamon Press, Oxford, 1975).
[38] E. Southern and R.W. Walker, A laboratory study of the friction of rubber on ice, in: Advances in polymer friction and wear, ed. L.H. Lee (Plenum Press, New York, 1974).
[39] R.G.C. Arridge, Mechanics of Polymers (Clarendon Press, Oxford, 1975).

Advances in Composite Tribology
edited by K. Friedrich

Chapter 3

Reciprocating Dry Friction and Wear of Short Fibre Reinforced Polymer Composites

A. SCHELLING * and H.H. KAUSCH

Laboratoire de Polymères, Ecole Polytechnique Fédérale de Lausanne, EPFL, 1015 Lausanne, Switzerland

Contents

* Present address: Du Pont de Nemours International S.A., 2 chemin du Pavillon, P.O. Box 50, CH-1218 Geneva, Switzerland.

Abstract

Most of the evaluations of friction and wear properties of materials are done under the conditions of continuous sliding movements, while some major applications are based on reciprocating mechanisms. Some theoretical assumptions and experimental results suggest that the continually reversing load which acts on the material during the reciprocating friction, could influence its wear rate and its friction coefficient.

The tribology of polyetheretherketone (PEEK), of carbon fibre reinforced PEEK (PEEK D450HF30) and of glass fibre reinforced PTFE (PTFE LUB904) is evaluated under dry reciprocating movement. A new high-speed reciprocating test rig had been designed and a numerical model simulating the mechanical stresses which affect the pin during reciprocating friction, was developed and used to analyze the tribological behaviour of the three materials. The friction coefficient of PEEK is studied under different reciprocating conditions. Various friction processes affecting this material are identified. The PEEK D450HF30 and the PTFE LUB904 are shown to have better friction properties than the PEEK. The wear rate of PEEK is higher under reciprocating than under continuous movement. This more severe wear is due to the existence of a mixed abrasive/adhesive contact condition. The wear of PEEK D450HF30 is caused by fibre fracture and pull-out on a ground counterface and mainly by adhesion when sliding on a smooth opposite part. The interface temperature highly influences the wear rate of this composite. The wear rate of the PTFE LUB904 is caused by one of its glass fibres abrading the steel counterface. For both composites, the type of movement has mainly a qualitative influence on the wear resistance.

1. Introduction

There are more and more industrial applications in which one tries to replace lubricated sliding contacts by pairs of materials which can run dry.

This type of requirement is justified for mechanisms which are used in the food and beverage industry (filling machine), where avoiding contamination of the product and the environment is an absolute necessity. Such a requirement is also justified when designing dry compressors which supply clean compressed gases for electronic, pharmaceutical, chemical and petrochemical purposes, as well as for hospitals and health care institutions. Removing the feed pump, filters and all kinds of similar equipment simplifies the design and helps to reduce the manufacturing and maintenance costs.

When changing the design of the lubricated mechanism is not feasible, materials used for parts like bearings, piston rings, vanes and gears have to be substituted by others to support the dry sliding contact. One looks in general for materials having low friction coefficients and wear rates under dry sliding conditions. Polymer composites have such properties.

In most industrial applications, a pure polymer cannot support the combination of high stress and high temperature which occurs at the contact points of a dry sliding contact. Adding reinforcing elements like fibres (carbon, glass or aramid) enhances the wear resistance, and solid lubricants (graphite, PTFE or MoS_2) reduce the friction coefficient.

Since thermoplastics are more convenient to process than thermosets, they are often used in designing mechanical parts with sophisticated shapes. Polytetrafluoroethylene (PTFE) is already known to have self-lubricating properties, but resins like polyetheretherketone (PEEK), polyimide (PI), polyamide (PA), polyethersulfone (PES) and polyphenylenesulfide (PPS) are potential candidates for applications like those mentioned above.

Materials intended to be used in friction and wear applications are in general submitted to selective tests on experimental devices like the pin on disc, pin on cylinder or the well known thrust/washer ASTM D3702-78. Generally these commercial test rigs simulate a continuous sliding movement between the two contacting parts. It has to be noted that a lot of industrial applications have to deal with friction and wear due to linear reciprocating or angular oscillating sliding movements. It will be shown in this chapter that some polymers and polymeric composites behave differently under continuous and reciprocating friction and, particularly, that friction and wear data collected on a continuous pin/disc device can highly underestimate the wear rate under reciprocating movement.

In the following some fundamentals on friction and wear of materials will be given with particular attention to the type of friction, i.e., continuous or reciprocating.

1.1. Friction and wear of polymers and polymeric composites

According to Amontons–Coulomb, the friction coefficient, μ (see eq. (1)), is defined as the ratio between the friction force, F_f, needed to create sliding and the normal force, F_n, which presses the two bodies against each other (fig. 1).

$$\mu = F_f/F_n. \tag{1}$$

In some cases it is usefull to differentiate the static friction coefficient, μ_s, from the dynamic friction coefficient, μ_d. The first one is measured just before sliding and the latter during sliding. In the litterature, in general, μ is used to express μ_d.

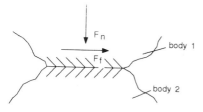

Fig. 1. The friction coefficient is defined as the ratio between the friction force F_f and the normal force F_n.

The wear rate, W, of a material corresponds to its volume loss per sliding distance (see eq. (2)). It is defined as [1]:

$$W = \Delta V / LA,\qquad(2)$$

where ΔV is the volume loss of the material, L the sliding distance, and A the area of contact.

One also talks about specific wear rate, W_s, when W is divided by the contact pressure, p.

The product pv [2] of the contact pressure, p, and the sliding velocity, v, is used to compare different testing conditions or to express the limit of application of a material.

The concept of *adhesive friction* between two contacting surfaces has been proposed by Bowden and Tabor [3]. The friction force is created by the breaking of microscopic forces on an atomic scale at the interface of the two contacting bodies. In the case of polymers the adhesive junctions are due to the Van der Waals forces and the diffusion mechanism between the two materials at the interface. This theory has been verified experimentally by Briscoe and co-workers [4] with polymers such as PMMA, LDPE, HDPE and PTFE.

The concept of *adhesive wear* has been developed by Archard [5]. The adhesive forces provoque fracture within the materials, which creates wear particles. According to this author, the worn volume generated from a material is defined as

$$\Delta V = k F_n L / H,\qquad(3)$$

where k is the wear coefficient and H the hardness of the worn material.

Friedrich [6] has shown that the surface of a brittle polymer, such as PPS, worn by adhesion is characterized as the presence of cracks and the material is removed as plates. A ductile polymer like PTFE is marked by large plastic deformations.

Abrasive friction occurs when the hardnesses of the two contacting materials are different by several orders of magnitude. When sliding, the hard particles create plastic deformations (micro-ploughing) in the softer counterpart, which causes the abrasive friction force. The corresponding *abrasive wear* of the softer material depends on the cutting properties of the hard particles (hardness and geometry) as well as on its resistance to penetration and micro-cracking. Abrasive wear is in general far more severe than adhesive wear. For example, in ref. [7] was shown that for similar sliding velocity and contact pressure, PEEK was wearing 5 orders of magnitude more when the counterface was covered with 15 μm abrasive grains than by pure adhesion on a smooth surface.

The *fatigue* or *delamination wear* concept has been introduced for metals [8]. A model has been developed and then adapted to other materials [9]. When asperities are rubbing on a counterface, cyclic stresses are generated in its subsurface. Such repetitive stresses acting on a defect located close to the surface can create cracks and then provoque their propagation. Finally, a wear particle is formed when the cracks arrive at the surface of the material. Such a wear mechanism has been observed in the case of a polyimide reinforced with glass fibres and hard particles rubbing against a smooth steel counterface [10].

The mechanisms presented above describe the friction and wear of pure polymers as well as of polymeric composites. When a composite is under stress, part of the load is transferred from the lower modulus (resin) to the higher modulus (fibres) component in order to respect the displacement continuity within the material. This has been demonstrated to be also valid when a composite is submitted to friction, i.e., the fibres close to the interface tend to support most of F_n and F_f [11]. By using an appropriate rule of mixtures [12], a fairly good approximation of the friction coefficient of a composite (carbon fibre/epoxy) has been obtained based on those of its components. Predictions of wear rates of PA 6.6 and PES composites have also been made using the wear rates of the components and these were verified experimentally [6]. A rule of mixture is in general valid for one friction/wear mechanism only.

When a polymer is wearing, the debris tends first to fill the cavities of the counterface, which creates progressively a so-called *transfer film*. After a period of time, a part of the polymer slides on itself. Under specific testing conditions, its notable influence on the friction coefficient and the wear rate have been shown for LDPE, PVC and PCTFE [13], and also for PTFE [14].

This chapter is dealing particularly with polymers and composites which can support dry friction above a temperature of 100°C, a sliding speed in the range of 5 m/s and a contact pressure of 5–10 bar. These are mainly the operating conditions of the piston ring in a reciprocating dry compressor.

PEEK is an aromatic semicrystalline polymer, which keeps its mechanical properties up to 200°C. It has the advantage to be processable by injection moulding. Friedrich and co-workers have studied its wear through adhesion [15] and abrasion [7] on both pure and composite forms and under different testing conditions and microstructures. It has been shown that the type of fibres (carbon, glass or aramid) and their orientation (normal or parallel to the friction plan) had to be defined in order to get the best friction and wear performances.

Briscoe et al. [16] have studied the friction and wear properties of the composite PEEK/PTFE. Adding PTFE tends to decrease the friction coefficient, but has a negative effect on the wear rate of the composite.

1.2. Reciprocating movement

When the amplitude of the reciprocating movement of the two parts is in the range of 0.1 mm, the wear is generally called *fretting wear* [17]. Such surface degradation is due to vibrations which occur at the contact points between elements like leaf springs, bolted joints, flanges and sealings. The wear mechanisms presented above can also be used to describe the fretting wear [17].

Other authors [18] have compared the friction coefficients and wear rates of polyimide under continuous movement or fretting contact. Tribological similarities and differences have been noted between the two movements. Basically the two kinematics create two different possibilities for the wear particles to escape the interface, which influences the morphology of the transfer film and thus the wear process.

As under continuous movement [19,20], fibre orientation with respect to the sliding surface and direction influences the friction coefficient and wear rate of the composite. Carbon fibre reinforced plastics under fretting on mild steel have shown optimum friction and wear properties when the fibres were located in the friction plane and had an orientation parallel to the sliding direction [21].

Fretting wear of glass, carbon and aramid fibre reinforced PEEK as well as carbon and glass reinforced epoxy have been studied under a sliding amplitude up to 600 mm [22]. For pure PEEK, these conditions created a more severe wear than the ones due to continuous movement, all other parameters being the same. The fibre contribution was clearly different from that under continuous movement [23].

Depending on the way the fretting force is applied to the part, significant cyclic stresses are generated and transmitted out of the contact area. Work on carbon fibre reinforced PEEK has shown [24] that these forces can reduce the lifetime of the element by fatigue (*fretting fatigue*).

When the amplitude of the reciprocating movement between the two sliding parts is in the range of 1–100 mm or higher, one talks about *reciprocating wear*.

Studies on some polymers [25,26] have indicated that each change of direction of the sliding velocity during the test tended to modify the friction coefficient. Adding a rotation to the linear continuous friction creates an increase in the friction coefficient of semicrystalline polymers like PTFE, HDPE and UHMWPE. This was attributed to a need for the macromolecules located at the contact surface to reorientate. Under the same conditions, this phenomenon has not been observed for other polymers, like LDPE, PP or PMMA. According to these observations, the friction coefficient of some polymers could be higher under reciprocating movement than under continuous movement, especially if the period of the movement leaves enough time for the macromolecules to change orientation.

A broad quantitative overview has been given in 1982 in order to select materials for the piston ring of the Stirling engine [27]. The stroke of the reciprocating movement was 40 mm, the counterface made of nitrited steel and the interface temperature was in the range of 80°C. A specific wear rate of 0.15×10^{-15} m^3/N m and a friction coefficient of 0.22 have been obtained with PTFE reinforced with carbon fibres and solid lubricants.

The maximum pv products of pure and reinforced PTFE and PI have been reevaluated for piston rings of dry compressors [28]. The specificity of the reciprocating movement has been introduced in order to be able to correct the pv value of the material when the one corresponding to continuous friction is known. Knowing the wear coefficient of the material under continuous friction K_w, the wear (mm h^{-1}) under reciprocating movement is (symbols as in ref. [28]):

$$W = 0.25 K_w K_t p_m v_m,$$

where
K_w = wear coefficient of the material (MPa^{-1}),
K_t = T_s/T_d = ratio between the suction and the discharge temperatures,

v_m = $2R\omega/\Pi$, average linear velocity of the piston,

$$p_m = \frac{k}{k-1} p_s \left\{ \left(\frac{p_d}{p_s} \right)^{(k-1)/k} - 1 \right\},$$

p_m = average pressure on the piston ring,
p_d = discharge pressure,
p_s = suction pressure,
k = c_p/c_v = ratio of the specific heats of the gas.

This study has established the relation between the wear rate of the piston ring and the mechanical characteristics of the compressor.

The *third-body* approach [29] is a powerful tool used to analyze and predict the tribological behaviour of materials. This concept is helpful when wear is due to oscillating or reciprocating movement. The wear behaviour of radial bearings made of glass fibre reinforced PTFE in contact with an oscillating steel shaft has been analyzed using the third-body approach [30]. In this particular case, it has been shown that the oscillating movement was less severe than the continuous one under similar testing conditions. The debris tended to stay at the interface and acted as rolls carrying the load. For small angles of oscillation, typically below $\pm 5°$, the transfer film on the steel shaft became irregular in thickness and affected the wear rate of the composite negatively. Similar work has been done on reciprocating linear movement [31]. With the use of a parameter called *Mutual Overlap Coefficient* (MOC), it had been demonstrated that the stroke of the movement for a given sliding distance was a parameter influencing the wear rate of the contacting parts.

1.3. Objectives

It has been shown above that on both the microscopic and macroscopic scale, the reciprocating movement can generate a more distinct friction and wear behaviour than that under continuous movement.

The first objective of this chapter will be to characterize the friction and wear of PEEK under dry reciprocating movement for different testing conditions.

The second part will consider the possibility for this material to be used for piston rings of reciprocating compressors. The experimental parameters, like sliding velocity, contact pressure, interface temperature and counterface, will simulate application conditions.

Finally, it will be demonstrated that a macroscopic model considering the deformation of the contacting parts during friction can help to understand their tribology.

2. Experimental details

2.1. Materials

The physical properties of semicrystalline polyetheretherketone (PEEK) have been described in details elsewhere [32]. It has a glass transition temperature $T_g = 143°C$ and a melting temperature $T_m = 334°C$.

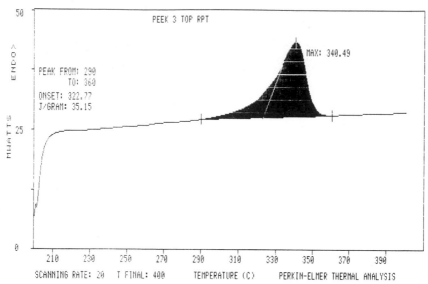

Fig. 2. Typical DSC curve of PEEK after injection moulding. Scanning rate 20°C/min.

The testing samples are cut from cylindrical injection molded bars having a diameter of 20 mm and a length of 50 mm. The neat resin is provided by ICI and has the commercial denomination "PEEK 450G". The processing parameters are those recommended by the supplier. The degree of crystallinity [33] of the specimen contact surface measured by DSC after processing is 28.7%. Figure 2 shows a typical DSC curve with the neat PEEK with a scanning rate of 20 K/min.

The PEEK composite tested here is also provided by ICI and has the commercial denomination "PEEK D450HF30". It contains about 10% carbon fibres, 10% PTFE and 10% graphite. The type of carbon fibre, i.e., high strength or high modulus, has not been revealed by the supplier. It is proccessed in the same manner as the neat resin. Figure 3 shows its microstructure after injection molding and before wear. The fibres are mainly oriented in the friction plane and have a diameter of 8–12 μm and an average length of 200 μm.

The PTFE composite is provided by Angst and Pfister and called PTFE LUB 904. It contains about 40% glass fibres having a diameter of 20–25 μm and a length of 300 μm. The microstructure of this composite (fig. 4) indicates that some fibres are already broken before wear.

The counterfaces are made of stainless steel DIN X20Cr13 (13% chromium). The Vickers hardness is 273.9 HV10. The wear tests are made with ground and polished counterparts. The surface parameters R_a, R_z, R_p and R_t (fig. 5) are measured before each trial by contact profilometry. The machining scratches are always oriented perpendicular to the sliding direction, whatever the type of test (continuous or reciprocating). Typical surface parameters for the ground and polished counterfaces are: $R_a = 0.66$ μm, $R_z = 5.97$ μm, $R_p = 3.83$ μm and $R_t =$

Fig. 3. Microstructure of the PEEK D450HF30 before testing: it contains 10 wt% of carbon fibres and 20 wt% of solid lubricants (PTFE and graphite).

7.94 μm and $R_a = 0.03$ μm, $R_z = 0.32$ μm, $R_p = 0.17$ μm and $R_t = 0.49$ μm, respectively.

2.2. Dry sliding reciprocating test rig (high speed–low pressure)

This machine simulates the friction and wear of a piston ring. The testing material is a fixed cylindrical pin, which has a circular contact surface with a diameter of 15 mm. The counterface is a reciprocating rectangular plate, moving in a vertical plane. The stroke is 71.5 mm.

2.2.1. Generalities

This device is composed of a four cylinder engine driven by an electrical 380 V DC motor. For the purpose of the test, the cylinder head was replaced by four test cells each located above one of the four cylinders. Figure 6 shows an open cell during the development stage.

A hollow rod is welded on the top of each piston upon which two exchangeable counterplates were fixed. Warm or cold oil can circulate through the rods and modify by conduction their temperature within a range of 20 to 150°C. Each cell tests two pins, and the complete device eight simultaneously. The humidity in the

A. Schelling, H.H. Kausch

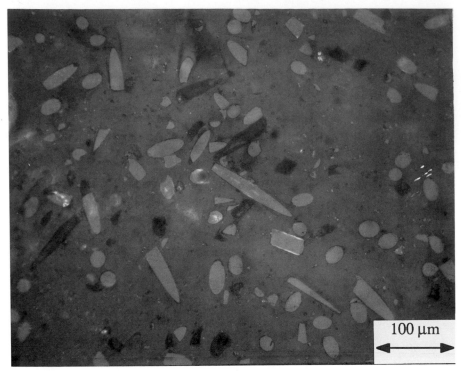

Fig. 4. Microstructure of the PTFE LUB904 before testing: it contains 40 wt% of glass fibres.

cells is constant during the experiments (50%). Contamination of the samples is avoided by using bellow rings, which connect the cell to the reciprocating shaft.

When the crankshaft is rotating the reciprocating sliding velocity of the counterplate is:

$$v(t) = \omega r \left(\sin \omega t + \tfrac{1}{2} \lambda_b \sin 2\omega t \right), \tag{4}$$

where ω is the angular velocity of the crankshaft, $2r$ the piston stroke, λ_b the connecting rod ratio, and the average sliding velocity during one period is defined as:

$$v = 2r\omega/\pi, \tag{5}$$

which is limited to 3.2 m/s because of the vibrations created by the bellow rings.

Each pin is mounted in a cylinder and pressed to the counterplate by pneumatic pressure (fig. 7). The contact pressure can reach 10 bar, which creates typically a normal force F_n on the pin of about 177 N for a contact surface area of 177 mm^2 (pin dia. 15 mm). Using this method, the average contact pressure between the pin and the plate was maintained constant during the duration of the test, regardless of the wear rate.

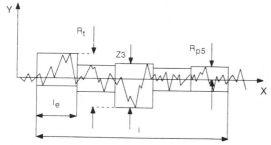

Fig. 5. The contact profilometry is used to define the roughness parameters R_a, R_z, R_p and R_t of the counterface before and after the friction and wear tests. The profile is measured along a 50 mm stroke (l) which is subdivided into five equal lengths of 10 mm (l_e) to calculate the parameters:

$$R_a = \frac{1}{l} \int_0^l |Y| \, dx,$$

$$R_z = \frac{1}{5} \sum_1^5 Z_i, \text{ where } z = \text{largest top–valley distance measured on } l_e.$$

$$R_p = \frac{1}{5} \sum_1^5 R_{pi}, \text{ where } R_{pi} = \text{largest top–middle line distance measured on } l_e.$$

$R_t = $ largest top–valley distance measured on the profile.

2.2.2. *Measurement of the friction coefficient* μ *(fig. 7)*

F_n is transmitted to the wall of the cell by a metallic web. A second web, perpendicular to the previous one and parallel to the friction plane, transmits F_f to a rigid beam. Strain gauges are glued on this web in order to be able to evaluate

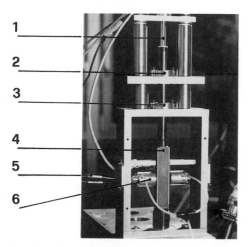

Fig. 6. Cell of the high speed–low pressure reciprocating test rig at the development stage. 1: heating/cooling nozzle, 2: bearing, 3: seal, 4: reciprocating plate, 5: thermocouple, 6: pin-holder.

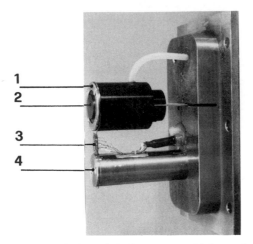

Fig. 7. Pin-holder of the low pressure–high speed reciprocating test rig. 1: pressurized cylinder, 2: pin, 3: strain gauges, 4: rigid beam.

F_f. The friction coefficient μ is defined by sampling during one period of the reciprocating movement:

$$\mu = \frac{1}{nF_n} \sum_{1}^{n} |F_{fi}|,$$

(6)

where $n = 100$.

2.2.3. Measurement of the interface temperature

The interface temperature resulted from the heat generated by the friction process a well as the heat conducted by the cooling/heating device (section 2.2.1.) or another cell working under different testing conditions. It is tried here to assess the average interface temperature, T_i, from the temperature measured in the pin during the test.

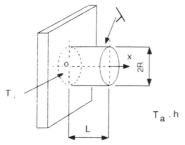

Fig. 8. The thermal model of the pin-on-plate contact is used to evaluate the interface temperature during friction, knowing the temperature in the pin.

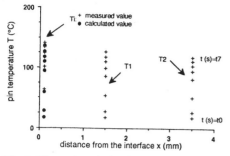

Fig. 9. Thermal analysis of PEEK: temperature profile along the axis of a cylindrical pin. +: interface temperature measured by contact, ●: interface temperature extrapolated from the model.

A simple thermal model of the pin-on-plate contact (fig. 8) has been developed elsewhere [34]. The interface temperature T_i is defined as:

$$T_i = [T(x) - T_a] \frac{\cosh mL}{\cosh m(L - x)} + T_a \text{ and } m = \sqrt{\frac{hC}{\lambda A}}, \qquad (7)$$

where $T(x)$ is the temperature in the pin measured at a distance x to the interface, T_a the ambient temperature in the cell, h the coefficient of thermal convection between the pin and the cell, $2R$, C and A, the diameter, circumference and contact area of the pin, λ the thermal conductivity of the pin, and L the length of the pin.

We assumed here that the relation between T_i and $T(x)$ is in first approximation not influenced by the sliding velocity.

The temperature profile along the axis of a PEEK pin is presented in fig. 9. The T_i extrapolated from the model is compared to T_{ic} measured with a thermocouple in contact with the surface of the plate at $v = 0$. The different temperature levels are obtained by heating the plate. T_1 and T_2 are measured in the pin by thermocouples located at 1.5 and 3.5 mm from the interface, respectively (fig. 10).

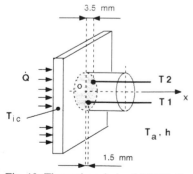

Fig. 10. Thermal analysis of PEEK: the temperature is measured in the pin by two thermocouples T1 and T2 located, respectively, at 1.5 and 3.5 mm from the contact surface. T_{ic} is measured by contact when the plate is stopped. The temperature of the counterplate can be changed by internal heating/cooling (Q).

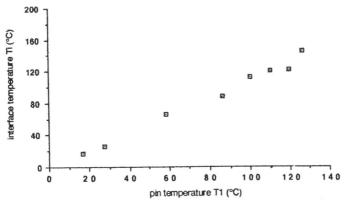

Fig. 11. Thermal analysis of PEEK: relation between the interface temperature extrapolated from the model and the temperature measured by T1. This chart is used to find the interface temperature during the friction and wear tests.

T_i is always lower than T_{ic} (fig. 9). This could be due to the fact that the heat is transmitted from the plate to the pin through the metallic asperities and not by the whole surface A as is assumed by the model. Consequently, the conductivity of the pin close to the interface is reduced. Figure 11 shows for PEEK the relationship between T_i and T_1.

Both friction coefficients and temperatures are recorded during the experiment and stored by a microcomputer.

2.2.4. Numerical model of the pin-on-plate contact

In order to have an adequate friction force measurement, the pin has to be able to move freely in the axial direction. Since the web which absorbs and measures the friction force is (0.1 mm) away from the friction plane, a tilting effort acts on the pin during sliding (fig. 7). Thus the contact pressure and the shear stresses at the interface are not symmetrical, i.e., those in front differ from those behind the contact.

The dry sliding conditions according to Amontons–Coulomb are: adhesion if $F_f < \mu F_n$ and sliding if $F_f = \mu F_n$.

Since these forces are non-linear, a numerical model is needed to define precisely the contact conditions between the pin and the plate and quantify the stresses (shear and pressure) at the interface during sliding.

The TACT [35] program is used to solve this problem. It combines the finite-element method (FEM) to describe the geometry of the parts and the linear iteration method to solve the equations. These well-known methods and their implementation are described elsewhere [36].

2.2.4.1. *Boundary conditions.*   The model considers a 2D plane-strain linear elastic calculation. Plane strain means that the deformation of the pin normal to the sliding direction is negligible, which is a reasonable assumption regarding its thickness.

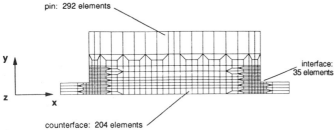

pin: 292 elements

interface:
35 elements

counterface: 204 elements

Fig. 12. Numerical model of the pin-on-plate contact: the pin, the interface and the plate are decomposed into 292, 35 and 204 finite elements, respectively.

The pin and the plate are decomposed into 292 and 204 elements, respectively, each element containing four nodes (fig. 12). A layer of 35 friction-gap elements connecting the two bodies is used to simulate the Amontons–Coulomb sliding conditions presented above. These require that no traction and no penetration occur perpendicular to the sliding direction and that the condition:

$$F_f \leqslant \mu F_n$$

is valid in the sliding direction.

The normal force, F_n, is, as in reality (pneumatic pressure), distributed over the whole cross section of the pin. F_f is distributed over 2 mm and corresponds to the action of the O-ring through which the friction force generated at the interface is equilibrated (fig. 13).

The elastic moduli, E, and the Poisson coefficients, ν, are the following:

Stainless steel: $E = 2.2 \times 10^{11}$, $\nu = 0.30$,
Neat PEEK: $E = 3.7 \times 10^{9}$, $\nu = 0.42$,
PEEK composite: $E = 1 \times 10^{10}$, $\nu = 0.42$,
PTFE composite: $E = 1.3 \times 10^{9}$, $\nu = 0.42$.

2.2.4.2. Results.   Figures 14 and 15 show the contact pressure and shear stresses in the PEEK at the interface, for the adhesion and sliding conditions, respectively

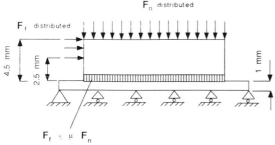

F_n distributed

F_f distributed

4.5 mm

2.5 mm

1 mm

$F_f \leqslant \mu F_n$

Fig. 13. Boundary conditions applied in the numerical model. The normal force F_n and friction force F_f are, as in reality, simulated by distributed loads.

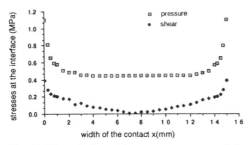

Fig. 14. Numerical model: contact pressure and shear stress repartition at the interface when the PEEK is in contact with the DIN X20Cr13 (adhesion). Contact pressure $p = 0.5$ MPa and $\mu = 0.35$.

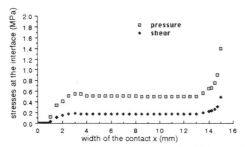

Fig. 15. Numerical model: same as fig. 14 but under sliding conditions.

($p = 0.5$ MPa and $\mu = 0.35$). Figure 16 indicates the corresponding deformed mesh during sliding.

The pressure repartition for adhesion (fig. 14) corresponds to historical values obtained when the elastic moduli of the two contacting materials are separated by several orders of magnitude [37]. The pressure becomes theoretically infinite at the extremities of the contact, which practically generates plastification of the polymer. The shear is proportionnal to the contact pressure at the place where micro-sliding occurs (fig. 14).

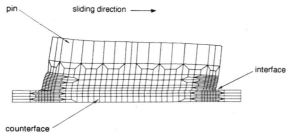

Fig. 16. Numerical model: deformed mesh corresponding to the friction PEEK/DIN X20Cr13. $p = 0.5$ MPa and $\mu = 0.35$. The interface contact pressure is equal to zero at the place where there is no more contact (fig. 15).

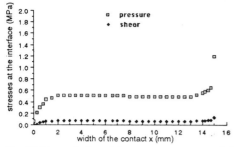

Fig. 17. Numerical model: contact pressure and shear stress repartition at the interface when the PEEK D450HF30 slides on the DIN X20Cr13. $p = 0.5$ MPa, $\mu = 0.13$. Conversely to PEEK (fig. 15) contact occurs everywhere.

Under sliding conditions, the pin tends to loose contact at the place where F_f is applied (fig. 16). At this place the contact pressure is equal to zero (fig. 15). The PEEK composite slides with a friction coefficient of 0.13 ($p = 0.5$ MPa). Figure 17 shows that no loss of contact occurs with the PEEK composite under these conditions. PTFE is also always everywhere in contact with the plate.

If we now consider $\beta = A_g/A$, where A_g is the contact surface during sliding, A the initial contact surface for $v = 0$ (adhesion), then fig. 18 shows how the friction coefficient influences the contact surface of the PEEK pin under sliding conditions. For $\mu > 0.26$, the loss of contact occurs locally, and consequently the friction is not reciprocating on the whole surface of the pin anymore. This observation will be used to analyze the friction and wear tests results presented in this chapter.

2.3. Dry sliding reciprocating test rig (low speed–high pressure)

Figure 19 shows this machine, which has been described in detail elsewhere [38]. Contrary to the previous device, friction occurs here in the horizontal plane. A fixed pin is mounted on a support having strain gauges to measure F_f and F_n. The contact pressure can reach 100 MPa. The countersurface is a mobile rectangular bar. Its velocity is constant along the stroke of 50 mm and can reach 0.037 m/s.

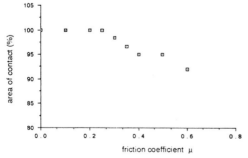

Fig. 18. Numerical model: influence of the friction coefficient of PEEK sliding on DIN X20Cr13 on the contact surface during sliding. Loss of contact occurs above $\mu = 0.26$.

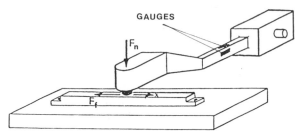

Fig. 19. Low speed–high pressure reciprocating test rig: the plate is moving horizontally and the pin is fixed and mounted in a support having strain gauges to measure F_f and F_n (from ref. [38]).

2.4. Dry sliding continuous test rig (Falex)

Like in the previous device, the friction occurs in the horizontal plane. Three rotating pins, each having a diameter of 6.3 mm, are pressed on a fixed disc, whose temperature can be adjusted internally by cooling/heating water. A thermocouple is located in the disc close to the interface and provides the average value of the interface temperature. The friction track has a diameter of 33 mm. The sliding velocity can reach 10 m/s. The friction force, F_f, is measured by strain gauges and the normal force, F_n, is applied using dead weights.

3. Friction under dry reciprocating movement

This section presents first friction tests of neat PEEK under high contact pressure (up to 100 MPa) and low sliding velocity (up to 0.037 m/s). Then, the neat PEEK as well as the PEEK and PTFE composites will be evaluated under high reciprocating speed (up to 2.5 m/s) and low contact pressure (0.5 MPa).

3.1. Reciprocating friction of PEEK under low speed–high pressure

In order to have a wide range of contact pressures, the contact sections of the pins varied from 1 to 276.5 mm^2. The degree of crystallinity of the neat PEEK was 28.7%. Three different materials have been used for the counterface, i.e., stainless steel DIN X20Cr13 (13% chromium), cast iron GG25 and neat PEEK (450G). The countersurfaces are ground and polished; table 1 lists their roughness parameters

TABLE 1
Surface roughness parameters measured for the stainless steel (DIN X20Cr13), cast iron (GG25) and neat PEEK 450G counterfaces before testing.

	DIN X20Cr13	GG25	PEEK 450G
R_a (μm)	0.16	0.08	0.26
R_z (μm)	1.50	1.20	1.91
R_p (μm)	1.19	0.75	2.16
R_t (μm)	2.70	2.38	4.79

TABLE 2

Influence of the normal force F_n (or the contact pressure p) on the friction coefficient μ of neat PEEK 450G sliding on stainless steel DIN X20Cr13 and cast iron GG25. Low-speed reciprocating friction ($v = 0.0016$ m/s).

F_n max (N)	p max (MPa)	μ/DIN X20Cr13	μ/GG25
100	100	0.15–0.11	0.15–0.18±0.05
100	0.53	0.17	0.20±0.05
100	0.39	0.18	0.20±0.05
100	0.36	0.17	0.20±0.05

before testing. Before each new test, the contact surfaces of the pins and the counterfaces are cleaned with trichloroethylene, followed by a running-in period. The ambient temperature and humidity are, respectively, 20°C and 50%. Considering the frequency of this reciprocating movement (1.67×10^{-2} s^{-1}), one can assume that the friction occurs under isothermal conditions.

3.1.1. Contact PEEK / DIN X20Cr13 and GG25

Table 2 shows, for different contact pressures, the friction coefficient of PEEK on DIN X20Cr13 and GG25 at 0.0016 m/s. Under high contact pressure, the friction coefficient changes, typically from 0.11 to 0.15 on stainless steel and from 0.15 to 0.18 on GG25.

Figure 20 shows the influence of the contact pressure on the friction force (or on μ) at 0.0016 m/s. The pressure starts initially at a minimum at one extremity of the stroke (0 MPa), increases constantly during sliding and reaches the maximum value at the end of the stroke. The reverse happens on the way back. The friction force (or μ) is measured during this reciprocating movement. Figure 20 shows that a change of μ occurs both for stainless steel and GG25 at $p = 45$ MPa. Figure 21 indicates that the friction coefficient is not influenced by the contact pressure when it varies from 0 to 0.4 MPa, which confirms in that case the second Amontons–Coulomb law.

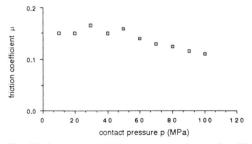

Fig. 20. Low speed–high pressure reciprocating friction of PEEK on DIN X20Cr13. Influence of the contact pressure on μ ($v = 0.0016$ m/s): above $p = 45$ Mpa, the friction coefficient decreases.

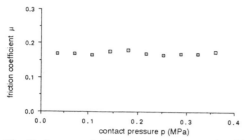

Fig. 21. Low speed–high pressure reciprocating friction of PEEK on DIN X20Cr13. Influence of the contact pressure on μ ($v = 0.0016$ m/s): for low values of p, μ stays constant.

Studies done with this device [39] have demonstrated that the friction coefficient of GG25 on GG25 was not influenced by the contact pressure in that range of sliding velocities.

According to the Mohr–Coulomb failure criterion:

$$\tau_c = \tau + \mu_i \sigma_n,$$

where τ_c is the shear limit of the material according to Mohr–Coulomb, τ the shear stress in the material, μ_i the internal friction coefficient, and σ_n the normal stress in the shear plane.

If we assume as a first approximation that $\mu_i = 0.05$, as is the case for most polymers, 45 MPa of normal pressure and 6.75 MPa (0.15×45 MPa) of shear create a Mohr–Coulomb shear stress level τ_c in the material of 9 MPa at the μ transition. In order to find out if this change in friction behaviour corresponds to the Mohr–Coulomb limit reached by PEEK, its behaviour under combined shear and compression stresses should be measured and friction tests with other polymers having similar mechanical properties should be done.

The sliding velocity, the maximum value of which is limited here to 0.016 m/s, shows no influence on the friction coefficient of PEEK under 50 MPa of contact pressure (fig. 22). This is probably due to the quasi-isothermal conditions under which this test is conducted.

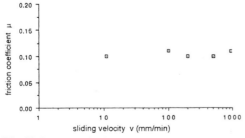

Fig. 22. Low speed–high pressure reciprocating friction of PEEK on DIN X20Cr13. Influence of a low sliding velocity on μ ($p = 50$ MPa).

$\mu_s = \mu_{max}$ et $\mu_d = \frac{1}{2}(\mu_{min} + \mu_{max})$

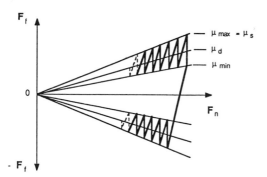

Fig. 23. Definition of μ_s (static) and μ_d (dynamic) for the low speed–high pressure reciprocating friction test.

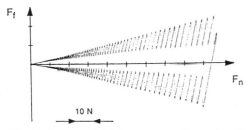

Fig. 24. Low speed–high pressure reciprocating friction of PEEK. Influence of the contact pressure on μ ($v = 0.0016$ m/s). Under low contact pressure, PEEK generates extensive stick–slip when sliding on itself. The pressure varies from 0 to 100 MPa.

3.1.2. Contact PEEK / PEEK

These conditions are characterized by a significant stick–slip, i.e., $\mu_s \neq \mu_d$. Figure 23 shows how we calculate μ_s and μ_d (section 1.2.). Figures 24 and 25 show the behaviour of PEEK when rubbing on itself at 0.0016 m/s under low and high contact pressure, respectively. Low pressure creates more instability than high

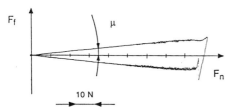

Fig. 25. Low speed–high pressure reciprocating friction of PEEK. Influence of the contact pressure on μ ($v = 0.0016$m/s). A high contact pressure generates less stick–slip than a low one (fig. 24). The pressure varies from 0 to 0.53 MPa.

TABLE 3
Influence of the normal force F_n (or the contact pressure p) on the static and dynamic friction coefficient, μ_s and μ_d, of neat PEEK 450G when sliding on itself. Low-speed reciprocating friction ($v = 0.0016$ m/s).

F_n max (N)	p max (MPa)	μ_s	μ_d
100	100	0.1–0.08	0.1–0.08
100	0.53	0.25 ± 0.02	0.18 ± 0.02
100	0.39	0.25 ± 0.02	0.18 ± 0.02
100	0.36	0.25 ± 0.02	0.18 ± 0.02

pressure. Under high contact pressure, Table 3 indicates a decrease of μ_d with p, like that observed when PEEK was sliding on a steel counterface (section 3.1.1.).

Stick–slip is due to the combination of a low rigidity of the pin support with the condition $\mu_s > \mu_d$ [3]. It has been demonstrated that the change from μ_s to μ_d is related to a vertical move (jump) of the pin [40]. In our case the change in rigidity of the counterface is the only parameter which has been modified from one contact (PEEK/steel) to the other (PEEK/PEEK), the surface parameters of the two counterfaces being almost the same. Such behaviour could indicate that PEEK would perform better on a metallic counterface than on itself and that its transfer film could affect its friction behaviour negatively.

μ_s tends to decrease with sliding velocity (fig. 26). This could be due to the fact that the low rigidity of the pin support gives at low displacement velocity more time to the the adhesive forces to develop at the interface. μ_d is not influenced by the sliding velocity, like when the PEEK was rubbing on steel (section 3.1.1.).

Figure 27 shows the worn surface of PEEK after about 10 m of rubbing on stainless steel. The extremities of the contact surface are more marked by the wear. This is probably due to a higher contact pressure at these places, as indicated by the numerical model (section 2.2.4.2.). The next section will show that the wear of the PEEK surface tends to be uniform when the sliding distance is more important.

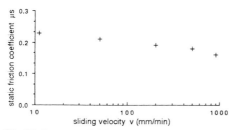

Fig. 26. Low speed–high pressure reciprocating friction of PEEK. Influence of the sliding velocity on μ_s when sliding on itself ($p = 50$ MPa): at high v, μ_s tends to decrease.

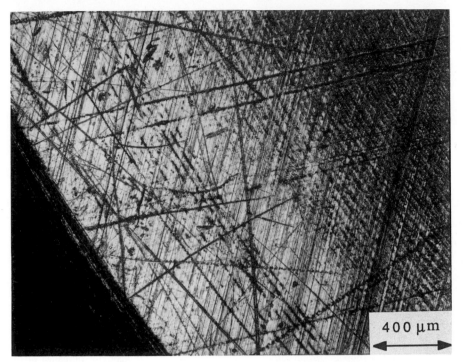

Fig. 27. Low speed–high pressure reciprocating friction of PEEK. Worn surface of the pin after 10 m of sliding on ground stainless steel DIN X20Cr13 ($p = 50$ MPa and $v = 0.0016$ m/s).

3.2. Reciprocating friction of PEEK and its composites at high speed–low pressure

The pins have a contact surface diameter of 15 mm and are rubbing on a stainless steel DIN X20Cr13 (13% chromium) counterface whose initial surface parameters are $R_a = 0.66$ μm, $R_z = 5.97$ μm, $R_p = 3.83$ μm and $R_t = 7.96$ μm. Its Vickers hardness is 273.9 HV10. The contact pressure is 0.5 MPa. The degree of crystallinity of the PEEK is 28.7% before testing. A running-in period precedes each measurement. Eight pins are tested simultaneously and the friction coefficients, wear rates and interface temperatures (T_i) are the average of these eight values.

3.2.1. Contact PEEK / DIN X20Cr13

Figures 28 and 29 show the influence of the sliding velocity on the interface temperature, T_i, and the friction coefficient, respectively. Figure 30 shows the shape of F_f during a reciprocating period at $v = 0.12$ m/s. At $v = 0.5$ m/s, μ reaches an average value of 0.55 during a reciprocating period of movement and locally even 0.7 to 0.9. Figure 31 indicates the influence of T_i on μ at a constant sliding velocity of 0.25 m/s.

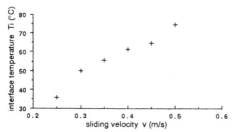

Fig. 28. High speed–low pressure reciprocating friction of PEEK on DIN X20Cr13. Influence of the sliding velocity v on the interface temperature T_i, $p = 0.5$ MPa.

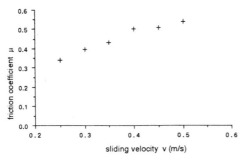

Fig. 29. High speed–low pressure reciprocating friction of PEEK on DIN X20Cr13. Influence of the sliding velocity v on μ, $p = 0.5$ MPa.

The increase in μ is generally related here to a degradation of the countersurface, i.e., the transfer film becomes irregular and consists of large (100 μm in length) and thick (20 μm) plates (fig. 32). The worn PEEK surface is marked by deep furrows as if it was worn by abrasive particles (fig. 33). This non-uniform countersurface corresponds to an irregular friction force (fig. 30). This poor friction behaviour correlates with the findings (extensive stick–slip) with the low speed–high pressure reciprocating device when the PEEK was sliding on itself (section 3.1.2.). This degradation of the contact conditions has also been seen elsewhere, with PTFE under oscillating movement [30].

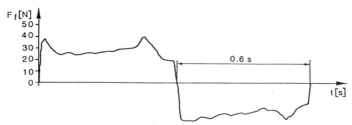

Fig. 30. High speed–low pressure reciprocating friction of PEEK on DIN X20Cr13. Friction force F_f during a period of movement at $v = 0.12$ m/s and $p = 0.5$ MPa.

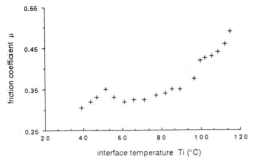

Fig. 31. High speed–low pressure reciprocating friction of PEEK on DIN X20Cr13. Influence of the interface temperature T_i on μ ($p = 0.5$ MPa, $v = 0.25$ m/s).

By comparing figs. 29 and 31, one can conclude that the interface temperature seems to be a major influencing factor. This remark is confirmed if we compare these results with the ones obtained under isothermal conditions where the friction coefficient was always below 0.15 (section 3.1.1).

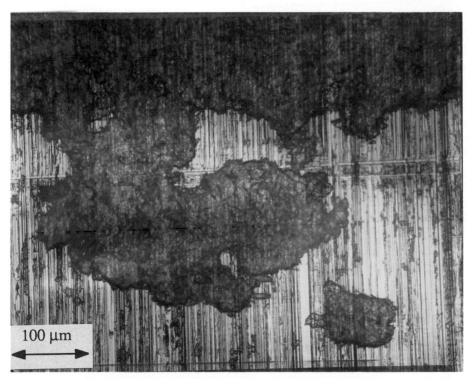

Fig. 32. High speed–low pressure reciprocating friction of PEEK on DIN X20Cr13. Counterface: the reciprocating friction causes an irregular and thick transfer film of PEEK influencing the friction negatively.

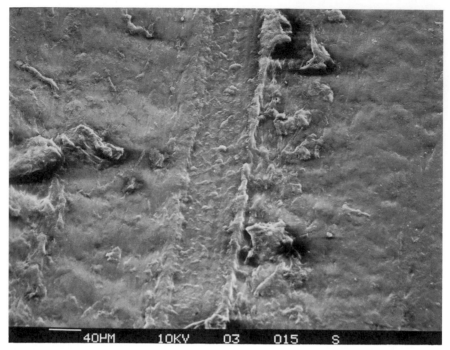

Fig. 33. High speed–low pressure reciprocating friction of PEEK on DIN X20Cr13. Worn surface of the PEEK: the irregular transfer film of the counterface causes a partially abrasive friction process.

A similar friction behaviour of PEEK (high μ and degraded counterface) has already been observed by Briscoe [16], but under continuous movement. In that case the interface temperature, which is about 170°C, is much higher than the one we measured here.

In our case we assume that the increase in μ is related to a change of the friction mechanism, i.e., from adhesion to a combination of adhesion and abrasion. The results obtained from the numerical model of the contact (section 2.2.4) can explain this transition. Figures 15 and 16 have shown a loss of contact of the PEEK at the place where F_f was applied. Since the friction is vertical, the debris can accumulate in the gap. When the reverse movement occurs, the debris is pressed on the counterface and generates after a few cycles pieces of agglomerated PEEK. The interface temperature becomes locally much higher than that deduced from the model (section 2.2.3). The mobility of the macromolecules at the PEEK–PEEK contact points is increased, which can also explain the strong influence of T_i on μ if we consider that the adhesion is still part of the friction process.

Under continuous movement [15], it has also been shown that the friction coefficient of PEEK could decrease with the sliding velocity. In that particular case this was explained by the creation of an hydrodynamic lubricating film of molten PEEK. Under reciprocating movement, these conditions should not occur because of the continual change in sliding direction.

The interface temperature is generally given by the relation [41]:

$$T_i = T_a + c\mu F_n v, \tag{8}$$

where T_i is the interface temperature, T_a the ambient temperature, c the thermal constant of the pin/plate system, μ the friction coefficient, F_n the normal force on the pin, and v the sliding velocity.

If we now consider (fig. 31):

$$\mu = \beta v, \tag{9}$$

where β is a constant, then

$$T_i = T_a + \beta c F_n v^2, \tag{10}$$

which is not the case according to fig. 28.

If we assume that the PEEK thermal properties are not changing too much in the considered range of temperatures, then a measured T_i lower than expected could be due to the fact that the contact between the pin and the plate is established by the asperities and not by the entire section of the pin, as it was assumed in the model. This phenomenon has already been explained above (section 2.2.3).

3.2.2. Contact PEEK composite / DIN X20Cr13

Under these conditions, the numerical model indicates that the pin is in contact with the plate everywhere (section 2.2.4). Figures 34 and 35 show the influence of the sliding velocity on the interface temperature and the friction coefficient of PEEK D450HF30, respectively. In contrast to the matrix, this material supports a sliding velocity of 2.5 m/s under a contact pressure of 0.5 MPa. The decrease of μ avoids the interface temperature to become too important when the sliding speed increases (see eq. (8)). Figure 36 shows the friction force F_f at $v = 0.6$ m/s during a period of the reciprocating movement. With this material, F_f is almost constant along the stroke. Figure 37 shows the influence of T_i on μ for $v = 0.25$ m/s and $p = 0.5$ MPa. This last behaviour has already been observed for this material, but under continuous movement [42].

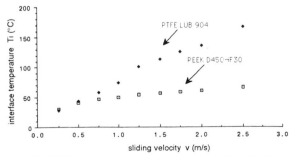

Fig. 34. High speed–low pressure reciprocating friction of PEEK D450HF30 on DIN X20Cr13. Influence of the sliding velocity v on the interface temperature T_i ($p = 0.5$ MPa).

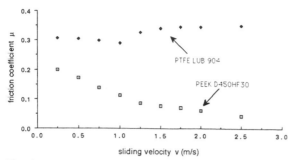

Fig. 35. High speed–low pressure reciprocating friction of PEEK D450HF30 on DIN X20Cr13. Influence of the sliding velocity v on μ ($p = 0.5$ MPa).

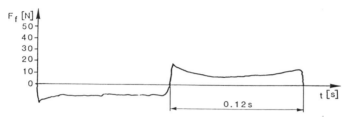

Fig. 36. High speed–low pressure reciprocating friction of PEEK D450HF30 on DIN X20Cr13. Friction force F_f during one period of movement, $v = 0.6$ m/s and $p = 0.5$ MPa.

By comparing our results with those [16] from the combination PEEK/PTFE, one can conclude that graphite reduces the friction coefficient of PEEK more effectively than that of PTFE. The results shown in figs. 35 and 37 indicate that T_i is again a factor which influences μ.

For an epoxy resin reinforced with long carbon fibres, it has been shown [12] that the friction coefficient can be given by:

$$1/\mu_c = V_f 1/\mu_f + V_m 1/\mu_m, \tag{11}$$

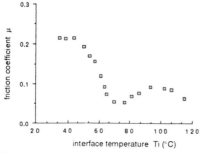

Fig. 37. High speed–low pressure reciprocating friction of PEEK D450HF30 on DIN X20Cr13. Influence of the interface temperature T_i on μ ($v = 0.25$ m/s and $p = 0.5$ MPa).

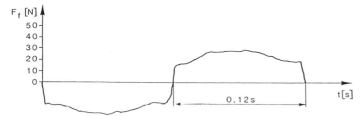

Fig. 38. High speed–low pressure reciprocating dry friction of PTFE LUB904 on DIN X20Cr13. Friction force F_f during one period of reciprocating movement ($v = 0.6$ m/s and $p = 0.5$ MPa).

where V_f is the relative volume content of fibre in the composite, μ_f the friction coefficient of the fibre, V_m the relative volume content of fibre in the matrix, and μ_m the friction coefficient of the matrix.

It has been shown above (section 3.2.1) that the friction coefficient of the neat PEEK (μ_m) increases with T_i, and others studies [43,44] have indicated that this is not the case for carbon fibres or solid lubricants. One can conclude here that the friction coefficient of this composite cannot be predicted by a simple rule of mixtures, such as the one proposed above (see eq. (11)). The fraction of solid lubricants is higher at the interface than in the volume of the composite. Such a low value of μ (0.06) as presented in figs. 35 and 37 may be obtained by a better spreading of the graphite particles at the contact surface, creating a solid lubricant film. Above $T_i = 65°C$, the increase in μ is probably due to the contribution of the matrix (section 3.2.1).

3.2.3. Contact PTFE composite / DIN X20Cr13

The numerical model (section 2.2.4) indicates that contact between the pin and the plate occurs everywhere. The influence of v on T_i and μ is indicated in figs. 34 and 35. The countersurface is highly marked by abrasion due to the glass fibres of the composite. The friction force is higher in the middle of the stroke than at the extremities (fig. 38), which corresponds also to a more severe wear at the place where the sliding velocity and interface temperature are higher (section 4.3). The fact that v does not influence μ on the average (fig. 35) could be due to the fact that the friction here is mainly controlled by the contact between the glass fibres and the steel. T_i can be rather well predicted by eq. (8). If we assume that the temperature limit of PTFE under 1 MPa of compressive stress is about 230°C, the highest sliding velocity this material could support under these conditions should be around 4 m/s.

4. Wear under continuous and reciprocating dry friction

This section will characterize the wear due to reciprocating friction and make a comparison with that created by continuous movement under similar testing

TABLE 4
Surface roughness parameters of the ground and polished stainless steel counterface DIN X20Cr13
before testing.

	Ground counterface	Polished counterface
R_a (μm)	0.66	0.03
R_z (μm)	5.97	0.32
R_p (μm)	3.83	0.17
R_t (μm)	7.94	0.49

conditions. The tests are done on the test rigs presented above (sections 2.2 and 2.3). The testing conditions, which are identical for both devices, are:
- contact pressure: 0.5 Mpa,
- sliding velocity (reciprocating and continuous): 0.25 m/s for pure PEEK and 1.2 m/s for the PEEK D450HF30 and the PTFE LUB904,
- test duration (after running-in period): 70 h, which makes 31500 cycles for the PEEK and 2.11 millions cycles for the PEEK D450HF30 and the PTFE LUB904.

The PEEK has an initial crystallinity of 28.7%. In order to increase its crystallinity and theoretically also its wear resistance [45], a heat treatment (250°C during 26 h) will be given to the samples before testing. Tests are done with stainless steel DIN X20Cr13 (13% chromium), ground and polished. Table 4 indicates the initial surface parameters of both discs and plates, measured in the direction of the sliding velocity.

4.1. Neat PEEK

Table 5 shows the wear rates W (section 1.2), the friction coefficients μ and the interface temperature T_i for both the continuous and reciprocating type of movement. W is higher under reciprocating movement and μ varies during the test period (4 h), i.e., $0.306 < \mu < 0.442$ (fig. 39). This change in μ can be due to the fact that the debris falls irregularly from the top of the plate and accumulates at the interface (section 2.2.4).

The heat treatment brings a small improvement versus the non-treated PEEK under reciprocating friction, but has an adverse effect under continuous movement. The change from 28.7 to 32% is probably not enough to be measurable

TABLE 5
Wear rate W, friction coefficient μ and interface temperature T_i of neat PEEK (annealed and untreated) under reciprocating and continuous friction.

	Reciprocating friction		Continuous friction	
	450G	450G annealed	450G	450G annealed
W (m/m)	2.70×10^{-7}	1.97×10^{-7}	3.30×10^{-9}	3.95×10^{-9}
μ	0.39	0.35	0.35	0.25
T_i (°C)	41.5	40	51	49.5

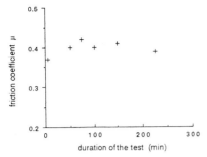

Fig. 39. High speed–low pressure reciprocating wear of PEEK on DIN X20Cr13. Variation of μ during a testing period of 4 h ($v = 0.25$ m/s, $p = 0.5$ MPa).

TABLE 6

Wear rate W, friction coefficient μ and interface temperature T_i of the PEEK composite D450HF30 when submitted to reciprocating friction on a ground/polished stainless steel DIN X20Cr13 counterface. The roughness parameters of the counterface are measured after testing. The initial values are put into brackets.

	Ground counterface		Polished counterface
	low T_i	high T_i	
W (m/m)	4×10^{-10}	1.95×10^{-9}	8.2×10^{-10}
μ	0.13	0.18	0.23
T_i (°C)	70	125	76.8
R_a (μm)	0.60 (0.66)	0.04 (0.66)	0.14 (0.03)
R_z (μm)	5.82 (5.97)	0.80 (5.97)	2.16 (0.32)
R_p (μm)	2.78 (3.83)	0.48 (3.83)	0.84 (0.17)
R_t (μm)	6.53 (7.94)	1.94 (7.94)	2.94 (0.49)

TABLE 7

Wear rate W, friction coefficient μ and interface temperature T_i of the PEEK composite D450HF30 when submitted to continuous friction on a ground/polished stainless steel DIN X20Cr13 counterface. The roughness parameters of the counterface are measured after testing. The initial values are put into brackets.

	Ground counterface		Polished counterface
	low T_i	high T_i	
W (m/m)	7.97×10^{-11}	1.18×10^{-9}	5.7×10^{-10}
μ	0.14	0.28	0.22
T_i (°C)	48.5	111	65
R_a (μm)	0.50 (0.66)	0.41 (0.53)	0.08 (0.03)
R_z (μm)	4.35 (4.94)	3.35 (5.96)	1.10 (0.32)
R_p (μm)	2.12 (3.22)	2.33 (4.25)	0.54 (0.17)
R_t (μm)	5.19 (6.54)	4.60 (7.37)	1.50 (0.49)

A. Schelling, H.H. Kausch

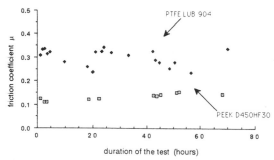

Fig. 40. High speed–low pressure reciprocating wear of PEEK D450HF30 on DIN X20Cr13. Variation of μ during a test period of 70 h.

quantitatively in this wear test. It nevertheless reduces the μ for both movements, what could indicate that the hardness of the polymer has still been increased by the heat treatment. The worn plate and pin surfaces after reciprocating sliding have already been presented above (section 3.2.1, figs. 32 and 33). This wear process resulted in the formation of flake-like debris with a typical length and

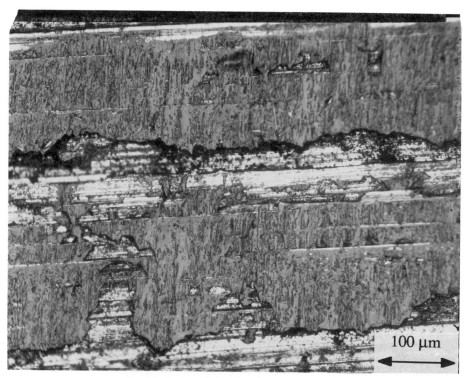

Fig. 41. High speed–low pressure reciprocating wear of PEEK D450HF30 on DIN X20Cr13. Counterface: the transfer film containing the solid lubricants has a positive effect on μ.

width of the order of 1 to 2 mm and a thickness of 0.7 mm. Such high wear rates for PEEK have been obtained under continuous friction by other workers, but at higher pv products (1.8 instead of here 0.13 MPa m/s). Since we know that the resistance to abrasion of PEEK is lower than to adhesion [7], the high wear rate we have measured here is probably due to the contribution of abrasion in the wear process, as already described before (section 3.2.1). Cyclic forces acting on the polymer during the reciprocating movement can also induce fatigue wear, but SEM observations made on subsurfaces did not reveal more cracks in the pin submitted to reciprocating friction.

The degree of crystallinity measured by DSC on the debris after continuous and reciprocating friction is, respectively, 41.25% and 40.5%. The value before testing was 28.7%. High compression and shear acting on the PEEK debris at the interface could be responsible for this further change in this physical property of the polymer. This observation had so far never been made with this material.

Under continuous movement, the disc is covered by a smooth and regular PEEK transfer film having a thickness of about a few μm. The pin surface is worn

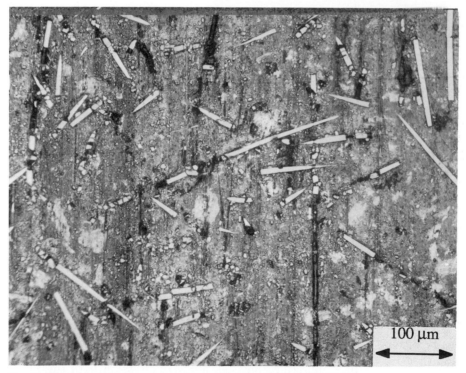

Fig. 42. High speed–low pressure reciprocating wear of PEEK D450HF30 on ground DIN X20Cr13. Worn surface of the PEEK D450HF30 after 70 h ($p = 0.5$ MPa, $v = 1.2$ m/s and $T_i = 70°C$): the metallic asperities of the counterface break the carbon fibres and generate abrasive wear on the PEEK matrix.

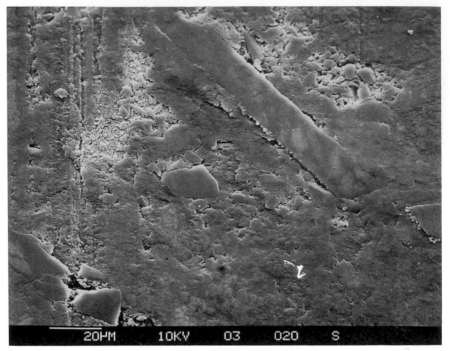

Fig. 43. High speed–low pressure reciprocating wear of PEEK D450HF30 on DIN X20Cr13 ($p = 0.5$ MPa, $v = 1.2$ m/s and $T_i = 70°C$). Worn surface of the PEEK D450HF30: debonding of a carbon fibre by the friction force.

uniformly by pure adhesion. Other workers have found similar wear rates under these conditions [1,15,45].

4.2. PEEK composite

Tables 6 and 7 show the wear rates, friction coefficients and interface temperatures for both types of movement. Figure 40 shows that μ changes during the testing period (70 h) when rubbing on a ground surface at $T_i = 70°C$, i.e., $0.111 < \mu < 0.146$. The variation of μ is smaller than with pure PEEK (section 4.1) and correlates with the better contact conditions (section 2.2.4).

Tables 6 and 7 indicate the roughness parameters of the counterfaces after testing, the initial values are put into brackets. At low T_i, the parameters of the ground surfaces are not much modified by the friction. The disc and plates are covered by a transfer film of PEEK D450HF30 (fig. 41) and the worn face of the pin (fig. 42) is characterized by broken fibres which stay in the polymer. The metallic asperities of the counterface have an abrasive effect on both the fibres and the matrix. Figure 43 shows a fibre before leaving the matrix.

At higher temperature (125°C), the reciprocating friction of the carbon fibres tends to polish the ground counterface. Its asperities have disappeared and the pin

Fig. 44. High speed–low pressure reciprocating wear of PEEK D450HF30 on polished DIN X20Cr13. Worn surface of the PEEK D450HF30 after 70 h ($v = 1.2$ m/s, $p = 0.5$ MPa and $T_i = 76.8°C$): severe wear characterized by extensive fibre debonding.

is mainly worn by severe adhesion. At this temperature, the strength of the PEEK D450HF30 starts to decrease, and such a drop in wear resistance has already been observed under continuous movement by other workers [15,45].

Under continuous movement and at higher temperature (111°C), the asperities of the ground counterface do not disappear, as was the case under reciprocating movement. The wear of PEEK D450HF30 also increases with T_i, but the wear mechanism remains the same as at lower temperature, i.e., fracture and debonding of the carbon fibres and abrasive/adhesive wear of the matrix.

A polished DIN X20Cr13 counterface has negative effects on both reciprocating and continuous wear. μ and T_i are higher than for ground counterfaces. The transfer film containing the solid lubricants does not stay at the interface, as was previously the case. Figures 44 and 45 show severe wear of the pin characterized by extensive fibre debonding, whatever the type of movement. Large plastic deformations of the PEEK cover the empty cavities of the debonded carbon fibres. This result confirms [2] that an optimum initial surface roughness exists for each pair of materials.

Finally fig. 46 shows the significant influence of T_i on the wear rate of the PEEK D450HF30 for the different tests done here. SEM analyzes of the subsur-

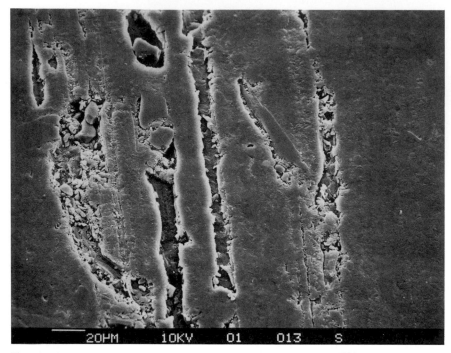

Fig. 45. High speed–low pressure reciprocating wear of PEEK D450HF30 on polished DIN X20Cr13. Worn surface of the PEEK D450HF30 after 70 h ($v = 1.2$ m/s, $p = 0.5$ MPa and $T_i = 76.5°C$): the large plastic deformation of the PEEK matrix covers the cavities where the fibres were located before debonding.

faces of the samples have not shown more cracks in the ones submitted to reciprocating friction. This could mean that here fatigue does not give a major contribution to the wear process.

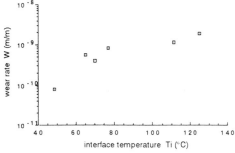

Fig. 46. High speed–low pressure reciprocating and continuous wear of PEEK D450HF30 on DIN X20Cr13. Summary of the wear results ($v = 1.2$ m/s, $p = 0.5$ MPa): the interface temperature T_i influences significantly the wear rate of the PEEK D450HF30.

TABLE 8

Wear rate W, friction coefficient μ and interface temperature T_i of the PTFE composite LUB 904 when submitted to reciprocating and continuous friction. The roughness parameters of the counterface are measured after testing. The initial values are put into brackets.

	Reciprocating friction	Continuous friction
W (m/m)	6.5×10^{-10}	8.22×10^{-10}
μ	0.30	0.33
T_i (°C)	91	124
R_a (μm)	0.08–0.21 (0.66)	0.10 (0.66)
R_z (μm)	0.52–2.9 (5.97)	0.52 (5.97)
R_p (μm)	0.47–0.9 (3.83)	0.40 (3.83)
R_t (μm)	0.7–3.94 (7.94)	0.73 (7.94)

4.3. PTFE composite

Table 8 indicates the wear rate, friction coefficient and interface temperature of PTFE LUB904 on DIN X20Cr13 for both types of movements. An X-ray analysis of the debris indicates a content of 36.6% (in weight) of metallic particles coming from the wear of the counterface. The amount of wear is almost the same for the counterface and for the pin. After 70 h of testing, a track of 15 μm depth has been made in the DIN X20Cr13 by the abrasion of the glass fibres of the PTFE LUB904 (fig. 47). The abraded profile of the counterface is partially reproduced on the PTFE matrix (fig. 48). The friction forces are not high enough to cause fibre debonding. The wear is here mainly dependent on the resistance of the glass fibres against steel, whatever the type of movement. Thus the difference in wear rate between the two movements is here probably due to the difference in temperature between the two devices [2].

The reciprocating friction causes non-uniform wear of the counterface, which corresponds also to a non-uniform friction coefficient along the stroke (section 3.2.3). The first roughness parameters in table 8 correspond to the centre of the plate (half stroke) and the second ones to the extremities. A higher sliding velocity

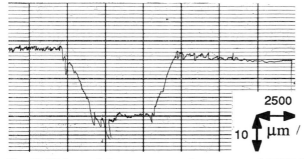

2500

10   μm /

Fig. 47. High speed–low pressure continuous wear of PTFE LUB904 on ground DIN X20Cr13. The counterface is highly abraded after 70 h (15 μm depth) by the glass fibres contained in the pin ($v = 1.2$ m/s, $p = 0.5$ MPa and $T_i = 124$°C).

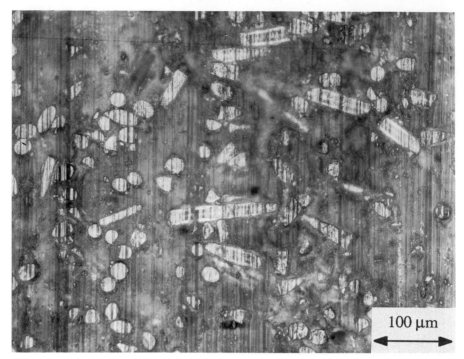

Fig. 48. High speed–low pressure reciprocating wear of PTFE LUB904 on ground DIN X20Cr13. Worn surface of the PTFE LUB904: the PTFE matrix is marked by the abraded counterface. The glass fibres stay in the matrix and their wear resistance controls the wear of the composite.

and interface temperature at the centre of the plate explains probably the more extentive wear at this location [46].

5. Summary and general conclusions

Neat PEEK: under isothermal conditions, the Amontons–Coulomb laws predict quite well its friction coefficient on smooth steel. Under more severe thermal conditions, μ and the wear rate increase drastically under reciprocating movement. A numerical model of the contact under reciprocating friction indicated that this was partly due to the testing conditions.

PEEK composite: adding carbon fibres, graphite and PTFE to the PEEK increases its pv factor by at least a factor five. A higher fibre content than the one considered here could even improve this performance. Nevertheless, the behaviour of the PEEK matrix is still a limiting factor at high temperature. The wear is controlled by fracture and debonding fibres when rubbing on ground steel and mainly by adhesion when the countersurface is (remains) polished. The wear rate is highly influenced by the interface temperature. The type of movement, i.e., reciprocating or continuous, has apparently only a qualitative influence on the wear.

PTFE composite: the glass fibres cause mainly the friction and wear of this material when sliding on steel. They have a marked abrasive effect on DIN X20Cr13 and cause a rather high friction coefficient (> 0.3) independent of the sliding velocity. The reciprocating movement generates a non-uniform wear of the counterface.

When the roughness parameters of the counterface are not affected by the test, the friction coefficients are almost identical under reciprocating and continuous movement.

Acknowledgements

The authors would like to thank the Swiss National Foundation for Scientific Research, the companies CERAC S.A. in Ecublens (CH) and Tetra Pak S.A. in Romont (CH) for their financial support of this work. Thanks are also given to Dr. A. Curnier (LMA, EPFL) for his help during this research.

List of symbols and abbreviations

E	elastic modulus
F_f	friction force
F_n	normal force
p	contact pressure
R_a	average surface roughness
R_p	average of five largest top–middle line distances measured along the profile
R_t	largest top–valley distance measured along the profile
R_z	average of five largest top–valley distances measured along the profile
T_a	ambient temperature
T_g	glass transition temperature
T_i	interface (contact surface) temperature
T_m	melting temperature
v	sliding velocity
W	wear rate
μ	friction coefficient
μ_d	dynamic friction coefficient
μ_s	static friction coefficient
ν	Poisson ratio

DIN X20Cr13	stainless steel with 13% chromium
GG25	cast iron
PEEK	polyetheretherketone
PEEK 450G	neat PEEK
PEEK D450HF30	PEEK composite containing 10 wt% short carbon fibres, 10 wt% PTFE and 10 wt% graphite

PTFE   polytetrafluoroethylene
PTFE LUB 904   PTFE composite containing mainly 40 wt% of short glass
                     carbon fibres

References

[1] M. Cirino, K. Friedrich and R.B. Pipes, Evaluation of polymer composites for sliding and abrasive wear applications, Composites 19(5) (1988).

[2] J.C. Anderson, The wear and friction of commercial polymers and composites, in: Friction and Wear of Polymer Composites, ed. K. Friedrich (Elsevier, Amsterdam, 1986) pp. 329–362.

[3] F.P. Bowden and D. Tabor, The Friction and Lubrication of Solids (Oxford University Press, Oxford, 1950).

[4] J.K.A. Amuzu, B.J. Briscoe and D. Tabor, Friction and shear strength of polymers, ASLE Trans. 20(4) (1977) 354–388.

[5] J.F. Archard, Contact and rubbing of flat surfaces, J. Appl. Phys. 24(8) (1953) 981–988.

[6] K. Friedrich, Wear of reinforced polymers by different abrasive counterparts, in: Friction and Wear of Composite Materials, ed. K. Friedrich (Elsevier, Amsterdam, 1986) pp. 233–287.

[7] M. Cirino, R.B. Pipes and K. Friedrich, The abrasive wear behaviour of continuous fibre polymer composites, J. Mater. Sci. 22 (1987) 2481–2492.

[8] N.P. Suh, The delamination theory of wear, Wear 25 (1973) 111–135.

[9] N.P. Suh, Tribophysics (Prentice Hall, Englewood Cliffs, NJ, 1986).

[10] M. Clerico and V. Patierno, Sliding wear of polymeric composites, Wear 53 (1979) 279–301.

[11] J.K. Lancaster, The effect of carbon fibre reinforcement on the friction and wear of polymers, Brit. J. Appl. Phys. 1(2) (1968) 549–559.

[12] T. Tsukizoe and N. Ohmae, Friction and wear performances of unidirectionally oriented glass, carbon, aramid and stainless steel fibre-reinforced plastics, in: Friction and Wear of Polymer Composites, ed. K. Friedrich (Elsevier, Amsterdam, 1986) pp. 205–231.

[13] N.S. Eiss Jr, The wear of polymers sliding on polymeric films deposited on rough surfaces, Trans. ASME 26/vol. 103 (1981).

[14] K. Tanaka, Transfer of semicrystalline polymers sliding against a smooth steel surface, Wear 75 (1982) 183–199.

[15] P.B. Mody, T.-W. Chou and K. Friedrich, Effect of testing conditions and microstructures on the sliding wear of graphite fibre/PEEK matrix composites, J. Mater. Sci. 23 (1988) 4319–4330.

[16] B.J. Briscoe, Lin Heng Yao and T.A. Stolarsky, The friction and wear of PTFE–PEEK composites, an initial appraisal of the optimum composition, Wear 108 (1986) 357–374.

[17] R.B. Waterhouse, Fretting, in: Treatise on Materials Science and Technology 13 (Academic Press, New York, 1979) pp. 259–286.

[18] A. Iwabuchi, K. Hori and Y. Sugawar, Effect of temperature and ambient pressure on fretting properties of polyimide, Wear 125 (1988) 67–81.

[19] N.-H. Sung and N.P. Suh, Effect of fiber orientation on the friction and wear of fiber reinforced polymer composites, Wear 53 (1979) 129–141.

[20] M. Cirino, K. Friedrich and R.B. Pipes, The effect of fiber orientation on the abrasive wear behavior of polymer composite materals, Wear 121 (1988) 127–141.

[21] N. Ohmae, K. Kobayashi and T. Tsukizoe, Characteristics of fretting of carbon fibre reinforced plastics, Wear 29 (1974) 345–353.

[22] O. Jacobs, K. Friedrich, G. Marom, K. Schulte and H.D. Wagner, Fretting wear performance of glass-, carbon-, and aramid-fibre/epoxy and PEEK composites, Wear 135 (1990) 207–216.

[23] K. Friedrich, O. Jacobs, M. Cirino and G. Marom, Hybrid effects on sliding wear of polymer composites, in: Proc. Conf. on Tribology of Composite Materials, Oak Ridge, TN, 1–3 May, 1990, eds P.K. Rohatgi, C.S. Yust and P.J Blau (ASM International, 1990) pp. 277–285.

[24] O. Jacobs, K. Friedrich and K. Schulte, Fretting fatigue of continuous carbon fibre reinforced polymer composites, Wear 145 (1991) 167–188.

[25] C.M. Pooley and D. Tabor, Friction and molecular structure: the behavior of some thermoplastics, Proc. R. Soc. London A 329 (1972) 251–274.

[26] B.J. Briscoe and T.A. Stolarsky, The influence of linear and rotating motions on the friction of polymers, J. Lubr. Technol. 103 (1981) 503–508.

[27] B. Bhushan and D.F. Wilcock, Wear behavior of polymeric compositions in dry reciprocating sliding wear, Wear 75 (1982) 41–70.

[28] Y. Wang, G. Chen and G. Jiao, Wear prediction for unlubricated piston ring, Wear 135 (1990) 227–235.

[29] M. Godet, The third-body approach, A mechanical view of wear, Wear 100 (1984) 437–452.

[30] J.K. Lancaster, R.W. Bramham, D. Play and R. Waghorne, Effects of amplitude of oscillation on the wear of dry bearings containing PTFE, J. Lubr. Technol. 104 (1982) 559–567.

[31] D. Play, Mutual overlap coefficient and wear debris motion in dry oscillating friction and wear tests, ASLE Trans. 28(4) (1985) 527–535.

[32] D.J. Blundell and B.N. Osborn, The morphology of poly(aryl-ether-ether-ketone), Polymer 24 (1983) 953–958.

[33] D.W. van Krevelen, Properties of Polymers, 2nd Ed. (Elsevier, Amsterdam, 1987).

[34] A. Schelling, Usure du polyétheréthercétone (PEEK) sous frottement sec alterné. Thèse de doctorat 839 (EPFL Lausanne, 1990).

[35] A. Curnier, TACT, A contact analysis program. Methodology Summary. LMA (EPFL Lausanne, 1983).

[36] O.C. Zinkiewicz, The Finite Element Method, 3rd Ed. (McGraw-Hill, London, 1977).

[37] K.L. Johnson, Contact Mechanics (Cambridge University Press, Cambridge, 1989).

[38] A. Curnier, F. Bonjour, J.A. Cheng, P.M. Kronenberg and M. Eichenberger, Document du tribomètre pion-table. Document. (Archives LMA-DME-EPFL, Lausanne, 1978).

[39] A. Curnier and J.A. Cheng, Etude expérimentale du frottement. Document (LMA/EPFL, Lausanne, 1989).

[40] T. Sukamoto, Normal displacement and dynamic friction characteristics in a stick-slip process, Tribol. Int. 20(1) (1987) 25–31.

[41] A.D. Sarkar, Friction and Wear (Academic Press, London, 1980).

[42] K. Friedrich and R. Walter, Microstructure and tribological properties of short fiber/thermoplastic composites, in: Proc. Conf. on Tribology of Composite Materials, Oak Ridge, TN, 1–3 May, 1990, eds P.K. Rohatgi, C.S. Yust and P.J. Blau (ASM International, 1990) pp. 217–226.

[43] K. Tanaka, Effects of various fillers on the friction and wear of PTFE-based composites, in: Friction and Wear of Polymer Composites, ed. K. Friedrich (Elsevier, Amsterdam, 1986) pp. 137–174.

[44] F.J. Clauss, Solid Lubricants and Self-Lubricating Solids, ch. 3 (Academic Press, New York, 1972).

[45] H. Voss and K. Friedrich, On the wear behavior of short-fibre PEEK composites, Wear 116 (1987) 1–18.

[46] K. Tanaka, Friction and wear of glass and carbon fiber-filled thermoplastic polymer, J. Lubr. Technol. (1987) 408–414.

Advances in Composite Tribology
edited by K. Friedrich
© *1993 Elsevier Science Publishers B.V. All rights reserved*

Chapter 4

Short-Fibre Reinforced, High-Temperature Resistant Polymers for a Wide Field of Tribological Applications

A.M. HÄGER

Institute for Composite Materials, University of Kaiserslautern, 6750 Kaiserslautern, Germany

M. DAVIES

ICI, Wilton Materials Research Centre, Middlesbrough, Cleveland TS6 8JE, UK

Contents

Abstract

This chapter presents numerous tribological data about high-temperature resistant polymers reinforced by short fibres and internally lubricated by various solid lubricants. Most formulation of tribologically used grades have been found empirically. To get a better understanding of such complex material systems some scientific approaches are introduced.

Beside today's most common testing techniques and evaluating methods the wear and friction influencing parameters and phenomenons are discussed.

Simple composite systems are investigated in order to get information about the filler's effect on the wear performance. A look is taken to the self-reinforcing effect of the matrix microstructure. The effect of solid lubricants is demonstrated and is discussed in more detail with PTFE. Therefore, some unique ESCA analysis are presented which show the development of the transfer film during the running in phase. It is demonstrated that not only the steel counterpart surface is modified but also the polymer composite surface undergoes some changes. And an attempt of a qualitative interpretation is undertaken.

Furthermore, the effect of various fibre reinforcements is discussed for a couple of model systems like LCP + GF or CF, GF and CF reinforced PEN or PC + GF. And a model to predict the wear behaviour of such fibrous systems is presented. It is demonstrated that the most effective wear reducing volume fraction of short fibre content is about 15%.

Finally, some recent investigations on more complex systems based on PAEK, PI, PBI matrices are discussed. These systems offer high wear resistance and low friction up to application temperatures of more than 200°C.

1. Introduction

The current trend in the development of polymers is to seek materials which retain usable properties at high temperatures. Examples of such recent developments are polyethersulphone (PES), polyetherimide (PEI) and polyether–ether ketone (PEEK) in the thermoplastic area and novel epoxies and bismaleimides in the thermosetting area.

In today's world where reliability is a key feature, it is necessary for engineers to design for wear integrity also. Polymers and polymer composites are now being used increasingly as engineering materials for technical applications in which tribological properties are of importance. There are a number of traditional fields,

TABLE 1
Tribological applications of polymer composite materials.

Field	Examples	Type of wear loading and requirements
Civil eng.	Conveyors	Abrasive, minimise wear
Mechanical eng.	Seals	Sliding, minimise friction and wear
Mechanical eng.	Bearings	Sliding, minimise friction and wear
Mechanical eng.	Gears	Sliding, minimise friction and wear
Mechanical eng.	Turbine or pump blades	Erosion, minimise wear
Medicine	Dental applications	Abrasive, minimise wear
Medicine	Hip replacements	
	– Shaft	Fretting, minimise wear
	– Cap	Sliding, minimise friction and wear
Automotive	Brakes	Sliding, minimise wear, optimise friction
	Tires	Abrasion, minimise wear, maximise friction

e.g. reinforced rubbers for tyres and transmission belts or filled thermosets for brake materials, but these are being augmented by applications which exploit the excellent sliding characteristics of polymeric materials [1] (see also table 1). The above-mentioned polymers are particularly interesting candidates for tribological applications because of their superior properties at the elevated temperatures which often occur during frictional heating.

Tribological applications, such as bearings, gears, cans, pumps, electrical contacts, prostheses and so on [1], often require high load capacities and low friction values. Therefore the polymers are reinforced by fibres and are internally lubricated with solid lubricants.

Thermoplastic based bearing materials offer a number of advantages over the more traditional metallic variants. They can exhibit excellent frictional and wear characteristics in the absence of external lubrication systems which can provide maintainance free operation, they can offer excellent corrosion resistance and often operate with low noise emission due to the good damping properties of polymers. Thermoplastics can also offer the economic advantages of quick fabrication via injection moulding.

The toughness and impact resistance of certain thermoplastic based systems has lead to many applications in chain conveyors and textile machinery. An application example is shown in fig. 1 (a slider block on a flat-bed knitting machine). Here, due to the risk of contaminating the garments under production, low friction and high wear resistance are necessary together with high stiffness and acceptable thermal stability. Another example is shown in fig. 2, which is a paper tractor for the accurate feeding of paper in printers. In this example a polymer composite system was the material of choice because of its low weight, excellent dimensional stability and wear resistance in the absence of potentially abrasive paper dust.

It is the case that most polymeric compositions intended for tribological applications are arrived at empirically. Reinforcements are added to increase the strength,

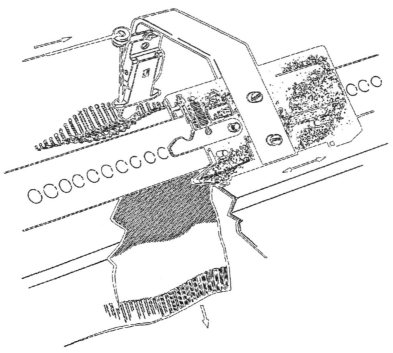

Fig. 1. Slider bock in flat-bed knitting machine.

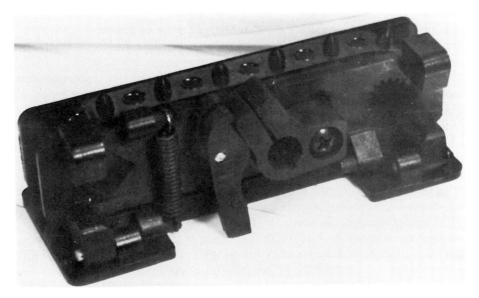

Fig. 2. Paper tractor.

TABLE 2
Composition of numerous commercial HT-thermoplastics.

Matrix	Abre-viation	Fillers (wt%)				
		PTFE	Silicone	Glass fibre	Carbon fibre	Graphite
Polyetherimide	PEI	–	–	30	–	–
		–	–	–	30	–
		–	–	–	40	–
		15	–	30	–	–
Polysulfone	PSU	15	–	–	–	–
		–	–	30	–	–
		15	–	30	–	–
		10	–	10	–	–
		–	–	–	30	–
		8	2	–	–	–
Polyethersulfone	PES	–	–	30	–	–
		15	–	30	–	–
		–	–	–	30	–
		15	–	–	30	–
		15	–	–	–	–
		10	–	–	10	10
Polyoxy-methylene	POM	5	–	–	–	–
		10	–	–	–	–
		15	–	–	–	–
		20	–	–	–	–
		25	–	–	–	–
		–	2	–	–	–
		18	2	–	–	–
		–	–	30	–	–
		15	–	30	–	–
		13	2	30	–	–
		–	–	–	20	–
Polyphenylene sulfide	PPS	20	–	–	–	–
		–	–	40	–	–
		15	–	30	–	–
		–	–	–	30	–
		15	–	–	30	–
		13	2	30	–	–
Polyetherether-ketone	PEEK	–	–	30	–	–
		–	–	–	30	–
		20	–	–	–	–
		15	–	30	–	–
		15	–	–	15	–
		10	–	–	10	10
		15	–	–	15	–
		15	–	–	–	15
		10	2	–	10	5
		15	2	–	–	10
Polyetherether-ketoneketone	PEEKK	10	–	–	30	–
		15	–	–	15	–
Polyarylether-ketone	PEKEKK (PAEK)	–	–	30	–	–
		–	–	–	30	–
Polyetherketone	PEK	10	–	–	10	10
Polyamide 6/6	PA66	13	2	30	–	–
Polycarbonate	PC	15	–	20	–	–

TABLE 3
Limiting *pv*-factors of numerous commercial HT-thermoplastics to be used at room temperature and sliding speeds of 1 m/s.

Matrix	Fillers (wt%)					Limiting PV (MPa m/s)
	PTFE	Silicone	Glass fibre	Carbon fibre	Graphite	
PEI	–	–	30	–	–	–
	–	–	–	30	–	–
	–	–	–	40	–	–
	15	–	30	–	–	–
PSU	15	–	–	–	–	0.175
	–	–	30	–	–	–
	15	–	30	–	–	–
	10	–	10	–	–	1.226
	–	–	–	30	–	–
	8	2	–	–	–	0.298
PES	–	–	30	–	–	–
	15	–	30	–	–	1.051
	–	–	–	30	–	0.350
	15	–	–	30	–	1.156
	15	–	–	–	–	–
POM	5	–	–	–	–	–
	10	–	–	–	–	–
	15	–	–	–	–	–
	20	–	–	–	–	0.438
	20	–	–	–	–	0.561
	25	–	–	–	–	–
	–	2	–	–	–	0.315
	–	2	–	–	–	0.420
	18	2	–	–	–	0.526
	18	2	–	–	–	0.631
	–	–	30	–	–	–
	15	–	30	–	–	0.613
	13	2	30	–	–	0.666
	–	–	–	20	–	0.701
PPS	20	–	–	–	–	–
	–	–	40	–	–	0.561
	15	–	30	–	–	1.226
	–	–	–	30	–	0.701
	15	–	–	30	–	1.226
	13	2	30	–	–	1.051
PEEK	–	–	30	–	–	–
	–	–	–	30	–	–
	20	–	–	–	–	–
	15	–	30	–	–	–
	15	–	–	15	–	1.401
high visc.	10	–	–	10	10	17.5
low visc.	10	–	–	10	10	4
PA66	13	2	30	–	–	0.701
PC	15	–	20	–	–	0.876
PEKEKK	–	–	30	–	–	> 16
	–	–	–	30	–	> 16
PBMI	–	–	–	–	20	> 20
PBI	–	–	–	–	–	> 10

Fig. 3. Influence of the counterface hardness on the composite wear.

hardness and wear resistance and internal lubricants are added to lower the coefficient of friction. Tables 2 and 3 list a large number of such empirical formulations. The last two materials listed are those used for the applications illustrated in figs. 1 and 2, respectively. Naturally, in order to be able to design materials for specific tribological applications, a detailed scientific understanding of the influence of the various additives is required.

Common internal lubricants are PTFE, silicone, graphite, MoS_2 and some oils. Generally it has been found that 10–20% (by weight) of these lubricants dispersed within polymer matrices can lead to considerable improvements in the tribological properties, particularly the coefficient of friction.

By their very nature, however, internal lubricants rarely reinforce the polymers into which they are dispersed. Fibrous reinforcement can often be used to off-set this disadvantage and generally it is found that the fibrous fillers also have a beneficial effect on the wear resistance. The simultanous presence of both a fibrous reinforcement and an internal lubricant gives rise to many complex interactions. Short glass fibres and PTFE particles are a common combination for compounds intended for dry sliding friction and wear applications. Evidently the better the comprehension of these complex interactions, the better are the chances of efficient and effective future composition developments.

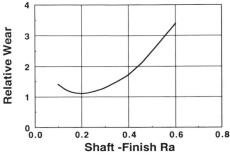

Fig. 4. Influence of the counterface roughness on the wear of composites.

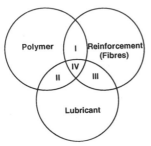

Fig. 5. Principal composition possibilities for tribologically used polymer composites [2].

A dry sliding polymer bearing consists not only of the polymeric component, but generally of a metallic component also. The hardness of the metallic component is of considerable importance, as is its surface roughness. Figures 3 and 4 show, respectively, in a schematic fashion how the relative wear of a dry sliding polymer–metal pair can be expected to vary with hardness and surface finish (as characterised by the R_a values) of the metallic component.

The vast majority of composites intended for tribological applications have three basic constituants, namely polymer matrix, some reinforcement (generally fibres) and a lubricating component. Figure 5 illustrates the compositional features and the in general best-performing commercial compositions are located in the central section of fig. 5 marked IV. Tables 2 and 3 give typical real compositional details for commercial compounds.

2. Experimental techniques for model friction and wear studies

It has been said that there are probably as many plastics wear tests as people performing plastics wear tests. This is because, despite the simplicity of the definition of the coefficient of friction and the specific wear rate, tribological properties are system properties and not material properties [4].

In reality, the tribological properties of a system depend on (a) the material components, (b) the operating variables and (c) the details of the tribological interactions. The operating variables are parameters such as the ambient temperature, the loads, the relative velocity of the contacting surfaces and the environment. The phrase "tribological interactions" covers the complications introduced by the detailed geometry of the tribocontact.

Since tribological characteristics are system specific it follows that different test methods will yield different values for the friction coefficient and rate of wear for nominally the same materials and the same set of conditions. It also follows that it is very difficult to apply test data to the design of real components in an engineering situation. In evaluating materials for a specific application, the most appropriate test method is the one which most closely mimics the application in question. It is mainly for this reason that there are as many tribology tests as people performing such tests.

2.1. Principles of measurement techniques

Whilst it is difficult to rely on the numerical values of the friction coefficient and specific wear rate resulting from laboratory test, such experiments can be of value in ranking materials and determining the mechanisms which give rise to the observed tribological phenomena.

There are many practicable ways of measuring friction and wear. Each technique has advantages and disadvantages. Here, only the most common techniques are presented and discussed.

The definition of the coefficient of friction is the rate of the frictional force to the normal load. In order to make the measurement, therefore it is necessary to measure two forces or two force-caused effects. The measurement of the normal load is generally not a problem. Often direct loading with a calibrated weight is used, but even if a load cell is not used, this provides a convenient solution. Measurement of the frictional force without interference from other effects is, however, more problematic. Common techniques involve the use of load cells, calibrated strain gauge assemblies or, in rotating equipment, rotary torque transducers. Problems arise usually due to the location of the friction measuring device in the test machine assembly, for example additional bearings between the point of friction and the measuring device can lead to additional sources of friction which cannot be neglected and cannot simply be eliminated by adjustment of the zero point. Usually attention should also be paid to the working temperature of the friction sensor, temperature shifts in the calibration are often not negligible.

There are a great many definitions of the wear rate. The German industry standard, DIN 50 321 [5], gives more than 23 definitions of the wear rate, but none of these are the same as the most common used in the bulk of the dry sliding tribology literature. The most common definitions which will be of relevance here, are:

Time-dependent wear rate (or depth wear rate):

$$\dot{W_t} = \frac{\Delta h}{t}. \tag{1}$$

Distance-dependent wear rate:

$$\dot{W_L} = \frac{\Delta h}{L}. \tag{2}$$

Dimensionless wear rate:

$$\dot{W} = \frac{\Delta V}{LA}. \tag{3}$$

Specific wear rate or wear factor:

$$\dot{W_s} = \frac{\Delta V}{F_n L} = \frac{\Delta m}{L \rho F_n}. \tag{4}$$

Here Δh is the height reduction of a specimen, t is the time of operation, L is the sliding distance, ΔV is the volume loss, Δm is the mass loss, A the apparent contact area, ρ the density of the material worn and F_n the normal load on the specimen. The wear rates are connected by the following equations:

$$\dot{W}_t = \dot{W}_s pv,$$ (5)

$$\dot{W}_t = \dot{W}v,$$ (6)

$$\dot{W}_t = \dot{W}_L p^2 v,$$ (7)

$$\dot{W} = \dot{W}_L p^2,$$ (8)

$$\dot{W}_s = \dot{W}_L p.$$ (9)

Wear rates can be measured either by weight loss or by height-reduction of the specimen. In both cases there are problems. The measurement of the weight loss has the advantage that only the wear of the investigated body is directly evaluated. Depending on the specimen geometry and the temperature conditions, different degrees of plastic flow of the material can occur, i.e. in some cases the material can be pressed out at the specimen edges and not be removed at the measurement. In addition, this method does not allow for easy computerised data collection. Changes in mechanisms can also not be detected. The common approach of taking measurements at intervals results in the problem of precise specimen location.

The height-reduction method also has its problems. Although the method allows an evaluation of the wear of the specimen–counterpart pair which approximates the wear of the softer body in a soft–hard body combination, viscoelastic deformation in polymer components can be mistaken for wear. This problem tends to exist, however, only for polymers with low moduli under the relevant regime of testing. The great advantage of the method, however, is that it is really adaptable to automation through the use of a transducer whose output is interfaced to a computer.

2.2. Typical apparatus configurations for sliding wear and friction

The most common laboratory-based tests in current use for dry sliding against hard counterparts are the pin-on-disc configuration (fig. 6), the thrust washer configuration (fig. 7) and the block-on-ring geometry (fig. 8). Surprisingly, journal bearing tests are little used for fundamental investigations throughout the literature, despite their similarity to many common applications.

A number of authors has given detailed accounts of equipment of these types. For example, Atkinson et al. [6] give details of a pin-on-disc apparatus. Anderson [7] has described a thrust bearing tester used in the evaluation of a wide range of polymeric compounds, and Grove and Budinski [8] and Briscoe and Ni [9] have discussed the journal bearing test. The only test which has been standardised at the time of writing by ASTM is the thrust bearing configuration (ASTM 3702-78) [10]. Despite the standard, however, it is not the only plastics wear test, for the reasons already mentioned.

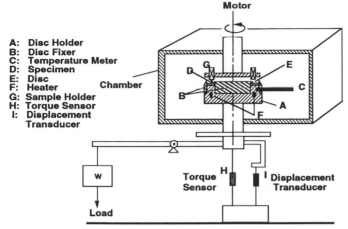

A: Disc Holder
B: Disc Fixer
C: Temperature Meter
D: Specimen
E: Disc
F: Heater
G: Sample Holder
H: Torque Sensor
I: Displacement
    Transducer

Fig. 6. Principal configuration of a pin-on-disk testing device.

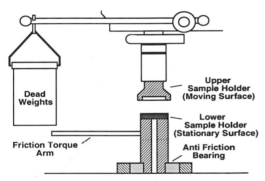

Fig. 7. Principal configuration of a thrust washer.

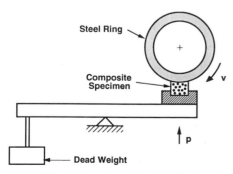

Fig. 8. Principal configuration of a block-on-ring testing device.

If one considers the simple pin-on-disc configuration, there are numerous variations on this theme. This means that any experimental methodology needs to be carefully detailed. The counterface material is normally a disc which is commonly of smooth hardened steel. The exact chemical composition of the counterface and its detailed surface topography are of great relevance. The pin can be loaded against the disc either from above, with the axis of the disc's rotation being vertical, or it can be loaded against the disc with the axis of rotation horizontal. In the first case the wear debris generated by the sliding process is more likely to remain in the contact zone, which, if the debris is abrasive in character, can lead to a significant difference in the wear rate.

The thrust bearing configuration is a much more conforming geometry than the pin-on-disc. The contact zone is a circular annulus which is perpetually under load. The wear debris tends to accumulate in the contact zone and, of course, has a profound influence on the results. Since the annular counterpart has a finite area of contact with the specimen, the wear track closest to the axis of rotation has a slower speed than that furtherst from the rotation axis.

No such problem is faced by using a block-on-ring configuration. The sliding velocity over the width of the specimen is constant. However, for constant pressure conditions, the curvature of the surface of the specimen has first to be adjusted to the shape of the counterface ring.

2.3. Influencing parameters / variable parameters

Because of the fact that tribological properties are not material properties but properties of the particular tribological system, there are many parameters which can have an influence. They can be divided into several groups, as illustrated fig. 9.

The commonly investigated parameters are discussed here.

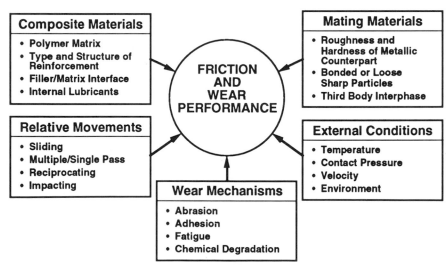

Fig. 9. Aeras of influence on the tribological performance of composite materials.

Load: the applied load is directly connected with the energy input or frictional work done on the material. A model suggested by Briscoe [11] demonstrates the energy flow during sliding. Elastic and plastic energy have to be distinguished. The plastic energy is creating wear debris by deformation and fracture, wheras the elastic energy is released in the form of thermal energy, which leads to frictional heating.

With polymeric materials, usually an increase in the applied load leads to an increase in the wear and a reduction of the coefficient of friction. Of course, these tendencies are influenced by fillers. Dramatic changes in the specific wear rate of neat polymers are found if the load leads to temperatures above T_g or T_m (see below).

Real surfaces are never truly smooth. Because of the finite roughness contact occurs only at certain points over the conforming surfaces. An increasing load leads inevitable to increasing levels of elastic and plastic deformation at the contacting asperities. This effect leads to a greater real area of contact. In many circumstances this increase in real area of contact due to asperity collapse can have a profound influence on the frictional coefficient.

Velocity: the sliding velocity is related to the frictional power. In this way it is responsible for frictional heating. An increasing velocity is often found to decrease the coefficient of friction and to increase the wear rate [12]. However, these trends are highly material specific.

Temperature: temperature has a significant influence on the properties of physics-based systems. The mechanical properties and damping characteristics in particular, change rapidly around the characteristic temperatures, such as T_g. Under non-dry conditions however, or when an internal solid lubricant such as PTFE is used, this effect is not observed, and the coefficient of friction tends to drop with increasing temperature.

Thermal softening of polymers can lead to a drop in surface hardness, which, as mentioned previously, can lead to increases in the real contact areas. This can lead to rapid increases both in the coefficient of friction and the rate of wear. These effects can be mitigated somewhat by the incorporation of mineral fillers.

Jaeger [13] has attempted to calculate analytically the temperature rises at real points of contact during sliding contact. The actual temperature in these zones is considered to be the sum of the ambient temperatures, the local average temperature is due to frictional heating and to flash temperatures due to individual transient asperity–asperity interactions [14]. Lancaster [15] has provided simplified versions of these equations for dry polymer–steel sliding contacts.

Counterpart topography: in addition to the above-mentioned factors the counterpart nature and the counterpart topography determine the details of the tribological behaviour. Investigations of the influence of the counterpart roughness are often complicated by the phenomenon of topography modification caused either by polymer transfer or by polishing through abrasive action of filler debris. Lancaster [16] showed in the case of polymers a rapid increase of the wear with increasing counterpart roughness.

A number of workers have reported minima in the wear rate versus counterpart roughness characteristics for polymer–steel sliding [17–20]. Chichos [20] explained the minimum as being a result of mechanistic changes. The reasoning is that at low roughness adhesive wear is dominant whereas at high roughness abrasive wear and fatigue are emphasised. The minimum is thought to correspond to a complex process which embodies characteristics of all the different Burwell mechanisms.

2.4. Data processing

A common technique for the evaluation of the rate of wear involves the determination of the mass loss of a specimen over a known period of time or equivalently distance of sliding at a constant velocity. Such data is commonly analyzed by taking the means of various interval wear rates where an interval wear rate is proportional to the mass loss divided by the time interval between specimen mass determinations.

An alternative method involves fitting a smooth polynomial function to the mass loss-time data followed by a differentiation of this function to yield a wear rate which is in general a function of time or equivalently sliding distance. The obtained wear rate is known as specific wear rate, often referred to as wear factor which is given by

$$\dot{W}_s = \frac{dm/dt}{F_n v \rho}. \tag{10}$$

F_n, v and ρ are, respectively, the applied normal load, sliding velocity and specimen density. dm/dt represents the gradient of the mass loss-sliding time record. Graphical representation of the mass loss-sliding time data can deviate significantly from linearity, particularly during the early stage of sliding process. For this reason a method was adopted whereby a polynomial function $m(t)$, was fitted to the experimental data using the least squares method. $m(t)$ was given by

$$m(t) - m_0 + m_1 t + m_2 t^2 + \cdots + m_n t^n, \tag{11}$$

where n is the order of the polynomial and $m_0 - m_n$ are the coefficients of $t^0 - t^n$ which give rise to the best fit.

Having determined $m(t)$, the specific wear rate (eq. (10)), now a function of time or equivalently sliding distance L ($L = vt$, where v is the velocity), is given by

$$\dot{W}_s(L) = \frac{1}{F_n v \rho} \left[m_1 + m_2 t + m_3 t^2 + \cdots + m_n t^{n-1} \right]. \tag{12}$$

Numerous investigations have observed this phenomenon of non-linearities in mass loss-sliding time characteristics.

Since the specific wear rate relates to the derivative of $m(t)$ then small inaccuracies in the definition of $m(t)$ will be magnified by the evaluation of the rate of wear. Because of this it clearly only makes sense to employ such a sensitive method of data analysis if the experimental data points themselves are subject to minimal uncertainty. By taking great care of experimental preparation the shapes of the derived wear rate versus sliding distance functions are found to be highly reproducible.

2.5. *Features*

2.5.1. *Different trends to steady state*

By using the above-outlined method to investigate the early stage of wear, different trends can be distinguished until steady-state conditions are reached. The experiments described in the following show that this different behaviour is attributed to a modification of the steel counterface by the polymers sliding against it.

In fig. 10 a graphical representation of a mass loss–time data set is shown for a newly prepared poly(ether sulphone) specimen sliding against a newly prepared steel counterpart using the procedure outlined above. From this figure it becomes clear that the early stages of the plot do not reveal a linear relationship between the mass loss and the sliding time. Shown superimposed on the data in fig. 10, is a sixth-order polynomial fit (i.e. $n = 6$ in eq. (11)), selected as the best representation of the experimental data.

Figure 11a shows the results of evaluating the specific wear rate using eq. (12) and the sixth-order polynomial fit to the data shown in fig. 10. Evidently, the non-linearity in fig. 10 has manifested itself as an increase in the wear rate with increasing sliding distance before a state of approximately constant wear rate is reached.

Clearly, the initial changes in the wear rate are due to either changes occurring in the specimen surface, changes occurring in the counterpart surface or a combination of both. In order to investigate this further, an experiment was conducted in which a newly finished and degreased counterface was loaded against a poly(ether sulphone) specimen which had previously been run against a newly prepared counterpart until a steady state had been achieved. The results of the experiment in the form of the specific wear rate versus sliding distance function are shown in fig. 11b. The fact that fig. 11b is closely similar to fig. 11a strongly suggests that the initial rise in the wear rate with sliding distance shown in fig. 11a, is not due to changes occurring in the polymer surface.

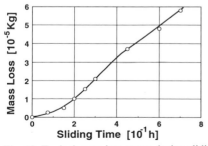

Fig. 10. Typical mass-loss curve during sliding wear experiments of PES.

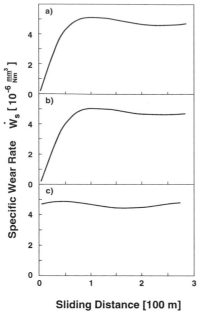

Fig. 11. Initial wear of PES running against smooth steel. (a) Newly prepared specimen, new counterface. (b) Worn specimen, new counterface. (c) Newly prepared specimen, used counterface.

Figure 11c shows the results of sliding a newly prepared poly(ether sulphone) specimen against a counterface which had been previously used in conjunction with a newly prepared specimen until steady-state conditions had been achieved. The figure reveals a specific wear rate which is approximately invariant with respect to the sliding time and at a level which is consistent with the steady-state rates of figs. 11a and b. This result, taken together with that of fig. 11b, suggests that counterface modification accounts for the initial rise in the specific wear rate of poly(ether sulphone) sliding against newly finished hard steel. Observations of the counterface by eye reveal that this modification takes the form of lumpy polymer transfer.

These experiments show that the presence of transfer films can have different influences on the wear behaviour. In the case of PES sliding on steel it is desirable to suppress the formation of such a transfer film in order to keep the initially low wear rate, which is more than one order of magnitude lower than the steady-state wear rate in the presence of a transfer layer.

2.5.2. Wear mapping

Where the initial stages of sliding are important for understanding the influences on the wear behaviour of the material investigated, different information is needed to design wear-loaded components. The designing engineer is interested in knowing the kind and magnitude of the wear mechanisms under given pv-conditions. Therefore, the desired goal is to create wear maps, showing the application limits under all possible working conditions.

3. Scientific approach to explanation of friction and wear trends

3.1. General idea

Modern, polymer-based materials, designed for tribological applications, usually contain three types of compounds. The most important one by the volume fraction is the matrix material, the polymer.

To improve the strength, stiffness and hardness of this matrix material either an orientation of the molecular structure is employed (only some degree of improvement) or in the case of heavy-duty applications a reinforcing compound is added. Commonly a reinforcement in fibre shape is selected, either of glass, carbon or aramid. Other shapes of the reinforcing materials, such as powders or small spheres, are e.g. used with glass or bronze. In the case of fibre reinforcement generally short-fibre and continuous fibre reinforcement has to be distinguished, but today most low friction-and-wear composite materials contain short fibres. These materials are easier to handle with respect to common manufacturing techniques.

The third type of component used in these materials is lubricants. Most time solid lubricants, such as PTFE, graphite or molybdenum disulphide, are selected, but in some cases liquid lubricants in fine dispersion are applied as well (fig. 5).

The most practical way to scientifically understand the influence of the components on the tribological behaviour of the composite is to look separately at the matrix behaviour and at simple composites of two components only. In fig. 5 these simple composites are represented by the areas marked as I and II. They are containing a matrix and a single reinforcing material or lubricant. Some investigations on such simple composites are presented further below. Later on, the interaction of two reinforcements and/or lubricants will be described.

In the following, three types of systems are discussed: self-reinforced matrices, lubricated polymers and fibre reinforced polymer systems.

3.2. Self-reinforcement and influence of microstructure

It has been mentioned above that reinforcement cannot only be reached by incorporating a second compound, but also by certain microstructures of the polymer itself. In principle all polymers are suitable for self-reinforcement. Their long molecules can act similarly as a fibre if the molecules are in a drawn condition.

This condition is reached either by drawing the part or by using a flow gradient during fabrication. The problem, however, is to freeze this condition in order that it persists over the lifetime of the part. Polymers with glass transition temperatures above the operation temperature will stay in this condition once it is generated.

Polymers which are built up from very stiff, stretched molecules are the liquid crystal polymers. These polymers can keep this structure up to temperatures as high as their melting point. Results on the sliding wear of LCP-based composites have been reported by Friedrich and Voss [21] some time ago. New data, recently

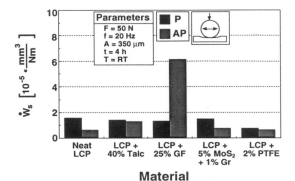

Fig. 12. Specific wear rate of LCP compositions in the parallel and the antiparallel direction with respect to the MFD, when tested under lubrication with a short chain alkylated polysiloxane.

generated by Schledjewski, with regard to the fretting wear behaviour, strengthen the fact that the wear of these materials is highly influenced by the molecular orientation in these systems (fig. 12) [22].

Similar trends can be found if normal high-temperature resistant, semi-crystalline polymers, such as PEEK, are oriented by some kind of a cold drawing process. A set of experiments on the sliding wear resistance has been conducted with drawn neat PEEK 450 G, a commercial grade from ICI, by Jacobs et al. [23].

It was available (a) as a compression moulded version and (b) in drawn forms with three different drawing ratios, DR = 3, 4 and 6, respectively. Figure 13 illustrates the principle of the drawing procedure, which is described in detail in ref. [22]. In this reference an increase in the crystallinity with the drawing ratio is reported. X-ray diffraction patterns, recorded with the incident beam (CuK$_a$-radiation) in three mutually orthogonal directions, show Debeye–Scherrer rings which arose from randomly oriented crystallites for the undrawn material (fig. 14). In contrast, the diffraction patterns of the drawn materials exhibited distinct intensity variations, indicating a rather high degree of alignment of the crystals (fig. 15). Figure 16 schematically illustrates the present knowledge of the microstructure of the drawn PEEK. The *c*-axis of the lamellae is preferably oriented parallel to the

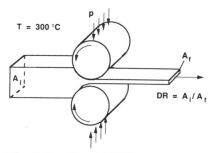

Fig. 13. Drawing of PEEK.

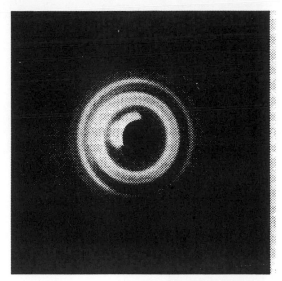

Fig. 14. X-ray diffraction pattern of undrawn PEEK.

drawing direction, where the *a*- and *b*-axis are more or less randomly oriented [24,25]. The molecules in the amorphous phases are also stretched and mainly oriented parallel to the drawing direction. Resulting from this microstructure, a strong and stiff bonding of the crystallites in the drawing direction is expected, but a poor bonding in the transverse direction.

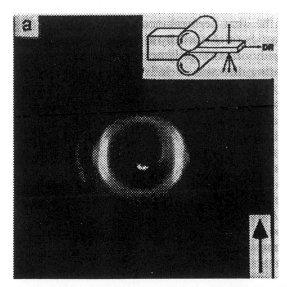

Fig. 15. X-ray diffraction pattern of drawn PEEK (DR = 6).

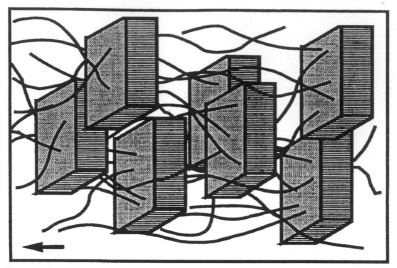

Fig. 16. Simplified representation of the morphology of drawn PEEK.

Sliding wear studies have been performed on a block-on-ring testing device as described above, over a period of 4 h with an apparent contact pressure of about 0.57 MPa and a sliding velocity of 3 m/s. From the mass loss the specific wear rate, as defined above, has been calculated.

Due to the anisotropic microstructure of the material, anisotropic wear behaviour is expected [26]. For this reason the wear tests with the drawn PEEK versions were carried out for three principal loading directions:
– with the sliding plane normal to the drawing direction (N);
– with the drawing direction in plane but transverse to the sliding motion (i.e. antiparallel, AP); and
– with sliding parallel to the drawing direction (P).

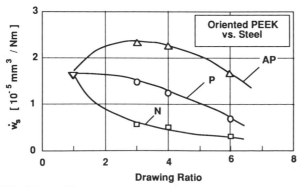

Fig. 17. Specific wear rates versus drawing ratio for different orientations of the sliding plane with respect to the drawing direction (conditions. $p = 0.57$ MPa, $v = 3$ m/s, T = room temperature).

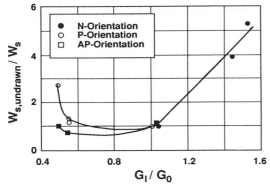

Fig. 18. Relative specific wear resistance versus tensile energy release rate for drawn PEEK.

The results are presented in fig. 17. The expected anisotropic wear behaviour has been found to exist. Other authors [27] published details of a similar dependence of the wear performance on the sliding direction for plastic reinforced by aramid fibres, which are also highly oriented polymers. The ranking order of the sliding directions relative to the molecular orientation generally is also in accordance with investigations on liquid crystal polymers [21] and oriented thermoplastics [28–30], respectively. However, the clear difference between the P- and AP-direction, found for PEEK, was less pronounced for the other polymers. Microscopic analysis revealed that undrawn PEEK shows indications of microabrasion, surface fatigue and adhesive wear. After sliding in the N-direction only some rather mild scratches existed. A surface worn in the P-direction shows even less damage by scratching, but more severe surface damage and flat holes from delamination of small chips. The wear surface of a specimen worn in the AP-orientation exhibited a totally different appearance. The surface is covered by a system of cracks parallel and striations transverse to the sliding direction.

The specific wear rate describes the volumetric material loss per unit of dissipated energy. Accordingly, the inverse of the specific wear rate, called specific wear resistance, is a measure of the energy dissipated during the removal of one volume unit of material. Therefore it seems reasonable to correlate the wear resistance with the material's strain energy release rate. Figure 18 presents a plot of the specific wear resistance versus the mode I strain energy release rate for crack propagation parallel to the sliding motion, i.e. transverse to the drawing direction for sliding in the N-orientation and parallel to the drawing direction for sliding in the P- or AP-direction. For both axes, the values obtained for the drawn materials were normalized to those of the undrawn PEEK. The assumption of linearity seems to apply in the case of N-orientation, although the mechanical value is derived under mode I conditions while under sliding the crack propagation parallel to the surface is driven by shear load. For the other sliding directions (AP and P) such a simple relationship could not be found. The probable reason is that the wear of the drawn material worn in the N-direction and of the undrawn material is dominated by the same abrasive mechanisms, while the wear mechanisms during sliding in the P- and AP-direction, respectively, are considerably different.

The anisotropic wear behaviour of the drawn materials, in general, can be explained by the morphological structure. When sliding takes place in the N-direction the molecules in the crystallized as well as in the amorphous regions preferably are orientated normal to the sliding plane; under this condition they are, even near the surface, strongly tied to the bulk. Therefore the material possesses a high resistance against the pulling out of the molecular bonds. Additionally, the relatively high flexibility of the polymer chains prevents bending fracture of the material near the surface, as was frequently observed for brittle glass or carbon fibre reinforcements under N-orientation [31].

Under sliding in the P-direction the molecules are subject to a tensile load more or less parallel to the fibre axis, e.g. the direction of the highest molecular bond strength. A high resistance against the initiation and development of surface cracks can be expected. However, the partly orientation of the molecules parallel to the surface diminishes the amount of covalent bonds in the perpendicular direction and thus reduces the cohesion of the surface layer to the bulk. As a consequence, the material becomes more susceptible to disruption under a load as it is produced by friction.

3.3. Effect of solid lubricants in different polymers

A variety of solid lubricants is employed in low-friction materials, due to their beneficial effect on the tribological characteristics when pre-dispersed into a base polymer before moulding. Perhaps currently the most common and important is polytetrafluoroethylen (PTFE). Silicones, molybdenum disulphide and graphites are also used. Reference [32] presents detailed experimental results on the effect of these substances on both the coefficient of friction and the wear rate for a wide range of base polymers. Lubricant packages decrease the coefficient of friction and the rate of wear. Since the frictional traction is reduced, flash temperatures decrease, and as such, lubricant additives generally allow polymers to tolerate higher rubbing intensities. The article of Gardos [33] presents a review of the little reported academic work on these so-called self-lubricating systems.

The good frictional properties of PTFE are based on the fact that PTFE tends to establish a thin transfer film on the counterface. This transfer film is of about 30 molecules thick [34]. During sliding the molecules are stretching out parallel to the direction of motion. The symmetrical, non-polar configuration of the PTFE chains is responsible for a weak intermolecular bonding. Once an orientated layer is established, slip in this layer becomes very easy (low shear strength). The result is a low coefficient of friction. This effect is used to improve the frictional properties of other materials by lubricating them internally with PTFE, which separates the sliding bodies by establishing such a thin film between them. The use of PTFE in high-temperature applications is limited to 250°C operation temperature.

The two other solid lubricants withstand higher temperatures but this temperature range is not interesting for polymeric materials. Their lubricating effect is

TABLE 4
Tribological investigated PEEK–PTFE blends.

No.	Material codes	Weight fraction (wt%)	Density (g/cm^3)		Volume fraction (vol%)	
			ICI	Authors	ICI	Authors
A	PEEK-150PF	0.0	1.32	1.303	0	0
B	RX-L-89-338 +PTFE-7C	8.0	1.34	1.334	5	5
B *	RX-L-89-338A +PTFE,KT300	8.0	1.32	1.301	5	5
C *	RX-L-89-339A +PTFE, KT300	22.5	1.36	1.346	15	14.8
D	RX-L-89-340 +PTFE-7C	52.0	1.625	1.588	40	39.5
E	RX-L-89-341 +PTFE-7C	71.0	1.81	1.798	60	59.5
F	RX-L-89-342 +PTFE-7C	90.5	2.03	2.036	85	85
G	RX-L-89-343 +PTFE-7C	97.0	2.12	2.122	95	95
H	PTFE-7C	100.0	2.13	2.169	100	100

based on the same reason, their molecular structure. The crystal structure of these materials is hexagonal and shows only weak bondings in the direction of the *c*-axis. Here van-der-Waals bondings are operating, where in the *c*-plane the atoms are covalently bonded. The great difference in the strength of the bindings causes easy slip of the *c*-planes. By incorporating either graphite or molybdenum disulphide powder into a material these particles will penetrate into the sliding contact zone, where the above described effect will lower the occurring frictional forces.

The effect of incorporating PTFE and graphite into polymers is well investigated. Some model polymers demonstrate this influence. Here a study on PEEK/PTFE-blends is presented [35]. The investigated materials are listed in table 4. The samples of pure PEEK and a PTFE content of 5 vol% have been injection moulded, the samples with higher PTFE contents were sintered. The specimens have been annealed in order to get the degree of crystallinity of PEEK at 30%. Figure 19 shows the specific wear rate which depends on the volume fraction of PTFE content. The experiments have been conducted using a pin-on-disk tester with 1 m/s sliding velocity and 1 MPa applied normal load. The best result is reached with a PTFE volume fraction of 5 to 10%. But the coefficient of friction provides a different point of view (fig. 20). The lowest coefficient of friction is obtained at about 10 to 20% PTFE content. For this system the optimum volume fraction of PTFE is about 15%. Both the wear rate and the coefficient of friction are in their lowest range (fig. 21). The reduction of friction at low PTFE contents is assumed to be due to the increasing efficiency of lubrication by the PTFE. The rise of the coefficient of friction beyond 20 vol% PTFE content can be explained by the microstructure of the materials. Some micrographs of the materials are shown in fig. 22. The micrographs of samples with a PTFE content up to 15 vol%

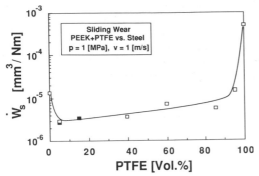

Fig. 19. Influence of PTFE content on the wear behaviour of PEEK. The black symbols represent results for materials with PTFE grade KT700 whereas the other symbols represent materials with PTFE grade 7C.

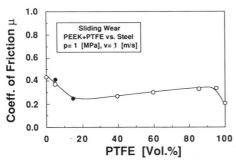

Fig. 20. Influence of PTFE content on the friction behaviour of PEEK. The black symbols represent results for materials with PTFE grade KT700 whereas the other symbols represent materials with PTFE grade 7C.

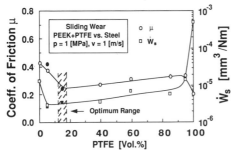

Fig. 21. Optimum range of PTFE content in PEEK composites with respect to the tribological behaviour. The black symbols represent results for materials with PTFE grade KT700 whereas the other symbols represent materials with PTFE grade 7C.

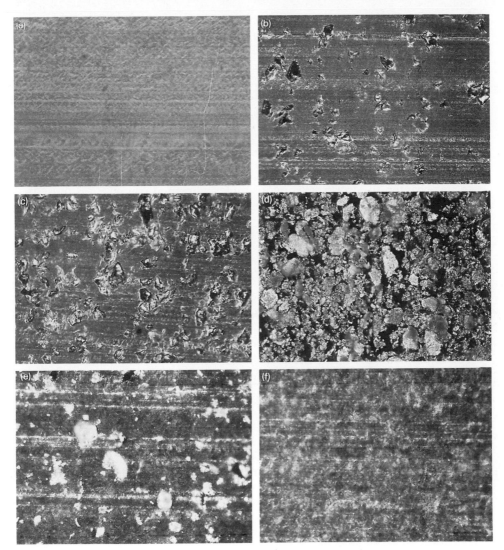

Fig. 22. Transmission light micrographs of PEEK–PTFE blends. (a) neat PEEK, (b) 5 vol% PTFE, (c) 15 vol% PTFE, (d) 40 vol% PTFE, (e) 95 vol% PTFE and (f) neat PTFE.

show a continuous phase of PEEK with dispersed PTFE phases. With 40 vol% PTFE content the appearance is different. PEEK and PTFE are present as two dispersed phases. With a higher PTFE content this phase becomes continuous. The change of tribological properties can be attributed for a very low PTFE content to the lubricating effect of this material and for contents higher than 20% to a change of the microstructure.

Similar tribological investigations have been performed by Briscoe and co-workers [36]. Opposite to the experiments reported here, he performed the investigations at lower pv-products. The results are very much different. The wear increases steadily at a low pv-product with increasing PTFE content. The resulting friction exhibits a strong reduction at low volume fractions, up to about 20%, and beyond this point a further but slight drop.

The studies make clear the great influence of the applied working conditions. Especially the wear behaviour of PEEK is improved by PTFE additions in the high pv-range and becomes worse in the case of low pv-products.

3.4. More detailed approach to the model system PC–PTFE

In this chapter an investigation on a simple, lubricated composite is presented. This composite, a polycarbonate lubricated with 20% spherical PTFE, is related to section II in fig. 5.

3.4.1. Experimental wear results

The experiments have been carried out on a block-on-ring apparatus as described above. The specimens took the form of cubes with a side of 6 mm, machined from injection mouldings. The counterparts were prepared to a pre-test surface roughness of $R_a = 0.08$ µm. All tests reported here were conducted at a sliding speed of 1 m/s under a normal load of 27 N ($= 0.75$ MPa pressure).

The initial wear rate in the previous case is considerably lower than that of the new specimen, but the final equilibrium wear rate is almost the same (see previous section).

In the present case the full curve in fig. 23, which is associated with a newly prepared counterpart and an equilibrated specimen, can be explained by material transfer to the metal counterpart. This transfer brings about an improvement in the wear resistance, and as such it is likely to be PTFE transfer. The dashed curve in fig. 23, which is associated with a newly prepared specimen, displays a far greater initial reduction of the wear rate with sliding distance. It follows, therefore, that the difference between the two curves must be associated with changes that occur in the specimen during prolonged sliding. It can be concluded that the high

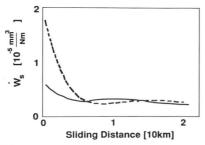

Fig. 23. Wear characteristics of 20% PTFE-filled PC sliding against newly prepared steel counterparts. Dashed curve. newly prepared specimen. Full curve. pre-worn specimen.

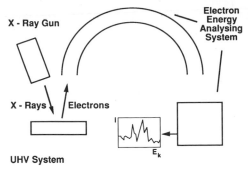

Fig. 24. Configuration of an ESCA analyzer.

equilibrium wear resistance of PTFE-filled polycarbonate can be attributed to two effects:
– transfer of polymer to the metal counterpart and
– modification of the compound surface as a result of sliding contact.
To study the current mechanisms the slid surfaces have been investigated by ESCA analysis.

3.4.2. ESCA

ESCA stands for electron spectroscopy for chemical analysis. With this method e.g. information about the distribution of elements on a surface is obtained. ESCA stands for a couple of testing principles based on the evaluation of the binding energy of electrons [37]. Usually the investigated surface is irradiated either with a monochromatic light beam or with an X-ray beam (XPS-method), using the characteristic, monochromatic X-ray irradiation of the source. The X-ray photons knock out electrons from the atoms of the investigated surface. The kinetic energy of these electrons is detected. Figure 24 shows schematically the experimental setup. By comparing the energy of the injected photon (known wavelength of the used light source) with the kinetic energy of the freed electrons the binding energy of the electrons can be derived.

$$E_k = h\nu - E_b - \Delta\phi, \tag{13}$$

where E_k is the kinetic energy of the electron, $h\nu$ is the photon energy of the X-ray beam, E_b is the binding energy of the electron and $\Delta\phi$ is the difference of the electric potential of the specimen and the detector. Since the binding energy is characteristic for each electron of each element, a comparison of the detected and known binding energies gives information about the element the electrons are emitted from. By setting the detector to a certain energy characteristic for one element and scanning the X-ray beam over the specimen a distribution map of this element is obtained (imaging XPS).

There are some other techniques available giving similar information. One alternative is TOFSIMS, but in comparison to ESCA TOFSIMS has the disadvantage of a much smaller field of view (about 0.2 mm compared with about 1 mm diameter). Beside a statistically more significant sampling, the larger area offers an easier handling of the samples (no charging problems to overcome) and inherently quantifiable data.

A VG "ESCASCOPE" was used for this work and this has been described fully elsewhere [38,39]. With this instrument specimens which either slid 0 or 15 km have been investigated.

A sample analysis area of about 900 μm diameter was chosen for the imaging work, using the 700 μm field of view aperture and the instrument "zoom" facility. The sub-10 μm objective aperture was selected to allow a resolution of about 5–8 μm sized features, along with 6 mm analyzer slits to optimize the signal levels. Al K_α radiation was used with the X-ray source set to 15 kV, 34 mA (510 W); the X-ray source–sample separation was approximately 16 mm.

XPS images were recorded for the F_{1s}, O_{1s} and C_{1s} photoelectron lines using 100 eV pass energy. The acquisition times were 4 h per element (comprising 2 h for the peak image (P) and 2 h for the background image (B)), acquired in "multiplex" mode where 1 min is spent alternately recording signal from the peak and then from the background energy (about 20 eV lower background energy than the peak).

The XPS images presented in fig. 25 are all overlay images with the F_{1s} component, shown in red, and the O_{1s} component, in green. F (red) corresponds to the PTFE and O (green) to the PC host material. In all cases, the images are "resultant" images which were produced by subtracting a background image from the peak image for each element.

The images in figs. 25a and b were recorded from unworn (0 km sliding distance) polymer samples, while those in figs. 25c to f were recorded from worn (about 15 km sliding distance) samples. Fluorine-rich "islands" of up to about 100 μm diameter can be seen at the surface of the unworn samples. This would be expected if the cut surface of the polymer compound reflected the bulk distribution, as revealed by TEM studies on thin cross sections through this material. It is interesting to note that the scale bar shown on the images would represent (typically for rough surfaces) the largest field of view available with the TOFSIMS technique; the larger field of view available in these XPS images serves to emphasize the heterogeneity in these samples.

It is evident that there is significant spreading of the fluoropolymer islands in the experimental sliding direction (right to left in fig. 25c, bottom to top for figs. 25d to f), leading to the formation of a thin film overlay (fig. 25f). The degree of spreading observed with a given set of wear testing conditions will depend on the initial distribution and particle size of the fluoropolymer at the surface, along with the sample surface roughness; these will affect the rate of formation of the film, as seen in the images shown in figs. 25c to f, which were all recorded after the same experimental sliding distance (15 km). Figure 25c shows a fluoropolymer distribution that is similar to an unworn sample (figs. 25a and b) with some spreading; figs. 25d and e show less resemblance to an unworn sample, with more extensive, virtually complete, formation of a fluoropolymer film over the compound surface.

The observation of film formation over the polymer compound specimen

Fig. 25. XPS overlay images showing the PTFE (red) and PC (green) distribution. (a) and (b) were recorded from unworn (0 km sliding distance) polymer samples while (c) to (f) were recorded from worn (about 15 km sliding distance) samples.

surface in this work appears to be unique; thin transfer films have between observed on metal or glass counterfaces used in wear experiments with pure PTFE and similar systems [34,40–44]. The formation and nature of such a film will have an important contribution to the interpretation of the wear rate against sliding distance data, the elucidation of friction and wear mechanisms and to the performance of polymer systems in tribological applications.

Purely spectroscopic observations indicate that there is no evidence of damage to the PTFE component, either due to the dry sliding wear experiment or due to the exposure to the X-ray beam during the imaging experiment. The observed F_{1s} and C_{1s} (CF_2 peak) binding-energy separation of 397.9 ± 0.2 eV is similar to that of pure PTFE (397.6 ± 0.1 eV, measured under the same conditions), and there is no evidence in the C_{1s} high-resolution spectrum of the existence of fluorinated carbon species other than $(-CF_2-CF_2-)_n$. It would seem likely, therefore, that the film in this case is indeed PTFE. A more traditional way of investigating overlayers using XPS would be to carry out an angle-resolved experiment. This was conducted in connection with this work, and the result supports the formation of a fluorine-rich overlayer as the $F_{1s}:O_{1s}$ ratio increases at low take-off angles (with respect to the sample surface).

However, the samples' surface roughness and slight curvature in the case of the worn samples combine to increase the experimental uncertainties. The imaging XPS experiment copes with the sample roughness and curvature (the depth of field with the experimental conditions used in this work was approximately ± 0.3 mm) and produces a clearer result.

The images in figs. 25a and d show several relatively large (about 20–200 μm size) dark features, where only very low background noise signals are observed; such features are absent in the worn samples (figs. 25c–f). This is an effect of the surface topography, where regions of relatively "high ground" are casting X-ray shadows over the sample surface. For the images presented in this work, the X-ray beam was incident from the bottom right to the top left of the image and at 35° to the plane containing the sample surface. Shadowing is absent for the worn samples, for which the surface has been smoothed extensively during the wear testing process.

To complete the understanding of the current mechanism, the counterface has been looked at too. Here the ESCA investigations have been combined with an argon-ion etching technique (fig. 26).

The results from the polymer specimens have demonstrated that some kind of an interfacial PTFE layer is formed during sliding. This PTFE layer is expected to be partly transferred to the metal counterpart, as reported by Tanaka et al. [34,45]. Indeed, the ESCA results demonstrate the existence of this PTFE transfer layer.

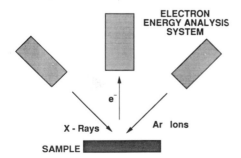

Fig. 26. Combination of ESCA analysis and argon-ion etching technique.

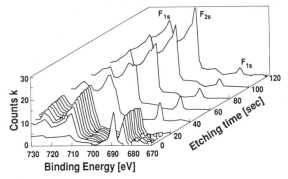

Fig. 27. Depth profile from ESCA analysis of the F_{1s} electrons on the transfer layer of PTFE-filled polycarbonate.

Here the question is how such a transfer film is structured. Therefore the transfer layer is stepwise removed by ion etching and after each step investigated by ESCA analysis.

The etching rate of polymers is in general about 100 nm/min. Therefore 12 s etch is sufficient to remove just about the transfer film; the thickness of this layer is assumed to be less than 10 nm. This value is in the same order of magnitude as reported by Tanaka et al. [34]. The overlayer thickness corresponds to about 30 molecule layers. ESCA analysis at the beginning of the etching and after several etching periods, up to 120 s, indicate that even after removing the transfer film there is still fluoride on the surface, but the binding-energy level of the F_{1s} electron has slightly shifted to lower energies. This fluoride remains over long etching periods on the counterpart surface (fig. 27).

The measured peaks are envelopes of several discrete peaks of similar energy. These peaks can be calculated by applying a Fourier transformation. Figures 28–30 give the results of such a transformation for the envelopes taken after 0, 12 and 120 s. Where the unetched specimen exhibits a peak of high intensity for the known binding energies for pure PTFE and a peak of very low intensity for the electron-binding energies found in metal fluoride, the intensities of these peaks

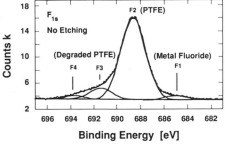

Fig. 28. ESCA analysis of the F1s electrons on the transfer layer of PTFE-filled polycarbonate prior to the etching process.

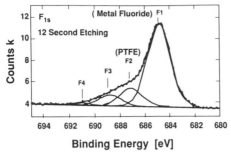

Fig. 29. ESCA analysis of the F1s electrons on the transfer layer of PTFE-filled polycarbonate after 12 s etch.

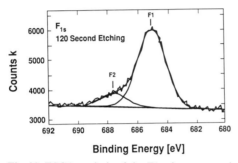

Fig. 30. ESCA analysis of the F1s electrons on the transfer layer of PTFE-filled polycarbonate after 120 s etch.

invert for the etched specimen. It can be concluded that the PTFE transfer film consists of two layers. The top layer is a thin cover of pure PTFE. The interface layer contains chemically bonded fluoride and is much thicker than the overlayer (fig. 31).

During a further experiment the energy analyzer was tuned to the electron-energy levels of bonded carbon (C_{1s}). The Fourier transformation of the enveloping counting curve reveals four peaks, of which one could be attributed to the

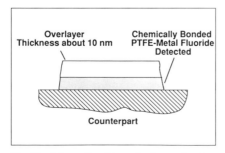

Fig. 31. Schematic of transfer-layer structure.

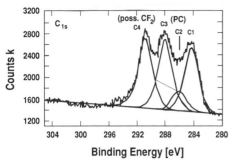

Fig. 32. Binding energies of C_{1s} electrons detected on the transfer film of polycarbonate sliding on a steel counterface.

polycarbonate and a second one represents possibly the CF2 bonding in neat PTFE (fig. 32). The other peaks are attributed to degraded PTFE or the carbon content of the steel counterpart. Finally, it can be concluded that the transfer layer of PTFE-filled polycarbonate contains mainly PTFE, but some polycarbonate is transferred as well.

3.4.3. Attempt at a qualitative interpretation

The wear behaviour of polymers internally lubricated with PTFE often does not follow a linear correlation with the volume fraction of PTFE. Such a linear relationship between the specific wear rates of the components can be written as

$$\dot{W}_c = (1 - V_{f\,PTFE})\dot{W}_{matrix} + V_{f\,PTFE}\dot{W}_{PTFE}. \qquad (14)$$

The wear drops below this linear relationship at some synergetic effect. The results presented in fig. 19 for PEEK–PTFE composites clearly show this synergism. This behaviour can be schematically described by an equation based on the specific wear resistance of matrix and lubricant. By assuming the synergetic effect being a sum of two currents, a volume fraction dependent and an oppositely acting

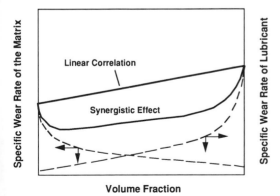

Fig. 33. Linear correlation and synergetic effect due to summation of two opposite trends.

mechanism, of which the principle is shown in fig. 33, the wear behaviour of the composite can be described as:

$$\frac{1}{\dot{W}_c} = (1 - V_{f\,PTFE})\frac{1}{\dot{W}^*_{matrix}} + V_{f\,PTFE}\frac{1}{\dot{W}^*_{PTFE}}, \tag{15}$$

where the specific wear rates marked with an asterisk represent the wear rates at the presence of PTFE.

$$\dot{W}^*_{matrix} = \dot{W}_{matrix}f. \tag{16}$$

These effective specific wear rates are related to the specific wear rates of the neat components by a factor describing the efficiency of the lubrication.

3.5. Effect of discontinuous fibre reinforcement

Discontinuous fibre addition is known to give considerable improvements on the mechanical properties of polymeric systems [46]. Since, loosely speaking, superior mechanical properties imply superior tribological properties, fibre reinforcement can be expected to bring about property advantages in dry sliding friction and wear situations. Early investigations [47–52] into the influence of short-fibre reinforcement on the wear properties indicated that such additives did bring about marked improvements in the tribological characteristics, particularly in the wear behaviour.

Clerico and Patierno [51] described the mechanism of the wear in short-fibre reinforced composites as being delamination initiated by cracking at the fibre–matrix interface, whereas Jain [52] described the mechanism as being one of pulling out of broken and worn fibres. Voss and Friedrich [53] recognised the significance of fibre orientation with respect to the sliding plane. They found that when the fibres in the core region of mouldings were aligned orthogonally to the sliding plane the wear resistance was significantly better than when the skin fibres were at the wear surface. These authors explained this as being due to the lower probability of fracture of fibres when normally oriented.

3.5.1. Study of model systems
3.5.1.1. Glass and carbon fibre reinforced LCP [21]. Pin-on-ring sliding wear experiments with filled LCP materials have been conducted at different pv-products. The LCP composites were delivered by Celanese Corp. as injection moulded plates. The composition and material codes are listed in table 5.

Figure 34 shows the results of these investigations. Two conclusions can be drawn. On the one hand, a fibre reinforcement of about 20% by volume fraction gives a minimum in the specific wear rate, on the other hand, carbon fibres are superior to glass fibres. This better performance of the carbon fibres can be attributed to their higher mechanical property profile and to the less abrasive action of their debris. By additions of mineral and lubricating graphite the wear characteristics of the glass fibre reinforced material is improved to an intermediate value between those of pure glass and pure carbon fibre reinforcement. The

TABLE 5
LCP materials investigated.

Grade code	V_f (%)	Filler	Density (g/cm³)
LCP	0	–	1.4
LCP-30GF	19	Glass fib.	1.61
LCP-50GF	35	Carbon fib.	1.79
LCP-30CF	23	Glass fib.	1.49
LCP-MFG	18	Glass fib.	1.89
	17	Mineral	
	4	Graphite	

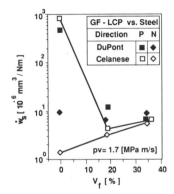

Fig. 34. Specific wear rates of LCP-composites as a function of the fibre content.

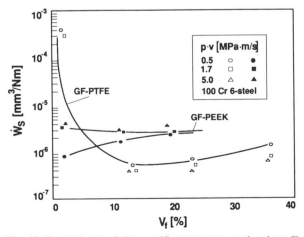

Fig. 35. Dependency of the specific wear rate on the short-fibre content for PTFE- and PEEK-composites.

TABLE 6
Sliding friction and wear results for short-fibre reinforced polycarbonate.

Grade code	V_f (%)	W_s (10^{-6} mm^3/N m)	μ_{dyn}
RXD-90370	0	1500.0	0.42
RXD-90132	5	8.3	0.32
RXD-90133	10	15.2	0.31
RXD-90134	20	20.7	0.34
RXD-90135	30	24.5	0.35

optimum volume fraction for fibrous additions has been found by numerous researchers in the glass fibre–PTFE and –PEEK system in the same manner, fig. 35.

3.5.1.2. Glass fibre reinforced polycarbonate.   LNP standard grade short glass-fibres were compounded into the polycarbonate matrix at volume fractions of 0, 5, 10, 20 and 30%. The tribological experiments were performed on a pin-on-ring wear test apparatus under a sliding velocity of 1.31 m/s and an 26.7 N applied load acting on the specimen ($pv = 0.97$ MPa m s^{-1}).

Table 6 gives the experimentally determined values of the equilibrium specific wear rate, W_s, and the dynamic coefficient of friction, μ_{dyn}, for the range of grades under test.

It would appear then that both the wear rate and the coefficient of friction dip to minimum values at a loading between 5 and 10%. In tribological terms, however, the differences in the friction values are not marked, with respect to the experimental error.

Figures 36a to d show the effect of the initial relative surface concentration of fibre and matrix. The plotted curves are "idealised" from the wear test data and may be used to describe the mechanistic changes occurring during sliding (particularly at "brake in"), when the volume of fibres is varied.

Figure 36a shows a sharp drop in the wear rate with time/distance, this then reaches some equilibrium value. The initial high value is due to the high level of

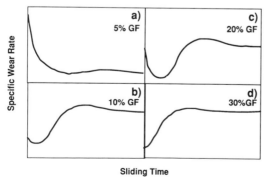

Fig. 36. Wear–sliding distance behaviour of polycarbonate filled with (a) 5%, (b) 10%, (c) 20% and (d) 30% glass fibres.

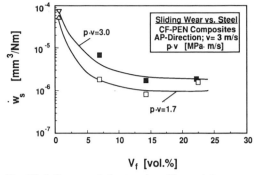

Fig. 37. Influence of short carbon fibre reinforcement on the sliding wear behaviour of PEN.

polycarbonate present at commencement of sliding. By raising the fibre content to 10% by volume (fig. 36b) a drop can be noted from the initial high rate of wear, followed by a significant rise again. The rise which is seen, is undoubtedly due to the onset of a higher glass–fibre induced "third-body" abrasive action. At even higher fibre contents, i.e. 20 and 30% (figs. 36c, d), the curve is essentially shifted to the left, i.e. in these grades the onset of third-body wear occurs rapidly and almost immediately after the commencement of sliding.

3.5.1.3. GF and CF PEN. In short-fibre reinforced PEN-based systems the increasing beneficial influence of increasing fibre content has been found again (fig. 37). But beyond a content of 20% neither for glass fibre nor carbon fibre additions further changes in wear rate were observed. The beneficial effect is attributed to a reduced ability of ploughing, tearing and other non-adhesive components of wear [54], provided a good interfacial adhesion between the matrix and the fibrous reinforcement exists.

For a 20 wt% GF–PEN composite no clear effects of the sliding direction with respect to the mould-filling direction could be revealed (fig. 38). An explanation for this behaviour is given by the isotropic character due to the quasi-random arrangement of the glass fibres. Principally the same characteristic was observed in the case of PAN-based composites with 20 and 30 wt% carbon fibre reinforcement (fig. 39), although the effect was less pronounced. Contradictory results have been reported by Shim et al. [55]. He investigated well-oriented CF–EP-composites and found a clear ranking of the sliding directions, with AP best and N worst. This effect was attributed to the different wear characteristics of PAN-based CF due to the crystallographic orientation with respect to the sliding direction.

3.5.2. Model approach

A number of workers have attempted to mathematically describe the wear process in short-fibre composite systems. Some use a rule-of-mixtures approach [56,57], in which the composite's tribological properties are expressed in terms of the properties of the constituents, whereas others [58,59] have used traditional approaches modifying the terms in the equations to reflect the properties of the

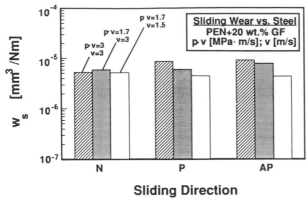

Sliding Direction

Fig. 38. Specific wear rate as a function of *pv*-product and relative sliding direction for PEN composites with 20 wt% short glass fibre reinforcement.

composite. In general, the former approach leads to a more fundamental under-standing of the composite.

The most comprehensive model to appear for the description of the wear behaviour of short-fibre reinforced systems is that due to Voss and Friedrich [21]. The model is based on observations of the variation of the specific wear rate with reinforcement fraction and on careful microscopic studies of the wear mechanisms. These workers observed that several processes were important; namely matrix sliding wear, fibre sliding wear, fibre cracking, fibre–matrix separation and third-body abrasion by fibre debris. By assuming that fibre–matrix separation occurs as a result of fibre cracking, Voss and Friedrich developed the following equation for the specific wear rate of the composite, $\dot{W}_s$, based on the rule-of-mixtures idea:

$$\dot{W}_c = \frac{1}{(1 - V_f)\dot{W}_m^{-1} + 0.5(1 - V_f^c)V_f\dot{W}_f^{-1}} + 0.5(1 - V_f^c)V_f\dot{W}_{fci}, \qquad (17)$$

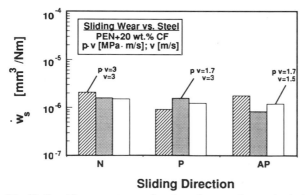

Sliding Direction

Fig. 39. Specific wear rate as a function of *pv*-product and relative sliding direction for PEN composites with 20 wt% short carbon fibre reinforcement.

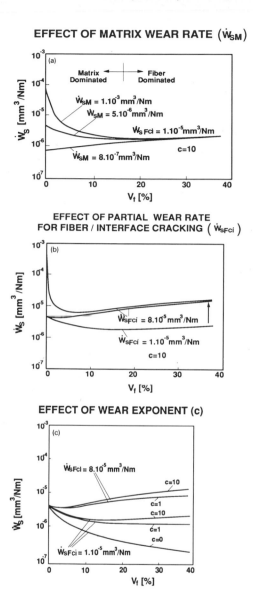

Fig. 40. Schematical demonstration of the influence of short-fibre reinforcement on the specific wear rate of a polymer composite.

where V_f is the fibre volume fraction, $\dot{W}_m$ and $\dot{W}_f$ are, respectively, the specific sliding wear rates of the matrix and the fibre and c is an empirical constant. $\dot{W}_{fci}$ represents the wear rate due to third-body processes.

Figure 40 schematically demonstrates the predicted variation of the wear rate of a polymer as a function of the volume fraction of short-fibre reinforcement according to eq. (17).

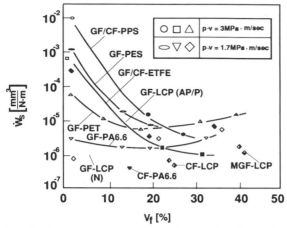

Fig. 41. Summary of specific wear rates as a function of the fibre volume fraction.

In fig. 41 the use of this equation to fit the functional dependence of the wear rate on the volume fraction is demonstrated. Very good agreement is found. Orientation differences in this model are accounted for by differences in the second term of eq. (17).

For low volume fractions of reinforcement the second term in the equation is small and as such the reductions in the wear rate with increasing reinforcement are due to the generally superior wear resistance of the fibrous phase. At larger volume fractions the amount of fibrous debris increases and as such the second term in the equation begins to have an influence. The minimum in the wear rate–fibre content function is therefore a result of these two competing processes. The model predicts that the minimum, which has clear technological significance can shift according to the nature of the fibrous addition. Fibres with greater intrinsic wear resistance in general will lower the minimum value, whereas the quality of the fibre–matrix bond will shift its position on the volume-fraction axis. Higher fibre–matrix bond strengths (lower values of c) will shift the minimum to higher volume fractions.

Hanmin et al. [60] have observed similar behaviour in the coefficient of friction as a function of volume fraction of carbon fibres in polyphenylene sulphide. No model has thus far been developed to express the friction coefficient in similar terms to that of eq. (17).

The relevance of this model has been demonstrated numerous times. The curves drawn in figs. 34, 35, 37 and 44 are theoretical values predicted by the model. Based on the Voss model a modified version is presented.

As before, the wear rate of the composite consists of two terms, where the first one now is:

$$\dot{W}_1 = \frac{1}{(1 - V_f)\dot{W}_m^{-1} + V_f\dot{W}_f^{-1}}, \qquad (18)$$

where as before, $\dot{W}_m$ is the intrinsic wear rate of the matrix. This time, $\dot{W}_f$ is the wear rate of the fibre associated with the particular system into which it is incorporated. This means that in this theory $\dot{W}_f$ will depend on factors such as the intrinsic wear resistance of the fibre, the fibre–matrix bond strength and, importantly, the influence that the matrix may have on the fibre wear behaviour via tribological-induced interface modification.

Thus we may write $\dot{W}_f$ in the following way:

$$\dot{W}_f = \dot{W}_{fi} f, \tag{19}$$

Where $\dot{W}_f$ is the intrinsic wear rate of the fibre itself and f describes the influence of incorporation into a composite system. For example, f can be expected to decrease with increasing fibre–matrix bond strength and decrease with increasingly beneficial sliding interface modifications.

The second term of this model, which describes the abrasive wear rate due to fibrous debris, is expressed as

$$\dot{W}_2 = \dot{W}_1 V_f \dot{W}_a(V_f), \tag{20}$$

Where $W_1 V_f$ is proportional to the rate of accumulation of fibrous debris and $\dot{W}_a(V_f)$ is the abrasive wear rate caused by the debris. Empirically it has been observed under abrasive conditions that

$$\dot{W}_a(V_f) = \dot{W}_{a0}(1 + aV_f), \tag{21}$$

i.e. the abrasive wear rate of fibre-filled composites increases linearly with fibre loading. Taking all together, eq. (22) represents an alternative to Voss' model with the same number of variables.

$$\dot{W}_c = \frac{1}{(1 - V_f)\dot{W}_m^{-1} + V_f \dot{W}_f^{-1}} + \left(1 + V_f(1 + aV_f)\dot{W}_{a0}\right). \tag{22}$$

Table 7 shows the experimental values of the wear rate for short-fibre reinforced polycarbonate, compared with those obtained using both Voss' model and eq. (22). In both cases the input parameters were adjusted until the most acceptable fits had been found.

TABLE 7
Comparison between experiment and predicted wear values by Voss' model and by the in eq. (22) presented model.

V_f (%)	Specific wear rate (10^{-6} mm^3/N m)		
	Experiment	Voss model	Eq. (22)
0	1500	1500	1500
5	8.3	8.9	8.5
10	15.7	11.9	11.8
20	20.7	20.7	18.2
30	24.5	30.3	24.1

4. Some recent investigations on HT-thermoplastics [62]

4.1. Materials

A couple of different PAEK-materials were studied. One set consisted of PEEK 450 G (ICI, Wilton, UK [61]) as the matrix material and various forms of lubricated and/or short-fibre filled PEEK 450 G versions (table 8). Small samples with a 3×4 mm sliding surface were cut from injection moulded plates, which were all delivered with 3 mm thickness. Sliding took place in the direction of the plate thickness, i.e. transverse to the injection moulding direction. In this way, possible effects due to fibre orientation were almost averaged out (because of the layer structure of different fibre orientation across the thickness [52]).

Another set of plates of neat PEEK-materials, which had different molecular weights, was made available by ICI. Finally, some doubly oriented PEKEKK versions, received from BASF were studied (set 3).

4.2. Counterface structure and time-related depth wear rate of PEEK

Under sliding condition with various pv-product levels the time-related wear depth of PEEK, reinforced with 10 wt% carbon fibres and lubricated by 10 wt% each of PTFE and graphite, was recorded. It was found that a dramatical increase beyond about 3 MPa m s^{-1} (fig. 42) takes place. This is a not very high limiting pv-product. However, in case of sliding against an oxide-layer covered counterpart (the oxide layer was build up in a 24 h heat treatment at 250°C) significantly less

TABLE 8
Commercial PAEK grades investigated under sliding wear conditions.

No.	Material code	Filler content	Density (g/cm^3)	Manufacturer
1	PEEK 450 G	neat matrix	1.294	ICI
2	PEEK-30GF	30 wt% short glass fibre	1.520	ICI
3	PEEK-30CF	30 wt% short carbon fibre	1.440	ICI
4	PEKEKK KR4177	neat matrix	1.300	BASF
5	PEKEKK KR 4177 G6	30 wt% short glass fibre	1.530	BASF
6	PEKEKK KR 4177 C6	30 wt% short carbon fibre	1.450	BASF
7/8	PEEK 450 D-HF 30 PEEK 450 D -FC 30	10 wt% short carbon fibre 10 wt% PTFE particles 10 wt% graphite filler	1.442	ICI
9	PEEK-LCL 4033	15 wt% short carbon fibre 15 wt% PTFE particles	1.446	ICI(LNP)
10	PEEK-VP 63/53	15 wt% PTFE particles 15 wt% graphite filler	1.448	LuV
11	PEK-VP 65/55	10 wt% short carbon fibre 10 wt% PTFE particles 10 wt% graphite filler	1.468	LuV

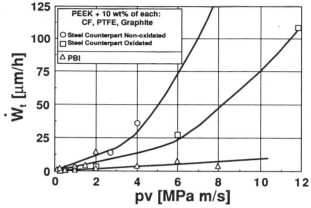

Fig. 42. Depth wear rate as a function of *pv*-factor and counterface structure for PEEK + CF + PTFE + graphite (10 wt% of each filler) and for polybenzimidazole (PBI).

wear occurs and the start point is shifted to higher values. This behaviour is conform with wear models based on the hardness of the contacting parts.

4.3. Apparent contact temperature, coefficient of friction and specific wear rate of PAEK-composites

Comparing the wear resistances of the short glass to the short carbon fibre reinforced PEEK- and PEKEKK-composites on the basis of different apparent contact temperatures, a significant decrease is obtained. Experiments have been carried out in a temperature range far below room temperature (here down to −40°C) and up to a testing temperature of 220°C (fig. 43). An improvement of the wear resistance of the carbon fibre filled version is possible if additional lubricants

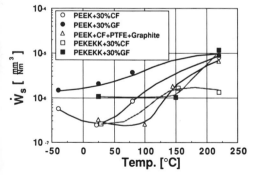

Fig. 43. Specific wear rate of three different PEEK-composites, as influenced by the testing temperature. Carbon fibres prove better than glass fibres, and even better results are achieved with additional PTFE- and graphite-lubricants (● = GF–PEEK, ○ = CF–PEEK at $pv = 5$ MPa m s^{-1}, △ = PEEK + CF + PTFE + graphite at $pv = 3.3$ MPa m s^{-1}, ■ = PEKEKK + GF, □ = PEKEKK + CF; in all cases $v \approx 3$ m/s). The corresponding values of the coefficient of friction of material no. 7 (PEEK + CF + PTFE + graphite) were: $T = 20°C$: $\mu = 0.149$; $T = 100°C$: $\mu = 0.156$; $T = 220°C$. $\mu = 0.150$.

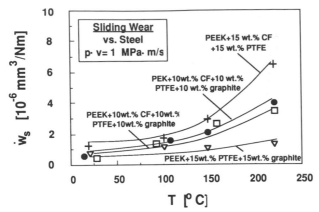

Fig. 44. The effects of various fillers (reinforcements, lubricants) on the wear of PEEK at different testing temperatures. For better clearness, only average data, achieved after two or three individual measurements, were used for the illustration. The scatter of the data is of the order of about ± 12 to 25% of the average values.

are added. These have a favourable action especially at elevated testing temperatures (but still below the glass transition temperature of PEEK, i.e. $T_g = 144°C$).

For the purpose of finding a PEEK-based composition with even lower friction and wear properties than the commercially available material no. 7 in table 8, especially for the use at temperatures above 150°C, other blends with different fibre/lubricant ratios were manufactured and tested as well (materials no. 9, 10 and 11 in table 8). The results are shown in figs. 44 and 9, with material no. 7 as the standard composition. For the same types and contents of fibre reinforcement and lubricating components, a change from an ICI-based PEEK matrix to a PEK-matrix (advertised by Hoechst Company, Frankfurt), did not result in drastic changes of the trends in specific wear rate versus testing temperature (fig. 44). The lack of lubricating graphite flakes in favour of a slight increase in both short carbon fibres and PTFE-particles, resulted in higher values of $\dot{W}_s$ for all temperature conditions evaluated. On the other hand, using no carbon fibres, but just graphite- and PTFE-filler lubricants, each as 15 wt%, clearly reduced the wear rates at elevated temperatures in comparison to the standard.

Differences in the coefficient of friction were, again, not as pronounced, although here material no. 10 seemed to be the most promising one, especially for the temperature of 150°C and above (fig. 45). In addition, it should be noted that the μ-values under the moderate pv-condition of 1 MPa m s^{-1} clearly decreased from a room-temperature level of about 0.23 to values below 0.1 at $T \geqslant 100°C$.

4.4. Molecular weight, degree of crystallinity and molecular orientation versus wear rate of PEEK

Davies et al. [62] studied the sliding wear performance of neat PEEK in much more detail. PEEK of three different molecular weights, as characterised by the

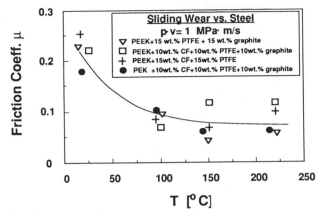

Fig. 45. Corresponding values of the coefficients of friction (materials and conditions as in fig. 44). The scatter here was determined to be between ±10 to 20%.

melt viscosity, and with similar crystallinities, after heat treatment, were worn against polished steel counterfaces using a block-on-ring methodology. The samples were worn under identical conditions, i.e. under a load of 19.1 N ($p \approx 0.5$ MPa), a sliding velocity of 1 m/s and a counterface roughness, R_a, of 0.085 ± 0.001 µm, at ambient temperature and laboratory humidity.

Figure 46 shows a representative wear curve for each molecular weight, plotted on the same axis. The shape of the curves consists of a characteristic initiation phase (up to about 105 m sliding distance) in which the specific wear rate changes from a lower to a higher plateau value. The equilibrium wear rate varied with molecular weight. The MP 2/3, low molecular weight, PEEK wore at twice the rate of the higher molecular weight MP 2/10 PEEK. However, the difference in equilibrium wear rate between the intermediate molecular weight, MP 2/8, and the high molecular weight material was small. This would suggest that under these conditions, mode I fatigue is not the dominant mechanism, since models based on this mechanism predict that the high molecular weight PEEK will be orders of magnitude more wear resistant than low molecular weight PEEK. The shape of the

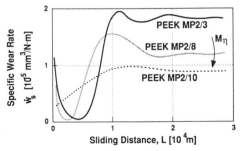

Fig. 46. Initiation phase and steady-state condition of specific wear rate versus sliding distance, as measured for PEEK with different molecular weights.

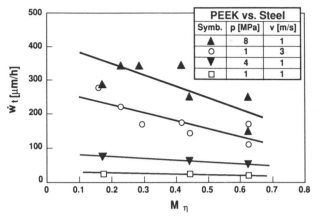

Fig. 47. Effect of molecular weight and testing conditions on the depth wear rate (in the equilibrium state) of PEEK.

curves undoubtedly contains information on the wear mechanisms occurring. However, further experiments will have to be carried out in order to extract this information.

Lu [63] did similar studies with these materials using a pin-on-disc device. The same trends were found with regard to the effect of molecular weight on the wear resistance. A plot of the depth wear rate as a function of molecular weight showed that the effects were greatest when the pv-level was high (fig. 47). For comparable pv-products, a higher velocity at the cost of lower pressure resulted in higher values of the depth wear rates than higher pressure/lower velocity conditions. In comparison to the results reported by Davies et al. [62] for the specific wear rate at $pv = 0.5$ Mpa m s^{-1}, Lu's results at $pv = 1$ Mpa m s^{-1} (both at $v = 1$ m/s) are even lower (MP $2/3 = M = 0.174$: $\dot{W}_s = 5.51 \times 10^{-6}$ mm^3/N m; MP $2/8 = M = 0.439$: $\dot{W}_s = 4.16 \times 10^{-6}$ mm^3/N m). The trends are, however, still the same. That differences in wear rate as a function of the testing procedure can occur is known from studies by Lancaster and co-workers [12], although in their case a pin-on disc configuration led for PEEK to higher values of $\dot{W}_s$ than a (rotating) line-contact configuration.

Results of wear measurements on annealed PEEK 450G-materials are compared with the unannealed ones in table 9. The wear resistance of the annealed materials was improved in all cases. This is probably attributable to the higher degree of crystallinity in the polymer (23% versus 31%) due to the annealing process of the as received material. Annealing was performed at 320°C for 1.5 h, followed by an isothermal heat treatment at 260°C for 50 h. This results in an increase in the material's hardness without reducing the toughness (expressed in terms of the product of strength and strain at fracture) or enhancing the frictional coefficient very effectively. As a combination of all of these parameters has been shown to reflect trends in the specific wear rate of polymer-based materials quite well [52], the experimental wear result is not surprising.

TABLE 9

Specific wear rates of unreinforced and glass fibre reinforced PEEK 450G, as received and after annealing, and the effect of c at a pv-level of 1.7 MPa m s^{-1}.

Material	$\dot{W}_s$ (mm^3 N^{-1} m^{-1})		
	0.6 m s^{-1}	1.5 m s^{-1}	3.0 m s^{-1}
PEEK (as received)	7.6×10^{-7}	3.5×10^{-6}	9.3×10^{-6}
18 vol% GF–EEK (as received)	2.2×10^{-6}	3.0×10^{-6}	4.1×10^{-6}
PEEK (annealed)	4.1×10^{-7}	1.2×10^{-6}	7.6×10^{-6}
18 vol% GF–PEEK (annealed)	1.9×10^{-6}	2.4×10^{-6}	2.3×10^{-6}

4.5. PI- and PBI-based composites at elevated temperatures

The results achieved with carbon fibre reinforced and PTFE and graphite lubricated polyether–ether ketone between RT and $T = 220°C$ in comparison to graphite-filled bismaleinimide and neat polybenzimidazole against a 100 Cr6 steel disc (the average roughness at the beginning of the test was about $R_a = 0.1–0.2$ μm and the hardness was around 63 HRC) are summarised in figs. 48 and 49. Both

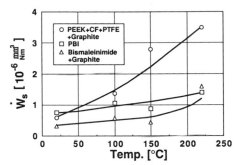

Fig. 48. Specific wear rate versus testing temperature for PBI, bismaleinimide + graphite and PEEK + CF + PTFE + graphite at 1 MPa and 1 m/s.

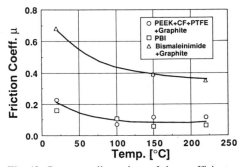

Fig. 49. Corresponding values of the coefficients of friction (material and conditions as in fig. 48).

the coefficient of friction (μ) and the specific wear rates ($\dot{W}_s$) were obtained after a running-in period of about 2–3 h, in which (1) a stable temperature had adjusted in the contact area and (2) the thermal expansion of the specimen had ended, and therefore could not superimpose the continuous measurements of height reduction due to wear. It becomes clear that the materials shown here have fulfilled the high-temperature requirement (220°C), and that the bismaleinimide-based compound, especially designed for low friction and wear applications, gives the best results. In the final choice of the right material for a particular application the following items may, however, also be of importance, e.g. how high the mechanical loading tolerance of the individual materials is and how difficult it will be to manufacture them into complicated shapes. In this sense, the injection mouldable, short-fibre reinforced thermoplastics (e.g. PEEK D450FC30) seem to be very superior to the PI-versions. Additional data recently published on the tribological performance of other thermoplastic composites at elevated temperatures can be found in ref. [64].

5. Conclusions

The results presented in this chapter indicate that many different polymer composites can be used for friction and wear applications. They can be tailored with regard to their load carrying capacity, their temperature and creep resistance, their coefficient of friction, and their resistance to wear under sliding against steel counterparts, but some more knowledge about the detailed interaction between the various components is required in order to obtain a full understanding of their tribological performance. Therefore, there are still further studies necessary.

Acknowledgements

Thanks are due to many companies, who kindly provided the different materials for our studies. Special thanks are due to Z. Lu for his considerable contributions to the experimental work reported here. Various parts of the works were financially supported by the Arbeitsgemeinschaft Industrieller Forschungsvereinigung (AIF D295), but also by the companies ICI (England) and Babcock (Germany).

List of symbols

a	empirical parameter in the composite wear equation
c	empirical parameter of the Voss model
f	efficiency factor
f	difference between the electric potential of specimen and detector
h	Planck constant
m	mass loss of specimen
m_n	coefficients of polynomial fit
μ	friction coefficient

n	frequency of photon
p	pressure
r	density
t	time
v	velocity
A	apparent contact area
C_{1s}	1s electron energy level of carbon
AP	antiparallel fibre orientation with respect to the sliding direction
DR	drawing direction
Δh	hight reduction
Δm	mass loss
ΔV	volume loss
E_b	binding energy of electrons
E_k	kinetic energy of electrons
ESCA	Electron spectroscopy for chemical analysis
F_n	applied normal force
F_{1s}	1s electron energy level of fluor
L	sliding distance
N	normal fibre orientation with respect to the sliding direction
O_{1s}	1s electron energy level of oxigen
P	parallel fibre orientation with respect to the sliding direction
T	temperature
T_g	glass transition temperature
T_m	melting temperature
V_f	fibre volume fraction
W	dimensionless wear rate
W^*	specific wear rate at the presence of another component
W_a	specific wear rate caused by fibre debris accumulation
W_c	specific wear rate of the composite
W_f	intrinsic specific wear rate of the fibres
W_{fci}	specific wear rate due to third body process
W_L	sliding distance related wear rate
W_m	specific wear rate of the matrix
W_s	specific wear rate
W_t	time related wear rate/depth wear rate
W_1	first term of the composites' specific wear rate
W_2	second term of the composites' specific wear rate

References

[1] M.P. Wolverton, J.E. Theberge and K.L. McCadden, Machine Design 55 (1981) 111.
[2] W. Bartholomeyzik, Techn. Rundschau 33 (1990) 52.
[3] K. Friedrich and R. Walter, in: Proc. ASM Int. Conf. on Tribology of Composite Materials, Oak Ridge, 1–3 May (1990).

[4]   H. Czichos, in: Friction and Wear of Polymer Composites, ed. K. Friedrich (Elsevier, Amsterdam, 1986).
[5]   DIN 50 321, Verschleiß-Meßgrößen (Beuth Verlag, Berlin, 1979).
[6]   J.R. Atkinson, K.J. Brown and D. Dowson, J. Lubr. Technol. 100 (1978) 208.
[7]   J.C. Anderson, Tribol. Int. October (1982) 255.
[8]   T.H. Grove and K.G. Budinski, in: Wear Tests for Plastics: Selection and Use, ASTM STP 701 (1978).
[9]   J.K. Lancaster, Tribology 12 (1973) 219.
[10] ASTM 3702-78, Wear weight of materials in self-lubricated rubbing contact using a draft washer testing machine (American Society for Testing and Materials, Philadelphia, PA).
[11] B.J. Briscoe, Tribol. Int. August (1981) 231.
[12] P.M. Dickens, J.L. Sullivan and J.K. Lancaster, Wear 112 (1986) 237.
[13] J.C. Jaeger, Proc. R. Soc. New South Wales 16 (1942) 203.
[14] J. Föhl, in: Reibung und Verschleiß, ed. K.H. Zum Gahr (DGM, 1983) p. 29.
[15] J.K. Lancaster, Tribology 5 (1971) 82.
[16] J.K. Lancaster, Proc. Inst. Mech. Eng. 183 (1968) 98.
[17] D. Dowson, J.M. Challen, K. Holmes and J.R. Atkinson, in: Wear of Non Metallic Materials, eds D. Dowson, M. Godet and C.M. Taylor (Mech. Eng. Publications, London, 1978).
[18] S. Bahadur and A.J. Stiglich, Wear 68 (1981) 85.
[19] B.J. Gillis, Ph.D. Thesis (University of Leeds, 1978).
[20] H. Chichos, Wear 88 (1983) 27.
[21] H. Voss and K. Friedrich, Tribol. Int. 19 (1986) 145.
[22] R. Schledjewski and K. Friedrich, Proc. European SAMPE Conf., May 11–13 (1992).
[23] O. Jacobs, T. Heitmann, H. Hoffmann and G.-M. Wu, Plast. & Rubber Process. & Appl. 14 (1990) 203.
[24] G.-M. Wu and J.M. Schultz, Polym. Eng. Sci. 29 (1989) 405.
[25] A.J. Waddon, M.J. Hill, A. Keller and D.J. Blundell, J. Mater. Sci. 22 (1987) 1773.
[26] E. Hornbogen, in: Friction and Wear of Polymer Composites, ed. K. Friedrich (Elsevier, Amsterdam, 1986).
[27] T. Tsukizoe and N. Ohmae, in: Friction and Wear of Polymer Composites, ed. K. Friedrich (Elsevier, Amsterdam, 1986).
[28] H. Voss, J.H. Magill and K. Friedrich, J. Appl. Polym. Sci. 33 (1987) 1745.
[29] H. Voss, K. Friedrich and J.H. Magill, J. Appl. Polym. Sci. 34 (1987) 177.
[30] J. Song, Cl.G. Maertin and G.W. Ehrenstein, in: Proc. ANTEC 88, Atlanta, April 18–21 (Soc. Plastic Engineering, 1988) p. 587.
[31] M. Cirino, R.B. Pipes and K. Friedrich, J. Mater. Sci. 22 (1987) 2481.
[32] Lubricomp: Internally Reinforced Thermoplastics and Fluropolymer composites, ICI Advanced Materials Bulletin No. 254-688 (1988).
[33] M.N. Gardos, in: Friction and Wear of Polymer Composites, ed. K. Friedrich (Elsevier, Amsterdam, 1986).
[34] K. Tanaka, Y. Uchiyama and S. Toyooka, Wear 23 (1973) 153.
[35] K. Friedrich and R. Scherer, International Conference of Advance Materials, E-MRS Spring Meeting 1991, Symposium A4: Composite Materials, Strasbourg, France, May 27–31 (1991).
[36] B.J. Briscoe, L.H. Yao and T.A. Stolarski, Wear 108 (1986) 357.
[37] N. Gurker, M.F. Ebel and H. Ebel, Surf. & Interface Anal. 5 (1983) 13.
[38] P. Coxon, J. Krizek, M. Humpherson and I.R.M. Wardell, J. Electr. Spectrosc. 52 (1990) 821.
[39] E. Adem, R. Champaneria and P. Coxon, Vacuum 41 (1990) 1695.
[40] B.J. Briscoe, Tribol. Int. August (1981) 231.
[41] R.P. Stein, Wear 12 (1968) 193.
[42] C.M. Pooley and D. Tabor, Proc. R. Soc. London A 329 (1972) 251.
[43] B.J. Briscoe, A.K. Pogosian and D. Tabor, Wear 27 (1974) 19.
[44] D.R. Wheeler, NASA Technical Paper 1728 (1980).
[45] K. Tanaka and T. Miyata, Wear 41 (1977) 383.
[46] M.J. Folkes, Short Fibre Reinforced Thermoplastics (Academic Press, 1984).

[47] J.K. Lancaster, J. Phys. D 1 (1968) 549.
[48] K. Tanaka, J. Lubr. Technol. 10 (1977) 408.
[49] B.J. Briscoe, Tribology 8 (1981) 231.
[50] G. Erhard and E. Strickle, Künststoffe 2(233) (1972) 282.
[51] M. Clerico and V. Patierno, Wear 53 (1979) 279.
[52] V.K. Jain, Wear 92 (1983) 279.
[53] H. Voss and K. Friedrich, Wear 116 (1987) 1.
[54] B.J. Briscoe and J. Tweedale, in: Proc. Conf. on Tribology of Composite Materials, Oak Ridge, TN, 1–3 May 1990, eds P.K. Rohatgi, P.J. Blau and C.S. Yust (ASM International, Materials Park, OH, 1990) p. 15.
[55] H.H. Shim, O.K. Kwon and J.R. Youn, Polym. Compos. 11 (1990) 337.
[56] H.M. Hawthorne, in: Proc. Int. Conf. Wear Mater. 1983 (ASME, New York, 1983) p. 576.
[57] S.V. Prasad and P.D. Calvert, J. Mater. Sci. 15 (1980) 746.
[58] T. Tsukizoe and N. Ohmae, J. Lubr. Technol. 10 (1977) 401.
[59] J.M. Thorp, Tribology 4 (1982) 59.
[60] Z. Hanmin, H. Guoren and Y. Guicheng, Wear 116 (1987) 59.
[61] Product Information, Ref. VKT 1/1184 (ICI Corporation, Wilton, UK, 1984).
[62] M. Davies and A. McNicols, Research Report (ICI Corporation, Wilton, UK, October, 1989).
[63] Z. Lu, unpublished results (University of Kaiserslautern, 1990).
[64] M.P. Wolverton, K. Talley and J.E. Theberge, in: 44th Annual Conf. Composites Institute, Society of Plastics Industry, February 6–9, 1989, Session 11-D/1 to Session 11-D/9 (1989).

Advances in Composite Tribology
edited by K. Friedrich

Chapter 5

Recent Developments in Tribology of Fibre Reinforced Composites with Thermoplastic and Thermosetting Matrices

U.S. TEWARI and J. BIJWE

ITMMEC, Indian Institute of Technology, Delhi, Hauz Khas, New Delhi-110016, India

Contents

Abstract

    In the first part of this chapter, various types of advanced polymeric composites are discussed. The rest of the chapter is devoted to the tribology of such advanced composites. The tribology of two types of polymeric composites, those with thermoplastics and thermosets, are separately discussed. The literature on the progress of such composites in recent years will be reviewed and discussed in separate sections in relation to the four wear modes, i.e. adhesive, abrasive, fretting-fatigue and erosive. Systematic step by step addition of solid lubricant and reinforcement of a polymer matrix could lead to an optimal combination of friction and wear characteristics. The influence of the various operating factors is also critically examined. The wear mechanisms of fibre reinforced thermoplastic composites are discussed in detail. The influence of various additives (powdery and reinforcing) and contents on the abrasive wear performance is also discussed. The wear and appropriate mechanical-property correlations are also examined. In the case of advanced thermosetting composites tribology, carbon, graphite, glass and aramide fibre composites with unidirectional and multidirectional reinforcement are discussed. Various operational factors, environmental conditions and types of fibre–matrix combination ultimately determine the actual performance of the composite.

1. Introduction

The continuous progress and innovations in modern technology are placing ever increasing demands on lubricants, which are not often met by conventional lubricants. Typical applications where solid lubricants or self-lubricating composites are the only solutions, are space and aeronautics, vacuum and cryogenic instruments, medical and food processing equipment, agricultural machines, motor transport, metal working, robots, computers, electronic and electrical devices, textile and chemical equipment, domestic devices, equipment in nuclear radiation areas, etc. The most common way to use solid lubricants is to apply them as films to metallic surfaces by various techniques such as burnishing, film spraying or vapour depositing (by sputtering or ion plating). They can also be used by fabricating a sliding pair of a component with a self-lubricating composite. A self-lubricating composite may be tailored by using a base matrix, reinforcement, solid lubricants and other various constituents which may be required for that typical application.

2. Polymers and composites in tribology

Modern triboengineering materials which offer attractive solutions as self-lubricating composites, are mostly compositions of polymeric, metallic or ceramic matrices and functional fillers providing one or several mechanisms of self-lubrication [1,2]. Among these, polymeric solid lubricants and/or fibre reinforced polymer composites form one of the most important classes of tribomaterials because of their unique property profile, such as self-lubricity, resistance to wear, corrosion and impact, low density, high specific strength, quiet operation due to vibration and noise absorption, ease in processing and low cost (in some cases). Several reviews and research papers on the application areas and tribology of polymers and composites are available [3–10].

2.1. High-performance polymers in tribology

Innovation in engineering technology demands materials with better stability at higher loads and temperatures along with superior tribological properties. Hence, high-performance engineering polymers which can dependably perform over extended periods of time under severe conditions, such as high thermal, mechanical and electrical stresses, in a harsh corrosive or extreme pressure environment, in biologically demanding conditions, weathering, etc., are continuously being synthesized and evaluated for triboapplications. Table 1 summarises some of such tribopolymers (thermoplastics and thermosets), additives and reinforcements. Much literature is available on the various aspects of tribology of these polymers [6–16].

Linear, covalently bonded polymers are called thermoplastics (TPs), because they can be reversibly softened by heat. Cross-linked polymers (three-dimensional network) are called thermosets (TS), because they are set (cured) as a function of temperature and cannot be remoulded by a temperature increase. TPs can be reheated, redissolved, remoulded, thermal or solvent welded or shaped by multi-step operations, unless there is some type of mechanical degradation [17]. They allow the production of complex shapes, principally by injection moulding, with relative ease and little or no finishing requirement. For thermosets, the starting materials are in liquid form, and during the moulding or shaping process a high degree of cross-linking, brought about by heat or ultraviolet light, causes the liquid system to turn into a rigid infusible solid. TS polymers have reasonable strength and stiffness, but relatively poor impact strength, and failure is essentially by brittle fracture [18,19].

2.2. Role of composites

Along with the special features of polymers as triboengineering materials there are, however, some inherent shortcomings, such as a high coefficient of thermal expansion, lower load carrying capacity, higher friction and wear rates than those with liquid-lubricated metals (not in boundary-lubricated condition), poor thermal conductivity and limited dimensional stability due to thermal expansion. Hence, virgin polymers are generally not adequate for the required end use. Most of the time reinforcement and/or solid lubrication leads to much enhanced triboperformance. Hence, polymer composites are nowadays invariably used in triboapplications rather than neat polymers.

2.3. Formulation of the composites

The friction and/or wear of a self-lubricating polymer can be reduced several-fold when proper additives are incorporated. Additives can be classified as shown in following flow chart.

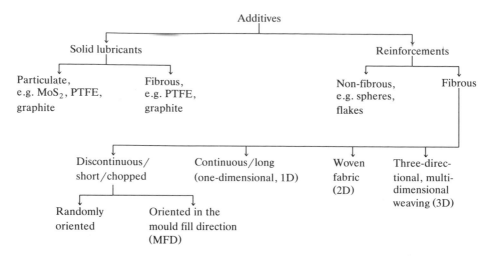

Some fibrous additives, such as graphite, can act in both ways, i.e. lubricating and/or reinforcing. It can improve the lubrication, reinforcement and thermal conductivity of the composite simultaneously. Some examples of various microstructures of composites depending on the arrangement of the constituents, are shown in fig. 1 [20].

2.3.1. Reinforcement

Fibres or spheres are able to reinforce a weaker polymer matrix if they themselves are stronger and of higher modulus and provided that the polymer matrix adheres to them. The degree of reinforcement depends on several factors, including whether or not the polymer can crystallise and if so, the amount of crystallinity developed. Semicrystalline polymers are more efficiently reinforced than amorphous polymers. This is in part due to the fact that the fibres act as nucleation sites for crystallisation and the fibres become surrounded by a micro-crystalline structure, which binds the fibre more firmly to the polymer. If the high-strength, high-modulus fibre is incorporated into the liquid system before hardening, the fibre reinforced product has greatly improved mechanical properties [18,19].

2.3.1.1. Reinforcement with glass microspheres. Glass microspheres, both solid and hollow, offer some advantages over glass fibres. Since they are spherical they have no orientation effect. Any anisotropy in a composite moulding results from the polymer itself. Their processability is better and surface finish is good. Solid spheres are more common at present, despite the advantage offered by hollow spheres of lower density. Because the aspect ratio of a sphere is unity, the reinforcing effect is much less than that of fibres. The lesser reinforcement is compensated to some extent by the possibility of using microspheres at higher loading (40–50%) than that for fibres. Mixtures of fibres and spheres can be used to optimise properties [18,19].

2.3.1.2. Reinforcement with fibres. Fibrous reinforcement generally imparts wear resistance. When higher load carrying capacity, wear resistance and better lubrica-

TABLE 1
Types of polymers and additives for tribocomposites [1].

Polymers				Additives		
Thermoplastics		Thermosets				
Polymer	Max. service temperature (°C)	Polymer	Max. service temperature (°C)	Friction and wear reducing	Reinforcing	Thermal conductivity increasing
PE (HMWPE, UHMWPE)	80	Phenolics	150	Graphite, graphite fluoride MoS_2, phthalocyanines metal oxides, silicon fluids	GF, CF, AF, graphite fibres, boron fibre, asbestos fibre, polyesters, Nomex, cotton, mica	Graphite, bronze silver
POM/polyacetal and its copolymers	125	Cresylics	150			
PA (PA6; PA6,6; PA11)	130	Expoxies	200			
PEEK	150	Silicones	250			
PPS	200	Polyimides	300			
PTFE	275					
Polyesters	150–200					
Polyoxabenzoates	300					
Polyimides	200–300					
PAI	230					
PEI	200					

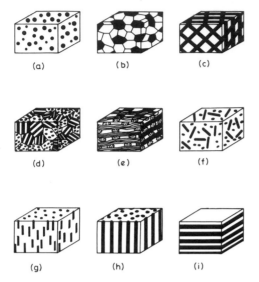

Fig. 1. Schematic representation of microstructures of composites: (a) dispersion structure, (b) polygonal dual-phase structure, (c) cell structure, (d) polygonal grain structure, with structural anisotropy of grains, (e) woven-fabric laminate, (f) randomly oriented short (discontinuous) reinforcement fibres, (g) unidirectional short (discontinuous) reinforcement fibres, (h) unidirectional long (continuous) reinforcement fibres and (i) laminate [20].

tion is desired, a combination of reinforcing fibres and particulate lubricants or a combination of lubricating and non-lubricating (only reinforcing fibres, e.g. glass fibres) fibres is formulated. Carbon fibres are more efficient for reinforcement than glass fibres at the same loading. Carbon-fibre composites have a lower density than glass-fibre composites with the same degree of reinforcement.

For fabricating advanced composites, the properties of the fibres and matrix which are most important, are as follows:

(1) The glass transition temperature (T_g) of the polymer, indicated by a sharp change in the load curve and which sets the maximum service temperature.
(2) The resin modulus, which controls the stiffness required to maintain rigidity between the fibres and allows the transfer of the load from fibre to fibre.
(3) The fracture toughness, which is related to the strain to failure and modulus which allows the composite to resist delamination, peel and shear failure.
(4) The strain to failure, which allows the fibres to exhibit their strain capability without matrix microcracking.
(5) The bonding capability of each polymer material to a selected fibre.

The fibre reinforcement starts with single filaments grouped to 1000 to 12 000 into a fibre bundle, which can be chopped into short fibres, woven into fabric or further combined into a fibre tow containing as many as 40 000 to 300 000 filaments. These can be further processed as shown below.

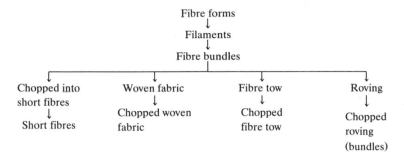

2.4. Short fibre reinforced polymers (SFRPs)

Depending on the processing technology, short/chopped fibre reinforcement can be done as a random distribution or predominently in the mould fill direction.

The fibres are most commonly incorporated into the polymer melt in a thin screw extruder fed by a metred mixture of fibre and polymer granules. The fibres fed to the machine are several millimeters in length but are broken down by mechanical forces during the mixing process. SFRPs with a fibre length of 0.3 to 0.6 mm or less account for the largest share, approximately 60%, of the FRP market. SFRPs include organic polymers with particulate fillers and chopped glass, carbon and aramid (aromatic amide) fibres. In the past ten years there has been a growing trend to increase the reinforcing efficiency of the fibres to the point where the mechanical properties of SFRPs approach at least 90% of the modulus and 50% of the tensile strength of the continuous fibre reinforced polymers (CFRPs).

One advantage of SFRPs over CFRPs is rapid, low-cost processability by injection/compression moulding or by extrusion. The performance of such composites depends on the alignment, uniform distribution and aspect ratio of the fibres and good adhesion between fibres and matrix. Equation (1) correlates the load carrying capacity and the aspect ratio of the fibres [5].

$$\sigma_f = \tau 2 l r^{-1} + \sigma_m, \tag{1}$$

where σ_f is the contact stress, l is the length and r the radius of the fibres, and σ_m is the compressive stress of the matrix polymer in the bulk composite pressed against the smooth counterface under a load W. τ is the tangential stress produced due to the difference in moduli of the fibre and the matrix. The higher the aspect ratio of the fibre (the aspect ratio is l/r), the more will the contact load be transferred from the matrix to the fibres.

Introduction of thermoplastic polymers reinforced with fibres longer (1–10 mm) than normal short fibres is a recent development. Products made by using long fibres have better impact resistance and higher stiffness than the conventional SFRPs. Long fibres also result in products with good surface finish (because of the orientation and because there are fewer fibre ends), better dimensional stability and enhanced creep and fatigue resistance [18].

2.5. Continuous fibre reinforced polymers (CFRPs)

Fibres spun from a melt or solution are produced as continuous filaments with a diameter in the range 5–10 μm. The filaments are collected in bundles of about 100 to form continuous strands. Several strands are combined together to form a continuous roving. It is used to produce unidirectional reinforcement in different planar directions. It is also used in filament winding and pultrusion. A schematic representation of the techniques for the fabrication of CFRPs is shown in figs. 2a, b and c. Maximum properties, such as a high modulus of elasticity, strength and elongation, are achieved by unidirectional lamination of the continuous fibre reinforcement in the 0° direction (fig. 2c(i)), i.e. parallel to the longitudinal axis of the fibre. The mechanical properties in the load direction (2 and 3 in fig. 2c) are much less good and are highly dependent on the matrix resin. This limitation can be overcome by bidirectional reinforcement, which is achieved by tailoring the directional placement of orthotropic (unidirectional) plies as shown in fig. 2c(ii). As the number of plies increases the isotropic strength is approached asymptotically [17].

2.6. Fabric reinforced polymers

There are two types, woven and non-woven, as described below.

2.6.1. Woven fabric reinforced polymers (WFRPs)

Woven material in laminate form is currently replacing more traditional structural forms, primarily because of the availability of fibres such as carbon and aramid, whose enhanced mechanical properties in composites surpass the property values of corresponding hardware of aluminium or steel with respect to the strength to weight ratio.

Yarn and roving can be woven in various ways to form cloth (fig. 3a). Fabric reinforcement (fig. 3b) results into bidirectional strength since the fibres are at right angles to each other, which is otherwise not possible with CFRPs or unidirectional placement of laminated composites (fig. 2c(i)). Combination of fibres with quite different properties, such as carbon warp with glass fill, can provide a fabric with good longitudinal strength/stiffness values as well as transverse (fill direction) (fig. 3b) toughness and impact resistance. Thus, many varieties of composites can be achieved by combining different yarns and weaves.

Weaving reduces interlaminar failure in the final product. However, the crimp in the fibre at the points of intersection reduces the reinforcement efficiency, which is at maximum when the fibre is straight. Various forms of weaves used for reinforcement are shown in fig. 4. Three types of weaves are common. Plain weave

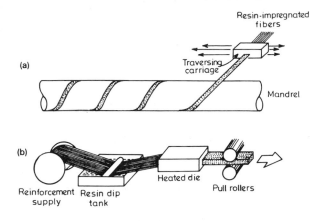

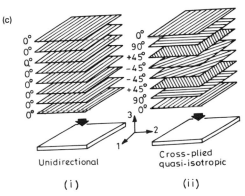

Fig. 2. Fabrication methods for advanced composites. (a) Filament-wound products are produced by winding resin impregnated fibres or tape onto a mandrel. (b) Sectional shapes such as rods and I-beams. The fibres pass through a resin tank and are then pulled through a heated die to cure and form final products. (c) Laminated composites are formed by laying up a number of prepreg mats or tapes and then cured. (i) unidirectional and (ii) multi-dimensional lamination.

(over and under, fig. 4a), twill (over and under every other fibre, fig. 4b) and satin (in various forms, all of which involve a fibre passing over or under more than two fibres, e.g. four, five or eight, see fig. 4c). Because there is less crimp in satin weaves, they give the best reinforcement and are preferred for high-performance composites. Woven rovings made from weaving-grade roving have excellent draping properties. They are cheaper than the conventional fabric since the expense of yarn preparation is eliminated and the heat treatment and finishing processes are avoided by treating the reinforcement with a size. In plain weave, the yarns do not drape easily, and it is used where single curvature with flat sheeting work of uniform strength through 360° in the fabric plane is required. The yarns are nearly

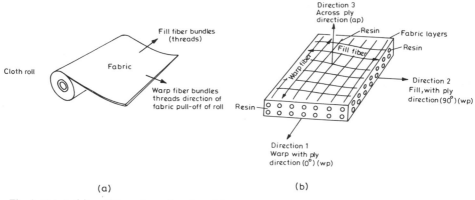

(a)   (b)

Fig. 3. Fabric (a) and its various directions (b).

straight in satin-weave woven fabric, where they pass over several cross yarns and crimp appears at longer intervals. In particular, eight-harness satin weave has good drape properties and is good for moulding with double curvature.

2.6.2. Non-woven fabric reinforced polymers

Non-woven fabric reinforced composites are made by either knitting or by stitching continuous rovings together. In the latter case the reinforcement in the stitched direction is less than when woven, but stitching avoids the problems of crimp. It allows for easy handling of the roving in sheet form and helps to reduce interlaminar shear. By layering the rovings in different directions before stitching, strength can be obtained in a preferred orientation. Three-directional reinforcement can be obtained by stitching layers of non-woven cloth together.

2.7. Multidirectionally reinforced fabric polymers (MRFPs)

The mechanical properties of bidirectionally reinforced fabric components (fig. 3b) or cross-plied quasi-isotropic laminates (fig. 2c(ii)) are not satisfactory in the third direction. Their values are matrix dominated and are more than one order of

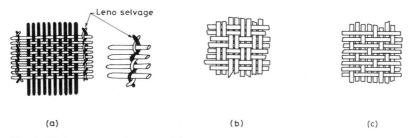

(a)   (b)   (c)

Fig. 4. Various types of weaves: (a) plain weave with leno selvage, (b) twill and (c) satin with five harness.

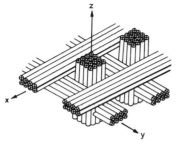

Fig. 5. Geometry of 3D-orthogonal weave.

magnitude less than in the reinforced directions. This problem becomes especially critical in applications involving high thermal stresses, such as a C–C-composite ballistic reentry nose and a solid rocket nose. The obvious solution to this is to add fibre reinforcement in the third direction, and in additional directions when necessary, to provide the composite materials with an isotropy approaching that of metal. Three-dimensional fabrics can be obtained by stitching layers of cloth together with polyester or aramid threads. There are more than twenty varieties of multidirectionally reinforced polymers. Figure 5 shows the geometry of one such (orthogonal) weave form. It has reinforced yarns arranged in an orthogonal (Cartesian) geometry, with all yarns intersecting at a 90° angle. The fibres can be introduced uniformly in each of the three directions to provide isotropic properties, or in unbalanced amounts, depending on the need. These woven blocks and parts are machined to the required size and shape. Such composites result in superior mechanical properties and erosion resistance after densification of the composite.

3. Friction and wear mechanisms of polymer composites

A composite must be able to support the dynamic stresses induced by an applied load and the tangential friction stresses. If it cannot do so, it will be prone to rapid wear by plastic deformation, brittle fracture or fatigue. In some cases this will cause catastrophic wear, and in some situations the composite will wear rapidly only until the contact area increases to the point where the polymer will support the decreased dynamic stresses and then transformation to a milder form of wear will occur. Various kinds of wear (e.g. adhesive, abrasive, fatigue, fretting fatigue, erosive, corrosive), alone or in combination, can occur, depending on the polymer, the additives and the sliding conditions. In order to provide the best lubrication, some sort of shear layer must develop between the sliding surfaces to reduce the adhesive and ploughing interactions between the moving surfaces. While reducing the friction, this layer can also reduce the stresses in the bulk of the polymer body. The thinner the layer, generally the better is the lubrication. PTFE is a well-known example of such a polymer. Depending on the application

TABLE 2
The use of different polymers and their composites in technical applications where friction and/or wear properties are of importance [21].

Group	Material	Applications	Maximum tribological potential
1	Unfilled and short fibre reinforced polymer composites	Seals, gears, slideways, bearings and abrasive applications	$pv < 15$ MPa m/s $v < 5$ m/s $T < 250°C$ $\dot{W}_s > 10^{-7}$ mm^3/N m $\mu > 0.03$
2	Bulk composites with continuous fibre reinforcements	Under-water or high-temperature conditions, aerospace seals and bearings	$pv < 100$ MPa m/s $v < 5$ m/s $T < 320°C$ $\dot{W}_s > 10^{-8}$ mm^3/N m $\mu > 0.09$
3	Thin-layer composites with metallic supports	High-pressure conditions, pivot bearings	$pv < 300$ MPa m/s $v < 1$ m/s $T < 320°C$ $\dot{W}_s > 10^{-9}$ mm^3/N m $\mu > 0.06$

friction and wear or both may be important. Table 2 [21] furnishes relevant data for technical applications.

Wear is not an intrinsic material property, but strongly depends on the following experimental parameters [21]:

(a) The wear mechanism, viz. abrasion, adhesion, fatigue and chemical degradation.
(b) The external conditions, viz. temperature, contact pressure, velocity and environment.
(c) The relative movements, viz. sliding, multiple/single pass, reciprocating and impacting.
(d) The contacting materials, viz. roughness and hardness of the metallic counterparts, bonded or loose sharp particles and third-body interface.
(e) The composite materials, viz. polymer matrix, type and structure of reinforcement, filler/matrix interface and internal lubricants.

The wear testing procedures differ for the various wear modes. A literature survey indicates that the methods used for similar wear modes also differ from researcher to researcher. Moreover, triboproperties are very sensitive to the environmental conditions (such as atmosphere, humidity and temperature), apart from the experimental parameters. Hence, it is very difficult to compare the data furnished by two different labs. This can only give an approximate idea of the wear performance. The composite must be tested in a tribosystem before an actual application for reliable results.

The following wear equations are frequently used to describe the wear behaviour of a polymer composite system [21]:

(a) The dimensionless wear rate, $\dot{W}$,

$$\dot{W} = \Delta m A^{-1} L^{-1} \rho^{-1} = \Delta V A^{-1} L^{-1} = \Delta h L^{-1}. \tag{2}$$

(b) The specific wear rate, $\dot{W}_s$ (in $m^3/N\,m$, or often called K_0 in $mm^3/N\,m$ in the literature)

$$\dot{W}_s = \dot{W} p^{-1} = \Delta V F_N^{-1} L^{-1}. \tag{3}$$

(c) The time-related depth wear rate, $\dot{W}_t$,

$$\dot{W}_t = \Delta h t^{-1} = \dot{W}_s p v, \tag{4}$$

where Δm is the mass loss, ρ the density, Δh the reduction in specimen height, ΔV the wear volume, t the sliding time, F_N the load in Newton, L the sliding distance and p the pressure.

As long as $\dot{W}_s$ is independent of p and v, this leads to a linear relationship between $\dot{W}_s$ and the product pv. In this range $\dot{W}_s$, i.e. K_0, can be considered to be some kind of material property of the composite tested in a well-defined system. But it may be noted that above a certain pv range, this is no longer true, and changes in p and/or v can cause remarkable changes in $\dot{W}_s$.

4. Tribology of short fibre reinforced composites

Several papers have appeared on the wear performance of short fibre reinforced composites containing various fibre–matrix combinations and fillers, in various wear modes. The influence of the various material properties (polymer, fibre type, preparation, alignment, bonding to the matrix) and experimental parameters, such as load, speed, sliding distance, counterface material and roughness, environment such as dry or wet, test temperature, atmosphere, humidity, etc., have been reported [11–13,16,20–24]. Especially for PTFE, addition of short glass or carbon fibres can reduce the high specific wear rate by several orders of magnitude. Generally, it is observed that inclusion of short fibres yields an improvement in wear performance. Short carbon fibres prove better in this respect than the more abrasive glass fibres, although a little degree of abrasiveness can be considered to be beneficial with respect to smoothening the counterface roughness. That is why high-strength carbon fibres with high abrasiveness but the same lubricating efficiency as high-modulus graphite fibres, are preferred for high-performance composites in sliding wear applications [13]. An increase in fibre percentage, however, does not necessarily improve the wear behaviour of a composite. The optimum fibre loading seems to be between 20 and 30 vol%. Generally, it is observed that, regardless of the polymer matrix, short fibre reinforcement results in a scatter band of the specific wear rate between 5×10^{-7} and 8×10^{-6} mm^3/N m [13].

TABLE 3
Details on selected PEI * composites.

	ULTEM 1000 (PEI)	ULTEM 4001	ULTEM 2200	**	ULTEM 4000
Designation	A	B	C	D	E
Composition ***	–	13–15% PTFE	20% GF	16% GF + 20% graphite	25% GF + 15% PTFE + 15% (graphite + MoS$_2$)
Specific gravity	1.27	1.33	1.42	1.45	1.7
Tensile modulus (1% secant) (N mm^{-2})	3000	2600	6900	–	7100
Flexural modulus tangent (N mm^{-2})	3300	3000	6200	–	9000
Compressive modulus (N mm^{-2})	2900	–	3500	–	3300
Rockwell hardness	M 109	M 110	M 114	–	M 85

* Catalogue by the General Electric Company.
** Composite prepared in the laboratory.
*** Data from the laboratory analysis.

4.1. Tribology of selected short fibre reinforced composites

With a view of understanding the role of filler/s in short fibre reinforced polymer composites, some typical combinations, shown in table 3, were selected [16,29–33]. Polyetherimide (PEI), a thermoplastic polyimide (PI), was selected as matrix material. Apart from outstanding thermal, mechanical and electrical properties, PIs have proved their excellency as high-performance tribopolymers [1,25–27]. Hence, new PI materials are continuously being synthesised and evaluated for triboperformance [15,25–27]. PEI is a comparatively recent developed polyimide [28] and literature on its friction and wear behaviour in various environments, with different fillers and under different experimental conditions, was not available. PEI is an amorphous, yet ductile (gardner impact strength 36 N m and ultimate elongation to break 60%) polymer. It was interesting to observe that this typical feature was indeed reflected in the friction and wear behaviour [29]. Though ductile, it did not wear by the film-transfer mechanism. Its wear debris was neither large, thick or chunky, nor its μ was very high. The dominant wear mechanism was fatigue. The wear rate was very high once fatigue was initiated after an incubation time.

4.1.1. Dry sliding wear against a smooth surface

Tribostudies of the selected composites were done against smooth, mild steel, giving rise to adhesive wear mainly, and on a three-pin-and-disc machine. Abrasive-wear studies with sliding against silicon carbide paper were performed on a one-pin-and-disc machine. Load, sliding speed, sliding distance and counterface roughness (grit size in case of abrasive wear) were the selected testing parameters.

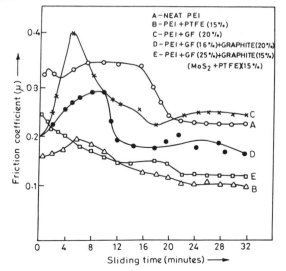

A−NEAT PEI
B−PEI +PTFE (15%)
C−PEI+GF (20%)
D−PEI+GF (16%)+GRAPHITE(20%)
E−PEI +GF (25%)+GRAPHITE(15%)
(MoS₂ +PTFE)(15%)

Fig. 6. Influence of fillers on the friction coefficient of selected PEI composites (table 3); the sliding speed is 2.1 m s^{-1}, $L = 43$ N (except for sample A, where $L = 25$ N), R_a of the counterface is 0.1–0.15 μm [31].

Details on the performance data found in these studies are published elsewhere [16,29–33].

4.1.1.1. Friction studies of selected composites. The salient features of friction behaviour of the selected composites against mild steel were as follows (figs. 6–9).

(1) The initial friction coefficient, μ, exhibited by all SFRPs was very high, which then stabilised to a low and steady value as the sliding time increased.

(2) Peaking in friction was displayed only by GF reinforced composites (C, D and E).

(3) The highest μ was recorded for the composite C, which did not contain any lubricant but GF.

(4) The lowest μ was found for the composite B, which did not contain any GF but the solid lubricant PTFE.

(5) As the amount of solid lubricant increased, both the peak friction (μ_p) and the steady-state friction (μ_s) decreased substantially. μ_p and μ_s showed dependence on the load, the sliding time, the sliding velocity, the speed and the fibre orientation for all the composites. With an increase in sliding time and load, the magnitude of μ_p and μ_s decreased. The time interval at which μ_p occurred reduced as the load increased (fig. 7). The two competing processes abrasion due to the GF and cushioning of the asperties due to the solid lubricants reached a steady state earlier as the load increased. The increase in the pin subsurface temperatures (thermocouples were inserted at 3 mm distance away from actual sliding surface) as a function of sliding time at two extreme loads is also plotted in fig. 7. In both cases the temperature rise was almost linear, with a steeper rise at higher load. The friction and wear data as a function of the sliding time at various

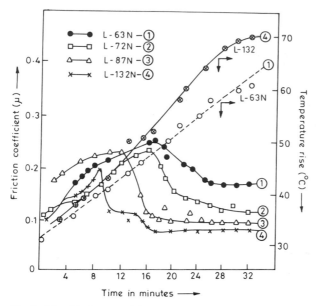

Fig. 7. The friction coefficient and pin subsurface temperature as a function of the sliding time at various loads for composite E; speed 2.1 m s^{-1}, $R_a = 0.2$ μm, the fibre orientation is N [31].

selected loads and sliding speeds are shown in figs. 8 and 9. At low speed (1.05 m/s) μ did not show stable value within the selected time interval. However, as the speed increased from 1.57 to 2.61 m/s, two major changes were recorded.

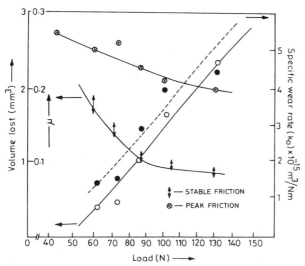

Fig. 8. Peak and steady-state friction coefficients, wear and specific wear rate as a function of load for composite E; speed 2.1 m s^{-1}, $R_a = 0.2$ μm, the distance slid is 4 km, the fibre orientation is N [31].

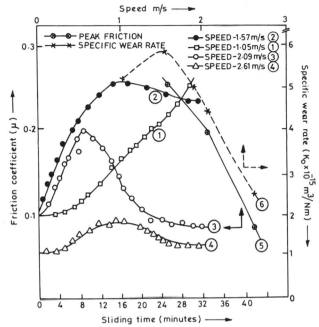

Fig. 9. Influence of the speed on the friction coefficient and the specific wear rate for composite E; $L = 132$ N, $R_a = 0.2$ μm [31].

(a) The magnitude of μ_p decreased rapidly with the increase in speed and the position of μ_p shifted towards the left side of the graph.

(b) μ_s decreased rapidly as the speed increased. A μ_s as low as 0.06 was recorded for the composite E under these high pv conditions ($p = 132$ N, $v = 2.61$ m/s).

4.1.1.2. Wear studies of selected composites.   The wear data of the composites C, D and E are plotted in fig. 10a, which shows the effective contribution of fillers and their combination with solid lubricants to enhancement in wear performance. The influence of successive incorporation of lubricants and fibres in the neat polymer on both friction and wear can be seen in fig. 10b. Composite E showed a good combination of friction and wear characteristics.

The wear studies of the fibre reinforced composites at various experimental parameters indicated that the specific wear rate (K_0) increased with increasing load. As the load increases, the frictional torque also increases, resulting in a higher flash temperature. This leads to surface melting and large transfer of molten polymeric material to the counterface. The deterioration in the fibre–matrix adhesion with the increase in temperature leads to easy peeling off or pulling out of short fibres from the polymer matrix, resulting in a higher wear rate. Such a deterioration in the fibre–matrix adhesion was also seen in SEM micrographs. The high flash temperature, however, was not solely responsible for the debonding of fibres. According to the delamination theory of Sung and Suh [34], when the fibres

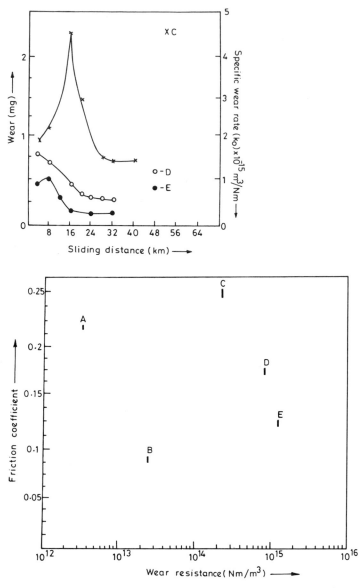

Fig. 10. (a) Steady-state wear studies for composites C, D and E; speed 2.1 m s^{-1}, $L = 43$ N, $R_a = 0.1$–0.15 μm [31]. (b) Friction and wear performance of PEI composites sliding under similar conditions; $L = 43$ N, speed 2.1 m s^{-1}, $R_a = 0.1$–0.2 μm (except for sample A, where $L = 25$ N).

are normal to the surface, because of the repetitive stresses such debonding occurring at the surface can propagate down along the length of the fibres. Since the magnitude of the repetitive stresses increases with increasing load, the debonding also increases and peeling off or pulling out of fibres or broken pieces becomes easy, resulting into higher wear. According to Tanaka [35], when the GF contact the metal surface directly while sliding, hot spots are produced, leading to degradation of the polymer. The higher the load, the higher the degradation, and the higher the wear. In the case of the GF reinforced composite C it was observed that the colour of the pin surface changed from light brown to black, indicating charring of the polymer as the load increased (the other composites were originally black in colour due to the included graphite) [16]. Thus, the increase in the wear rates of the GF reinforced composites as a function of the load was attributed to the following two processes: (i) increase in pulverization of the fibres and, hence, the load-supporting action of the fibres being less effective and (ii) thermal degradation of the PEI matrix. The same phenomena are responsible for the increase in wear rate with increasing sliding time. It was observed that composites with the fibres normal to the sliding surface exhibited lower friction and wear values as compared to those with parallel fibres. According to the delamination theory [34], with F_N (load when the fibres are normal to the surface) the debonding will occur at the surface and will propagate only a finite distance along the length of the fibres, because the tensile-stress component which causes debonding diminishes as the distance from the surface increases. Repeated sliding will not change the length of debonding until the fibre length is reduced by wear. The fracture of fibres due to the mechanical pulling action at the interface or due to buckling, occurs with difficulty since lateral deformation of normally oriented fibres will be constrained by the presence of the neighbouring fibres. For F_p (load when the fibres are parallel to the surface) initiation of cracks may occur at the surface or at the fibre–resin interface, at a finite depth from the surface where the tensile-stress component perpendicular to the surface is maximum. Once cracks are nucleated, they will propagate under cyclic loading and the debonding length will increase, resulting in large-scale fibre separation. The separated individual fibres will fracture more readily since bending increases the wear rate. A schematic representation of this is shown in fig. 11.

The friction and wear of composites B and C as a function of the surface roughness of the counterface revealed interesting results. Surfaces with R_a equal to 0.1 to 0.2 μm were more suitable for low and steady friction and wear than very rough or very smooth surfaces.

The composites showed optimum surface roughness values for the lowest friction and wear. The friction and wear minima did not match for composite B, while for composite C they did. A number of papers are available reporting similar studies for various polymers and composites. It is a well-established fact that there exists an optimum surface roughness of a counterface at which the friction and/or wear is minimal. However, the reason why for some composites the minima of friction and wear match at a typical R_a and for some composites there is complete mismatching, is not yet well understood.

It was observed that, of the selected composites, composite E showed the best range of friction and wear. Its behaviour was compared with that of the commercially established wear-resistant PI composite SP22 [25] sliding under identical conditions [33] (fig. 12). It was observed that the latter exhibited marginally

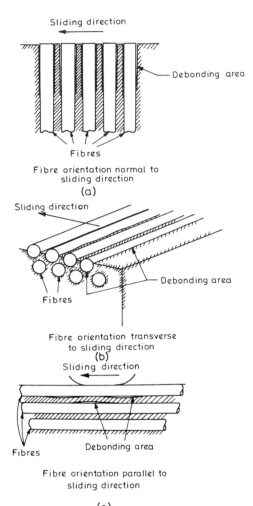

Fig. 11. Schematic representation of the failure modes in uniaxial continuous fibre reinforced composites while sliding, with fibre with different orientations [34].

superior tribobehaviour, $\mu_E = 0.12$, $\mu_{SP22} = 0.09$, $K_{0E} = 8.5 \times 10^{-16}$ m^3/N m and $K_{0SP22} = 5 \times 10^{-16}$ m^3/N m, than the former, but at the expense of easy mouldability.

4.1.1.3. SEM microscopy and wear mechanism. The wear mechanism of neat polymer and PTFE-filled polymer (composite B) was very much different from that of fibre reinforced composites. It was confirmed that, while wearing, PEI did not transfer a film but fine powder on the counterface. The wear was dominated by fatigue [29], while that of composite B was dominated by immediate film transfer of PTFE on the counterface [32].

While wearing, composites C, D and E did not transfer PEI film on the disc, instead, composites D and E transferred a thin film of graphite and PTFE,

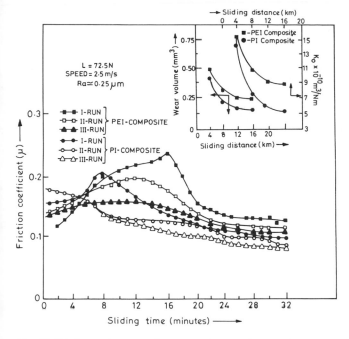

Fig. 12. Friction and wear performance of composite E and SP22 Vespel composite, worn under similar conditions; speed 2.1 m s^{-1}, $L = 72.5$ N [33].

respectively, on the counterface, which was responsible for the rapid decrease in μ as compared to composite C. Under severe sliding conditions conducive to melting of the polymer, composites C and D transferred layers of molten polymer on the

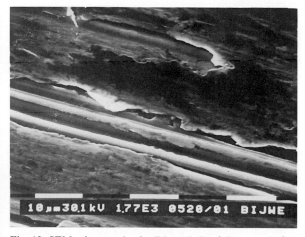

Fig. 13. SEM micrograph of mild steel disc (after wearing) showing a transferred molten layer of PEI due to the severe sliding conditions; speed 2.1 m s^{-1}, $L = 72$ N [16].

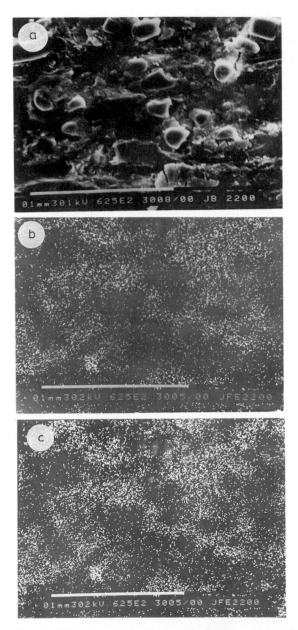

Fig. 14. SEM micrograph of composite C worn under high pv-conditions ($v = 2.1$ m s^{-1}, $L = 72$ N), showing (a) severe damage to the polymer matrix and fibres, pulled-out fibres and cracked tips, (b) and (c) corresponding EDAX studies: (b) silicon dot mapping, here a higher dot density indicates the existence of GF and pulverized particles, (c) Fe dot mapping, showing back-transfer of iron due to abrasion from the mild steel disc to the pin surface where fibres are absent [16].

Fig. 15. SEM micrograph of a disc covered with a transferred layer (from composite E) after wearing; speed 2.1 m s^{-1}, $L = 72$ N [31].

disc (fig. 13). While sliding under high pv conditions, because of the high frictional heating, the polymer gets softened. Since GFs are very abrasive, the counterface gets abraded and back-transfer of Fe particles to the softened polymer matrix occurs. Particles are transferred only at those spots where GFs are absent. Figure 14 shows SEM micrographs of such an event. The extent of abrasiveness of GFs reduced as the amount of lubricant increased in composites D and E. In fact, composite E transferred PTFE film immediately on the disc (fig. 15), exhibiting low friction.

Along with adhesive and abrasive (due to the GF) wear modes, the prominent wear mode of composites C, D and E was fatigue (fig. 16). Comparison of micrographs of these composites with that of graphite fibre reinforced ductile polyurethane composite [3] indicates similarities in the wear mechanism (fig. 16). In a brittle matrix such as epoxy, cracks generated in the matrix propagate through the fibre when the bonding between fibre and matrix is strong. In a highly ductile matrix, such as PU (polyurethane) reinforced with graphite fibres, cracks cannot propagate through the matrix and the fibres. Therefore, the fibres bend with the matrix under contact with the asperities and the wear rate of this kind of composite is controlled by the wear rate of the fibres. Experimental results have shown that the ends of the fibres in such composites are extremely well polished and elliptical rather than circular, although initially the fibre tips are perpendicular to the surface. These facts indicate that the fibres were worn gradually with very small increments when they bend while sliding [3]. In figs. 16a–d, the arrays of well-polished elliptical ends of fibre tips indicate the similarity in wear mechanism. The influence of a high load on the surface topography of composite D is shown in fig. 17. The prominent features of the worn pin-surface of the GF reinforced composites (figs. 17–19) were microcracking, fibre pulverization, wear thinning of the fibres and deterioration in the fibre–matrix adhesion, which is responsible for

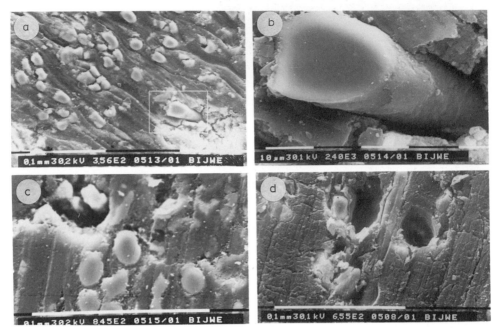

Fig. 16. SEM micrographs of composite D at different locations (speed 2.1 m s^{-1}, $L = 112$ N), showing (a) severe melt flow of polymer in the sliding direction, maximum fibres normal to the surface, cracks generated in the sliding direction and (a′) pulled-out fibres. (b) Magnified view of pulled-out fibres from the matrix with elliptical and polished tips. (c) Elliptical fibre tips, crater from where molten material is chipped off with the array of unworn fibres. (d) Multiple parallel microcracks perpendicular to the sliding direction, indicating fatigue, cavities due to fibre consumption, deteriorations in the fibre–matrix adhesion [30].

peeling off or pulling out of fibres from the matrix. Figure 19 shows some micrographs of the worn pin-surface of composite E.

4.1.2. Sliding wear against abrasive (silicon carbide) papers

Abrasive-wear studies on a pin-on-disc machine abrading polymer pins against water-proof silicon carbide (SiC) abrasive papers were done with a view to observe the effect of short fibres on the abrasive wear performance [36,37]. An additional, similar series of composites (polyamide (PA), Nylon 6,6 filled with short carbon fibres (supplied by Hysol Grafil, UK)), as shown in table 4, were selected with a view to have a better insight in the abrasive-wear characteristics of SFRTPs. These studies were done under single pass (speed 5 cm/s) [37] as well as multipass (0.43 m/s) [38] conditions. The other experimental parameters were load, abrading distance and grit size. The effects of fibre concentration and fibre orientation (in

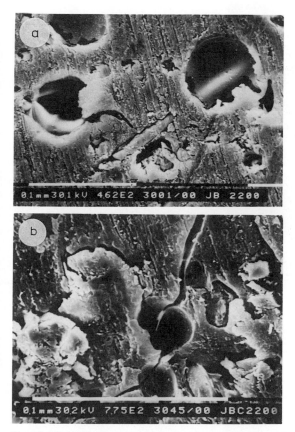

Fig. 17. Micrographs of composite D worn under high load and speed (2.1 m s^{-1}, $L = 132$ N), showing (a) fibre–matrix debonding, wear thinning of the longitudinal fibres and cavities due to fibre consumption, (b) deep cracks initiating and propagating from fibre to fibre, pits formed due to graphite extraction and fibre consumption, back-transfer of molten polymer from the disc to the pin surface (patches in the left portion of the micrograph) [30].

typical cases) were also studied. Efforts were also made to correlate the wear with appropriate mechanical properties.

In the case of the composites A to E (table 3), polymer pins with a diameter of 7 mm and for the PA composites (table 4) square samples (12 × 12 mm with 4 mm thickness) were used. The selected loads were in the range 4 to 14 N for sliding against 220 grade abrasive SiC paper (grit diameter of about 52 μm). The important observations in the wear studies relating to the material properties were as follows.

Neat nylon 6,6 showed a lower wear rate than that of PEI. The parent polymers (both PEI and PA) displayed better wear behaviour than SFRTPs. The wear

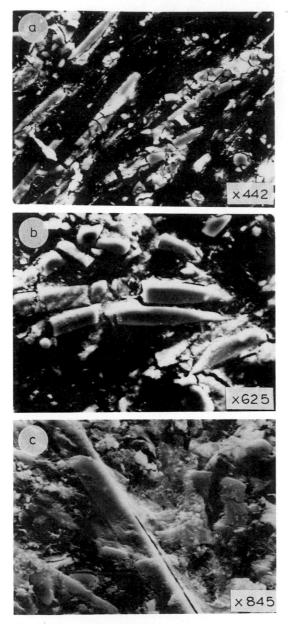

Fig. 18. Micrographs of composite E (fibres parallel to the surface), worn under $L = 72$ N and $v = 2.1$ m s^{-1}, showing (a) microcracking of fibres, (b) breaking and pulverisation stages of fibres and cracks developed in the polymer matrix and (c) deterioration in the fibre–matrix adhesion [31].

increased as the percentage of filler increased (fig. 20). Incorporation of fibres and combinations of fibres with powdery fillers (PTFE) deteriorated the wear performance substantially. The composites which displayed the best combination of

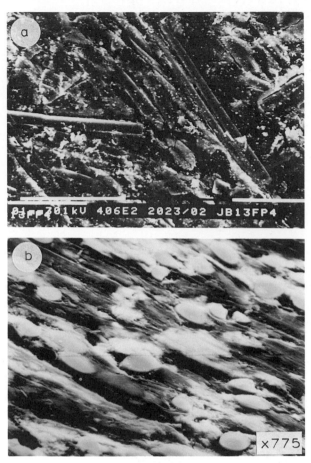

Fig. 19. Micrograph of composite E (fibres P) worn under high pv-conditions ($v = 2.1$ m s^{-1}, $L = 132$ N), showing (a) wear thinning of fibres with still more deterioration in the fibre–matrix adhesion and (b) (fibres N) elliptical fibre tips burried in the matrix, thick cracks propagating in the sliding direction and severe melt flow of polymer [31].

triboproperties in adhesive-wear applications (E and E') were the lowest in ranking for the abrasive-wear performance. Composites with fibres normal to the surface displayed slightly lower wear rates than those with parallel fibres (fig. 21a). In the case of PA composites, however, since the contact area of the samples was different (because of the experimental constraints) no clear cut pattern emerged (table 5). Compared to short fibres, powdery fillers or their combination with fibres were more detrimental for abrasive wear (figs. 20, 21); composites having a lower filler concentration, i.e. B, with 15% PTFE, exhibited more wear than those containing 20% GFs. The wear increased linearly with the load for both types of composites (fig. 21). The specific wear rates (K_0) showed a linear decrease with

TABLE 4
Details on selected composites of PA 6,6 [+].

Property	Unit	Nylon 6,6	Nylon 6,6 * 20% CF	Nylon 6,6 * 30% CF	Nylon 6,6 * 40% CF	Nylon 6,6 * 30% CF + 15% PTFE
Designation		A'	B'	C'	D'	E'
Tensile strength	MPa	81	193	241	276	209
Tensile elongation	%	10	3–4	3–4	3–4	3–4
Flexural strength	MPa	103	289	351	413	278
Flexural modulus	GPa	2.8	16.5	20	23.4	15.2
Ultimate tensile strength **	MPa	29.9	98.2	144.9	162	89.1
Ultimate elongation to break **	%	92	9.67	6.9	6.6	10.33

[+] Data sheet by Hysol Grafil.
* Fillers and fibres are by weight.
** Data generated in the laboratory.

load for both the series of composites, which is in accordance with the wear equation based on the crack-propagation theory [23].

Efforts were made to correlate the wear performance with various appropriate mechanical properties, such as the ultimate elongation to break (e), the ultimate tensile strength (S), the hardness (H), the $(He)^{-1}$ and $(Se)^{-1}$ (i.e. a Ratner–Lancaster plot). It was observed that wear versus e^{-1} showed a more linear relationship than any of the other properties (fig. 22). For multipass conditions, however, this did not hold true. Other factors such as material transfer and abrasive surface modification (such as pulling out of abrasive grains), seemed to

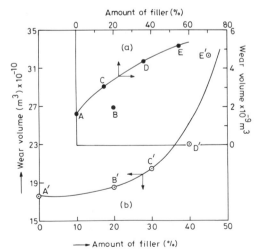

Fig. 20. The wear as a function of the amount of filler when abrading against 220 paper ($v = 5$ cm s^{-1}, $L = 8$ N), (a) for PEI composites (distance abraded is 4 m) and (b) for PA composites (distance abraded is 1.72 m [37]).

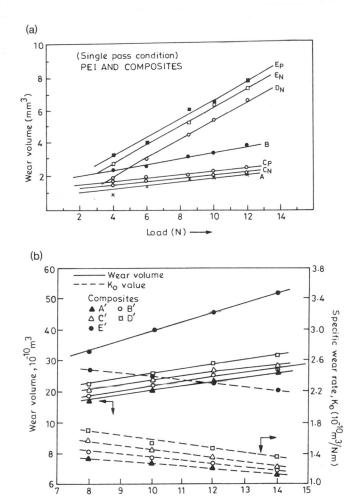

Fig. 21. The wear as a function of the load, (a) for PEI composites ($L = 8$ N, $v = 5$ cm s^{-1}, distance abraded is 4 m) and (b) for PA composites ($L = 8$ N, $v = 5$ cm s^{-1}, distance abraded is 1.72 m) [37].

TABLE 5
Effect of fibre orientation in the composite on the abrasive wear of CF-PA composites [37].

Composite	Contact area (mm^2)	Trend in wear	
		Single pass	Multipass
B'	0.76	N > P	N > P
C'	0.36	N < P	N < P
D'	0.36	N < P	N < P
E'	0.76	N > P	N > P

(a)

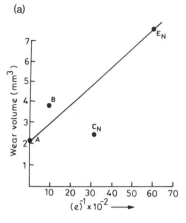

(b)

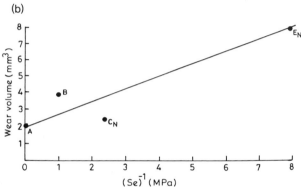

(c)

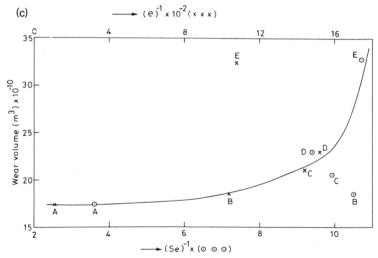

Fig. 22. The wear as a function of $(Se)^{-1}$ (Ratner–Lancaster plot) and e^{-1}. (a) and (b) For PEI composites and (c) for PA composites [37].

play a vital role in the wear performance of the composite. It was thus concluded that in the selected composites, elongation to break played a prominent role in determining the wear performance, rather than other properties.

The wear as a function of the abrading-particle size was also studied, and the well-known "size effect" [39] was observed very clearly in the case of PEI composites [36,40]. In the case of PA composites, the wear increased slowly after showing an initial rapid increase with increasing grit size. The typical size effect (i.e. saturation in wear after a typical grit size), however, was not observed in the case of PA composites.

5. Tribology of continuous fibre reinforced composites

The developments in tribological research on continuous fibre/fabric rein-forced composites in recent years have been collected in this section. The available material is divided into four sections depending on the dominant wear mode involved.

5.1. Adhesive wear mode

Wear studies involving sliding against smooth metallic surfaces, referred to adhesive wear studies, are collected in this section.

Lhymn [41] selected a uniaxial laminate (16 plies) composite with 60 wt% carbon fibre in a polyphenylene sulfide (PPS) matrix with the motivation to investigate the influence of fibre orientation on the friction and wear properties both in the adhesive (sliding against smooth stainless steel) and the abrasive (abrading against abrasive paper) wear mode. The sliding speed and load were the two varying operational parameters. Figure 23 shows various composites with various fibre orientations with respect to the sliding direction. The wear versus sliding velocity relation indicated that the wear reached a maximum around a speed of 1 m s^{-1}. The effect of fibre orientation did not, however, disclose any clear pattern. With increasing load, μ showed a decreasing trend. The fibre orientation influenced the friction coefficients in the order $\mu_{AP} < \mu_P < \mu_N$ at a

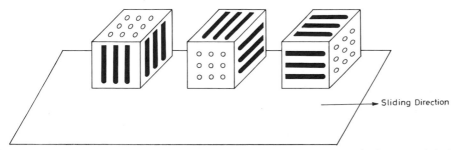

Fig. 23. Various fibre orientations with respect to the sliding direction: (left) normal (N), (middle) antiparallel (AP) and (right) parallel (P).

constant velocity. The friction data could be explained on the basis of the existing molecular–mechanical model. The wear data, however, could not be correlated either with the mechanical properties or with the microstructure.

Ovaert and Cheng [42] selected PEEK and its 15 ply unidirectional carbon-fibre (type IM-6) reinforced composites for studying the dry-wear performance against mild steel under light-load conditions. The wear was studied as a function of the counterface roughness (rms R_a) at two pv-values. It was observed that the wear rate of neat PEEK showed a minimum at $R_a = 0.15$ μ_m as the roughness of the counterface increased. Similar studies for CF-PEEK composite (fibres parallel to the sliding direction) exhibited a minimum around 0.25 μ_m. Both the wear and the friction coefficient decreased as a result of CF addition. For polished (very smooth) surfaces, the PEEK wear debris was in the form of sheets or platelets, indicative of a fatigue-delamination wear mechanism. The region of relatively low wear was characterised by the formation of interfacial films, while for the CF-PEEK composite no evidence of film formation on the counterface was seen in the minimum-wear region. (While at rougher surfaces, a film was transferred.) The wear debris was transferred to the counterface regardless of the counterface roughness, and its morphology reflected the addition of CF. Attempts were made to apply the model of Hollander and Lancaster [43] to the wear data. The application of the model appeared to be limited by the extent of material transfer and counterface modification.

An in-depth study of the triboperformance of PEEK and its two graphite-fibre reinforced composites (unidirectional or 1D composite with AS_4 graphite fibre, $V_f = 60\%$ and two-directional or 2D composites with woven HTA7 graphite fibre, five-harness satin weave, $V_f = 60\%$) was carried out by Mody et al. [44]. The fibre orientation, temperature and sliding velocity were the three varying parameters in the dry-sliding studies against steel on a pin-on-disc machine. The salient features of the investigations were as follows.

(a) *Neat polymer*: the wear rates increased with increasing pv and temperature. The coefficient of friction increased with increasing pv, but decreased with increasing temperature.

(b) *1D composite*: the coefficient of friction showed little dependence on pv and temperature. For all orientations (fig. 23), it was higher than that for neat polymer. The wear rates, however, were definitely much smaller than that of the parent polymer, with the exception of the AP orientation. The wear rates increased with increasing pv and temperature for all orientations and followed the sequence $\dot{W}_P < \dot{W}_N \ll \dot{W}_{AP}$. Although the wear for the P orientation was slightly less than that for N, the damage was smaller in the former case.

(c) *2D composite*: the wear rates for this composite were an order of magnitude smaller than those for the 1D composite. While they increased markedly with both pv and temperature, the difference between the different orientations was not significant (fig. 24). However, from the damage point of view, the P orientation, which comprised of 80% in-plane fibres parallel and 20% in-plane fibres transverse to the direction of sliding, constituted the most wear-resistant grade, while the (N, AP) orientation was prone to considerable fibre damage. At 50°C, a high

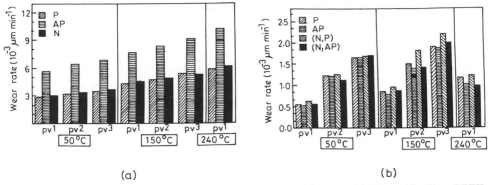

(a)   (b)

Fig. 24. Variation of the wear rates with temperature and pv, for (a) 1D and (b) 2D graphite-fibre–PEEK composites. ($pv_1 = 0.3$, $pv_2 = 0.6$, $pv_3 = 0.9$ MPa m s^{-1}; $p = 0.6$ MPa and constant) [44].

degree of isotropy was found for all the pv-values. The coefficients of friction exhibited larger variations between the orientations, and generally increased with increasing pv-values.

The average variation between the materials was as follows.

$$\dot{W}_{1D\text{-}PK} \cong 0.52\ \dot{W}_{PK},$$

$$\dot{W}_{2D\text{-}PK} \cong 0.18\ \dot{W}_{PK},$$

$$\mu_P \cong \mu_{AP} \cong 0.8\ \mu_N \quad \text{(for 1D composite)},$$

$$\mu_P \cong \mu_{AP} \cong \mu_{N,P} \cong \mu_{(N,AP)} \quad \text{(for 2D composite)}.$$

In fact, the simultaneous existence of parallel and antiparallel oriented carbon fibres in woven form resulted in a synergistic effect on the wear performance of the composite. Cirino et al. have proposed 1D and 2D wear models [45].

Sliding-wear studies of various unidirectionally reinforced PEEK and epoxy composites, as shown in table 6, against steel on a pin-on-ring machine were carried out by Cirino et al. [45]. Abrasive-wear studies with abrasion against SiC papers were also performed, with a view to design an ideal composite material with good wear resistance against a variety of wear loads (extreme surfaces, i.e. severe abrasion to smooth sliding). The following salient points emerged from the investigations.

(1) The ductile thermoplastic PEEK exhibited wear rates an order of magnitude smaller than the more brittle thermosetting EP matrix.

(2) The wear rates of PEEK composites were reduced by a factor of 1/5 to 1/10 as compared to the unreinforced PEEK, with the exception of CF-PK (N), while in EP composites a greater reduction of about two orders of magnitude compared to neat EP was observed. The fact that the EP and PEEK composites ultimately resulted into similar ranges of the wear rates indicate the strong domination of the associated wear mechanisms.

(3) GF reinforcement did not prove beneficial in both the matrices (in the case of the abrasive studies, however, it produced a greater reduction in wear).

TABLE 6
Details on materials selected [45].

Sl. No.		Abbreviation	Material	Fibre volume fraction	Density (g cm^{-3})	Manufacturer
1	Thermoset	CF-EP (1)	epoxy resin 3501-6	58%	1.62	BASF
		EP (2)	epoxy resin	–	1.28	Hercules
		CF-EP (2)	AS4 carbon-fibre epoxy	62%	1.59	Hercules
		GF-EP (2)	E-glass fibre/epoxy	58%	2.01	Fiberite
		AF-EP (2)	K49 aramid fibre/epoxy	60%	1.34	Fiberite
2	Thermoplastic	AF-PK	K/49 aramid fibre/PEEK	60%	1.38	ICI
		GF-PK	E-glass fibre/PEEK	61%	2.04	ICI
		CF-PK	AS4 carbon-fibre/PEEK	55%	1.56	ICI
		PK	Polyether-etherketone (PEEK)		1.27	ICI

(4) The fibre orientation definitely influenced the wear rates. However, the trends were not uniform for the two matrices. The general trend was that the P orientation for all types of composites appeared to be more favourable, with the exception of the AF-PK and AF-EP composites, for which the N direction was more beneficial.

For studying the fretting-fatigue wear behaviour of CF-epoxy laminates, Schulte et al. [46] prepared a laminate A (with stacking sequence $(\pm 45, 0, \pm 45_4, 90, \pm 45)_a$ and a laminate B with stacking sequence $(0_2, 90_2, 0_2, 90_2)_b$. These contained the

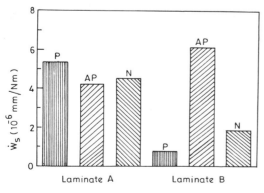

Fig. 25. Specific wear rate, $\dot{W}_s$, of laminate A and B and the influence of the sliding direction [46].

unidirectional carbon fibres T300 and HTA and different epoxy matrices. Sliding wear studies against a steel ring in three fibre orientations were also done and are shown in fig. 25. Laminate B in the P orientation exhibited the lowest wear rate.

Friedrich et al. [47] carried out a more systematic research on hybrid composites, with the objective to find fundamental information on the tailoring of hybrid composites and on their superiority over monolithic ones. A little is reported on these efforts in the literature [48,49]. For this, initially the friction and wear performance of composites reinforced with different fibres, like CF, GF and AF, for various fibre orientations is tested. A hybrid composite containing two or more types of reinforcement in various forms, like continuous long fibres or woven fabric, is fabricated. Generally two types of structures (sandwich and layer) are selected for these studies. It was proven that such hybrid composites result in a synergistic effect and that a layer hybrid structure is superior to a sandwich structure. Details on tailoring the various factors affecting the performance of such hybrid composites and their wear mechanisms are reported elsewhere in this book.

For studying the effect of the various experimental parameters, such as sliding speed, distance and load, on the dry wear of continuous E-glass-fibre reinforced epoxy composites against steel, Zamzam [50] prepared various composites of EP resin (resin HY 956 + hardener LY554) with varying amounts of GFs. The wear versus sliding distance curves of EP composite (42% GF) at various sliding speeds indicated that the wear increased almost linearly with increasing sliding distance. The wear resistance of the composite increased slowly and linearly up to a critical point ($v = 4$ m s^{-1}) with increasing sliding speed. Above this, the wear resistance showed a steep rise. This effect was thought to be due to the high adhesion between the contacting surfaces at lower speeds resulting in higher wear. With increasing speed, the frictional heat increases and, consequently, the adhesion decreases. Moreover, at higher speeds oxidised iron particles were thought to be transferred to the pin surface, resulting in low wear. The wear resistance of the composite decreased linearly with increasing contact pressure, which agreed well with the literature data. For six EP composites the wear resistance versus fibre fraction curve showed a peak at 30% GF fraction. The decrease in wear resistance after this peak was attributed to insufficient bonding between the fibres, due to the decrease in resin with increasing fibre fraction. It was proposed that at low fibre fractions the friction and wear of the composites were dominated by that of the matrix, whereas at higher fibre fractions, the wear mechanisms associated with the fibres became more operative. The following five individual mechanisms were observed to be dominating the material removal process: matrix wear, fibre sliding wear, debonding of the fibres from the matrix, plastic deformation of the matrix up to fracture and severe sliding wear of the composite.

Odi-owei and Schipper [51] selected POM and its three composites containing 15% PTFE, 15% PTFE + 20% CF and 15% PTFE + 20% GF. Both the carbon and the glass fibres were long (diameter 0.01 mm) and randomly dispersed. Friction and wear tests were performed on a pin-on-disc machine against steel discs having two types of surface roughnesses. It was observed that POM + 15% PTFE proved to be superior to all other composites (fig. 26). It is clearly seen that

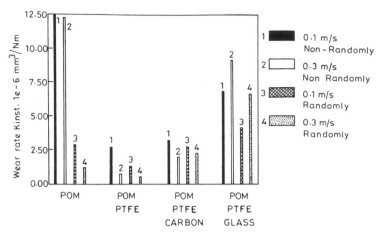

Fig. 26. Instantaneous wear rates of POM composites as a function of the surface preparation and sliding velocity [51].

POM or POM + PTFE + GF did not prove to be suitable for bearing material, while POM + PTFE + CF exhibited satisfactory behaviour only under certain operating conditions. There are very few cases in the literature where fibrous reinforcement has reflected into less enhancement in the wear resistance as compared to that with particulate-filled composites. It was observed that, except for neat polymer, a discontinuous or continuous film of the materials was transferred to the counterface. The film was arbitrarily affected by the sliding velocity. It was also seen that the surface preparation (topography) influenced both the friction and the wear enormously. A randomly prepared counterface was preferable for a low value of the wear. Herrington and Sabbaghion [52] studied the static and kinetic friction coefficients between standard washers of different metals and a graphite-fibre ($V_b = 64.4\%$) reinforced epoxy composite with the view of examine the influence of clamping at different outer-ply angles. The used eight-ply composite sheet (0/90/45–45) was quasi-isotropic. The laminate was cut at three different angles (0°, 45° and 90°). In most of the friction pairs the kinetic friction coefficient was slightly lower than the static one. The washer material was shown to have significant influence on the friction coefficient. Among the selected metallic washers, the galvanised washer showed the minimum friction coefficient while an increased normal force had little effect on the friction coefficient. However, it raised the likelihood of ploughing, which, when it occurs, causes a transition from a friction phenomenon to a wear phenomenon. It was also seen that the friction phenomena were definitely dominated by the surface preparation of both the washer and the composite. Surface contamination caused by routine handling of as-received washers resulted in a 50% lower value of the friction coefficient than found for those washers which were carefully degreased. The directionality of the outer fibres showed a much smaller effect on the friction coefficient, which did not reach its maximum when the direction of the applied force was transferred to the outer plies. Similarly, a minimum friction coefficient was not reported for an outer-ply angle of 0°. These different results were attributed to the manufacturing process of the composite. Thus, the friction coefficients were more dependent on the resin surface than on the direction of the outer layer of fibres.

With the view to study the influence of the various experimental and material parameters on the wear characteristics of fabric reinforced polymer composites, Vishwanath et al. [53] selected two different phenolic resins (Novalak and Resol) modified with softer poly(vinyl-butyral) (PVB) with varying amounts of butyralde-hyde, since such a modification results in enhanced impact resistance. The laminates were prepared from E-glass plain-weave woven roving fabric (80–85%). Wear tests were performed on a pin-on-disc machine under dry conditions against cast iron. The warp fibres in the specimen were perpendicular to the sliding surface. The results indicate that the wear of the selected composites increased non-linearly with the normal load, speed and amount of modifier. The material-transfer phenomenon was observed in composites having a butyraldehyde content $\geqslant 15\%$ and at a speed $\geqslant 10$ m s^{-1}. The resin matrix with a butyraldehyde content of 10% gave the minimum wear rate and a high mechanical strength. Not only the modifier but also the resin type influenced the wear performance of the composite. Resol matrix in the composites gave high wear rates at slow speeds as compared to those with the Novalak type. The disproportionate (excessive) increase in wear with increasing speed and modifier content was attributed, firstly, to the glass fibres in the matrix, which break after being exposed to the rubbing surface and change the surface characteristics of the specimen, and secondly, that the fact that the modifier was softer than the phenolic resin. Increasing sliding speed increases the frictional heat and, thereby, results in more thermal penetration as well as softening of the matrix, which in turn deteriorates the fibre–matrix bonding. This ultimately reflects in higher wear at higher speed and modifier content. The combined effects of thermal softening, abrasion due to glass particles and the charring of resin resulted in increasing wear rates at high speeds. Resol-type composites showed a greater tendency for thermal softening than Novalak-type composites, which resulted in delamination of plies and, hence, a higher wear.

In a subsequent work, Vishwanath et al. [54] fabricated various composites of phenolic resin modified with PVB reinforcing with E-glass plain-weave woven roving (composite A), plain-weave woven fabric (composite B) and satin-weave woven fabric (composite C), the amount of resin being 20%. Dry sliding wear tests of these against cast iron were carried out on a pin-on-disc machine, keeping the warp fibres in the specimen normal to the rubbing disc, while the weft fibres were parallel to the disc. The sliding speed was the only variable test parameter. The general trend in the friction studies at various sliding speeds indicated that μ decreased with increasing speed, which was thought to be because of formation of liquid phase at the interface due to thermal degradation of the resin. At higher speeds, the μ of composite A was lower than that of the other composites. The interface temperature of this composite was higher than that with other composites, confirming the above statement of higher thermal degradation at high speeds. Composite A showed a greater tensile, flexural, impact and interlaminar shear strength. The less good mechanical properties of composites B and C were due to the damage introduced in the yarn at the microlevel while twisting and doubling during the spinning of the yarn. GFs, being very brittle by nature, are prone to damage during weaving, resulting in a reduced strength. The minimum wear rate ($\dot{W_s} = W_v v_s^{-1} F_n^{-1}$, where W_v is the wear volume, v_s is the speed and F_n is the load), however, was recorded for composite B (plain-weave woven fabric). Thus, the weave geometry was observed to be a very important material parameter, controlling the wear performance of the composites. It was thus confirmed that composite A, with the best combination of mechanical properties, showed the lowest μ but not the smallest wear rate. The wear rate was in the order C > A > B and the bulk surface temperature was in the order A > C > B. In the plain-weave woven fabric composite B, showing the lowest wear rate, yarn cross-over interlacing appears alternatively, resulting in a closer weave and better holding of the weft fibres in position. Possibly, at cross-over points wear debris may accumulate and act as a protective layer. Thus, the larger the number of interlacings, the greater is the crimp, the higher is the possibility of wear debris accumulation and the lower is the wear rate. In composite C, the smaller number of interlacings resulted in poor holding of the weft fibres, while in composite A, due to the loose weave, the weft fibres were loosely held. This ultimately resulted in lower wear rates due to increasing debonding of the fibres. It was also observed that the structural integrity of composites B and C was lost at a speed $\geqslant 13$ m s^{-1} due to severe debonding and charring of the resin, while in composite A only charring was observed at speeds around 18 m s^{-1} without any loss of structural integrity. SEM studies of the worn surfaces at low speeds indicated delamination cracks, fibre debonding and deposition of powdery wear debris. The size of the wear debris was finer on the surface of composite A than that on B and C. With increasing sliding velocity, the delamination, debonding and fibre exposure increased. The higher the speed, the coarser was the wear debris. The debonding in the weft fibres was higher than that in the warp fibres, because the weft fibres were parallel to the sliding surface and in the direction of sliding. The overall effect of localised charring, fibre debonding and fibre breakage was least in composite B.

Recently, Vishwanath et al. [55] selected two types of polymers, i.e. PVB modified phenolic resin with 10 wt% butyraldehyde content and commercially available polyester resin. Various composites of these polymers with plain-weave woven roving E glass in different proportions were tested against cast iron, with the view to study the influence of the type and amount of polymer on the friction and wear properties. Tribostudies of six composites at various sliding speeds revealed interesting results. It was observed that the wear rate of glass/phenolic composites, which maintain their standard integrity even at high sliding velocities, was less than that of glass/polyester composites. The friction coefficient of the former system was lower than that of the latter. The minimum friction coefficient and wear rates were observed at 30 wt% of the matrix resin in both systems, while a composition of 20 wt% of the matrix resin gave the highest tensile and flexural strength. At the low matrix content of 20 wt% the fibre exposure to the counter-face was high, resulting in increased fibre fracture. Moreover, a low matrix content reduced its impact load absorption capability resulting in load transmission from matrix to adjacent subsurface fibres, leading to increased debonding of fibres. The debonded fibres become loose and break as a result of frictional thrust, giving rise to high wear rates. The amount of entrapped wear debris, which is higher for low matrix content composite, also increased the abrasion and, hence, the wear. These conclusions were supported by SEM studies of worn surfaces.

Wear studies at various sliding speeds revealed that there exists a critical velocity for which minimum wear is recorded in both types of the composites. The critical velocity of glass/phenolic composites at low resin content was lower than that of the glass/polyester composite. However, at high resin content, both composite systems showed the same order of critical velocity. During sliding, both the adhesive and the abrasive wear mechanism were operating, resulting in prolific powdery wear debris at sliding velocities below the critical value. The frictional heat increased with increasing sliding velocity, which reduced the brittleness of both the matrix and the reinforcing glass fibres. On the other hand, the rate of impact-type repeated loading caused by the hard asperities present in the counter-face, increased with increasing sliding velocity. This enhanced delamination cracks in the matrix in addition to debonding and breakage of the fibres. At the critical sliding velocity the effect of the impact-type repeated loading was neutralised by the drop in brittleness of both the fibre and resin due to the frictional heat. Moreover, the wear debris due to the abrasive and adhesive wear was further rubbed against the counter-surface, making it smoother. Some of this debris becomes embedded in the matrix, forcing a protective layer. The combined effect led to minimal wear. At higher speeds the effect of the reduction in brittleness of the matrix and fibres was less pronounced compared to that of the rate of repeated cyclic loading, resulting in increased debonding and microcracking of the matrix resin and, hence, high wear.

5.2. Abrasive wear mode

By abrading a polymeric pin against an abrasive surface, such as silicon carbide paper, one studies abrasive wear. The various parameters affecting the abrasive-wear performance of composites and the related wear mechanisms are discussed in detail by Friedrich [11].

Recently, a significant contribution to the abrasive-wear investigations on unidirectional continuous fibre reinforced polymer composites was made by Cirino et al. [56]. With the view to study the influence of the fibre orientation on the wear, apart from the regular orientations N, P and AP (fig. 23), various intermediate orientations of the fibres with respect to the abrading surface were selected. EP(2), AF-EP, PK and CF-PK were the composites selected for these studies; they are included in table 6. The wear studies were done against non-waterproof SiC abrasive paper with grit 220, velocity 300 mm min^{-1}, pressure 2.2 MPa and the distance over which was abraded was 500 mm (single-pass condition) on a pin-on-flat apparatus [56]. The wear proved to be strongly dependent on the type of material and fibre orientation. In both the composites CF-PK and AF-EP, the absolute minimum occurred in the N direction, although the distribution was different. The wear rates of CF-PK were strongly influenced by the direction of sliding relative to the fibres as well as by the fibre orientation relative to the sliding surface, while in the case of the AF-EP system the wear rates did not significantly change with fibre orientation up to 30° around the N direction. The results were described by discussing the fibre–abrasive-particle interactions on a microscopic level. It was observed that the material removal was higher for the cases where the fibres were oriented in the plane of the sliding surface and smaller where the fibres were oriented out of the sliding-surface plane. The dominating wear mechanisms in the CF-PK composite were of the microcracking and microcutting type, whereas in the AF-EP system microploughing was the predominant wear mechanism. This system behaved in a ductile way, resulting in a relatively smoother surface appearance. The microploughing resulted in lower wear. This was especially evident for AF-EP with fibre orientations between normal and parallel. In addition, empirical wear equations were developed for predicting the dry abrasive wear behaviour of these two systems with arbitrary fibre orientation.

In subsequent work, see refs. [45,57], additional efforts were made, investigating the influence of the operational parameters (pressure, sliding velocity, apparent contact area on the pin sample and SiC papers of two grades) on the friction and wear of the composites as shown in table 6. The CF-EP system was selected for studying the influence of the operational parameters. With increasing contact pressure the coefficient of friction did not show significant change, while the dimensionless wear rate increased linearly. The sliding velocity did not prove to be a sensitive parameter both for friction and wear. The friction coefficient did not change significantly with increasing apparent contact area, but the wear rate slightly decreased. The effect of the two types of paper (waterproof and non-waterproof) for the AF-EP and CF-PK systems for three fibre orientations was also studied. On waterproof papers the wear rates of the AF-EP system were higher and on non-waterproof paper those of the CF-PK system were higher. The influence of the fibre orientation and the material–fibre combination on the wear against two grades of paper (grit 70 μm and 15 μm) was also studied. The wear rate of PK was lower than that of EP by a factor of about two. Fibre addition to the EP decreased the relative wear rate as well as the friction coefficient. CF addition to PK proved beneficial for 15 μm grit only. The composites showed a smaller degree of anisotropy while sliding against the 15 μm abrasive surface as compared to that with the 70 μm abrasive surface. In general, the N orientation generated the lowest wear rates and the AP orientation resulted in the highest wear rates. The minimum wear resistance were shown by the reinforced systems with N orientation. The wear resistance ratios ($\dot{W}_{C}^{-1}/\dot{W}_{EP}^{-1}$) of the composites with respect to the material parameters are shown in table 7. By examining the worn surfaces and subsurfaces with SEM, the basic wear mechanisms were identified as a function of the fibre orientation. The varying ability of each mechanism to remove material resulted in anisotropic wear behaviour. The dominating wear

TABLE 7

Wear-resistance ratios of the composite materials in the abrasive and sliding systems. The wear-resistance ratio is equal to the wear resistance of the composite material divided by the wear resistance of the neat epoxy matrix [45].

		A *			Neat	B **			Neat
		Reinforcement				Reinforcement			
		CF	GF	AF		CF	GF	AF	
EP	N	1.8	2.3	7.9		65.4	0.5	122.4	
	P	1.7	1.4	1.1	1	99.8	2.1	44.1	7
	AP	0.9	0.8	1.2		44.1	1.1	46.3	
PK	N	1.7	2.4	15.2		4.9	5.1	61.4	
	P	1.6	1.8	0.9	1.8	158.1	5.5	46.1	11.8
	AP	1.0	1.2	1.3		94.8	2.0	44.1	

* A – wear data on abrasion against 70 μm SiC.

** B – wear data while sliding against smooth steel.

mechanism could be differentiated by observing the fibre–grit interactions. When the fibres were normal to the abrading surface, their ends were cut in a slicing manner, due to the abrasion. When they were parallel to the sliding direction, ploughing and cutting of the matrix material was the predominant wear mechanism. In-plane bending of the fibres caused fibre fracture and cracking, resulting in a relatively higher wear when the fibres were in the AP orientation. The extent of wear by these mechanisms strongly depended on the fibre–matrix combination also. Finally, network data were compiled on the wear behaviour in terms of three material parameters, the fibre orientation, the fibre type and the matrix material. This enabled a systematic selection of low-wear composite materials, which could consist of a ductile matrix with high fracture toughness, such as PEEK reinforced with aramid fibres oriented normal to the contacting surface. GF or CF oriented in-plane and parallel to the wearing direction (P and AP), can be more beneficial. This could be an ideal design of a composite material with an overall optimal wear resistance against a wide variety of wear loads (severe abrasion to smooth sliding). Additional beneficial components could be PTFE fibres interwoven in the structure, PTFE particles or any lubricating agent (e.g. silicon fluids or graphite flakes).

In order to compare the wear rates of the different systems (adhesive and abrasive conditions), the surface roughness (R_t) of the steel rings and the average particle diameter (D) of the SiC particles were plotted on the same axis as the asperity size. It was seen that the wear rates differed by about six orders of magnitude (the wear rates in the abrasive system being higher) [45]. Moreover, the effect of fibre reinforcement on the wear rate of the neat matrix was much greater under sliding-wear conditions, where a decrease up to two orders of magnitude was observed. Table 7 shows similar data on the wear performance of composites in both wear modes.

For CF-PPS uniaxial laminate (16 plies, CF 60%) composite, Lhymn [41] selected various speeds and loads for studying the influence of the fibre orientation on the abrasive-wear behaviour. The following points emerged from wear studies against abrasive paper.

(1) The values of the specific wear rate were in the order of fibre orientation as N < AP < P, regardless the variation in speed or load.

(2) The wear rates were almost insensitive to the sliding-velocity variation, while they decreased with increasing normal load.

(3) The coefficient of friction (at higher load) showed the trend P ≪ AP < N. The friction coefficient did, however, not show much variation with sliding velocity. At low road, however, this relation showed substantial variation. The N orientation showed the highest friction coefficient, which increased with speed. The other two orientations did not show a definite pattern.

(4) The friction coefficient decreased with increasing load at higher velocity. At lower load no definite pattern emerged.

(5) The correlation between the wear rate and the friction coefficient was very poor.

The molecular–mechanical model of the frictional force was found suitable for explaining the friction data for the various orientations. The molecular part is controlled by the yield strength and bond-rupture energy for the different fibre configurations. The mechanical part is controlled by the indentor penetration depth and the asperity radius. The effect of the sliding velocity on the friction coefficient was analyzed on the basis of an equation derived from crack-propagation theory. The developed model-equation fitted quite well at high velocities, where the thermally activated process is operative. However, this was not valid for low speeds, due to the fact that the model was developed on the basis of heat generation at local spots and the subsequent thermally activated process, and at low speeds thermal effects are negligible. For explaining the general trend observed in the wear performance of composites with different fibre orientation $((\dot{W}_s)_N < (\dot{W}_s)_{AP} < (\dot{W}_s)_P)$, a previously derived phenomenological equation was used. The variation in the mechanical properties, such as elastic modulus, fracture strain and interlaminar shear strength (I_s), were correlated with the wear-rate equation for different fibre orientations. The unique PAN carbon-fibre morphology (onion skin) was correlated with the friction and wear behaviour to explain the difference between P- and AP-type specimens. Because the carbon fibres had a circumferentially onion-like graphite sheet structure, the bonding between the two sheet layers is weak. Hence, $(I_s)_N > (I_s)_P$ results in $(\dot{W}_s)_N < (\dot{W}_s)_P$. Moreover, $(I_s)_{AP} > (I_s)_P$ due to the lamellar platelet morphology of the graphite sheets and the lamellar crystalline platelets surrounding a fibre in a cylindrical fashion in a polymer matrix. This was thought to be the reason for $(\dot{W}_s)_{AP} < (\dot{W}_s)_P$.

Mcgee et al. [58] selected three graphite fibre (Thornel 300) reinforced (one with continuous fibres and two with chopped fibres) composites for studying abrasive-wear behaviour. The load, abrading distance and grit diameter were the three selected operational parameters. Graphite-fibre (unidirectional) reinforced $(V_f = 62\%)$ polyimide (bismaleimide) (referred to as BM composite) was tested for three fibre orientations. The other two composites were reinforced with nylon $(V_f = 40\%)$ and PPS $(V_f = 40\%)$. In addition, wear studies on steel and aluminium were included for comparison. The results indicated that the wear rates for these composites were significantly higher than those for metals. The wear rates of the composites followed the sequence $NY_{Gr-F} < BM_{Gr-F}(N) < PPS_{Gr-F} < BM_{Gr-F}(P) \lll BM_{Gr-F}(AP)$. Thus, the matrix material and the fibre orientation $(N < P \lll AP)$ were observed to be influencing factors. SEM studies of the worn surfaces indicated better fibre–matrix adhesion in the case of the nylon composite. Apart from the higher ductility, this was thought to be an additional factor for the better wear performance. Fibre debonding, buckling and fracturing were observed in SEM studies. Since the weak fibre–matrix interface is not exposed in the case of normal orientation, this results in better wear performance. In the AP direction (leading to maximum wear), the fibres are subjected to shear, bending and torsion loading by the abrasive grits, which could tear away the fibres, resulting in the highest wear. While varying the test parameters, the wear versus grit-size relation, however, showed a size effect. Above a critical grit size (in this case 180 mesh), the wear rate became constant or increased very slowly, similar to that in metals.

5.3. Fretting-fatigue wear mode

Relative cyclic sliding motion of two surfaces in intimate contact constitutes a special wear process, called fretting. This results in significant reduction in the fatigue resistance. Fretting fatigue occurs when repeated loading of a part causes a sliding movement at the material interfaces in the design. This movement may induce at or near the points of contact between the faces stresses of sufficient intensity to cause cracking [59]. This situation often leads to material failure in an unpredictable manner. The common examples are multilayer leaf springs, bolted joints, flanges and seals. In many of these applications FRPs are increasingly replacing metals. However, only little is reported on the fretting-fatigue wear behaviour of FRPs in the literature [59–63]. Recently, Friedrich and co-workers have done significant research in this area [46,61–63], which is reported elsewhere in this volume.

5.4. Erosive wear mode

If solid particles impinge on a target surface and cause local damage with material removal, the wear is known as erosive wear. Studies to develop understanding of the mechanism of erosive wear have been motivated by the reduced lifetimes and the failure of mechanical components used in erosive environments, e.g. in pipelines carrying sand, coal slurries in petroleum refining, aircraft gas turbine compressor blades, etc. [64–66]. The effects of the various testing parameters on the erosive wear of different polymeric materials have been discussed by many authors [67,68]. Solid particle erosion on polymeric composites, however, has not yet been extensively investigated. With the view to understand the erosive-wear characteristics of fibre reinforced composites, and the related mechanism, some papers have appeared in the literature. The performance of the following materials have been investigated: nylon 6,6 and its graphite-fibre reinforced composite with

TABLE 8
Materials selected for erosive wear studies [64].

Composite	Polymer	Fibre
Unidirectional continuous graphite-fibre reinforced polyimide laminate, A	Hexel's F-173 bismaleimide (BM) PI resin	Thornel 300 graphite fibre
Woven (0/90) graphite-fibre reinforced epoxy laminate, B	Fiberite's 934 epoxy resin	Thornel 300 graphite fibre
Woven aramid-fibre reinforced epoxy laminate in a quasi-isotropic (0/90/+45) symmetric layup, C	Fiberite's 934 epoxy resin	Kevlar fibre
Chopped graphite-fibre reinforced thermoplastic, D	LNP corporation's PPS (injection moulded)	40% graphite fibre (chopped)

quartz particles as erodent [66], nylon, epoxy and polypropylene reinforced with carbon, glass and steel fibres [69], quartz–PI, quartz–polybutadiene composite [70] and glass-fibre reinforced epoxy composite [71]. The erosive-wear performance of some advanced composites is discussed in the this section.

Pool et al. [64] selected four composites (table 8) for erosive wear studies against sand particles of various sizes. The results are shown in fig. 27. The volume erosion rates in the composites tested were greater than those of low-carbon steel by at least an order of magnitude. The wear rates of the materials (gram removed/gram of abrasive) followed the sequence (table 8):

$$A_\perp > A_\parallel \gg B > C > D.$$

The maximum wear rate occurred at an angle of impingement of 90° for composites A and B, indicating brittle-type erosion behaviour, while composite C exhibited wear maxima in the range of angles 34°–45°, indicating semi-ductile

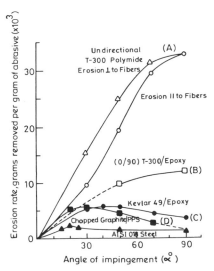

Fig. 27. Erosion rate versus the angle of impingement for various composite materials ($v = 31$ m s^{-1}) [64].

behaviour. Material D showed a maximum at 25°, indicating ductile behaviour. The preliminary investigations on the erosive wear of the selected composites indicated that the brittleness of the fibres and the interfacial bond strength between fibre and matrix are the main factors governing the erosion rates of composites, apart form the type of polymer, i.e. thermoplastic or thermoset. The well-bonded ductile fibres in a thermoplastic matrix should ideally exhibit the lowest erosion rate.

Lhymn and Wapner [72] studied slurry erosion (33% sand particles in tap water) of 40% glass-fibre (short) reinforced PPS composite. A linear increase in the specific-weight loss, w_e ($w_e = \Delta m / At$, A being the sample's exposed area and t the time of erosion), as a function of erosion time and sand-particle size was observed. The eroded volume increased faster with increasing particle size than increasing particle velocity. This composite showed a much lower slurry-erosion rate than cast iron. SEM studies indicated that the failure process appeared to be controlled by subsurface crack propagation and subsequent work-debris formation. Utilizing the crack-propagation theory concept, a phenomenological equation was derived to explain the effect of the particle size and the velocity on slurry erosion. It indicated that the erosion rates increased with increasing impacting particles' mass, velocity and fracture toughness. It also indicated that the wear rate was inversely proportional to the fracture energy.

Mathias et al. [73] studied BM composites reinforced with unidirectional carbon fibre (65–75%) using angular shaped alumina abrasives of various sizes as erodent at different speeds and angles. It was observed that for all selected experimental conditions, the erosive-wear resistance of the composite was smaller than that of the parent polymer. There was a small dependence of the wear on the orientation of the fibres, but, in general, composites with fibres parallel to the erodent stream had a higher erosion resistance than those with the fibres perpendicular to the particle stream direction. The wear versus velocity (v) and particle size (D) showed an exponential relation, while the wear and α' (angle of impingement) relation proved to be complex and varied with both D and v. The composite exhibited a larger wear for $\alpha' = 90°$ than for $\alpha' = 30°$ in all cases, except for low v and high D. Thus, when the kinetic energy of the particles was low, they could not remove enough matrix material to reveal the underlying fibres. For neat polymer the wear was independent of α', with few exceptions. That the maximum wear was at an angle of $\alpha' = 90°$ under almost all the conditions indicated that the brittle nature of the graphite fibres dominated the wear process. The wear occurred in stages: first, erosion and local removal of the matrix and, second, erosion of the exposed fibres. When the size of the damage zone was of the order of the size of the fibre diameter, the matrix gave a significant contribution to the material removal. When the damage zone was large, the material removal was dominated by the brittle fibres. The SEM studies revealed the damage mechanism as a mixture of brittle fracture and ductile tearing.

6. Conclusions

Developments in the tribology of high-performance polymer composites in recent years have been discussed. Various operational parameters, such as load, speed, counterface type, surface topography and sliding duration, influence the wear performance of a composite. However, the extent of the influence differed not only from composite to composite, but also from wear mode to wear mode. Stress was given to understanding the influence of the material parameters, such as type of matrix, fibre, their combination, fibre orientation relative to the sliding surface and fibre concentration, on the friction and wear performance of advanced composites. It was concluded that for a type of matrix and fibre, a typical fibre orientation or a typical fibre–matrix combination, a fabric weave which exhibits excellent behaviour in one wear mode need not necessarily display the same performance in another one. Generally, carbon fibre in parallel orientation and aramid fibre in normal orientation were observed to result in enhancement in the wear performance of the composite. Glass fibres were found to be acceptable in a few systems. Highly ductile thermoplastics were observed to be more suitable in a wide range of triboapplications (not in fretting fatigue) than brittle thermosets. The concept of hybrid composites, their tailoring and performance is discussed in detail. In hybrid composites, the synergistic effect due to woven-fabric reinforcement in two or three dimensions, is probably the result of fibre interlocking in the contact area.

List of symbols

AP	antiparallel
AF	aramid fibre
α'	angle of impingement
BM	bismaleimide
CF	carbon fibre
CFRP	continuous-fibre reinforced polymer
EP	epoxy
FRP	fibre reinforced polymer
GF	glass fibre
HMWPE	high molecular weight polyethylene
K_0, $\dot{W}_s$	specific wear rate
MFD	mould fill direction
N	normal
P	parallel
PA	polyamide
PAN	polyacrylonitril
PE	polyethylene
PEEK, PK	polyetheretherketone
PEI	polyetherimide
PI	polyimide
POM	polyoxymethylene
PP	polypropylene
PPS	polyphenylene sulphide
PTFE	polytetra fluoroethylene

PU   polyurethane
PVB   polyvinylbutyral
SEM   scanning electron microscopy
T_g glass transition temperature
TP   thermoplastic
TS   thermoset
UHMWPE   ultra-high molecular weight polyethylene
μ friction coefficient
v velocity
WFRP   woven fabric reinforced polymer

References

[1] R.L. Fusaro, Self lubricating polymer composites and polymer transfer film lubrication for space application, Tribol. Int. 23 (1990) 105–122.
[2] A.I. Sviridyonok, Self lubrication mechanism in polymer composites, Tribol. Int. 24 (1991) 37–43.
[3] N.P. Suh, Tribophysics (Prentice Hall, Englewood Cliffs, NJ, 1987).
[4] J.K. Lancaster, Polymer based bearing materials, the role of fillers and fibre reinforcement, Tribol. Int. 5 (1972) 249–255.
[5] J.K. Lancaster, Composites for aerospace dry bearing application, in: Friction and Wear of Polymer Composites, Composite Materials Series, Vol. 1, ed. K. Friedrich (Elsevier, Amsterdam, 1986) ch. 11, p. 363.
[6] B.J. Briscoe, Wear of polymers. An essay on fundamental aspects, Tribol. Int. 14 (1981) 231–243.
[7] H. Bohm, S. Betz and A. Ball, The wear resistance of polymers, Tribol. Int. 23 (1990) 399–406.
[8] T. Tsukizoe and N. Ohmae, Friction and wear of advanced composite materials, Fibre Sci. Technol. 18 (1983) 265–286.
[9] G.A. Cooper and J. Berlie, New bearing materials for high speed dry operation, Wear 123 (1988) 34–50.
[10] B.J. Briscoe, Tribology of polymers: state of an art, in: Physico-chemical Aspects of Polymers Surface, Vol. 1, ed. K.C. Mittal (Plenum Press, New York, 1983) p. 387.
[11] K. Friedrich, in: Friction and Wear of Polymer Composites, Composite Materials Series, Vol. 1, ed. K. Friedrich (Elsevier, Amsterdam, 1986) ch. 8, p. 233.
[12] H. Voss and K. Friedrich, Wear performance of bulk liquid crystal polymer and its short fibre composites, Tribol. Int. 19 (1986) 145–156.
[13] K. Friedrich, Polymeric matrix composites, high wear resistance, in: Int. Encyclopedia of Composites, Vol. 4, ed. S.M. Lee (VCH, New York, 1990).
[14] K. Tanaka and Y. Yamada, Effect of temperature on friction and wear of some heat resistant polymers, in: Polymer Wear and its Control, ACS Symp. Series, Vol. 287, ed. L.H. Lee (ACS, Washington, DC, 1985) p. 103.
[15] M.N. Gardos, Self lubricating composites for extreme environmental conditions, in: Friction and Wear of Polymer Composites, Composite Materials Series, Vol. 1, ed. K. Friedrich (Elsevier, Amsterdam, 1986) ch. 12, p. 397.
[16] J. Bijwe, U.S. Tewari and P. Vasudevan, Friction and wear studies of short glass fibre reinforced polyetherimide composite, Wear 132 (1989) 247–364.
[17] H. Mark, N. Bikales, C. Overberger and G. Menges, eds, Encyclopedia of Polymer Science and Engineering, Vol. 3, Chief ed. I.I. Kroschwitz (Wiley–Interscience, New York, 1985).
[18] M.S.M. Alger and R.W. Dyson, Thermoplastic composites, in: Engineering Polymers, ed. R.W. Dyson (Blackie and Sons, Glasgow, 1990) ch. 1, p. 1.
[19] R.W. Dyson, Long-fibre reinforced thermoset composites, in: Engineering Polymers, ed. R.W. Dyson (Blackie and Sons, Glasgow, 1990) ch. 5, p. 101.
[20] K.H.Z. Gahr, Microstructure and Wear of Materials, Tribology Series, Vol. 10 (Elsevier, Amsterdam, 1987).

[21] K. Friedrich, Z. Lu and R. Scherer, Wear and friction of composite materials, in: Proc. Int. Conf. on Advance materials, EMRS spring meeting, 1991, Sympo. 44, Composite Materials (Strasbourg, France, May, 1991) pp. 27–31.

[22] J.W.M. Mens and A.W.J. deGee, Friction and wear behaviour of 18 polymers in contact with steel in environments of air and water, Wear 149 (1991) 255–268.

[23] C. Lhymn, Effect of normal load on the specific wear rate of fibrous composites, Wear 120 (1987) 1–27.

[24] C. Lhymn, Tribological properties of polybutyleneterephthalate glass composites, Adhesive wear, Mater. Sci. & Eng. 80 (1986) 93–100.

[25] K. Friedrich, Sliding wear performance of different polyimide formulations, Tribol. Int. 22 (1989) 25–31.

[26] U.S. Tewari, S.K. Sharma and P. Vasudevan, Friction and wear studies of bismaleimide resin, Tribol. Int. 21 (1988) 27–30.

[27] M.N. Gardos and H.E. Sliney, Graphite fibre reinforced PI composites, in: Tribology in the 80's, Vol. 2 (NASA, Lewis Centre, Cleveland, Washington, 1984) p. 829.

[28] M.S.M. Alger, High temperature and fire resistant polymers, in: Speciality polymers, ed. R.W. Dyson (Blackie and Sons, Glasgow, 1987) ch. 3 p. 38.

[29] J. Bijwe, U.S. Tewari and P. Vasudevan, Friction and wear studies of bulk polyetherimide, J. Mater. Sci. 25 (1990) 548–556.

[30] U.S. Tewari and J. Bijwe, Tribological investigations of polyetherimide composite, J. Mater. Sci. 27 (1992) 328–334.

[31] J. Bijwe, U.S. Tewari and P. Vasudevan, Friction and wear studies of polyetherimide composites, Wear 138 (1990) 61–76.

[32] J. Bijwe, U.S. Tewari and P. Vasudevan, Friction and wear studies of internally lubricated composite, J. Synth. Lubr. 6 (1989) 179–202.

[33] U.S. Tewari and J. Bijwe, Comparative studies on sliding wear of polyimide composites, Composites 22 (1991) 204–210.

[34] N.-H. Sung and N.P. Suh, Effect of fibre orientation on friction and wear of fibre reinforced polymeric composites, Wear 53 (1979) 129–141.

[35] K. Tanaka, personal communication (1988).

[36] J. Bijwe, C.M. Logani and U.S. Tewari, Influence of fillers and fibres reinforcement on abrasive wear resistance of some polymer composites, Wear 138 (1990) 77–92.

[37] U.S. Tewari, J. Bijwe, J.N. Mathur and I. Sharma, Studies on abrasive wear of carbon fibre (short) reinforced polyamide composites, Tribol. Int. 25 (1992) 53–60.

[38] J. Bijwe, J.N. Mathur and U.S. Tewari, Influence of reinforcement and lubrication on abrasive wear performance of polyamide, J. Synth. Lubr. 8 (1991) 177–195.

[39] A. Misra and I. Finnie, On the size effect in abrasion and erosion, Wear 65 (1981) , 359–373.

[40] J. Bijwe, Friction and wear of engineering polymer: Studies on polyimides, composites, PhD Thesis (IIT Delhi, 1989).

[41] C. Lhymn, Tribological properties of unidirectional polyphenylene sulfide–carbon fibre laminate composites, Wear 117 (1987) 147–159.

[42] T.C. Ovaert and H.S. Cheng, Counterface topographical effects on the wear of polyether-ether-ketone carbon fibre composite, Wear 150 (1991) 275–287.

[43] A.E. Hollander and J.K. Lancaster, An application of topographical analysis to the wear of polymers, Wear 25 (1973) 155–170.

[44] P.B. Mody, T.W. Chou and K. Friedrich, Effect of testing conditions and microstructure on the sliding wear of graphite fibre/PEEK matrix composites, J. Mater. Sci. 23 (1988) 4319–4330.

[45] M. Cirino, K. Friedrich and R.B. Pipes, Evaluation of polymer composites for sliding and abrasive wear applications, Composites 19 (1988) 383–392.

[46] K. Schulte, K. Friedrich and S. Kutter, Fretting fatigue studies of carbon fibre/epoxy resin laminates, Part II. Effects of a fretting component on fatigue life, Composite Sci. & Technol. 30 (1987) 203–219.

[47] K. Friedrich, O. Jacobs, M. Cirino and G. Marom, Hybrid effects on sliding wear of polymer composites, Oral presentation, Conf. on Tribology of Composite Materials (Oak Ridge, TN, 1–3 May, 1990).

[48] H.M. Hawthorne, Wear in hybrid carbon/glass fibre epoxy composites materials, in: Proc. 3rd Int. Conf. on Wear of Materials, ed. K.C. Ludema (ASME, Washington DC, April 8, 1983) p. 576.

[49] T. Tsukizoe and N. Ohmae, Friction and wear performance of undirectionally oriented glass, carbon, aramid and stainless steel fibre-reinforced plastics, in: Friction and Wear of Polymer Composites, Composite Materials Series, Vol. 1, ed. K. Friedrich (Elsevier, Amsterdam, 1986) ch. 7, p. 205.

[50] M.A. Zamzam, The wear resistance of glass fibre reinforced epoxy composites, J. Mater. Sci. 25 (1990) 5279–5283.

[51] S. Odi-owei and D.J. Schipper, Tribological behaviour of unfilled and composite polyoxymethylene, Wear 148 (1991) 363–376.

[52] P.D. Herrington and M. Sabbaghion, Factors affecting the friction coefficient between metallic washers and composite surface, Composites 22 (1991) 418–426.

[53] B. Vishwanath, A.P. Verma and C.V.S.K. Rao, Wear study of glass woven roving composites, Wear 131 (1989) 197–205.

[54] B. Vishwanath, A.P. Verma and C.V.S.K. Rao, Effect of fabric geometry on friction and wear of glass fibre-reinforced composites, Wear 145 (1991) 315–327.

[55] B. Vishwanath, A.P. Verma and C.V.S.K. Rao, Effect of matrix content on strength and wear of woven roving glass polymeric composites, Composite Sci. & Technol. 44 (1992) 77–86.

[56] M. Cirino, K. Friedrich and R.B. Pipes, The effect of fibre orientation on the abrasive wear behaviour of polymer composite materials, Wear 121 (1988) 127–141.

[57] M. Cirino, R.B. Pipes and K. Friedrich, The abrasive wear behaviour of continuous fibre polymer composites, J. Mater. Sci. 22 (1987) 2481–2492.

[58] A.C. Mcgee, C.K.H. Dharan and I. Finnie, Abrasive wear of graphite fibre reinforced polymer composite materials, Wear 114 (1987) 97–107.

[59] P.J.E. Forsyth, Occurrence of fretting fatigue failures in practice, in: Fretting Fatigue, ed. R.B. Waterhouse (Applied Science Publishers, London, 1981) p. 99.

[60] N. Ohmae, K. Kobayashi and T. Tsukizoe, Characteristics of fretting of carbon fibre reinforced plastics, Wear 29 (1974) 345–353.

[61] K. Friedrich, Fretting fatigue failure of polyester resin and its glass fibre mat composites, J. Mater. Sci. 21 (1986) 1700–1706.

[62] O. Jacobs, K. Friedrich, G. Marom, K. Schulte and H.D. Wagner, Fretting wear performance of glass, carbon and aramid fibre epoxy and PEEK composites, Wear 135 (1990) 207–216.

[63] O. Jacobs, K. Friedrich and K. Schulte, Fretting fatigue of continuous carbon fibre reinforced polymer composites, Wear 145 (1991) 167–188.

[64] K.V. Pool, C.K.H. Dharan and I. Finnie, Erosive wear of composite materials, Wear 107 (1986) 1–12.

[65] C.E. Smeltzer, M.E. Gulden and W.A. Compton, Mechanisms of material removal by impacting dust particles, J. Basic Eng. 19 (1970) 639–654.

[66] G.P. Tilly, Erosion caused by airbourne particles, Wear 14 (1969) 63–79.

[67] S.M. Walley, J.E. Field and P. Yennadhiou, Single particle impact erosion damage on polypropylene, Wear 100 (1984) 263–280.

[68] K. Friedrich, Erosive wear of polymer surfaces by steel ball blasting, J. Mater. Sci. 21 (1986) 3317–3332.

[69] G.P. Tilly and W. Sage, The interaction of particle and material behaviour in erosion process, Wear 16 (1970) 447–465.

[70] J. Zahavi and G.F. Schmitt, Solid particle erosion of reinforced composite materials, Wear 71 (1981) 179–190.

[71] A.M. Latifi, Solid particle erosion in composite materials, Masters' Thesis (Wichita State University, Wichita, KS, December, 1987).

[72] C. Lhymn and P. Wapner, Slurry erosion of polyphenylene sulphide glass fibre composites, Wear 119 (1987) 1–11.

[73] P.J. Mathias, W. Wu, K.C. Goretta, J.L. Routbort, D.P. Groppi and K.R. Karasek, Solid particle erosion of a graphite fibre reinforced bismaleimide polymer composite, Wear 135 (1989) 161–169.

Advances in Composite Tribology
edited by K. Friedrich
© 1993 Elsevier Science Publishers B.V. All rights reserved

Chapter 6

Wear Models for Multiphase Materials and Synergistic Effects in Polymeric Hybrid Composites

K. FRIEDRICH

Institut für Verbundwerkstoffe GmbH, Universität Kaiserslautern, Erwin-Schrödinger-Strasse 58, W-6750 Kaiserslautern, Germany

Contents

Abstract

This chapter presents some of the existing approaches for describing mathematically the friction and wear behaviour of multi-component systems. The relationship between the size of the reinforcing phase and the asperity size plays an important role in friction and wear phenomena. Depending on this parameter, different types of wear behaviour can occur.

The abrasive wear of particle-reinforced polymer composites can be described by a simple additive equation based on a series model. The apparent filler wear rate in the composite is higher than that measured for the same material in sheet form and this increase is highly dependent on the type of particle-surface treatment. If these composites are, however, subjected to sliding wear against smooth steel counterparts, the simple rule-of-mixtures description does not apply anymore, due to a higher complexity of the wear mechanisms.

From investigations on the unlubricated friction and wear of unidirectionally oriented fibre reinforced polymers (FRP) sliding against carbon steel, the following results were obtained:
(1) The law of mixture for the friction coefficient of FRP was deduced, and the validity of this law was assured by the experimental results.
(2) Carbon-FRP gave a small specific wear rate and a low friction coefficient. By contrast, glass-FRP provided a large specific wear rate and a high friction coefficient.
(3) A model was proposed in which the wear of FRP proceeds by wear-thinning of the fibre with subsequent breakdown of the fibre and by peeling-off of the fibre from the matrix. Using this model, an experimental equation for the wear of FRP was deduced.

Further, work concerning the tribological behaviour of hybrid composites has led to the following conclusions:
(1) The steady-state wear rates of hybrid composites sliding against stainless steel often lie considerably below the values linearly interpolated between those of the single-fibre composites. Thus, wear synergism is exhibited by the hybrid composites.
(2) Fibre orientation affects a composite's wear behaviour, but both the extent of its significance and the optimum orientation for wear resistance depend on the composite's composition.
(3) Using the best wear results achieved with monolithic fibre composites, i.e. with carbon fibres oriented in-plane and parallel to the sliding direction, or with aramid fibres in normal orientation, would lead to a model composite with a 3D-hybrid structure, which is supposed to have optimum wear resistance.

(4) Approaching this model by producing 2D-hybrid composites shows synergistic effects between different fibre materials and directions chosen.
(5) A mathematical, semi-empirical model is suggested that allows one to describe these hybrid effects by (a) starting from a rule-of-mixture prediction on the wear resistance of the hybrid composite, and (b) subtracting a reduction term of the wear rate, resulting from the interactive protection against wear of one component by the other. If the hybrid efficiency factor is negative, the wear of the hybrid composite is, on the other hand, greater than that predicted by the rule-of-mixtures approach.

1. Introduction

Polymer composites with continuous-fibre reinforcement of high volume fraction and perfect alignment are known to have very high values of their specific strength and stiffness. Their properties can be tailored to the load system acting on a structural part made from these materials. Besides these advantages, the wide variety of different fibre and matrix materials permits the design of composites with unique properties for different kinds of application. This is especially true if different types of fibres are used to develop the microstructure of a particular composite material (hybrid composites). The majority of material developments in this respect have been carried out to improve the mechanical properties of the composites, e.g. their resistance to fatigue conditions [1–4]. Little has yet been published, however, on how hybridization influences secondary properties, as, for instance, the composite friction and wear behaviour.

Fibre reinforced thermoplastics and thermoset resins are used as seal, gear and dry bearing materials. Thermoset resin based composites, and especially those with continuous carbon fibres, can give better service (e.g. lower wear rates, higher strength) than those with thermoplastic matrices and short-fibre reinforcements when sliding against metals under severe conditions [5–7]. This is because the high fibre contents attainable with thermosetting composites and the preferential load-bearing by the fibres ensures that their tribological properties are mainly determined by those of the reinforcements [6,7].

The wear of some unidirectional carbon fibre composites sliding on steel was found to be little affected by the fibre orientation relative to the sliding direction [5,8]. Yet other studies report sizeable wear dependence on fibre orientation for carbon [9,10] and glass, aramid and steel fibre composites [11] on similar counterfaces. Again, correlations between the wear rates and both Young's modulus and interlaminar shear strength for composites with different carbon fibres have been reported [11]. However, it has also been shown that comparisons of wear behaviour between different composites may be questionable because each composite generates its own wear-controlling counterface topography during sliding [6,12]. Nevertheless, in general, carbon fibre reinforced thermoset resins exhibit much greater wear resistance than similar glass fibre composites [6,11–13].

Mixed fibre or hybrid composites can provide optimal tailored mechanical properties (e.g. toughness) since addition of a second fibre can often improve one

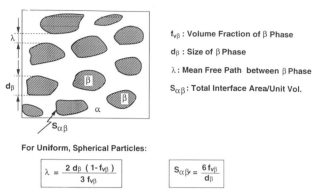

For Uniform, Spherical Particles:

$$\lambda = \frac{2\,d_\beta\,(1\text{-}f_{v\beta})}{3\,f_{v\beta}}$$

$$S_{\alpha\beta} = \frac{6\,f_{v\beta}}{d_\beta}$$

Fig. 1. Microstructural parameters in a two-phase isotropic material [22].

property of a single fibre composite without much sacrifice of others [14]. Syner-gism * has even been found in hybrids such that some composite properties exceed those expected based on a simple "rule-of-mixtures" behaviour [15].

The objectives of this chapter are (a) to summarize some of existing models for describing the wear behaviour of multiphase materials in general and of particle filled polymer composites in particular, (b) to especially focus on those papers on wear which have studied fibre reinforced composites and which have taken into account the special condition that more than one type of reinforcing component was used in a certain polymeric matrix, (c) to report recent works for developing hybrid composites with high wear resistance, and (d) to establish a model which describes the positive or negative synergistic wear effects resulting from hybridiza-tion of different fibre materials in one particular composite system.

2. Rule-of-mixtures approaches to wear of multi-component materials

2.1. Abrasive wear models

Quite recently, a quantitative approach to the wear behaviour of multiphase systems has been reported in the literature by Simm and Freti [16]. In particular, these authors discussed models describing the wear rate as a function of the volume fraction of a second phase [17–20]. Phase interaction effects and interface phenomena have also been observed [19,21]. Figure 1 shows the important param-eters for quantitative microstructural analysis which have been established to describe two-phase materials [22]. The wear behaviour depends on various factors, as discussed in the previously referenced literature [17–21]. The wear mechanism(s) of each phase is the first important aspect. Under defined conditions, a given phase shows a specific wear mode and wear rate, which depends on its individual

---

* Synergism is the joint action of discrete agencies (as e.g. various types of fibres), in which the total effect is greater than the sum of their effects when acting independently (from "The Meriam-Webster Dictionary", 1974).

properties. Consequently, when various phases are combined, forming a multi-phase material, it is expected that the overall behaviour will be a function of the respective contribution of each phase. Based on this approach, the wear resistance has been mathematically described by Khruschov as a linear function of the volume fraction of the phases present [18]. In terms of the total wear rate, i.e. the inverse of the wear resistance, the model is called the inverse rule of mixtures (IROM):

$$W^{-1} = \sum V_i W_i^{-1}, \tag{1}$$

where W is the total volume wear rate, W^{-1} is the total wear resistance, W_i is the wear rate of the ith phase and V_i is the volume fraction of the ith phase.

Garrison [18] demonstrated that Khruschov's model predicts the wear behaviour of multiphase materials only when each phase shows a wear rate proportional to the applied load. That is the case for ductile metal systems. However, when hard phases and ceramic materials are involved, the IROM cannot be applied because the wear rate is non-linearly related to the load [18,23].

The other existing model is the linear rule of mixtures (LROM):

$$W = \sum V_i W_i. \tag{2}$$

The relationship gives a balanced value of the wear rate without a predominant effect of any one phase. In general, it was pointed out that LROM is suitable to describe a combination of very similar phases, while IROM better describes structures which consist of phases with very different properties [19].

Another important point is the particle–matrix interfacial bond strength. It can have the most significant effect in multiphase systems [19,20]. The above models are based on no phase interaction. In fact, important micro-mechanisms, like cracking, wear debris formation, decohesion and pulling out of the reinforcing phase, can occur at interfaces.

The relationship between wear rate and second-phase volume fraction is sum-marized in fig. 2, where the above-described models are compared with previously reported experimental results [19,20,24]. The LROM was confirmed by Zum Gahr

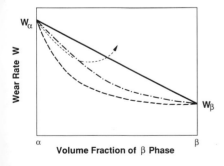

Fig. 2. Qualitative wear behaviour in various two-phase structures (schematic representation): ferrite–pearlite, LROM (———) [19]; ferrite–martensite, IROM (— — —) [19]; chromium cast irons–M_7C_3 carbides (·—·—·) [19]; chromium cast irons–M_7C_3 carbides (······) [24]; NiCrBSi W_2C (······) [20].

[19] to describe ferrite–pearlite behaviour in a two-body pin-on-paper abrasion test. In the same work, it was found that a ferrite–martensite system conforms better with IROM. Considering M_7C_3 carbides in high chromium white irons, Zum Gahr's results showed a modified IROM-type relationship in which an additional term related to interaction and interface effects is included. However, De Mello et al. [24], working with the same material, have found a minimum in the wear rate in a three-body wet abrasion test. As one can see in fig. 2, the qualitative representation of the two results shows a great difference in wear rate versus the second-phase volume fraction. This may be explained by different test conditions. Studying a NiCrBSi alloy reinforced with various carbides and oxides, Sandt and Krey [20] obtained a minimum-type relationship for fine silicon carbide abrasive in the pin-on-paper test, which is similar to De Mello's results. In this case, pulling out of the reinforcing phases probably became a predominant factor.

All these results are mainly correlated with the second-phase volume fraction. Although abrasive size effects and microstructure size effects were observed [20,21], they have not been specifically reported.

In order to better understand the mechanisms involved in the total tribological system, the interacting effects between the geometrical factors must also be analyzed. The total abrasive action on the multiphase system has to be considered as the macroscopic sum of all the microscopic effects produced by the individual abrasive grains. Each of these wear micro-events generates a groove in the material. It is expected that the reinforcing role of a dispersed phase may change as the groove size is smaller or bigger than the microstructural size.

This concept reflects the intensity of the abrasion phenomena. It takes into account abrasive properties such as hardness, shape and size, and it includes the load which the abrasive grain is subjected to on the surface. The relationship between microstructural detail and abrasive micro-events is shown in fig. 3 in schematic form. Condition 1 represents a small groove in comparison with the size

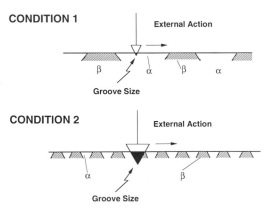

Fig. 3. Relationship between microstructure and groove size, considering an abrasive micro-event. Condition 1: groove size smaller than the microstructure. Condition 2: groove size equal to or larger than the microstructure.

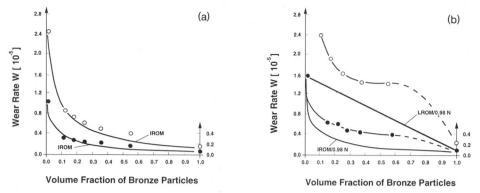

Fig. 4. Dimensionless wear rate versus reinforcing-phase volume fraction in epoxy–CuAl system (CuAl-particle diameter about 100 μm): (a) Fine silicon carbide abrasive; (b) Coarse silicon carbide abrasive (● = 0.98 N; ○ = 2.94 N in (a) and (b)).

of the microstructural phases (α and β). The abrasive particle wears both phases alternately in a quite independent manner. Interfacial effects are not expected to be involved to a great extent. The specific wear mode of each phase would essentially contribute independently to the total wear rate.

In condition 2, the microstructural phases are smaller than the groove size. Abrasive action is produced simultaneously on α and β and also on the interface. In this case, the interface may play an important role in the overall wear behaviour.

2.2. Abrasion of the epoxy–CuAl composites by SiC papers [16]

The choice of a polymer–metal system as experimental material was guided by several criteria. The first consideration was the ease of sample preparation with complete microstructural control. In addition, such a system is interesting because it combines very different materials with dissimilar properties and wear resistances, such as is normally the case for wear facing materials. Finally, the bond strength between the phases is relatively low (20–50 MPa). It was believed that because of this low strength, pulling out of the metal particles would occur when the interfaces are stressed. The wear rate W is reported versus Cu–Al volume fraction in figs. 4a and b for fine (46 μm grain size, surface grain density 235 cm^{-2}) and coarse (160 μm grain size, surface grain density 17 cm^{-2}) SiC abrasive, respectively. In the same graphs, the LROM and IROM are represented.

All the results show that the Cu–Al particles had a high reinforcing effect at low volume fraction, up to about 0.3. Furthermore, the wear rate decreased slowly, tending to approach a constant value. For the fine abrasive and a low testing load, the measured wear rates are in good agreement with the calculated IROM (see fig. 4a). It should be noted that the differences between the experimental results and the IROM curves increase slightly with increasing the second-phase (Cu–Al) volume fraction. For the coarse abrasive, large discrepancies with the IROM were observed, even for the lowest testing load (see fig. 4b). The measured wear rates were higher than predicted by the IROM model.

Analysis of the results at a given volume fraction shows that the load and abrasive size act in the same way. In fact, this is directly related to the type of the test used. For the pin-on-paper abrasion test, increasing the abrasive size diminishes the grain density on the paper. Consequently, for the same load condition, the load on each individual particle increases, allowing deeper penetration of the abrasive just as it occurs when the testing load is increased.

To understand the role of the second phase on a microscopical level, observations of abrasive micro-events were made. These showed that in case of the fine abrasive and the low load, the size of the grooves was smaller than both the Cu–Al particles and the mean free path λ. Each phase was worn as a result of its own individual response to the abrasive grain and no pulling out phenomena were observed in spite of the low interface bond strength. Therefore, these conditions were such that the results were in good agreement with the IROM.

When the depth of penetration increases up to the point where the groove size reaches the microstructural size, it was observed that pulling out of the reinforcing particles occurs as an additional contribution to the wear. In this case, the testing conditions were a coarse abrasive and a 2.94 N load; this led to a large discrepancy with the IROM in the overall test.

Thus the results obtained are qualitatively supported by the observations made from the short time tests. When no interaction or additional mechanism was involved, the results are in good agreement with the IROM. Alternatively, when the testing conditions cause the reinforcing particles to be pulled out, the wear rate is higher than predicted by the IROM. In this case, it was also observed that when the second-phase volume fraction was increased, the amount of pulled out particles also increased. This explains why the deviation from the IROM becomes greater when the second-phase volume fraction increases. Zum Gahr [21] obtained a similar behaviour with a steel fibre reinforced polyester composite, where the use of fibres led to an anisotropic microstructure. The modified rule of mixtures equation which described the composite's abrasive wear resistance best, was found to be:

$$W^{-1} = \left[V_{St}W_{St}^{-1} + V_P W_P^{-1} - V_{St}V_P \left(W_{St}^{-1} + W_P^{-1} \right) \right]. \tag{3}$$

2.3. Model validity and practical aspects

Like any other wear phenomenon, abrasive behaviour cannot be described using fundamental material properties. It results from the interaction between such properties and the wear environment [25]. This explains why experimental work is often only comparative and cannot be used to predict the wear behaviour in practice. In multiphase structures, the situation is complicated by the combination of phases and/or constituents with different properties. Starting from the wear rate of each constituent, the rules of mixtures only describe the contribution of each phase to the total behaviour. This can be considered as a simplified concept which assumes that the wear rate of each constituent was obtained under the same experimental conditions. Thus the rules of mixtures are not useful in the quantitative prediction of industrial wear rates, but instead are useful as a laboratory tool in the development of abrasion-resistant materials.

In addition, these rules of mixtures only describe the volume fraction effect without considering the microstructural size. The work by Simm and Freti [16] has shown that the relationship between the size of the reinforcing phase and the groove size which results from an abrasive micro-event, is an important parameter, which must also be considered. From a practical point of view, the reinforcing effect of a reinforcing phase seems to be more effective when the microstructure is larger than the groove size. This allows the hard phases to act as real barriers instead of only reducing the plastic deformation on the overall material surface. This is also in very good agreement with results obtained on the abrasive wear of (a) polyethyleneterephthalate (PET) composites filled with glass beads of various sizes [26], and (b) short-fibre reinforced PET with different fibre/matrix bond quality [27]. Furthermore, if the bond strength between the phases is a potential weak point, a coarse microstructure avoids high stresses at the interface.

Since the groove size depends on external factors, its practical prediction will not be easy. Nevertheless, the size of the reinforcing phase should be maximized in relation to the abrasive size. This should avoid the situation where the groove size approaches the size of the microstructure.

2.4. Abrasion of particle-filled polymers for dental applications [28]

2.4.1. Material compositions

Reinforcement of polymers by high-modulus fibres results in composite materials of high modulus with a tensile strength and toughness which can be greater than those of the component materials. Hard-particle reinforcement of glassy or crystalline polymers gives an increase in modulus but no improvement or a reduction in tensile strength and impact strength. Particle-filled polymer composites are thus suited to uses where a high modulus is needed, where loads are mainly compressive or where high-temperature creep resistance is important.

Dental filling materials are one such application. The requirements for a suitable material are that it can be inserted into a cavity as a viscous liquid or paste and will rapidly set. It should provide a good match to the natural tooth in modulus, thermal diffusivity and thermal expansion. Also, the material should be wear resistant and dimensionally stable, so that it fits tightly into the cavity without leakage. Amalgam is generally used to fill cavities in the occlusal (grinding) surfaces of posterior molar teeth, but has the disadvantage of being unaesthetic in front teeth. Silicate cements are used for filling front teeth but these have poor mechanical properties and slowly dissolve, so that the filling lifetime varies between 1 and 10 years but typically is only about 5 years. In the 1950s polymeric filling materials were tried but were unsatisfactory, as they wore rapidly, showed excessive polymerization shrinkage and had a coefficient of thermal expansion

TABLE 1
Properties of dental materials [28].

	Tensile strength (MN m^{-2})	Compressive strength (MN m^{-2})	Modulus (GN m^{-2})
Enamel	10	100–400	50
Dentin	50	200–350	12
Amalgam	60	400	
Silicate cement	3–7	150–200	20
Unfilled resin	2–3	75	2
Composite	2–3	150–250	5–12

which was much greater than that of teeth [28]. Subsequently, particle-filled polymer composites were developed with much improved properties. Current dental composites are mostly based on BIS-GMA [29], a condensate of bisphenol A and glycidyl methacrylate, developed by Bowen [30]. This dimethacrylate monomer, which has a viscosity of 1.4 N s m^{-2}, is mixed with other methacrylates, such as as glycol dimethacrylate, to reduce the viscosity, together with quartz and amine accelerator. Typically, the inorganic phase is 75 wt% or 50 vol%. Table 1 compares the properties of composites with other materials [28–30].

The wear properties of composite filling materials are of concern for two reasons. Firstly, it is preferable that a filling should have a smooth surface so that deposits and bacteria do not easily collect. To achieve this, the filling may be initially moulded against a smooth surface or polished with a fine abrasive, in either case subsequent abrasion may lead to a surface roughening. Secondly, composites have not yet proved satisfactory for use in the occlusal surfaces of molars and premolars (class I and II restorations) since they wear too rapidly. Craig and Powers [31] have reviewed clinical observations of the relative wear rates of filling materials and compared these with the results of laboratory wear tests. In general, laboratory tests correlate with clinical experience within any group of materials, but are misleading when comparing different classes, for instance composites, unfilled resins and amalgams.

2.4.2. Wear processes and model

The main problems with dental composite resins, as deduced from an analysis of the reasons given for replacement, are (a) insufficient wear resistance, (b) insufficient marginal integrity or sealing ability, which is suspected to lead to secondary caries formation, and, to a lesser extent, (c) insufficient colour stability. In vivo studies have shown that surface erosion and abrasion occurs in the contact-free areas (CFA) as well as in the occlusal-contact areas (OCA) of posterior composites. There are several methods available to analyze the wear and abrasion resistance in vitro, including: two-body abrasion, three-body abrasion, oscillatory wear and chewing simulation. A three-body abrasion apparatus capable of varied sliding action at the interface has produced good correlation with in vivo data [32]. The results of these studies were dependent upon the choice of the third body abrasive and the mechanical settings employed for speed and load. This method for evaluating the wear resistance of posterior composites shows promise as a predictive tool for different types of composites. However, the results are presently inconclusive due to the complexity of the wear process in the oral cavity.

Wear processes take many forms, but the processes most important for dental fillings presumably involve combinations of indentations and ploughing. This means that one is concerned with abrasive wear in which a hard particle ploughs or cuts a surface. In two-body abrasion the abrasive is in the form of a grit bonded into a matrix. Rabinowicz [33] models two-body abrasion of ductile materials as a ploughing process, which gives rise to an expression for the specific wear rate, R, i.e. volume removed per load and unit length of movement, as:

$$R = Kp/H, \tag{4}$$

where H is the indentation hardness, p is the applied contact pressure and K, the wear factor, depends on the particle shape and on the proportion of the material displaced from the ploughed furrow which is actually removed. K thus depends on the toughness and work hardening of the material abraded. Brittle materials apparently wear as a result of surface cracking. This process has been treated by Lawn [34] and gives a final equation similar to eq. (4).

In three-body abrasion the abrasive is in the form of loose particles between the sample and a backing plate. The wear rates also follow eq. (4) but the K values for metals are found to be one to two orders of magnitude lower [33]. Three-body abrasion is experimentally simpler in that the abrasive can be continually replenished and so is less subject to blunting and clogging, which can reduce the wear rates in two-body abrasion tests. Three-body abrasion of brittle materials has been extensively studied by Wilshaw and co-workers [35]. In the work by Prasad and Calvert [28] a simple equation was developed to describe abrasive wear of composite materials in terms of the wear properties of the individual components. A full understanding of the abrasive wear of composites would, however, require a full understanding of the abrasive wear of homogeneous materials. In the absence of this the authors formulated a rule of mixtures which can be expected to give, at least, approximate values for composite wear rates in terms of the wear rates of the separate phases.

The volume of material removed, ΔV, on abrasion of a composite should be the sum of the wear of the filler and matrix components:

$$\Delta V = \Delta V_f + \Delta V_m. \tag{5}$$

Wear studies on many systems have shown that the volume lost is proportional to the load and, after an initial stage, to the distance moved by the sample [33–35].

Given an apparent contact pressure p on a sample moved over a sweep distance L and wear rates per unit pressure R_f and R_m, the volume losses will be:

$$\Delta V_f = R_f L p A_f, \tag{6a}$$

$$\Delta V_m = R_m L p A_m, \tag{6b}$$

where A_f and A_m are the exposed areas of filler and matrix (as normalized to the total area), which will vary depending on the relative wear rates. It was assumed

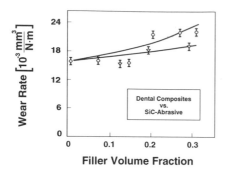

Fig. 5. Wear rates of quartz–PMMA composites by loose 4.5 μm silicon carbide abrasive. ($\triangledown$) Silane-treated quartz filler, (○) untreated filler. Full lines (———) fitted using eq. (9).

that at steady state the ratio of the loss rates from the filler and matrix is equal to the ratio of the volume fractions, $\chi/(1 - \chi)$:

$$\frac{\chi}{1 - \chi} = \frac{R_f A_f}{R_m A_m} = \frac{R_f A_f}{R_m (1 - A_f)}, \tag{7}$$

$$A_f = \chi R_m / [(1 - \chi) R_f + \chi R_m]. \tag{8}$$

Putting this into eqs. (6a) and (6b), one gets an expression for ΔV and hence for R_C, the wear volume of the composite per unit load and per unit sweep length:

$$1/R_C = (1 - \chi)/R_m + \chi/R_f. \tag{9}$$

This expression is equivalent to a series model for wear (IROM; cf. section 2.1) in which the two components are exposed successively to abrasion. This expression is not expected to be a complete description of composite wear. In particular if one must interpret the experimental observations in terms of particle pull-out and enhanced interfacial wear. However, eq. (9) does provide a base-line with which to compare experimental results; it is an "ideal" wear rate.

2.4.3. Wear results and model predictions

The wear rate of quartz-filled polymethylmethacrylate (PMMA with 45 to 75 μm quartz particles) against 4.5 μm silicon carbide abrasive increases with filler volume fraction as shown in fig. 5. If the particles are silane treated the increase is less marked. Crimes [36] and Hartley [37] found that the wear rate of quartz plates against 5 μm silicon carbide under similar circumstances was $5-9 \times 10^{-3}$ mm³/ (N m). Under the same conditions the wear rate of PMMA is 16×10^{-3} mm³/(N m). Thus according to eq. (9) a decrease in wear rate with filler content would be expected. Equation (9) can be fitted to the data as shown in fig. 5, using an apparent filler wear rate which is five times the measured rate for treated fillers and sixteen times the measured rate for untreated fillers.

Scanning electron micrographs suggest that the components are both wearing by chipping, since both are covered with small pits. The silica particles are somewhat recessed with respect to the surface as if they were wearing faster. The non-

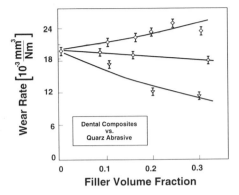

Fig. 6. Wear rates of composites by 10 μm quartz abrasive. (▽) Silan-treated quartz filler, (○) untreated quartz, (◇) untreated glass-bead filler. Full lines fitted using eq. (9).

silanated particles also show more gaps between the particles and the matrix and more places where particles have been pulled out. Thus silane treatment does seem to improve the particle–matrix bond.

Wear rates for composites against quartz abrasive are shown in fig. 6. In this case the wear rate drops markedly as silane-treated quartz particles are introduced into the resin. Wear rates of PMMA and quartz against the same quartz abrasive were 19.9 and 1.6×10^{-3} mm³/(N m), respectively. Using these values and eq. (9) one can predict an even more marked decrease in wear rate. Non-silanated quartz fillers give little improvement in wear rate and glass beads (diameter of glass-bead filler between 4 to 44 μm) give an increased rate. The wear rate for glass sheet against abrasive was 25×10^{-3} mm³/(N m), so that eq. (9) predicts a slight increase in wear rate with filler content, but not as great as is observed. Micrographs of the silane-treated composites show that the particles stick out above the matrix by a few μm whilst in the untreated composites the particles tend to be flush with the surface and occasionally are pulled out. The particle surfaces appear smoother than the matrix, except close to the edge of the particle. The surface-profile results confirm this. Glass bead composites again showed numerous fractures and pull-outs.

With calcium carbonate abrasive the wear rates decrease markedly with filler content and the effect is again greatest with the silanated quartz filler and least with the glass beads (fig. 7). In this case, the reinforcing quartz particles stand out clear of the matrix, since they are essentially unworn by the abrasive, therefore the predicted composite's wear rate by eq. (9) approaches a value of zero.

The wear results presented here show two regions of behaviour. With silicon carbide and silica abrasives the wear rate of the filler material is comparable to that of the matrix, and the predominant wear process is chipping. With the relatively soft calcium carbonate abrasive the filler wear rate is very small and composites wear by matrix loss and particle pull-out.

Using eq. (9) with the measured matrix wear rate it is possible to obtain an apparent filler wear rate. Values obtained for this apparent wear rate from the

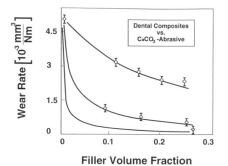

Filler Volume Fraction

Fig. 7. Wear rates of composites abraded by loose 3 μm calcite particles. ($\triangledown$) Silane-treated quartz filler, ($\bigcirc$) untreated quartz filler, ($\diamondsuit$) untreated glass filler. Full lines fitted using eq. (9).

fitted curves shown in figs. 4, 6 and 8 are given in table 2. The form of eq. (9) is in reasonable agreement with the filler volume fraction dependence of the composite's wear rate, but the apparent filler wear rates are greater than the rates measured on silica plates by a factor of about four for silane-treated silica filler and ten or more for untreated silica. Glass-bead fillers seam to behave similarly. The most likely explanation for this enhanced wear rate is that it is associated with more rapid chipping of the particles and matrix at the interfaces. With glass beads the wear seems to take the form of particles fracturing completely and being pulled out. The role of the silane in reducing composite wear is clearly to improve the

TABLE 2
Matrix, filler and composite wear rates [28].

Abrasive system	Particle size (μm)	Composite system	Wear rates (10^{-3} mm³/Nm)			
			Com-posite [a]	Matrix	Filler [b]	Filler in bulk
3-body, SIC	4.5	Untreated quartz/PMMA	23	16	130	5–9
3-body, SIC	4.5	Treated quartz/PMMA	19	16	35	5–9
3-body, quartz	10	Untreated quartz/PMMA	18	20	16	1.6
3-body, quartz	10	Treated quartz/PMMA	11	20	6	1.6
3-body, calcite	3	Untreated quartz/PMMA	0.4	5	0.1	
3-body, calcite	3	Treated quartz/PMMA	0.1	5	0.03	
3-body, quartz	10	Untreated glass beads/PMMA	24	20	80	25
3-body, calcite	3	Untreated glass beads/PMMA	2	5	0.8	
2-body, SIC	10	Sevitron, PMMA		259		
2-body, SIC	10	Smile, glass/bis GMA	1150	331	97	
2-body, SIC	10	Untreated glass/bis GMA	268	331	226	
2-body, SIC	10	Adaptic, quartz/bis GMA	75	366	42	
2-body, SIC	10	Untreated quartz/bis GMA	110	366	64	
SIC abrader	100	Adaptic, quartz/bis GMA	86	61	80	

[a] Results from this work, composite 30 vol% filler and results calculated from ref. [37], composite 50 vol% filler.
[b] Calculated using eq. (9).

TABLE 3
List of nominal and actual filler volume fractions in the different dental materials used for the sliding wear studies.

Material code	Filler particle size d (mm)	Nominal volume fraction V_f (%)	Actual volume fraction V_f (%)	Surface treatment	Symbols
55	–	–	–	Matrix	□
60 A	≈ 10	60	58.6	No silane	
60 B	≈ 10	50	47.9	No silane	
60 C	≈ 10	40	35.2	No silane	◇
60 D	≈ 10	30	25.4	No silane	
60 E	≈ 10	20	16.1	No silane	
60 F	≈ 10	10	5.3	No silane	
62 A	≈ 10	60	59.1	With silane	
62 B	≈ 10	50	49.4	With silane	
62 C	≈ 10	40	38.4	With silane	○
62 D	≈ 10	30	27.9	With silane	
62 E	≈ 10	20	18.5	With silane	
62 F	≈ 10	10	7.7	With silane	
64 A	2	55	55.3	No silane	
64 B	2	50	47.8	No silane	
64 C	2	40	39.6	No silane	
64 D	2	30	27.3	No silane	△
64 E	2	20	13.8	No silane	
64 F	2	10	3.4	No silane	
66 A	2	55	55.3	With silane	
66 B	2	50	49.9	With silane	
66 C	2	40	39.6	With silane	
66 D	2	30	30.6	With silane	▽
66 E	2	20	18.0	With silane	
66 F	2	10	9.2	With silane	
68 A	2	55	53.6	With silane [a]	▼
68 B	2	20	17.8	With silane [a]	

[a] Plus antioxidant (Trigonox 21).

particle-to-matrix bond so that the matrix supports the particle at the boundary and this reduces the tendency of these regions to chip off when loaded.

2.5. Particulate dental composites under sliding wear conditions

2.5.1. Material compositions

For the dental composites in this study, a polymethylmethacrylate (PMMA) based matrix (special composition DPMA/TEGDM = 67/33) was filled with different amounts of glass (Gd 83-type) particles. Two monomodal sizes, $d = 2$ and slightly less than 10 μm, respectively, were used [38]. A further subdivision is that

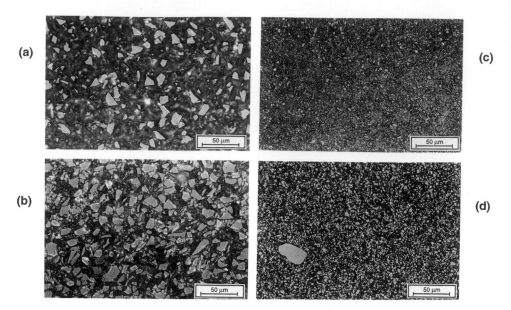

Fig. 8. Polished cross sections of dental composites containing different amounts of particles (V_f) with different sizes (d): (a) $d = 10$ μm, $V_f = 30\%$ nominal, (b) $d = 10$ μm, $V_f = 60\%$ nominal, (c) $d = 2$ μm, $V_f = 20\%$ nominal, (d) $d = 2$ μm, $V_f = 50\%$ nominal.

groups with and without silane coupling agent on the filler were tested. Table 3 shows the formulation details, i.e. composite identification code, nominal and actual filler volume fraction (as determined from density measurements with $\rho_{Filler} = 3.2$ g/cm^3 and $\rho_{Matrix} = 1.18$ g/cm^3), filler particle size and type of surface treatment. Cross sections of specimens having different amounts and sizes of filler particles are presented in fig. 8. These figures allow a comparison of the actual distances between the particles and the mean free paths between them, as calculated according to the equation for λ as given in fig. 1. The calculated values are listed in table 4, and they are in a reasonably good agreement with the average distances estimated from the polished cross sections.

2.5.2. Specific wear rates and sliding wear mechanisms

The sliding wear experiments with these dental composites were performed in a block on ring configuration against 100 Cr 6 steel with surface roughness value of $R_a = 1.6$ μm and $R_z = 7$ μm. As the composites were rather brittle (about $K_c = 2.5$ MPa $\sqrt{m}$), the loading conditions were chosen such that fracture of wear specimens under the applied pressure and surface shear due to the sliding counterpart could be prevented ($p = 0.1$ MPa, $v = 4$ m/s). Figure 9 illustrates the course of the specific wear rates of all the materials listed in table 3 as a function of filler volume fraction. Both, the 2 and 10 μm particle filled composites show, relative to the unfilled matrix, a little reduction in wear rate at low filler content (up to 10 vol%). Above this value, the composites' wear rates exceed that of the

TABLE 4

Relationship between particle distance λ and filler diameter as a function of filler volume fraction, according to the equation: $\lambda = 2d(1 - V_f)/3V_f$.

Filler volume fraction V_f (%)	Particle distance λ (μm)	
	$d = 2$ μm	$d = 10$ μm
5	25.30	126.50
10	11.97	59.90
15	7.54	37.74
20	5.32	26.64
25	3.99	20.00
30	3.10	15.50
35	2.47	12.40
40	2.00	9.99
45	1.63	8.14
50	1.33	6.66
55	1.06	5.45
60	0.89	4.44

PMMA-matrix and run through a maximum, which is very much higher and steeper for the 2 μm particulate composite than for the 10 μm system. After these maxima (which are at about $V_f = 20\%$ for the 10 μm and about 28% for the 2 μm composite) the values for both multiphase materials approach each other at a level comparable to that of the neat polymer (in the range above $V_f > 45\%$). Regarding the surface treatment of the fillers and the addition of an antioxidant, there was no systematic effect detectable in the case of the 2 μm particle system. Only in the maximum region of the 10 μm composite, the silane-treated materials resulted in slightly better values of the wear resistance.

An analysis of the worn surfaces, as performed by optical light and scanning electron microscopy, gives an idea about the types of mechanisms involved in material removal. Figure 10 illustrates for the 2 μm system that the macroscopic

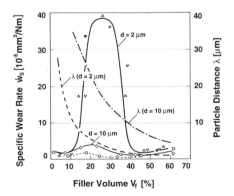

Fig. 9. Specific wear rates of dental composites as a function of actual particle content. In addition the course of the particle distances $\lambda(V_f)$ in the different systems is illustrated.

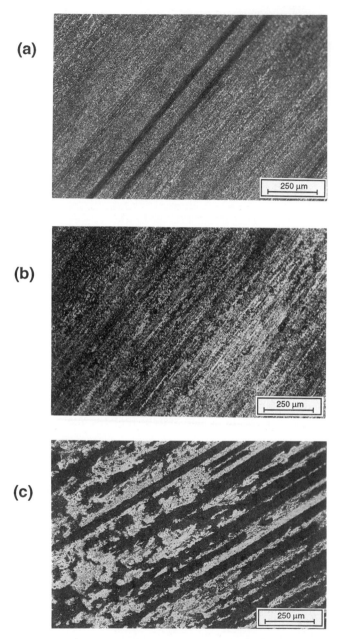

Fig. 10. Reflected-light micrographs of the worn surfaces of the 2 μm composites demonstrate the transition from striation patterns ((a) for $V_f = 20\%$ nominal and (b) for $V_f = 40\%$ nominal) to a striation/patchwork pattern ((c) $V_f = 55\%$ nominal).

appearance of the worn surfaces is dominated by a striation pattern (with orientation parallel to the sliding direction). This pattern is clearly visible up to a filler content of about 40% (figs. 10a and b). Above this range the striations are superimposed by a patchwork pattern consisting of regions (fig. 10c) which become more frequent and larger with further increase in filler content. The same features appeared also on the worn surfaces of the 10 μm system, except that the transition from striations to patches occurred already between 30 and 40% (nominal) for the non-silane-treated and between 20 and 30% for the silane-treated composite (figs. 11a to c). Higher magnifications of the different features by the use of SEM showed that the striation pattern consists of regions being very smooth, and others having a rough structure with longitudinal and transverse cracks at their edges with the smooth striations (figs. 12a and b). The latter also appear to be slightly higher in elevation that the rougher, severely worn ones.

An analogy to this observation was found in case of the patchwork pattern (figs. 12c and d). The patches consisted of flat, smooth regions, between which a fine grain structure of worn material was found. The more of these patches are formed and the better they get connected to each other, the more of this grainy material remains attached to the wear surface between these patches. The patches themselves seem to consist of back-transferred wear debris which got entrapped by the worn surface underneath and which in this way protects it temporarily. The fine cracks in these smooth layers indicate that part of the material will, after some time, separate again from these patches, become new wear debris, and eventually become back-transferred at another location. In case of the striations, which run from one end of the sample to the other, wear debris formed in the rougher regions gets almost totally removed from the surface at the end of the sample. Only at the edges of the smooth striations some of it can be temporarily collected, before it will be cracked and delaminated again. This leads, therefore, to a higher wear of the material than in the case where a dense patchwork pattern has built up on the material to be worn.

2.5.3. Model description of wear

The rather complex shapes of the wear curves are a result of various mechanisms which dominate the materials wear behaviour in the different ranges of filler volume fraction. The contribution of the individual mechanisms to the total wear of the composites is further controlled by the size of the filler particles and their adhesion to the polymer matrix. In case of the 2 μm system, the composite seems to follow in the range up to $V_f = 10\%$ an inverse rule of mixture (IROM), i.e. it is dominated by the simultaneous wear of the matrix with some improvement due to the harder, more wear-resistant filler particles. Since their distance λ is much greater than the maximum roughness value R_z, there is no particle/particle interaction. The wear surface mainly looks as the smooth striations shown in fig. 12a. When V_f becomes greater than 10%, λ is equal to or from 15% on smaller than R_z. At this point, strong interaction between the roughness profile of the steel counterpart and the densely packed, fine particles $(d < R_z)$ takes place. Ploughing of particles out of the surface starts to dominate the wear process, and

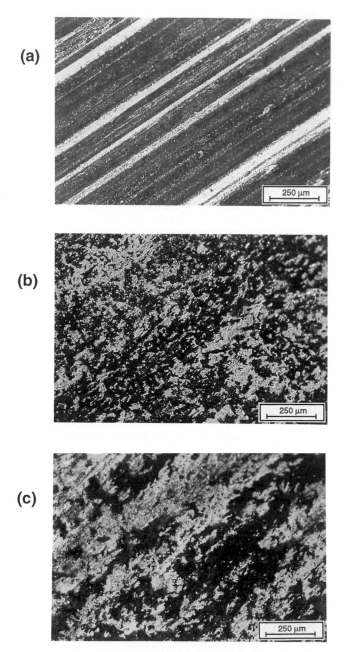

Fig. 11. Striations (a) on the worn surface of a 10 μm silane-treated composite at $V_f = 10\%$ (nominal) change to a pronounced patchwork pattern at $V_f = 40\%$ (nominal) (b). The size of patches increases with increasing filler content (e.g. $V_f = 50\%$, (c)).

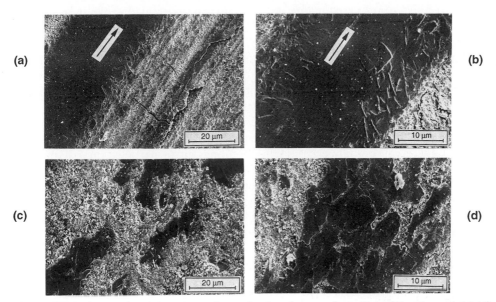

Fig. 12. SEM micrographs of the worn surfaces of a dental composite at $V_f = 10\%$ with rough (right) and smooth (left) striations ((a) and (b): arrow indicates sliding direction). In comparison, the patchwork pattern of smooth patches with grainy areas around (as it occurs at higher values of V_f) is shown in (c) and (d).

these particles can then act as third body abrasives to the material itself. Once this mechanism is locally initiated, it leads soon to the formation of the rough striations. The probability for this effect is the more pronounced, the more particles there are in the system. In addition, cracks formed at sharp edges of the particles can easily grow under some vibrational motion of the rotating steel ring. They can connect with each other due to the shorter particle distances, thus leading to surface fatigue effects and disconnection in the form of larger patches from the surface and their transport out of the striation at the end of the contact surface, i.e. enhanced wear of the composite. Near a V_f-value of 25%, however, a counter-mechanism starts to built up, which can be referred to as "patching" or "back-transfer" of worn material [39]. This is now possible, because the particle distance approaches the particle diameter $(2d > \lambda > d)$, which makes it more difficult for the counterpart asperities (roughness profile) to penetrate the surface and to initiate the ploughing mechanism. This results in the fact that worn matrix material, temporarily removed from the spaces between the particles, gets entrapped between the harder particles sticking with their tips out of the surface. At this point, the macroscopic appearance of the wear surface tends to change from a striation to a patchwork pattern. As the patches on the one hand protect the underlaying composite material from momentary wear, and at the same time create a free distance between the steel surface and the areas not covered with patches of back-transferred material, the wear is drastically lower than in V_f-cases where the ploughing mechanism dominates the process. Above $V_f = 45\%$ an equilibrium situation is reached, which consists of (a) simultaneous filler and matrix wear, as indicated in fig. 3a, (b) some loosening of particles due to surface fatigue, thus a small contribution of third body abrasion, and (c) the formation of patches on the surface for temporary wear protection. The result in this case is the reaching of a wear rate level similar to that of the neat matrix.

In case of the 10 μm-system, the same statements as made for the 2 μm system are valid for the region of low filler content ($V_f = 10\%$). Above this level ($10\% < V_f < 20\%$), there will be only occasionally a pulling out of particles (mainly because $d > R_z$ and $\lambda > R_z$); this can, however, also here create the formation of striations, which are rougher in nature than the smooth parts where the IROM is applicable. But the increase of the wear by this effect is very much lower than that seen for the 2 μm particles. Opposite to the statement made by other authors [16] in connection with fig. 3b, it is especially under this condition, i.e. where the size of the microstructural details (λ, d) is greater than the asperity size (R_z), of importance that the bonding between the components is well organized. Under strong filler/matrix bonding, pull out of harder particles by the roughness tips of the steel counterpart is much more difficult, and therefore the formation of third body abrasives as well (fig. 13a). This is the reason why the silane treatment in this part of the 10 μm curve showed a clear effect. On the other hand, it can be expected that a better bond quality is rather ineffective when the aspertiy size is much greater than the particles and when there is enough space between them, so that the asperities can locally penetrate the matrix deep enough so as to dig out the full particle with matrix around it (fig. 13b).

Due to the greater particle size and the remaining of most of them in the wear surface, the ploughing mechanism in the 10 μm-system is earlier replaced by the patching mechanism. Hard particles, which stick out of the worn PMMA matrix easily begin to capture worn material at their edges, thus forming a patchwork pattern under the condition that λ approaches the particle size ($2d \geqslant \lambda > d$) (fig. 13c). The optimum protection is reached when $\lambda \approx d$. This is again combined with a reduction in wear rate until an equilibrium level is reached, which is slightly lower than that seen for the 2 μm material.

In summary, the different trends in specific wear rate versus filler volume fraction due to the action of various mechanisms are schematically illustrated in fig. 14, along with the actual curves resulting from them. The complexity does not allow to describe these curves by a simple rule of mixtures; the only attempt which can be made to approach these trends mathematically is a semi-empirical relationship which includes (besides the rule of mixture components) special terms which account for the individual mechanisms dominating the composites' wear process in the various regions of V_f. A more simple development of such an equation has been performed by Friedrich and Voss [40,41] for sliding wear of short-fibre composites, and it is referred to this approach in section 4.3.2 of this chapter.

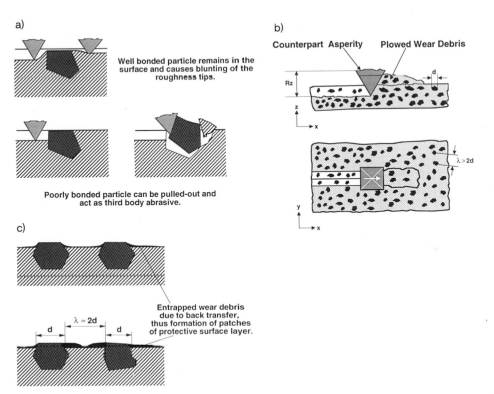

Fig. 13. Schematic illustration of the individual wear mechanisms as a function of filler size, bond quality and particle content: (a) effect of silane treatment on the interaction between counterpart asperities and large (10 μm) particles ($d > R_z$; $\lambda > R_z$), (b) ploughing of fine (2 μm) particles out of the surface ($d < R_z$; $\lambda > 2d$), (c) formation of a protective layer between embedded hard particles of short distance ($2d \geqslant \lambda > d$).

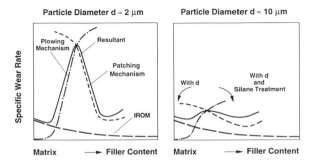

Fig. 14. Schematic illustration of the various trends due to the different wear mechanisms and the actual courses of curves resulting from them.

3. Models for sliding wear of continuous fibre reinforced composites

3.1. General remarks

Advanced composite materials, such as fibre-reinforced plastics (FRP) or fibre-reinforced metals (FRM), have received great technological and industrial attention in recent years. Among these composite materials, FRP has been potentially useful in many fields of industry because of its good mechanical properties, especially because of its high specific strength. The friction and wear properties of FRP have not been fully understood, although pioneering researches by Lancaster and Giltrow have shown a number of significant factors affecting the friction and wear of FRP [7,42–47]. Tsukizoe and Ohmae [48–56] have studied the friction and wear of unidirectionally oriented FRP in contact with carbon steel, and discussed the influences of volume fraction of fibres, kind of fibres, and the tribological anisotropy with respect to the sliding direction.

3.2. Rule of mixtures for the friction coefficient [54]

When a countersurface slides against an FRP surface, both, normal load, P, and the tangential force, F, are supported by the fibres and the matrix; therefore the friction coefficient μ can be given by

$$\mu = \frac{F}{P} = \frac{F_f + F_m}{P_f + P_m}, \tag{10}$$

where the suffixes f and m denote fibre and matrix, respectively. It is known further that

$$A_f = V_f A, \tag{11}$$

$$A_m = V_m A = (1 - V_f) A, \tag{12}$$

where A is the nominal area of contact, and V_f and V_m are the volume fractions of fibres and matrix, respectively. When a peeling-off of the fibres from the matrix, under the action of shear deformation, does not occur, one can assume that the shear strain γ_f is equal to γ_m:

$$\gamma_f = \gamma_m. \tag{13}$$

If G_f and G_m, the moduli of rigidity of the materials underneath the contacting surface, are equal, then the shear stress, τ, becomes constant, that is

$$\tau_f = \frac{F_f}{A_f} = \tau_m = \frac{F_m}{A_m}. \tag{14}$$

From eqs. (11), (12) and (14), then is obtained:

$$F_f = V_f F, \tag{15}$$

$$F_m = V_m F. \tag{16}$$

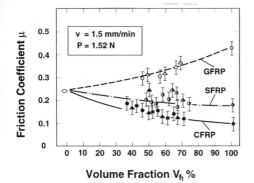

Fig. 15. Influence of the volume fraction of fibres on the friction coefficient of polyester composites (orientations □ = N, △ = AP, ○ = P, see fig. 19).

By taking into account the relation $P = P_f + P_m$, it follows:

$$\frac{F}{\mu} = \frac{F_f}{\mu_f} + \frac{F_m}{\mu_m}, \tag{17}$$

and finally

$$\frac{1}{\mu} = V_f \frac{1}{\mu_f} + V_m \frac{1}{\mu_m}. \tag{18}$$

From this equation Tsukizoe and Ohmae [54] were able to calculate the friction coefficient, μ, of FRP when the friction coefficients of the fibre, μ_f, and of the matrix, μ_m, are given.

When the FRP is hybrid-reinforced with two fibres f_1 and f_2, the law of mixtures for the friction coefficient is given by

$$\frac{1}{\mu} = V_{f_1} \frac{1}{\mu_{f_1}} + V_{f_2} \frac{1}{\mu_{f_2}} + V_m \frac{1}{\mu_m}, \tag{19}$$

where

$$V_{f_1} + V_{f_2} + V_m = 1. \tag{20}$$

The relationships between friction coefficient and volume fraction of fibres are shown in fig. 15. Figure 16 shows typical results of hybrid FRP. The thick solid, dotted and chain lines in these figures show the calculated friction coefficient either by eq. (18) or (19). Since there exists a good agreement between the theoretical and the experimental results, the assumption of eq. (13) might be reasonable for the friction of FRP at low sliding speeds and light normal loads, in which case fracture of FRP at the sliding surface does not occur.

From the results in figs. 15 and 16, it is clear that carbon fibre is the best reinforcement as far as the friction of FRP is concerned. The friction anisotropy depending on the fibre orientation relative to the sliding direction, cannot be

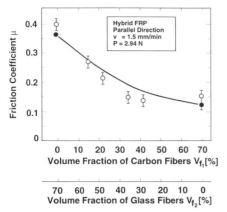

Fig. 16. Influence of the volume fraction of fibres on the friction coefficient of hybrid epoxy composites.

recognized in these figures. In fig. 16 when one utilized a medium hybrid FRP ($V_{f_1} = 35\%$, $V_{f_2} = 35\%$) instead of a carbon fibre FRP ($V_{f_1} = 70\%$, $V_{f_2} = 0\%$), there is only a slight increase in the friction coefficient. Thus for a practical application of FRP where friction becomes an important problem, the use of hybrid carbon FRP can be recommended with regard to the performance/cost ratio.

3.3. Wear properties

Tsukizoe and Ohmae's results of corresponding wear tests [49–53] have shown that the wear volume increases linearly with the sliding distance; no running-in period of wear has been apparent in their case. Thus it was possible to character-ize the wear in terms of a specific wear rate $\dot{w}_s$, which has as unit $\text{mm}^3/(\text{N m})$. Figure 17 summarizes wear performances of seven kinds of unidirectionally ori-ented FRP, the volume fraction of which was approximately 70% (see table 5). The figure shows the relationships between the specific wear rate and the friction

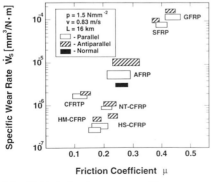

Fig. 17. Relation between specific wear rate and friction coefficient (P, AP, N, see fig. 19).

TABLE 5
Constitution of composites tested.

no.	Symbol of FRP	Fibre reinforcements	Resin matrices
1	HS-CFRP	High-strength carbon fibre	Epoxy resin or polyester resin
2	HM-CFRP	High-strength carbon fibre	Epoxy resin
3	NT-CFRP	High-strength carbon fibre (no surface treatment)	Epoxy resin
4	GFRP	E-glass fibre	Epoxy resin or polyester resin
5	SFRP	Stainless steel fibre	Epoxy resin or polyester resin
6	AFRP	Aramid fibre	Epoxy resin
7	CFRTP	High-strength carbon fibre	PTFE
8	Hybrid FRP	High-strength carbon fibre and E glass fibre	Epoxy resin

coefficient. As for the tribological anisotropy, it seems that every FRP has good wear-resistance in the normal and the parallel, but poor wear-resistance in the antiparallel sliding direction.

From the results in fig. 17, it is evident that HS-CFRP and HM-CFRP have a small specific wear rate of the order of 10^{-7} mm^3/(N m) and a low friction coefficient of 0.2. In contrast, GFRP and SFRP show a large specific wear rate of the order of 10^{-4} mm^3/(N m) and a high friction coefficient of 0.4. The lowest friction coefficient, of 0.1, was obtained for CFRTP. The good tribological properties of the carbon fibre reinforced plastics group (CFRTP, NT-CFRP and HM-CFRP) may be caused by the good mechanical properties of FRP, for instance, high Young's modulus and high interlaminar shear strength, as well as by the good tribological properties of the fibres, e.g., self-lubricating ability and high strength.

3.4. Wear mechanisms and correlations with other mechanical properties

From the experimental results and discussions in the previous sections, and from the scanning electron microscopical observations of worn FRP surfaces [9,48,50,53], Tsukizoe and Ohmae proposed the following model of wear processes: "The wear of FRP proceeds by
(a) wear thinning of the fibre reinforcements,
(b) subsequent breakdown of the fibres,
(c) peeling-off of the fibres from the matrix,
and a sequential occurrence of these processes governs the wear of FRP".

The essential factors affecting these three processes may be as follows:
(a) in the wear thinning of the fibre → load P and sliding distance L,
(b) in the breakdown of the fibre → strain $\mu p/E$ of FRP caused by the friction force, load P and sliding distance L,
(c) in the peeling-off of the fibre from the matrix → interlaminar shear strength I_s of FRP, strain $\mu p/E$ of FRP, load P and sliding distance L.

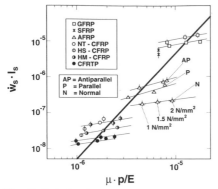

Fig. 18. Relation between $w_r I_s$ and $\mu p / E$: (P, AP, N, see fig. 19).

Therefore, the wear volume ΔV can be given by

$$\Delta V = f\left(\frac{\mu p}{E}, \frac{1}{I_s}, P, L\right). \tag{21}$$

As the first-order approximation, one can assume

$$\Delta V = k\frac{\mu p}{E}\frac{1}{I_s}PL, \tag{22}$$

where k is a dimensionless constant. The the specific wear rate $\dot{w}_s$ can be written as

$$\dot{w}_s = k\frac{\mu p}{E}\frac{1}{I_s} \tag{23}$$

or

$$\dot{w}_s I_s = k\frac{\mu p}{E}, \tag{24}$$

where $\dot{w}_s I_s$ and $k\mu p/E$ are both dimensionless quantities. Figure 18 shows the relation between $\dot{w}_s I_s$ and $k\mu p/E$ under normal pressures p of 1, 1.5 and 2 N/mm² for seven kinds of FRP. The solid line in this figure shows the experimental values for $p = 1.5$ N/mm², and the wear equation at a constant normal pressure p is given by

$$\dot{w}_s I_s = \alpha\left(\frac{\mu p}{E}\right)^\beta. \tag{25}$$

The values of α and β may be calculated form the experimental line in fig. 18, and the experimental wear equation for FRP can be written as

$$\dot{w}_s = 1.40 \times 10^{10}\left(\frac{\mu p}{E}\right)^{3.08}\frac{1}{I_s}. \tag{26}$$

TABLE 6
Composition and mechanical properties of composite specimens.

Ply fraction [a]	Total fibre volume fraction [b] (V_F)	Individual fibre volume fraction [c]		Flexural strength (GPa) [d]	Tensile modulus (GPa)	Apparent interlaminar shear strength (MPa) [e]
		V_g	V_c			
100% g	0.55	–	–	2.02	53	82
2/1 g/c	0.59	0.39	0.20	1.07	86.5	88.6
1/1 g/c	0.57	0.295	0.257	1.10	111	84.7
100% c	0.55	–	–	1.34	156	56.5

[a] g – glass fibre composite, c – carbon fibre composite.
[b] Measured by HNO_3 dissolution.
[c] By combustion of residue from footnote b.
[d] Maximum stress in 3-point bending, beam span/depth = 25.
[e] As for footnote d but beam span/depth = 4, width/depth = 1.

When Young's modulus, E [MPa], and the interlaminar shear strength, I_s [MPa], of FRP are given and the friction coefficient, μ, between the FRP and a carbon steel is known, one can estimate the specific wear rate $\dot{w}_s$ in mm³/(N m) under the normal pressure $p = 1.5$ N/mm² from the wear eq. (26).

3.5. Sliding wear of hybrid carbon / glass fibre epoxy composite materials [57]

3.5.1. General ideas and experiments

The objective of the study by Hawthorne [57] was to determine if hybridization can provide benefits for composite tribological properties. Consequently, investigations have been made of the friction and wear behaviour of unidirectional single-fibre and mixed carbon and glass fibre/epoxy materials when sliding, with various fibre orientations, against a steel counterface.

Composite plates were fabricated by laying-up multiple plies of either unidirectional glass or carbon fibre "prepreg" tapes, or both together in a sequence to give the required hybrid, and curing in an autoclave. The tapes consisted of 60 wt% of either high-strength glass fibres (S-2 glass, $E \approx 86$ GPa) or high-modulus carbon fibres (Celion GY 50, $E \approx 330$ GPa) preimpregnated with the same epoxy resin. The composition and mechanical properties of the four composites are summarized in table 6. Test specimens were cut from the fabricated plates having the geometry and fibre orientations at the sliding surface as illustrated in fig. 19.

A pin-on-disc tribometer with a self-aligning pin holder was used to push three identical specimens at a time with equal load against the flat surface of a rotating counterface. The specimens were spaced 120° apart on a (mean) circle diameter of 3.8 cm, and their initial unsupported length was 2.4 mm. The initial roughness of all the ground 304 stainless steel counterfaces was 0.12 ± 0.05 mm perpendicular to the lay and 0.10 ± 0.04 mm parallel to the lay. The sliding speed was 0.5 m/s and the total normal load was 93 N (1.94 N/mm² nominal bearing stress). The wear was determined by weight loss measurement on cleaned and dried pin

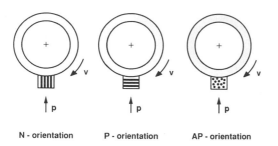

N - orientation   P - orientation   AP - orientation

Fig. 19. Designation of fibre orientation relative to the sliding direction.

specimens at 5 km sliding distance intervals. The friction torque was monitored continuously and the pin temperatures were sensed by fine thermistors (0.6 mm diameter) at 2.3, 1.5 and 0.7 mm (nominal), respectively, from the original bearing surfaces of the three specimens. Extrapolation of these three temperatures provided estimates of the mean interface temperatures.

3.5.2. Experimental results

The friction coefficients of most composites on the ground steel surfaces increased from < 0.2 to relatively steady higher values in less than 1 min (i.e. < 30 m sliding or < 250 cycles). Some, especially those with normally oriented fibres, gave more erratic results, and there was considerable variation of values for similar specimens. Typical friction result trends for equilibrium pin temperatures near the sliding interface were approached in about an hour (≈ 1.8 km or 15 000 cycles). The estimated mean interface temperatures for all specimens tested ranged between 50–80°C, but no consistent correlation with composite fibre orientation or composition was discerned. Fine wear debris collected on the counterface wear track and, for the faster wearing specimens, as loose material adjacent to the track.

Opposite to the results from Tsukizoe and Ohmae [54], as reported on in one of the previous sections (3.3), all composites in Hawthorne's study [57] exhibited initial "running-in" wear behaviour, although to differing extents, progressing towards steady-state wear rates. Prolonged testing of some specimens, especially those with fibres oriented normal to the counterface, showed a tendency to develop unstable sliding with vibration. This produced pin surface break-up and consequently unreliable assessment of the wear, so most tests were discontinued around 20 km. Wear rates of all the composites at 15 km sliding distance are shown plotted against their fibre composition (i.e. the relative proportion of carbon and glass fibres) in fig. 20.

Optical photomicrographs of some typical worn pin surfaces illustrated the progressive severity of wear for all three fibre orientations in the series carbon → hybrid → glass fibre composites. For carbon fibre specimens with any orientation, the sliding wear tracks showed only slight alteration of the original ground surface. The more abrasive glass fibre composites completely removed the original grinding

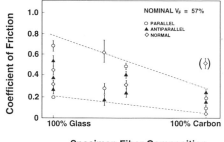

Fig. 20. Sliding friction results for specimens with different fibre composition.

marks while the hybrids produced intermediate wear tracks with some scoring (varying in extent with fibre orientation). Examinations by scanning electron microscopy (SEM) have shown many of the detailed features reported by others with single-fibre composites [9,11,16] such as wear-thinned fibres, transferred and back-transferred layers, nature of debris, etc., and are not discussed further. The main feature typically observed by SEM on the worn surfaces of hybrid composites is a great disparity in the extent of wear of the glass and carbon fibre laminae for the parallel-fibre specimens. Antiparallel-fibre specimen surfaces were relatively smooth, with some scoring and discrete worn patches (the latter more numerous on the glass fibre laminae) both evident. No clear distinction between laminae could be seen for hybrid specimens with the fibres oriented normal to the counterface. Surfaces of these specimens typically exhibited rough worn and scored areas besides much smoother areas.

3.5.3. Hybrid wear model based on the elastic properties of the constituents

The carbon fibre specimen wear rates, at $\approx 10^{-7}$ mm^3/(N m), agree well with results reported for CFRP sliding against various steel counterfaces [8,10,11]; the friction values in fig. 20 are also compatible with previous results [8,11]. Present glass fibre composite wear rates, however, are an order of magnitude smaller than values obtained for sliding against a rougher mild steel counterface [11]. There are also small wear rate differences between the three fibre orientations, but their ranking is not the same as those reported for sliding on other counterfaces [8,11].

For both single-fibre and hybrid-fibre composites the decrease from high initial wear rates to limiting values is consistent [16] with the evidence for alteration of the counterface topography along the wear tracks by asperity removal and transferred layers. For the hybrids, those with fibres oriented parallel to the sliding direction exhibit the least amount of "run-in" and the total wear and their steady-state wear rates are closer to the CFRP rates than those of GFRP. This could be explained by early formation of the wear pattern where the carbon fibre laminae become the effective bearing surface and thus mainly determine the hybrid sliding wear characteristics. When, or precisely how, this duplex pattern

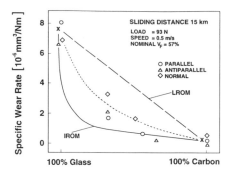

Fig. 21. Variation of the specific wear rate.

develops has not been ascertained, but the known superior fatigue resistance and less abrasive nature of CFRP compared with GFRP are probably the cause.

After "run-in" the transverse oriented 1/1 hybrid-fibre specimens have the lowest equilibrium wear rates, not much larger than those of the 100% carbon composites. Together with the fact that the relatively smooth wear surfaces resemble closely those typical of worn CFRP specimens, this points to the dominating influence of the carbon fibre layers once again. Since the low wear rate of CFRP is closely associated with the formation of lubricating transferred films, it is suggested that this also applies to the transverse-fibre hybrids. The small areas of attrition of these hybrid surfaces are most likely the result of sliding fatigue wear [58]. The poorer fatigue resistance of GFRP would then account for the greater incidence of these areas on the glass fibre laminae.

The hybrids with fibres oriented normal to the counterface exhibit the largest wear rates, but their worn surface appearance suggests only that wear may involve mild abrasion or surface fatigue processes [16,59]. Transfer films from the carbon fibre laminae may play a role here also, since many smooth areas were found, with no back-transferred layers, which is characteristic of thin transfer film formation in CFRP [16]. An additional factor may be important with these hybrids, however, as discussed in the mechanical analysis below.

The results in fig. 21 indicate that all the hybrid carbon/glass fibre composites exhibit lower wear rates than those expected from a linear interpolation (LROM) between the single fibre composite values. For example, substituting 33% of the glass fibres in GFRP by carbon fibres reduces the wear rates with between 55–80%, depending upon fibre orientation. Similar comments apply to the initial wear rates, suggesting that such large reductions are determined more by the type of composite than by the sliding-generated topography on the counterface. This greater reduction of wear rates could thus be regarded as a positive effect in the wear behaviour of these materials (although the reduction is not as high as one would expect from the inverse rule of mixtures IROM, i.e. the full drawn line in fig. 21).

Since the wear in these composites is mainly determined by the fibres [6,7], the dashed lines in fig. 21 would define hybrid composites wear rates if these were made up of sequential contributions from the CFRP and GFRP materials in simple proportion to their fractions in the hybrid, i.e.

$$W_H = V_c W_c + V_g W_g. \tag{27}$$

Clearly this is not the case, and another possibility is considered that proportions the wear rate contributions from the two boundary single-fibre composites in a manner defined by the constant axial strain conditions existing at the composite bearing surface. Thus for the three-phase hybrids,

$$\varepsilon_H = \varepsilon_c = \varepsilon_g = \varepsilon_m. \tag{28}$$

For normal fibre orientation $\varepsilon = \sigma/E$, and since $P^* = A\sigma$, the load any component of length l is given by

$$P = V\varepsilon E = P^*l. \tag{29}$$

The total load on the composite can be described as

$$P_H + P_c + P_g + P_m \tag{30}$$

or

$$P_H = \varepsilon(V_c E_c + V_g E_g + V_m E_m). \tag{31}$$

Since $E_m \ll E_c$ or E_g, eq. (31) reduces to

$$P_H = \varepsilon(V_c E_c + V_g E_g), \tag{32}$$

i.e. all the load is borne by the fibres.
Then

$$P_c/P_g = V_c E_c/V_g E_g, \tag{33}$$

or

$$P_c = (V_c E_c/V_g E_g)P_g, \tag{34}$$

leads to

$$P_H = P_c + P_g = P_g(1 + V_c E_c/V_g E_g). \tag{35}$$

Thus

$$P_g = \left(\frac{V_g E_g}{V_g E_g + V_c E_c}\right)P_H \tag{36}$$

and

$$P_c = \left(\frac{V_c E_c}{V_g E_g + V_c E_c}\right)P_H, \tag{37}$$

i.e. the load on the fibre phases is distributed according to the bracketted fraction of the total load on the composite. Now the assumption is made that the wear of both CFRP and GFRP has about the same load dependence over the range appropriate to this analysis, so that the same proportioning can also be applied to the wear rates of the two fibre phases:

$$W_H = \left(\frac{V_c E_c}{V_g E_g + V_c E_c}\right) W_c + \left(\frac{V_g E_g}{V_g E_g + V_c E_c}\right) W_g. \tag{38}$$

That is, the higher Young's modulus of the carbon fibres ensures that this phase bears most of the normal load and, thus, also provides the main contribution to the hybrid composite wear rate. A more or less similar analysis has been applied to the sliding friction behaviour of single-fibre composites [13].

Theoretical hybrid composite wear rates, calculated from eq. (38) using data for the normal oriented single-fibre composites, are shown by the dotted curve in fig. 21. Comparison with the experimental data suggests that this model accounts quantitatively reasonable for the wear of these hybrids. It does not, of course, apply to the unidirectional hybrids with fibres lying in the sliding plane. However, as discussed earlier, the greater influence of the carbon fibres in these hybrids is qualitatively accounted for by other factors.

No evidence for any sub-surface plastic deformation, such as has been reported in metal matrix composites [59], was observed on polished cross sections of the worn specimens of normal oriented fibres after stable sliding tests. Specimen chatter due to stick–slip behaviour and some surface "brooming" of normal oriented fibre composite pins did occur after unstable sliding developed. Again no sub-surface cracking, which might lead to delamination wear [10], has been detected in any specimens examined.

It is noted that composite wear rates bear little relationship to either their measured flexural or interlaminar shear strengths. Since the composite's Young's modulus is linearly dependent upon the specimen composition (table 6), wear rates do relate to the modulus in the same way as does the composition (fig. 21). Finally, the above discussion presents many simplifications of what are complex wear phenomena and further study is required, particularly in the early stages of sliding, to clarify the wear mechanisms of such hybrid composites. The behaviour of intimately mixed fibre composites, with these or other fibres, may also be worth investigating.

4. Systematic development of polymeric hybrid composites for high wear resistance

4.1. Ideas and objectives

Hawthorne [57] reported on the sliding wear performance of glass- and carbon fibre hybrid composites against steel. He found that a reasonable percentage of the more expensive carbon fibres could be replaced by the much cheaper glass fibres without losing much of the high wear resistance measured for the composites reinforced with only carbon fibres. A similar result was found by Tsukizoe and Ohmae [54], who tried to describe the coefficient of friction and the wear behaviour of hybrid composites by a modified rule of mixtures. In addition to these studies of fibre hybrid composites, there exists a small number of papers that deal with the effects of different polymer/polymer combinations on the specific wear rate of polymeric hybrid composites sliding against steel counterparts (e.g. Briscoe et al. [6]) or with the influence of different types of fibres in a particular polymeric matrix on the wear performance of monolithic composites [61].

All these papers have in common that some types of fibres with particular orientation can result in greater improvements of the composites' wear resistance than others, which may be more favourable, however, when they are positioned under a different orientation relative to the wear direction. Bringing the favourable orientations and types of fibres together, may even lead to further improvements or to so-called "synergistic" effects. Since this had, however, been shown only for a few fibre/matrix combinations, a systematic study was performed quite recently, which had as objectives

(a) to study the wear performance of quite a number of the major fibre/matrix combinations in a more systematic was as it was done before,

(b) to develop from this information hybrid composites, which expected to have, if possible, better wear resistance than that which could be expected from the simple rule of mixtures of the wear rates of the different monolithic constituents.

The efforts were concentrated only on the wear rate of these materials because, in most tribological situations, it is the wear life of a component rather than its coefficient of friction that determines its applicability.

TABLE 7
Unidirectional composites employed in wear tests.

Set	no.	Abbrev.	Material	Fibre Vol%	Density (g/cm^3)	Manufacturer
1	1	EP	3501-6 Epoxy	–	1.28	Hercules
	2	CF-EP$_1$	AS4 Carbon fibre-EP	62	1.59	Hercules
	3	GF-EP$_1$	E-Glass fibre-EP	58	2.01	Fiberite
	4	AF-EP$_1$	K49 Aramid fibre-EP	60	1.34	Fiberite
2	5	PEEK	Polyetheretherketone	–	1.27	ICI
	6	CF-PEEK	AS4 Carbon fibre-PEEK	55	1.56	ICI (APC-2)
	7	GF-PEEK	E-Glass fibre-PEEK	61	2.04	ICI
	8	AF-PEEK	K49 Aramid fibre-PEEK	60	1.38	ICI
3	9	am.PA	am.Polyamide (J-polymer)	–	1.17	Du Pont
	10	CF-PA	AS4 Carbon fibre-PA	54	1.55	Du Pont
	11	AF-PA	K49 Aramid fibre-PA	67	1.37	Du Pont
4	12	CF-PA66	AS4 Carbon fibre-PA66	20	1.27	ICI (Plytron)
	13	GF-PA66	E-Glass fibre-PA66	31	1.57	ICI (Plytron)
	14	AF-PA66	K49 Aramid fibre-PA66	40	1.26	ICI (Plytron)
5	15	CF-EP$_2$	Carbon fibre-Epoxy	63	1.61	Hebrew U.
	16	AF-EP$_2$	Aramid fibre-Epoxy	62	1.38	Hebrew U.

4.2. Basic materials and testing procedures

Different sets of unidirectionally oriented, continuous-fibre composites were used to obtain a fundamental approach to find the wear rates of monolithic composites (table 7). Materials in sets no. 1 and 2 were 3 mm thick plates from a previous study. Details of the manufacturing procedure are described in ref. [62]. Thin tapes of continuous fibres impregnated with a thermoplastic matrix (materials in sets no. 3 and 4) were manufactured into plates of 3 mm thickness by hot pressing. A typical cross section of such a plate, made of carbon fibres in an amorphous-polyamide matrix (J 2-polymer, Du Pont), is shown in fig. 22.

The monolithic composites listed in table 7 as set no. 5, were directly obtained as pre-cured plates [63]. They were post-cured in our laboratory over a period of 2 h at a temperature of 180°C.

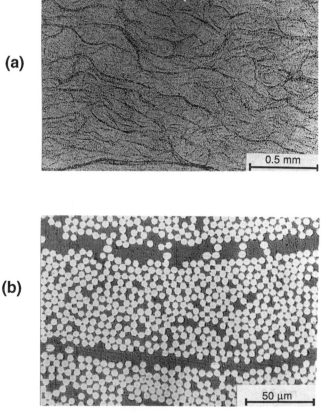

(a)

0.5 mm

(b)

50 μm

Fig. 22. Cross section of a unidirectional carbon fibre/amorphous PA composite after press-forming of 5 mm, preimpregnated tows (upper photograph). The thickness of the tows (about 0.1 mm) is reflected in the enlargement of the cross section (lower photo).

A description of the multiple-pass sliding wear studies, as performed with a standard pin-on-ring apparatus, is given in several previous papers [64–66]. Briefly, the unidirectional (UD) composite samples are pressed against rotating steel rings (German Standard 100 Cr 6; initial roughness $R_a = 0.14$ μm) with three principal fibre orientations: normal (N) to the sliding direction, in-plane and parallel (P) to the sliding direction, and in-plane but perpendicular (AP) to the sliding direction (fig. 19). All tests were carried out at room temperature in a laboratory environment and under pressure/velocity conditions of $pv = 1.7$ MPa m s^{-1} ($v = 1.5$ m s^{-1} in sets no. 1, 2, 3; $v = 3$ m s^{-1} in sets no. 4 and 5). This particular pv-condition was chosen for two reasons:

(a) so as to have a better basis for comparing the results of this study with similar tests on a wide number of short-fibre reinforced thermoplastics carried out in the past [41,67,68] and

(b) because the wear testing facility used allowed only a limited range of pressure/velocity combinations when operating in a satisfactory manner, e.g. without vibration effects, which might have added to the actual material removal due to sliding wear contacts.

Furthermore, it can be assumed that, at the chosen pv-level, the tribological performance of the matrix is the same with and without fibre reinforcement. This is perhaps not true at much higher pv-conditions. Here, the matrix may wear differently without any fibre reinforcement (owing to the higher probability of creep effects).

The specific wear rates, $\dot{w}_s$, were derived (after reaching steady-state wear conditions) as:

$$\dot{w}_s = \frac{\Delta m}{AL\rho p}, \tag{39}$$

where Δ_m is the mass loss, A the apparent contact area, p the apparent contact pressure, L the sliding distance and ρ the material density. The inverse of the specific wear rate is referred to as the material wear resistance.

In addition to the sliding wear studies, some of the materials listed in table 7 were also subjected to severe abrasive wear loading conditions by the use of SiC papers with different grain size. Typical abrasive wear testing procedures are described in ref. [69].

4.3. Wear rates of monolithic composites

4.3.1. Sliding wear against steel

Reinforcement of a polymeric matrix system by continuous fibres normally results in a reduction of the material wear rate. The relative improvement in wear resistance is, however, highly influenced by both the type of polymer matrix and the type of fibre used as reinforcement. This fact is demonstrated in fig. 23 for the first three sets of materials listed in table 7 [70]. The wear rates of the different

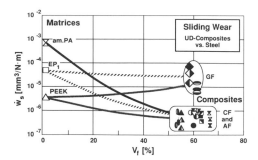

Fig. 23. Matrix and fibre effects on the sliding wear of unidirectional composites. The various symbols refer to different fibre materials and orientations in the different matrices. The exact values are given in the following figures.

compared matrix materials vary over a range of almost three orders of magnitude. The incorporation of carbon or aramid fibres brings the wear rates closer together (within one order of magnitude). In all cases, both types of fibre resulted in an improvement in wear behaviour, although this was much more pronounced for the rapidly wearing amorphous polyamide than for the more wear resistant poly(ether ether ketone) (PEEK). Glass fibre composites exhibited 10- to 100-fold higher wear rates than the carbon or aramid fibre composites, and they wore even faster than the unreinforced matrix when the latter possessed a high wear resistance (PEEK, for instance).

The different symbols used for the composites in fig. 23 refer to the different fibre orientations tested. The results for the various materials are presented in the histograms given in figs. 24–26. Regardless of the type of matrix used, it became evident that the carbon fibre composites tested in the P orientation exhibited the lowest wear rates (fig. 24). The reverse situation occurred in the case of the aramid fibre composites, which resulted in minimum wear rates when the fibres were placed normal to the sliding plane and direction (fig. 25). Figure 26 illustrates that the glass fibre composites possessed wear rates that were more than one order of magnitude higher than those of the other materials; they also showed a slight

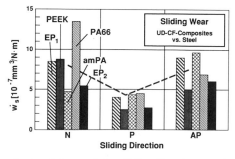

Fig. 24. Effects of the sliding direction on the wear of unidirectional carbon fibre composites with different polymer matrices.

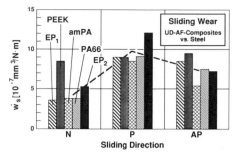

Fig. 25. Effects of the sliding direction on the wear of unidirectional aramid fibre composites with different matrices.

superiority of the P direction relative to the AP and N directions. It should be noted at this point that, within one block of fibre orientation (N, P, or AP) in each of the three different histograms, the different fibre volume fractions in the various composites tested are not considered, although it is known that this parameter can very effectively influence the wear performance of a polymer composite as well [13]. On the other hand, systematic studies of the effect of the fibre volume fraction on the sliding wear of short-fibre reinforced thermoplastics have shown that the greatest effects resulting from fibre reinforcement occur within a range of 0–20 vol% of fibre loading, whereas above this value the effects are much less pronounced [64,71].

In table 8, all results obtained with the different matrices and composites, worn under different orientations, are summarized in terms of their wear resistance ratios (wear resistance of the material normalized to that of the unfilled epoxy). The higher this quantity is above unity, the better is the composite in this particular direction in comparison to the isotropic standard (unreinforced epoxy). All ratios above an arbitrarily chosen value of 90 (i.e. 90 times higher sliding wear resistance than epoxy) are printed in large characters for better comparison. With the exception of aramid fibre reinforced PEEK in the N orientation (for which some fibre or interface damage during the high-temperature treatment of PEEK in

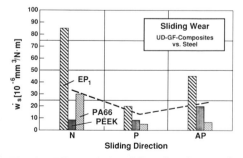

Fig. 26. Effects of the sliding direction on the wear of unidirectional glass fibre composites with different matrices.

TABLE 8
Wear-resistance ratios as normalized to that of neat epoxy resin (EP$_1$: $\dot{w}_s^{-1} = 1.79 \times 10^4$ N m/mm^3).

Matrix	Orientation	CF	GF	AF	Unreinforced
EP$_1$	N	65.4	0.5	122.4	1
	P	99.8	2.1	44.1	
	AP	44.1	1.1	46.3	
PEEK	N	65.1	5.1	61.4	11.8
	P	158.1	5.5	46.1	
	AP	94.8	2.0	44.1	
amPA	N	117.7	–	145.5	0.07
	P	128.6	–	65.1	
	AP	57.0	–	102.4	
PA66	N	41.3	1.8	145.5	27.7
	P	122.9	16.3	60.8	
	AP	80.2	8.9	73.7	
EP$_2$	N	102.4	–	104.3	1
	P	197.5	–	45.7	
	AP	90.7	–	75.8	

the manufacturing process cannot be excluded [80]), the following conclusions can be drawn from table 8:

(a) carbon fibres in parallel orientation and/or aramid fibres in normal orientation have resulted in the best wear resistance values of the composites (regardless of the type of matrix);

(b) carbon fibres in the AP or N direction may also sometimes lead to good results;

(c) glass fibres are ineffective relative to carbon or aramid fibres; if, however, glass fibres should be incorporated in the composite for other reasons, their arrangement should be in the P orientation, or, less ideally, in the AP direction, but not in the N orientation;

(d) among the unreinforced matrices, tougher semicrystalline thermoplastics (e.g. PEEK and polyamide 6.6) result in higher wear resistances than amorphous PA or a cross-linked more brittle epoxy resin.

4.3.2. Sliding wear model for monolithic fibre composites

Recent papers by Voss and Friedrich [64,71] have shown that one can assign partial rates to all of the wear processes described in fig. 27, as discussed already in section 3.4. It is assumed that (3) and (4) occur sequentially and can therefore be considered as a combined process ($W_{s,Fci}$). The removal of fibres also caused an additional matrix wear process because these fibre particles can act as third body abrasives, as can be deduced form the wear surface micrographs. All these mechanisms which are not due to pure sliding, are included in $W_{s,Fci}$. Hence, the wear rate of the composite ($W_{s,C}$) is the sum of a wear rate which accounts for the sliding processes ($W_{s,s}$) and one which accounts for the additional wear mecha-

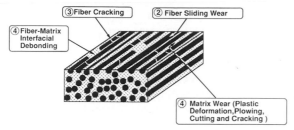

Fig. 27. Schematic illustration of the microscopic wear mechanisms observed on the wear surfaces of the composite materials under sliding wear conditions.

nisms ($W_{s,Fci}$). Each partial wear rate is assumed to be a function of the fibre volume fraction, V_F.

$$W_{s,C} = W_{s,s}(V_F) + W_{s,Fci}(V_F). \tag{40}$$

The sliding part, $W_{s,s}$, is in itself a mixture of matrix (M) and fibre sliding (Fs). Their contributions are proportional to the exposed area of these components relative to the area of the whole sliding surface:

$$\frac{1}{W_{s,s}} = \left(1 - \frac{A_F}{A}\right)\frac{1}{W_{s,M}} + a\left(\frac{A_F}{A}\right)\frac{1}{W_{s,Fs}}, \tag{41}$$

where A_F is the exposed area of fibre and A is the total sliding surface area ($A_F/A = V_F$). The inverses of the wear rates are summed since the same wear mechanism acts simultaneously on two separate phases of the material [72]. A factor a has been introduced, which describes the fraction of the fibre surface worn by sliding. A second factor, b, accounts for the part of the fibre surface which si removed by fracture and interfacial debonding. The following conditions must hold:

$$a + b = 1, \tag{42}$$

$$W_{s,Fci}(V_F) = b\frac{A_F}{A}W_{s,Fci}. \tag{43}$$

By substituting eqs. (41) and (43) into eq. (40) one can write:

$$W_{s,C} = \left[(1 - V_F)\frac{1}{W_{s,M}} + aV_F\frac{1}{W_{s,Fs}}\right]^{-1} + bV_fW_{s,Fci}. \tag{44}$$

The factors a and b are, in general, also functions of V_F, and the relative contributions of the mechanisms can be modelled in the following way:

$$a = 0.5(1 + V_F^c), \tag{45}$$

$$b = 0.5(1 - V_F^c). \tag{46}$$

The exponent c describes the rate of transition from one mechanism to the other; it should be a function of fibre geometry and fibre–matrix bonding. For simplicity, a and b can be assumed to be equal to 0.5. The only partial wear rate which contributes to the overall wear rate of the composite which is not directly accessible is $W_{s,Fci}$ which accounts for the post-sliding wear processes of mechanisms (3) and (4). However, $W_{s,Fci}$ can be estimated by considering eq. (44). From the results of this study, the wear rate of the composite ($W_{s,C}$) is found, as well as the wear rate of the unfilled matrices ($W_{s,M}$). Furthermore, appropriate values for the fibre sliding wear rates ($W_{s,Fs}$) have been successfully determined in previous studies [26,41,57,64]. The glass fibres possess the highest wear factor (glass fibre: $W_{s,Fs} = 1 \times 10^{-7}$ mm^3(N m)$^{-1}$), followed by the aramid fibres with $W_{s,Fs} = 5 \times 10^{-8}$ mm^3(N m)$^{-1}$. The slowest sliding wear rate can be expected for the carbon fibres ($W_{s,Fs} = 1 \times 10^{-8}$ mm^3(N m)$^{-1}$). When one takes into consideration the properties of the fibres, such as lubrication effectiveness, thermal conductivity, toughness and strength, the above ranking of the wear factors for the fibres appear to be appropriate. Using the above values of $W_{s,Fs}$, the measured wear rates of unfilled EP and PEEK matrices, and $a = b = 0.5$, the following specific wear rates for the post-sliding wear process (i.e. $W_{s,Fci}$: fibre fracture, fibre-matrix interfacial debonding pulverization, fibrillation, etc.) can be approximated:

(a) glass fibre composites:

$$W_{s,Fci} \approx 3 \times 10^{-5} \text{ to } 3 \times 10^{-4} \text{ mm}^3(\text{N m})^{-1},$$

(b) aramid fibre composites:

$$W_{s,Fci} \approx 4 \times 10^{-7} \text{ to } 3 \times 10^{-6} \text{ mm}^3(\text{N m})^{-1},$$

(c) carbon fibre composites:

$$W_{s,Fci} \approx 7 \times 10^{-7} \text{ to } 3 \times 10^{-6} \text{ mm}^3(\text{N m})^{-1}.$$

The wide ranges for the above values resulted from the effects of fibre orientation relative to the sliding direction (excluding the unexpected high value of CF-PEEK in the N orientation). The magnitude of the partial wear rates of the glass relative to the carbon and aramid fibre composites illustrates the deleterious effect of the abrasiveness of glass fibres. This resulted in a large contribution ot the wear of glass fibre composites by fibre fracture, debonding and pulverization, and enhanced abrasion of the matrix material.

4.3.3. Abrasive wear against SiC paper

A previous study conducted by Cirino et al. [73] investigated the wear resistance of some of the composite materials under severe abrasive conditions, i.e. rubbing against silicon carbide paper of various grain sizes. The results are summarized in histogram form in figs. 28a and b. Anisotropic wear behaviour is evident under both extreme wear conditions (i.e. abrasive and sliding wear systems); however, the dependence of the wear rate on the fibre orientation is not consistent within each composite material. For example, the GF-EP system exhibited its lowest wear rate in the AP orientation under the severe abrasive condition; whereas, for the sliding wear condition, the P orientation resulted in the least and the N orientation in the greatest wear rate.

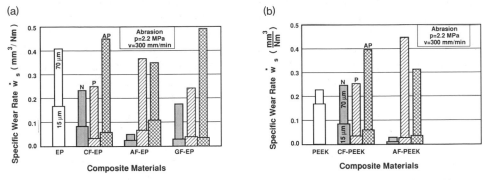

Fig. 28. Average specific wear rates in abrasive wear systems: (a) epoxy matrix materials and (b) PEEK matrix materials ($p = 2.2$ MPa; $v = 300$ mm min^{-1}).

In order to compare the wear rates of the different wear systems, the surface roughness (R_t) of the steel ring and the average particle diameter (D) of the silicon carbide particles are plotted on the same axis as the asperity size. Figure 29 shows that the wear rates differ by about six orders of magnitude between the abrasive and sliding wear conditions. Furthermore, the effect of fibre reinforcement on the wear rate of the neat matrix is much greater under sliding wear conditions, where a decrease of up to two orders of magnitude was observed.

For a general comparison of the effectiveness of the different fibre materials and orientations in improving the wear performance of the two different polymer matrices, it is helpful to relate the specific wear resistances of the composites to that of a reference material ($W_s^{-1}/W_{s(standard)}^{-1}$, the wear resistance ratio). Table 9 illustrates the results of this procedure for all composites and the PEEK matrix, with neat EP as the chosen standard. This database can be utilized in the design of a composite material where maximum resistance in a specified wear situation is required. For example, under severe abrasion it would be of great advantage to have a ductile polymer matrix (with high fracture toughness, such as PEEK) and an aramid fibre reinforcement oriented normal to the surface which is to be worn.

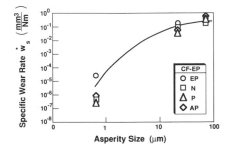

Fig. 29. Comparison of the specific wear rates of the CF-EP material under various wear conditions. (○) EP; (□) N; (△) P; (◇) AP.

TABLE 9

Wear-resistance ratio of the composite materials in the abrasive wear systems. The wear-resistance ratio is equal to the wear resistance of the composite material divided by the wear resistance of the unfilled epoxy matrix ($W^{-1}_{s(standard)}$). Note that a higher number indicates better wear resistance against wear loading.

Matrix	Reinforcement			Neat polymer
	CF	GF	AF	
EP				
N	1.8	2.3	7.9	1
P	1.7	1.4	1.1	
AP	0.9	0.8	1.2	
PEEK				
N	1.7	2.4	15.2	1.8
P	1.6	1.8	0.9	
AP	1.0	1.2	1.3	

Additional benefits can be expected by incorporating glass or carbon fibres oriented in-plane and parallel to the wearing direction.

It could be shown further [74] that deviations from the three major directions (AP, P and N) result in very drastical changes in case of AP- and P-orientation of aramid fibres in a polymeric matrix, whereas deviation from the favourable N orientation does not change the wear characteristics of the composite very much (figs. 30a to c). This was however not the case for the more brittle carbon fibres, for which the surface of the composites' wear rates as a function of fibre orientation angles α and β remained at a rather uniform level (fig. 30d).

4.4. Wear studies on hybrid composites

4.4.1. Concept for hybrid composites with good wear resistance

From the conclusions drawn above, a model composite can be designed that combines all the favourable fibre types and orientations in one structure. A composite with an overall good wear resistance could be made, for example, by three-dimensional hybridization, with interwoven carbon fibres in-plane (XY-plane) and aramid fibres in the Z-direction. The matrix should be a tough, highly wear-resistant thermoplastic polymer (fig. 31).

Of course, such a hypothetical 3D hybrid structure cannot very easily be verified. But it is at least possible to approach such a structure along two lines. One is to study 2D woven-fabric composites, e.g. carbon-fibre woven-fabric reinforced PEEK. In fact, this has been done in previous work [75,76]; it was shown that the woven structure resulted in a better wear resistance and a lower coefficient of friction than the two basic orientations of the unidirectional material (figs. 32a and b). This type of wear synergism is illustrated further in fig. 32c. Regardless of the orientation and the actual amount of fibres in the weft and warp direction of the 2D-fabric relative to the sliding direction applied, the woven surface gave better wear performance than the unidirectional materials. The explanation of this

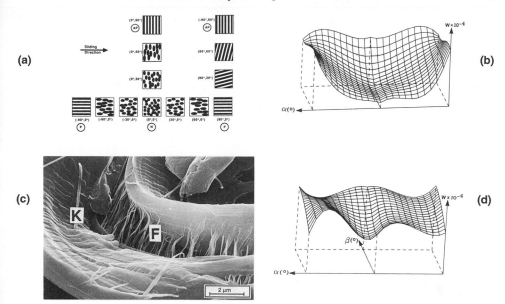

Fig. 30. (a) Illustration of the fibres on the wear surfaces for orientations between the N, P and AP orientations. (b) Graphic display of the constructed surface using the empirical relationships obtained from the wear data (fig. 28) of the AF-EP composite material. (c) SEM photograph of the worn surface of the AF-EP in the (90, 30) orientation. The bent aramid fibre shows fibre fibrillation (F) on the tensile side and kink bands (K) on the compression side. (d) Graphic display of the constructed surface using the empirical relationships obtained from the wear data of the CF-PEEK composite material (fig. 28).

phenomena is based on the fact that the cross-over points in the woven-fabric wear surface act as locations of wear debris accumulation. As long as patches of the condensed wear debris are kept entrapped at the points, the actual surface of the

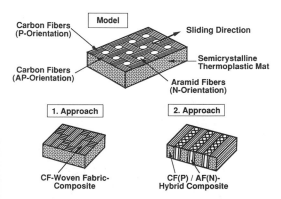

Fig. 31. Design concept for a hybrid composite with good wear resistance.

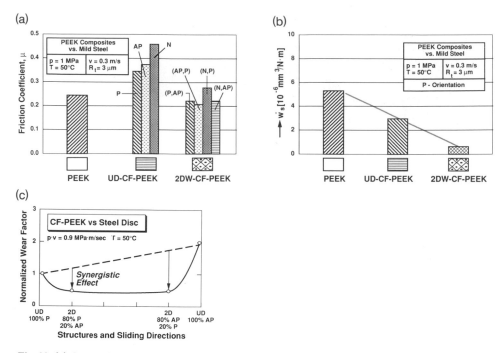

Fig. 32. (a) Comparison between the coefficients of friction of the neat PEEK matrix, the unidirectional (UD) CF-PEEK composite and the two-dimensional CF woven fabric-PEEK composite, respectively. (b) As fig. 32(a) for the specific wear rates. (c) Wear synergism of CF-PEEK composite due to 2D woven-fabric fibre arrangement (relative to UD).

composite material in between these points gets temporarily protected from further wear (fig. 33).

Another approach to the hybrid model composite is to combine layers of different fibre types and orientations within one laminate. This can result in either a sandwich or a layer structure (fig. 34). Table 10 includes all the hybrid materials made from the basic components mentioned as sets no. 3–5 in table 7. The majority of them contain carbon and aramid fibres in a cross-ply sandwich form.

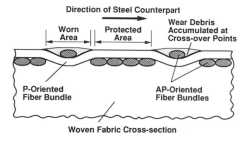

Fig. 33. Wear protection mechanisms due to accumulation of wear debris at cross-over points in the 2DCF woven-fabric/PEEK composite.

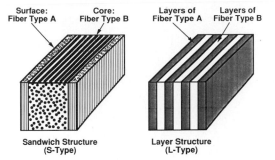

Fig. 34. Possible structures for layered hybrid composites.

Some of the combinations of sliding directions applicable to these hybrids are listed in a separate column. Figure 35 shows a cross section of one of the amorphous PA matrices, having carbon fibres in the two surface layers and aramid fibres with a transverse orientation in the core region. The higher magnification of the transitional region from one fibre type to the other clearly shows that no crack formation has occurred during the manufacturing process of these hybrids (in spite of differences in the thermal expansion coefficients of the fibres, and, therefore, the formation of internal stresses when cooling form the molten state of the polymer matrix).

4.4.2. Sliding wear data

The following selection of diagrams (figs. 36–39) shows a variety of the different trends observed when two monolithic composites are brought together to form a

TABLE 10
Hybrid composites employed in wear tests.

Set	Matrix	no.	Abbreviation	Total V_f (A & B)	A/B (%)	Hybrid type	Sliding direction applied
3	amPA	1	CF(0)-AF(90)-CF(0)	59–65	variable	S	P–N–P
		2	AF(0)-CF(90)-AF(0)	61	50/50	S	N–P–N
		3a	CF(0)-AF(0)-CF(0)	61	50/50	S	P–P–P
		3b					N–N–N
4	PA66	4a	CF(0)-AF(90)-CF(0)	20–40	variable	S	P–N–P
		4b					N–P–N
		5	CF(0)-AF(90)-CF(0)	20–40	variable	L	P–N–P
		6a	CF(0)-GF(90)-CF(0)	25	50/50	S	N–P–N
		6b					N–AP–N
		7a	GF(0)-AF(90)-GF(0)	35	50/50	S	P–N–P
		7b					AP–N–AP
5	EP$_2$	8a	CF(0)-AF(0)-CF(0)	62	50/50	S	P–P–P
		8b					N–N–N

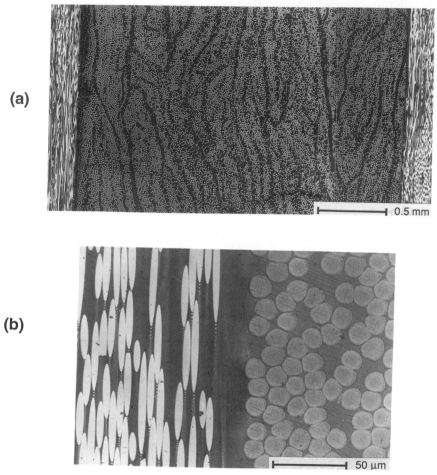

Fig. 35. Cross section of a hybrid sandwich structure, made of an aramid-fibre/amorphous PA core and two surface layers with carbon fibres in 90° orientation relative to the fibres in the core (upper photo). The lower micrograph shows a magnification of the interfacial region between the carbon (left) and the aramid fibres (right).

new hybrid structure. The shape of the curves depends, in general, on the following:

(a) The difference in wear rate between the two limiting compositions (which are in turn a function of the particular fibre material, the fibre orientation and the polymer matrix); the structures of the limiting compositions are indicated in each of the diagrams by the X-axis labelling e.g. the X-axis of fig. 36a has to be read as follows: (1) the far left extreme represents 100% aramid fibre reinforcement, i.e. the carbon fibre $V_f = 0$, with the aramid fibres oriented normal to the sliding plane; (2) the far right extreme represents 100% carbon fibre reinforcement, i.e. 0% of aramid fibre, with the carbon fibres oriented in-plane and parallel to the

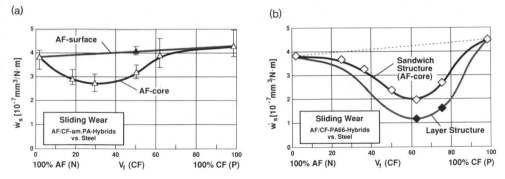

Fig. 36. (a) Specific wear rates of aramid-fibre/carbon-fibre/amorphous PA hybrids. (b) Specific wear rates of aramid-fibre/carbon-fibre/PA 6.6 hybrids.

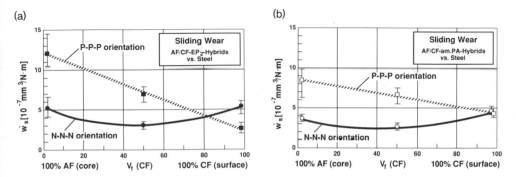

Fig. 37. (a) Specific wear rates of unidirectional aramid-fibre/carbon-fibre/epoxy$_2$ hybrids. (b) Specific wear rates of unidirectional aramid-fibre/carbon fibre/amorphous PA hybrids.

sliding direction; (3) any data point between these two, e.g. carbon fibre $V_f = 0.60$, means that the reinforcement of the hybrid composite consists of 40 vol% aramid fibres in normal orientation and 60 vol% carbon fibres in parallel orientation; besides this X-axis labelling each diagram contains information about the type of

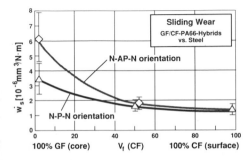

Fig. 38. Specific wear rates of glass/carbon/PA 6.6 hybrids.

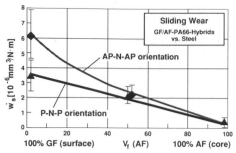

Fig. 39. Specific wear rates of glass/aramid/PA 6.6 hybrids.

polymer matrix used (e.g. amorphous polyamide in fig. 36a), and about the type of laminate studied (e.g. sandwich laminate in the case of fig. 36a, with aramid fibres either in the core or in the two surface layers).

(b) The structure of the hybrid composites formed out of the two extremes (sandwich or layer structure).

(c) The stacking sequence (e.g. in the case of sandwich hybrids, the composites with carbon fibres positioned in the surface layers and aramid fibres in the core are considerably superior to the inverted structure with aramid fibres in the surface layers and carbon fibres in the core).

(d) The efficiency with which one or the other fibre material can cause, in a given polymer matrix, effects that result in wear behaviour of the hybrid that is superior to that of the basic, monolithic composites.

Figure 36a shows the specific wear rates of hybrid composites consisting of normal oriented aramid fibre and parallel carbon fibre in an amorphous PA matrix. The arrangement of the aramid fibre in the sandwich core resulted in a positive hybrid effect, i.e. the wear rates for the hybrid were lower than that expected from a simple rule of mixtures (IROM) based on the wear rates of the two components when being tested separately. The best results were achieved when more than 50% of the fibres were aramid. For the 50/50 mixture it became evident that a change in the stacking sequence, i.e. the aramid fibre now positioned in the surface layers, does not result in the same positive hybrid effect. Instead, the wear data were found to be on the linear connection between the two limits.

A similar tendency as that seen before is also found for the aramid/carbon fibre hybrids with PA 6.6 as the matrix (fig. 36b). The only difference is that, in spite of the same fibre arrangements in the sandwich, the minimum in the curve of the hybrid values is shifted towards higher volume fractions of carbon fibres. This may be due to the other type of matrix, and to resulting effects of the bond quality between fibres and matrix, and to changes in the wear mechanisms caused by these effects. It is shown further that a layer hybrid yields better improvements in the wear resistance than a simple sandwich structure.

Arranging both aramid and carbon fibres in the same direction leads to very different curves when testing such hybrids in various directions. As illustrated in

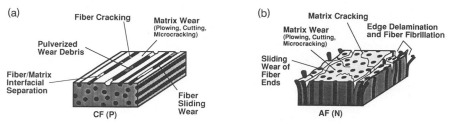

Fig. 40. (a) Schematic of wear mechanisms in a monolithic composite with carbon fibres in-plane and parallel to the sliding direction. (b) As (a), for aramid fibres normal to the sliding plane. Further consequences due to hybridization of carbon and aramid fibres are described in fig. 43d.

fig. 37a for aramid/carbon/epoxy hybrids (and in fig. 36b for aramid/carbon/amorphous PA hybrids), a parallel orientation of all the fibres results in a linear connection between the two extremes and the 50/50-hybrid composite. For the normal fibre arrangement, on the other hand, the 50/50 hybrid proves better than the two monolithic composites at both ends of the curve.

If the hybrid composites are made of glass and carbon fibres, the same trend is visible, as was reported by Hawthorne [57] for such a system. Replacing up to 50% of the carbon fibres by glass fibres does not result in a dramatic increase in the composite wear rates (fig. 38). Above this composition, however, the values seem to increase more rapidly towards the high wear rates as measured for the monolithic glass/PA 6.6 composite. A similar observation was made when mixing glass and aramid fibres in a PA 6.6 matrix, as shown in fig. 39.

4.4.3. Sliding wear mechanisms

A comprehensive microscopical analysis of the wear mechanisms occurring during the sliding of monolithic composites versus steel counterpart (as listed in table 7, sets 1 and 2) has been reported elsewhere [62]. Only those details important for interpreting the positive hybrid effects, especially in the aramid(N)/carbon(P) composites, will therefore be illustrated in this context. Figures 40a and b schematically give an impression of the wear mechanisms of the monolithic carbon- and aramid fibre composites under the relevant orientations.

Besides material removal due to fibre and matrix sliding, a great amount of wear debris is generated by fibre cracking and fibre/matrix interfacial separation in the carbon fibre composites (fig. 41). Aramid fibres under normal orientation result in a microscopically rather smooth surface, with some degree of fibre/matrix debonding. Moreover, there are two effects occurring that have a more macroscopic dimension, one of which is the formation of individual, long cracks that tend to run across the whole width of the specimens, the other being the delamination of specimen edges, combined with fibrillation of these regions (fig. 42a). If aramid fibres are oriented in-plane and parallel to the sliding direction, they may get peeled off the surface due to their poor bonding to the surrounding matrix (fig. 42b).

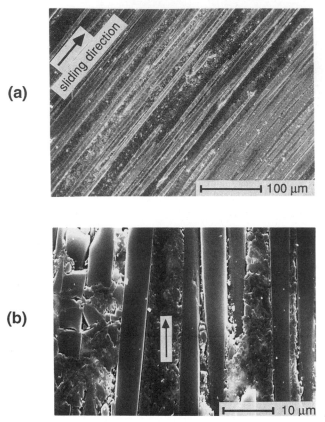

(a)

sliding direction

⊢————————⊣ 100 µm

(b)

⊢————————⊣ 10 µm

Fig. 41. SEM micrographs of the worn surface of a carbon-fibre/PA 6.6 sample in the P-orientation: (a) overview, (b) details of fibre thinning, interfacial debonding and fibre fracture (arrow indicates sliding direction).

The favourable interactions between the aramid- and the carbon fibre layers that are formed when carbon fibres are placed in parallel orientation in the surface regions of the sandwich hybrids, are as follows:

(a) Cracks that may develop in the aramid fibre core of the sandwich due to the poor fibre/matrix bonding are stopped as soon as they encounter the strong barrier of carbon fibres in the surface layers oriented transversely to the aramid fibres.

(b) Edge delamination and fibrillation of the aramid fibres are strongly prevented by the stiffer carbon fibres, which are better bonded to the matrix.

(c) Simultaneously, the normal oriented aramid fibres in the core, especially in the vicinity of the carbon/aramid fibre interfacial region, are able temporarily to hold back worn material, consisting of compounded pulverized fibre particles in smeared out polymer matrix debris, as flat patches on their surface; this third body, once formed, partly separates the counterpart surfaces and supports the

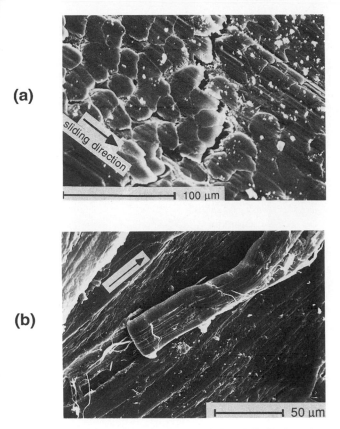

Fig. 42. SEM micrograph of edge cracking and fibrillation of an aramid-fibre/PA 6.6 sample in N orientation (a) and pulling of a single aramid fibre out of the surface, in case of in-plane, parallel orientation (b).

load-carrying capacity [39]; thus, the actual specimen surface is to some extent protected against further material removal due to sliding wear.

Figure 43a shows such a situation for the worn surface of a carbon (P)/ aramid(N)/PA 6.6 hybrid composite, and fig. 43b illustrates this hybrid effect schematically.

4.5. Hybrid wear model

A description of the wear rates of hybrid composites has been attempted by Hawthorne [57] using a rule-of-mixtures approach for the wear-rate values of the two hybrid components and weight factors that were a function of the elastic moduli of the different fibre materials used. Although this approach led to good agreement between experimental data and predicted values, it does not consider that the components constituting the hybrid may mutually interact and cause a

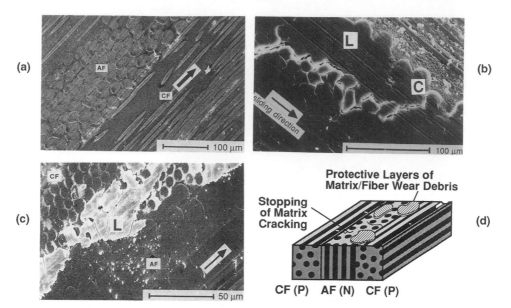

Fig. 43. (a) Worn surface of an AF(N)/CF(P)-layer composite. (b) Worn surface of a carbon-fibre (P)/aramid-fibre (N)/PA 6.6 sandwich composite, showing crack-stopping (C) at the boundary between the normal oriented aramid fibre and the parallel carbon fibre and the protective layers of matrix/fibre wear debris (L). (c) Build-up of a protecting patchwork on the wear surface of a CF(N)/AF(N) hybrid. (d) Schematic illustration of the wear-reducing mechanisms due to hybridization.

deviation of the hybrid wear resistance from the simple rule-of-mixtures calculations.

The hybrid model developed in this study, therefore, contains not only the wear contributions of the two components (when measured separately), but also a wear reduction term, which accounts for possible changes due to hybridization:

$$\dot{w}_{s,H} = \dot{w}_{s,AB} - \dot{w}_{s,red},\tag{47}$$

where $\dot{w}_{s,H}$ is the specific wear rate of the hybrid composite, $\dot{w}_{s,AB}$ is the specific wear rate due to simultaneous sliding of the two components A and B against the same counterpart and $\dot{w}_{s,red}$ is the wear reduction due to interactive protection of one component by the other (to be arranged sequentially with $\dot{w}_{s,AB}$ because interaction between A and B builds up with time).

The main term in eq. (47), $\dot{w}_{s,AB}$, is calculated from the wear resistances of the two components A and B, since the wear of A and B starts simultaneously:

$$\dot{w}_{s,AB} = \left[(1 - V_B)\dot{w}_{s,A}^{-1} + V_B\dot{w}_{s,B}^{-1}\right]^{-1},\tag{48}$$

where V_B is the relative contribution of component B to the hybrid ($V_B + V_A = 1$).

The reduction term, $\dot{w}_{s,red}$, is a more complex quantity because it has to account for different factors: (a) the absolute amount of reduction of the wear rate as a

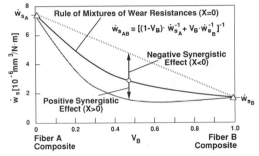

Fig. 44. Schematic of the effects of a positive and negative hybrid-efficiency factor on the course of the hybrid wear curve relative to the rule-of-mixtures prediction.

function of hybrid composition and (b) the relative dominance of one or other fibre in the hybrid effect. A general expression taking these factors into consideration is:

$$\dot{w}_{s,red} = X \dot{w}_{s,A} f(V_A, V_B),\tag{49}$$

where X is the hybrid efficiency factor (see fig. 44) ($X = 0$, no synergistic effect due to hybridization, $X > 0$, positive synergistic effect, $X < 0$, negative synergistic effect); $f(V_A, V_B)$ is the shape function, which accounts for the relative dominance of component A or B in the hybrid effect.

In eq. (49), the second factor simply brings the wear-rate dimensions into play and is easily accessible from the basic experiments with the monolithic material type A. The two other factors are more difficult to estimate, because they depend on a variety of different parameters. The influence of some of these, e.g. the matrix used or the fibre/matrix bond quality, can somehow be estimated. But there are others that are almost impossible to quantify. Among these are the occurrence or suppression of certain types of wear mechanisms due to hybridization, their efficiency in improving the wear resistance of the hybrid composite, and the dominance of one or the other fibre type in the hybrid effects. The magnitude of the hybrid efficiency factor, X, may therefore vary over a wide range (e.g. for positive hybrid effects, between 0 and about 10), and feeling for its approximate value for a given fibre A/fibre B system can only be accomplished after sufficient experimental experience.

In a similar way, the variables in the shape function are rather unpredictable, unless there exists enough experimental evidence that, for certain fibre/matrix combinations, special fibres are more efficient than others in acting in a synergistic manner. Only the general form of this shape function can be approached on a reasonably reliable basis, either as simple power functions or by the use of more complex hyperbolic functions:

(a) simple shape function:

$$f(V_A, V_B) = (1 - V_B)^a V_B^b,\tag{50}$$

where $a > 1$, $b < 1$ if fibre type A dominates the hybrid effect; the opposite is found for dominance of B (e.g. $0.5 \leqslant a$; $b \geqslant 3$).

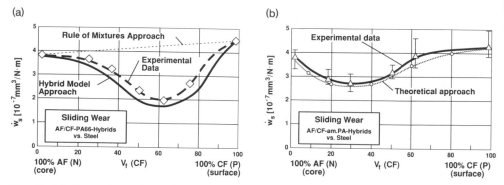

Fig. 45. (a) Comparison of experimental data and theoretical approach of hybrid wear effects in aramid-fibre/carbon-fibre/PA 6.6 composites. The variables in the wear-reduction term had the following values: hybrid efficiency factor $X = 0.98$; complex hyperbolic shape function with $a_1 = 0$, $b_1 = 1.3$, $c_1 = 1$ and $a_2 = 0.964$, $b_2 = 4$, $c_2 = 0.5$. (b) The data for the aramid-fibre/carbon-fibre/ amorphous PA composites were modeled by: hybrid efficiency factor $X = 1.54$; complex hyperbolic shape function with $a_1 = 0.664$, $b_1 = -4$, $c_1 = 0.2$ and $a_2 = 0.984$, $b_2 = 2$, $c_2 = -0.2$.

(b) complex shape function:

$$f(V_A, V_B) = [a_1 + \tanh b_1(c_1 - V_B)][a_2 + \tanh b_2(c_2 - V_B)], \qquad (51)$$

where a_i, b_i and c_i have to be adjusted according to the dominance of fibres A or B in the hybrid effect.

Using the latter equation, the experimental data given in fig. 37b for the aramid-fibre(N)/carbon-fibre(P)/PA6.6 hybrid with sandwich structure, were approached iteratively. Figure 45a shows that this can be done with quite good agreement, and improvements can even be achieved when adjusting a_i, b_i and c_i more precisely. But, despite this, it must be clearly stated at this point that the model suggested so far can only be used to estimate the trends in wear behaviour of a hybrid system. To make this possible, it is necessary to known the wear performance of the basic components and to have at least a rough idea about the size of the hybrid efficiency factor and the dominance of one type of fibre relative to the other used in the hybrid system under particular consideration.

Figure 45b illustrates the same type of approach for the aramid-fibre(N)/ carbon(P) hybrids in the amorphous PA matrix. Since the minimum in wear resistance is here at a different position from that in the previous case, the variables in the shape function are, of course, very different from those used before. This illustrates even more the difficulty of modelling wear of highly anisotropic materials.

An example of the situation of a negative hybrid efficiency factor is given for the glass/carbon fibre system (fig. 46). The wear rate of the 50/50 hybrid is much higher than that expected from the rule of mixtures for the wear resistance of the two components under separate consideration. Table 11 summarizes the results for most of the systems presented, where hybridization was favourable or not.

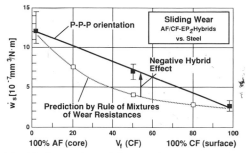

Fig. 46. Comparison of the experimental wear data and the rule-of-mixtures prediction of aramid-fibre (P)/carbon-fibre (P)/aramid-fibre (P)/EP$_2$.

4.6. Outlook

There are also some indications in the literature which testify that composites with a 3D fibre arrangement and a high-temperature polymer matrix with lubricants as inclusions have excellent wear resistance (table 12 [77–79]). In the particular case mentioned glass or carbon fibres were used as reinforcements, and the 3D structure was verified in different geometries of the bearing elements. Flat sliding blocks and filament wound bearings with fibres in the radial, tangential and axial directions were shown (fig. 47).

A comparison of the wear-rate data, especially those of 3D CF-PI with the majority of fibres in the contact plane, with the results of a 3D CF-PEEK material, as given in table 13, reflects a slight superiority of the PEEK matrix composite. This is especially true when the materials are compared on the basis of their wear

TABLE 11
Hybrid efficiency factors for different hybrid composites tested at various orientations.

Set	Matrix	no.	Abbreviation	Hybrid type	Sliding direction	Hybrid efficiency factor X	$\dot{w}_{s_{red}}$ at A/B = 50/50 [a]
3	amPA	1	CF(0)-AF(90)-CF(0)	S	P–N–P	> 0	21%
		2	AF(0)-CF(90)-AF(0)	S	N–P–N	0	–
		3a	CF(0)-AF(0)-CF(0)	S	P–P–P	< 0	22%
		3b			N–N–N	> 0	27%
4	PA66	4	CF(0)-AF(90)-CF(0)	S	P–N–P	> 0	45%
		5	CF(0)-AF(90)-CF(0)	L	P–N–P	> 0	64%
		6a	CF(0)-GF(90)-CF(0)	S	N–P–N	0	–
		6b			N–AP–N		
		7a	GF(0)-AF(90)-GF(0)	S	P–N–P	< 0	135%
		7b			AP–N–AP		127%
5	EP$_2$	8a	CF(0)-AF(0)-CF(0)	S	P–P–P	< 0	63%
		8b			N–N–N	> 0	36%

[a] $\dot{w}_{s_{red}}$ expressed in % of rule of mixtures value $\dot{w}_{s_{AB}}$.

TABLE 12
Depth wear rates of 3D CF- and -GF-PI composites worn at different testing temperatures and
pressure times velocity conditions [77].

Fibre of 3D weave		Wear rate (m/s)		
		No heat applied	316°C	
		$6.07\,\text{MPa}\times0.76\text{m}\times\text{s}^{-1}$	$2.76\,\text{MPa}\times0.51\text{m}\times\text{s}^{-1}$	$6.07\,\text{MPa}\times0.76\text{m}\times\text{s}^{-1}$
CF-PI	‖	2.5×10^{-8}	1.1×10^{-8}	$3.2\ \times10^{-8}$
	⊥	1.9×10^{-8}	1.1×10^{-8}	$8.0\ \times10^{-8}$
GF-PI	‖	9.9×10^{-9}	5.6×10^{-9}	2.23×10^{-7}
	⊥	2.3×10^{-8}	4.9×10^{-9}	$7.8\ \times10^{-8}$

‖ – Majority of fibre lay is parallel with the plane of sliding.
⊥ – Majority of fibre lay is normal to the plane of sliding.

factor, i.e. under simultaneous consideration of the differences in the external
testing parameters [$\dot{w}_s = \dot{w}_t/(pv)$]:
  (a) 3D CF-PI at room temperature
$$\dot{w}_s = 2.5 \times 10^{-8}\ (\text{m/s})/[6.07\ \text{MPa}\ 0.76\ \text{m/s}] = 5.4 \times 10^{-6}\ (\text{mm}^3/\text{N m}).$$

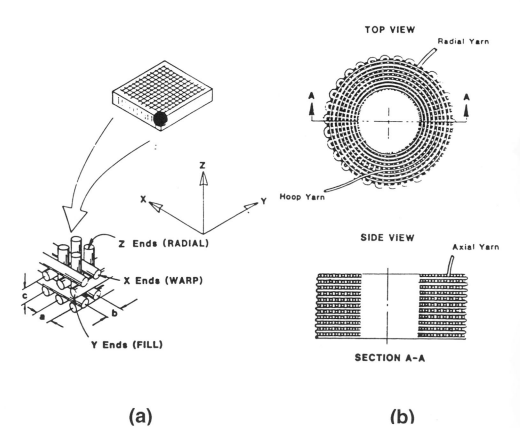

Fig. 47. Schematic of the 3D fibre arrangement in a flat sliding block (a) and a joint bearing (b) [93].

TABLE 13
Influence of the testing temperature on the coefficient of friction, time related wear rate and specific wear rate of a three-dimensional woven CF-PEEK composite material, as manufactured using a commingle technique (courtesy L.F. Taske II, CCM-Report no. 89-02, Centre for Composite Materials, University of Delaware, USA, 1989).

Material code	$T = 20°C$			$T = 220°C$		
	μ	$\dot{w}_t$	$\dot{w}_s$	μ	$\dot{w}_t$	$\dot{w}_s$
PEEK+50%-3D graphite fibre	0.491	0.036	1.52	0.189	0.086	3.62

$\dot{w}_s$ in 10^{-6} mm^3/Nm; $\dot{w}_t$ in mm/h; $p = 2$ MPa; sliding direction: normal to MFD; $v = 3.3$ m/s.

(b) 3D CF-PEEK at room temperature

$$\dot{w}_s = 0.036 \ (\text{mm/h})/[2 \ \text{MPa} \ 3.3 \ \text{m/s}] = 1.52 \times 10^{-6} \ (\text{mm}^3/\text{N m}).$$

The same trend is in principle visible for the data obtained at elevated temperatures.

5. Conclusions

It was the objective of this chapter to present some of the existing approaches for mathematically describing the friction and wear behaviour of multi-component systems, in particular of polymeric matrix composites with different types of reinforcement (e.g. particle fillers or continuous fibres of one or various kinds, i.e. hybrids). The relationship between the size of the reinforcing phase and the groove size plays an important role in abrasion phenomena. Depending on this parameter, different types of wear behaviour can occur. From a practical viewpoint, this concept can guide the selection of multiphase materials facing abrasive wear conditions. In this case, the size of the dispersed phase must be maximized relative to the abrasive grains.

The abrasive wear of particle-reinforced polymer composites can be described by a simple additive equation based on a series model. The apparent filler wear rate in the composite is higher than that measured for the same material in sheet form and this increase is highly dependent on the type of particle-surface treatment. Surface-profile measurements and electron microscopy suggest that the enhanced wear rate is due to enhanced chipping of the filler at the interface.

If these composites are, however, subjected to sliding wear against smooth steel counterparts, the simple rule-of-mixtures description does not apply anymore, due to a higher complexity of the wear mechanisms. However, also here the statements concerning size of the reinforcing phase, the quality of the interfacial strength and the filler fraction are still valid.

From investigations on the unlubricated friction and wear of the unidirectionally oriented fibre reinforced polymers (FRP) slid against carbon steel, the following results were obtained:

(1) The law of mixture for the friction coefficient of FRP was deduced, and the validity of this law was assured by the experimental results.

(2) Carbon-FRP gave small specific wear rate and low friction coefficient. In contrast, glass-FRP provided a large specific wear rate and a high friction coefficient.

(3) A model was proposed in which the wear of FRP proceeds by wear-thinning of the fibre with subsequent breakdown of the fibre and by peeling-off of the fibre from the matrix. Using this model, an experimental equation for the wear of FRP was deduced. It was found, by critically examining this wear equation, that the most significant parameters for the wear of FRP and Young's modulus and, in some cases, the interlaminar shear strength.

Further work concerning the tribological behaviour of hybrid composites has led to the following conclusions:

(1) The steady-state wear rates of hybrid carbon/glass fibre epoxy composites sliding against stainless steel lie considerably below the values linearly interpolated between those of the single-fibre (CFRP and GFRP) composites, regardless of the fibre orientation relative to the sliding direction. Thus, wear synergism is exhibited by the hybrid composites.

(2) Fibre orientation affects the composite wear behaviour but both the extent of its significance and the optimum orientation for wear resistance depends on the composite composition.

(3) In addition to the composite composition and fibre orientation, the wear behaviour of these hybrid composites appears to be influenced by the relative abrasiveness, fatigue resistance, film transfer forming characteristics and relative load bearing capability of the two constituent fibre phases.

(4) Hybrid wear rates correlate with the Young's moduli of the different fibres used, but not with their flexural or interlaminar shear strength.

(5) Using the best wear results achieved with monolithic fibre composites, i.e. with carbon fibres oriented in-plane and parallel to the sliding direction, or with aramid fibres in normal orientation, would lead to a model composite with a 3D-hybrid structure, which is supposed to possess optimum wear resistance.

(6) Approaching this model by producing 2D-hybrid composites shows synergistic effects between different fibre materials and directions chosen.

(7) Good results were achieved with aramid fibres in N-orientation in the core of a sandwich hybrid composite and carbon fibres in a parallel arrangement in the surface layers of the sandwich.

(8) A model is suggested that allows one to describe these hybrid effects by (a) starting from a rule-of-mixtures prediction on the wear resistance of the hybrid composite and (b) subtracting a reduction term of the wear rate, resulting from the interactive protection against wear of one component by the other. If the hybrid efficiency factor is negative, the wear of the hybrid composite is, on the other hand, greater than that predicted by the rule-of-mixtures approach.

Finally, it can be stated from the results of this chapter that many different polymer composites can be used for friction and wear applications. They can be tailored with regard to their load carrying capacity, their temperature and creep resistance, their coefficient of friction and their resistance to abrasion and wear under sliding against steel counterparts. But in spite of (a) the wide variety of existing material systems for these applications and (b) the flexibility in tailoring polymer composites for a particular property profile, there are still further studies necessary in order to obtain a full understanding of their tribological performance. A basic help to reach this objective can be achieved from some further references not yet mentioned in the previous context [80–86]. These include information on the role of fatigue crack propagation in the wear of polymeric tribo-systems [82], on how to select polymer composites for friction and wear applications [80,81,83–85], and on the construction of wear mechanism maps helpful to understand the wear rate transitions and their relationship to wear mechanisms [86].

Acknowledgements

I would like to acknowledge the help and cooperation of students, technicians, research assistants and secretaries at the Institut für Verbundwerkstoffe, University of Kaiserslautern, for their experimental, typing and graphic design work. Thanks are due to the many companies who kindly provided the different materials for our studies. Especially mentioned should be in this respect ICI, Wilton, UK, and Du Pont de Nemours, Wilmington, USA. Various parts of our works were financially supported by the Deutsche Forschungsgemeinschaft (DFG FR 675/11-1: Impregnation Mechanisms During Manufacturing of Thermoplastic Composites) and by the Arbeitskreis Industrieller Forschungsvereinigungen (AIF, Forschungskuratorium Maschinenbau, Proj.Nr. D 295: High Temperature Thermoplastic Bearing Materials). Finally, thanks are expressed to Professors T. Tsukizoe and N. Ohmae, Japan, Dr. H.M. Hawthorne, Canada, Drs. P.D. Calvert and S.V. Prasad, USA, and Drs. W. Simm and S. Freti, Switzerland, for kindly accepting the use of some of their earlier published work in this article.

List of symbols

A	area [mm^2]
a	factor
b	factor
c	exponent
d	particle size [μm]
D	particle diameter [μm]
E	Young's modulus [GPa]
F	tangential force [N]
G	shear modulus [MPa]
H	indentation hardness [MPa]
I_s	interlaminar shear strength [MPa]
K	wear factor [mm^3/(N m)]
K_c	fracture toughness [MPa $\sqrt{m}$]

L sliding distance [m]
l length of a component [m]
P normal load [N]
p apparent contact pressure [MPa]
R specific wear rate ($\hat{=} \dot{w}_s = \dot{w}/p = W_s$) [mm^3/(N m)]
R_a average roughness [μm]
R_t total roughness [μm]
R_z maximum roughness [μm]
V volume fraction ($\hat{=} \chi$) [%]
v velocity [m/s]
W total (dimensionless) wear rate ($\hat{=} \dot{w}$)
W_i (dimensionless) wear rate of the ith phase
W^{-1} wear resistance
α, β orientation angles [°]
γ shear strain [%]
ρ density [g/cm^3]
Δm mass loss [g]
μ coefficient of friction
σ tensile strength [MPa]
τ shear stress [MPa]
γ particle distance [μm]

Subscripts
C   composite
c   carbon
F   fibres
f   fibres or fillers
g   glass
H   hybrid
i ith phase
m   matrix
P   polymer matrix
St   steel fibres

References

[1] S. Fischer and G. Marom, Compos. Sci. & Technol. 28 (1987) 291.
[2] K. Hofer Jr, M. Stander and L. Bennett, Polymer Eng. Sci. 18 (1978) 120.
[3] K. Schulte, E. Reese and T.W. Chou, in: Proc. of ICCM-VI/ECCM-II, Vol. 4, eds F.L. Mathews et al. (Elsevier Applied Science Publishers, London, 1987) p. 489.
[4] G. Marom, H. Harel, K. Friedrich, K. Schulte and H.D. Wagner, Composites 6 (1989) 537.
[5] J.P. Giltrow and J.K. Lancaster, The role of the counterface in the friction and wear of carbon fiber reinforced thermosetting resins, Wear 16 (1970) 359–374.
[6] J.K. Lancaster, Polymer-based bearing materials – the role of fillers and fiber reinforcement, Tribology 5 (1972) 249–255.
[7] J.P. Giltrow, Friction and wear of self-lubricating composite materials, Composites 4 (1973) 55–64.

[8] J.P. Giltrow and J.K. Lancaster, Properties of CFRP relevant to applications in tribology, in: Proc. Int. Conf. on Carbon Fibers (Plastics Institute, UK, 1971) 251–257.

[9] T. Tsukizoe and N. Ohmae, Wear performance of unidirectionally oriented carbon fiber reinforced plastics, Tribol. Int. 8 (1975) 171–175.

[10] N.-H. Sung and N.P. Suh, Friction and wear of fiber reinforced polymeric composites: effect of fiber orientation of wear, in: Proc. 35th Ann. Tech. Conf. Society of Plastics Engineering (1977) 311–314.

[11] T. Tsukizoe and N. Ohmae, Wear mechanisms of unidirectionally oriented fiber reinforced plastics, Proc. 1st Wear of Materials Conf. ASME (1977) 518–525.

[12] J.K. Lancaster, Geometrical effects on the wear of polymers and carbons, Trans. ASME F (JOLT) 97 (1975) 94–197, 216.

[13] T. Tsukizoe and N. Ohmae, Tribomechanics of carbon fiber reinforced plastics, Ind. Lubr. & Tribol. 28 (1976) 19–25.

[14] N.L. Hancox, ed., Fiber Composite Hybrid Materials (Applied Science Publishers, London, 1981).

[15] J. Summerscales and D. Short, Carbon fiber and glass fiber hybrid reinforced plastics, Composites 9 (1978) 157–166.
J. Summerscales and D. Short, Hybrids – a review. Part 2, physical properties, Composites 11 (1980) 33–38.

[16] W. Simm and S. Freti, Abrasive wear of multiphase materials, Wear 129 (1989) 105–121.

[17] E. Hornbogen, in: Proc. Int. Conf. on Wear of Materials, Vancouver, April 14–18 (American Society of Mechanical Engineers, New York, 1985) pp. 477–484.

[18] W.M. Garrison Jr, Wear 82 (1982) 213–220.

[19] K.-H. Zum Gahr, in: Proc. Int. Conf. on Wear of Materials, Vancouver (American Society of Mechanical Engineers, New York, 1985) pp. 45–58.

[20] A. Sandt and J. Krey, Metall 39(3) (1985) 233–237.

[21] K.-H. Zum Gahr, Microstructure and Wear of Materials, Tribology Series, Vol. 10 (Elsevier, Amsterdam, 1987).

[22] E.E. Underwood, Quantitative Stereology (Addison Wesley, Reading, MA, 1970).

[23] D.A. Rigney, ed., Fundamentals of Friction and Wear of Materials (American Society for Metals, Metals Park, OH, 1981).

[24] J.D.B. De Mello, M. Durand-Charre and T. Mathia, Colloque International sur les Matériaux résistant à l'usure, St.-Etienne, November 23–25 (1983) pp. 21.4–21.11.

[25] C.S. Yust, Int. Met. Rev. 30(3) (1985) 141–154.

[26] K. Friedrich, ed., Friction and Wear of Polymer Composites, Composite Materials Series, Vol. 1 (Elsevier, Amsterdam, 1986).

[27] K. Friedrich and M. Cyffka, On the wear of reinforced thermoplastics by different abrasive papers, Wear 103 (1985) 333–344.

[28] S.V. Prasad and P.D. Calvert, Abrasive wear of particle-filled polymers, J. Mater. Sci. 15 (1980) 1746–1754.

[29] D.C. Watts, Dental restorative materials, in: Materials Science and Technology, series eds R.W. Cahn, P. Haasen and E.J. Kramer, Vol. 14, Medical and Dental Materials, ed. D.F. Williams (VCH-Publishers, New York, 1992) ch. 6, pp. 209–258.

[30] R.L. Bowen, J. Am. Dent. Assoc. 66 (1963) 57.

[31] R.G. Craig and J.M. Powers, Int. Dent. J. 26 (1976) 121.

[32] A.W.J. de Gee, P. Pallav and C.L. Davidson, J. Dent. Res. 65 (1986) 654–658.

[33] E. Rabinowicz, Friction and Wear of Materials (Wiley, New York, 1965).

[34] B.R. Lawn, Wear 33 (1975) 369.

[35] K. Phillips, G.M. Crimes and T.R. Wilshaw, Wear 41 (1977) 327.

[36] G.M. Crimes, Ph.D. Thesis (University of Sussex, 1973).

[37] N.G. Hartley, Ph.D. Thesis (University of Sussex, 1971).

[38] T.A. Roberts, ICI Chemicals & Polymers, private communication (1990).

[39] M. Godet, The third-body approach, a mechanical model of wear, Wear 100 (1984) 437–452.

[40] H. Voss and K. Friedrich, On the wear behavior of fiber reinforced thermoplastics sliding against smooth steel surfaces, Proc. Int. Conf. Wear of Materials (ASME, Vancouver, Canada, April, 1985) pp. 751–764.

[41] H. Voss, Aufbau, Bruchverhalten und Verschleißeigenschaften kurzfaserverstärkter Hochleistungs-thermoplaste, Fortsch.-Ber. VDI-Zeitschr. Reihe 5, Nr. 116 (VDI-Verlag, Düsseldorf, 1987).

[42] J.K. Lancaster, Composite self-lubricating bearing materials, Proc. Inst. Mech. Eng. 182(1, 2) (1967/1968) 33.

[43] J.P. Giltrow and J.K. Lancaster, Carbon-fiber reinforced polymers as self-lubricating materials, Proc. Inst. Mech. Eng. 182(3N) (1967/1968) 147.

[44] J.K. Lancaster, The effect of carbon fiber reinforcement on the friction and wear of polymers, Br. J. Appl. Phys. Ser. 2, 1 (1968) 549.

[45] J.K. Lancaster, Composite for aerospace dry bearing applications, in: Friction and Wear of Polymer Composites, ed. K. Friedrich (Elsevier, Amsterdam, 1986) ch. 11, pp. 363–396.

[46] J.K. Lancaster, Lubrication of carbon fiber-reinforced polymers, part I – Water and aqueous solution, Wear 20(3) (1972) 315.

[47] J.K. Lancaster, Lubrication of carbon fiber-reinforced polymers, part II – Organic fluids, Wear 20(3) (1972) 335.

[48] N. Ohmae, K. Kobayashi and T. Tsukizoe, Characteristics of fretting of carbon fiber reinforced plastics, Wear 29(3) (1974) 345.

[49] T. Tsukizoe and N. Ohmae, Friction properties of advanced composite materials, Proc. JSME-ASME Joint Western Conference on Applied Mechanics (Hawaii, March, 1975) JSME paper No. D-2.

[50] K. Tanaka, Effects of various fillers on the friction and wear of PTFE-based composites, in: Friction and Wear of Polymer Composites, ed. K. Friedrich (Elsevier, Amsterdam, 1986) ch. 5, pp. 137–174.

[51] T. Tsukizoe and N. Ohmae, Friction and wear of with parallely oriented fiber reinforced plastics – Tribological assessment for CFRP, GFRP and SFRP, J. Jpn. Soc. Lubr. Eng. 21(5) (1976) 330. In Japanese.

[52] T. Tsukizoe and N. Ohmae, Friction properties of composite materials, Trans. Jpn. Soc. Mech. Eng. 43(367) (1977) 115. In Japanese.

[53] T. Tsukizoe and N. Ohmae, Wear mechanism of unidirectionally oriented fiber-reinforced plastics, Trans. ASME 99 F(4) (1977) 401.

[54] T. Tsukizoe and N. Ohmae, Friction and wear performance of unidirectionally oriented glass, carbon, aramid and stainless steel fiber-reinforced plastics, in: Friction and Wear of Polymer Composites, ed. K. Friedrich (Elsevier, Amsterdam, 1986) ch. 7, pp. 205–231.

[55] M. Yukumoto, T. Tsukizoe and N. Ohmae, Systems approach to the wear of fiber-reinforced plastics, J. Jpn. Soc. Lubr. Eng. 23(12) (1978) 881. In Japanese.

[56] N. Ohmae, M. Yukumoto and T. Tsukizoe, Analysis of system structure in the wear of FRP, Proc. 2nd. European Tribology Congress (Düsseldorf, October, 1977) Band II/III, 57/1.

[57] H.M. Hawthorne, Wear in hybrid carbon/glass fiber epoxy composite materials, in: Proc. Int. Conf. on Wear of Materials, Reston, VA, April 11–14, 1983, ed. K.C. Ludema (ASME, New York, 1983) pp. 576–582.

[58] V.K. Jain and S. Bahadur, Experimental verification of a fatigue wear equation, Wear 79 (1982) 241–253.

[59] Z. Eliezer and V.D. Khanna, On the effect of fiber orientation on the wear of composite materials, Wear 53 (1979) 387–389.

[60] B.J. Briscoe, L.H. Yao and T.A. Stolarski, Wear 108 (1986) 357.

[61] J.K. Lancaster, In: Proc. Int. Conf. on Tribology in the '80s, Vol. 1 (NASA Conference Publication, 1983) p. 333.

[62] M. Cirino, K. Friedrich and R.B. Pipes, Composites 5 (1988) 383.

[63] H. Harel, J. Aronhime, K. Schulte, K. Friedrich and G. Marom, J. Mater. Sci. 25 (1990) 1313–1317.

[64] H. Voss and K. Friedrich, Wear 116 (1987) 1–18.

[65] J. Krey, K. Friedrich and O. Jacobs, VDI-Berichte 734 (VDI-Verlag, Düsseldorf, 1989) 179.

[66] O. Jacobs, K. Friedrich, G. Marom, K. Schulte and H.D. Wagner, Wear 135 (1990) 207.

[67] K. Friedrich, Friction and Wear of Polymer Composites, Fortschr.-Ber. VDI-Z, Series 18, No. 15 (VDI-Verlag, Düsseldorf, 1984).

[68] K. Friedrich and J.S. Wu, Polymer composites with high wear resistance, in: Encyclopedia of Composite Materials, Vol. 4, ed. S.M. Lee (VCH Publishers, New York, 1991) pp. 255–279.

[69] K. Friedrich, in: Delaware Composite Design Encyclopedia, eds L.A. Carlsson and J.W. Gillespie Jr., Vol. 4, Failure Analysis of Composite Materials (Technomic Publ., Lancaster, USA, 1990) pp. 139–206.

[70] R.M. Turner and F.N. Cogswell, In: Proc. 18th Int. SAMPE Technical Conference, Seattle, Washington, USA, 7–9 October (1986) pp. 32–44.

[71] H. Voss and K. Friedrich, Wear performance of a bulk liquid crystal polymer and its short fiber composites, Tribol. Int. 19 (1986) 145–156.

[72] E. Hornbogen, Friction and wear of materials with heterogeneous microstructure, in: Friction and Wear of Polymer Composites, ed. K. Friedrich (Elsevier, Amsterdam, 1986) pp. 61–88.

[73] M. Cirino, R.B. Pipes and K. Friedrich, The abrasive wear of continuous fiber polymer composites, J. Mater. Sci. 22 (1987) 2481–2492.

[74] M. Cirino, K. Friedrich and R.B. Pipes, The effect of fiber orientation on the abrasive wear behaviour of polymer composite materials, Wear 121 (1988) 127–141.

[75] P.B. Mody, T.-W. Chou and K. Friedrich, J. Mater. Sci. 23 (1988) 4319.

[76] P.B. Mody, T.-W. Chou and K. Friedrich, in: ASTM STP 1003, ed. C.C. Chamis (American Society for Testing and Materials, Philadelphia, PA, 1987).

[77] M.N. Gardos and B.D. MacConnell, Development of a high-load, high-temperature self-lubricating composite, Parts I–IV, ASLE-Prepr. 81-3A-3, 81-3A-6 (1981).

[78] M.N. Gardos, A.E. Castillo, J.W. Herrick and R.A. Sonderlund, Solid lubricated turbine bearings, in: ASLE Proc. SP-14 Int. Conf., Denver, CO, August 7–10 (1984) pp. 248–257.

[79] Tribo-Comp TDF, Brochure (Tiodize Company, Huntington Beach, CA, 1985).

[80] G.A. Cooper and J. Berlie, New bearing materials for high speed dry operation, Wear 123 (1988) 33–50.

[81] G.A. Cooper and J. Berlie, Stiff and strong bearing materials for high speed dry operation, Wear 123 (1988) 51–58.

[82] M.K. Omar, A.G. Aktins and J.K. Lancaster, The role of crack resistance parameters in polymer wear, J. Phys. D 19 (1986) 177–195.

[83] K. Holmberg and G. Wickström, Friction and wear tests of polymers, Wear 115 (1987) 95–105.

[84] H.L. Price, Selecting polymer composites for friction-wear applications, Modern Plastics 17 (1987) 114.

[85] Y.T. Wu, How short aramid fibers improve wear resistance, Modern Plastics 18 (1988) 67.

[86] S.C. Lim and M.F. Ashby, Wear mechanism maps, Acta Metall. 35(1) (1987) 1–24.

PART III

Tribology of Ceramic, Glass and
Metal Matrix Composites

Advances in Composite Tribology
edited by K. Friedrich

Chapter 7

Tribology of Ceramic Matrix Composites against Metals

KANAO FUKUDA and MASANORI UEKI

Advanced Materials & Technology Research Laboratories, Nippon Steel Corporation, 1618 Ida, Nakahara-ku, Kawasaki 211, Japan

Contents

Abstract

The sliding of silicon nitride matrix–silicon carbide whisker composites against bearing steel has been investigated, and it was observed that the wear of the ceramic specimens decreased with increasing amount of incorporated whiskers. In hot-pressed silicon nitride based ceramics, the appearance of anisotropy in the mechanical properties has generally been observed. Such an anisotropy was also found in the tribological properties of those ceramics. A series of investigations on ceramics sliding against pure iron was also carried out, with as result that composite ceramics did not always show better wear resistance than monolithic ones. The tribological compatibility must be considered with first priority when the sliding countermaterial is selected. In this chapter, the discussion is based on the viewpoint of the adhesive-wear mechanism with special emphasis on the tribological compatibility of the sliding materials

1. Introduction

At present, ceramics used practically which are candidates for use as tribological parts have three distinct features which no other materials have. Namely: (1) their mechanical properties are the major controlling factors in their selection for tribological use, (2) they are applied taking their characteristics, such as heat and corrosion resistance, into account before considering their tribological properties, and (3) the partners which they are sliding against, are metals and alloys in most cases. The features described above express the present status of the development and understanding of ceramics. Primarily, the reason why the tribological properties of ceramics are discussed based on their mechanical properties [1], is that there is fear of abrupt fracture of the ceramics in use. In many cases, the tribological parts are at the same time parts of machinery, and that their reliability prevents abrupt fracture during operation is an essential requirement for their application. Therefore, it is more important that the materials to be used have appropriate strength and toughness than good tribological properties. One of the advantages of ceramics over other materials is their excellent chemical stability at high temperature and/or in a corrosive atmosphere. Therefore, it may happen that ceramics are used for the tribological parts for the only reason that the environment at usage is the above-described atmosphere. However, it is well known that a tribo-chemical reaction, which is an enhanced chemical reaction, occurs in a sliding environment. Under these circumstances, since the compatibility [2] of the sliding pair also changes, there is the possibility to have progressive wear of the ceramics. Ceramics are applied in tribological parts as substitution of the metallic parts presently used, and, therefore, in many cases their sliding partners are metallic materials. Since the ceramics are still in a transient (incubation) period of development for practical application, the optimum method to utilize them has not yet been developed, as, for example, what is the best material for the sliding partner of ceramics.

In this chapter, the incorporation of a second phase into ceramics as method to improve their reliability and its effect on the tribological properties in sliding with metallic materials are described. There are two ways to improve the reliability of ceramics, which are the elimination of small pores (defects), mainly through process innovation, and increasing the fracture toughness through the control of crack propagation. The method to increase the toughness described here is the incorporation of particulates, whiskers and fibres in the ceramic matrix. Using this method, toughening can be achieved by preventing rapid propagation of cracks in the ceramics. Such prevention is exerted by absorbing the energy applied to the parts through deflection and/or branching [3] of the cracks and pulling-out [4] of whiskers and fibres when the cracks meet obstacles like particulates, whiskers, and fibres. As result of the toughening effort by the incorporation of second phases, practical commercial materials such as Al_2O_3(alumina)–SiC(silicon carbide)-whisker (SiC_w) and Al_2O_3–TiC(titanium carbide) composite materials have been developed.

Adhesion, cutting, and fatigue are proposed and discussed so far as the major wear mechanisms of the materials, as schematically shown in fig. 1. Of these, adhesion is considered as the controlling mechanism for sliding wear (fig. 2). The wear through adhesion is determined by the shear strength of the materials composing the sliding system and by the intensity of the adhesive force between the materials in sliding. Figure 2 shows a model of the sliding of ceramics against

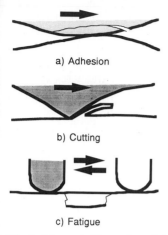

a) Adhesion

b) Cutting

c) Fatigue

Fig. 1. Schematic illustration of the proposed wear mechanisms.

metals. When both materials adhere intensively at the sliding contact interface, wear of materials occurs if the adhered junction breaks either at the metal (A) or the ceramic (C) side of the interface. However, if the adhesive force at the interface is small, the materials do not wear, since the junction breaks exactly at the contact interface when the shear stress is applied to the sliding system [5]. The shear fracture at the sliding interface occurs not always in the side of the material with lower strength. It also depends on the geometry of the contact zone. However, as a result, the amount of wear of the materials composing the sliding system depends on the probability of occurrence of breaking inside one of the materials, i.e. depends upon the shear strength of the materials.

When we consider the mechanism of wear based on the aforementioned adhesion, it is apparent that in order to minimize the wear, the shear fracture has to be made to occur exactly at the contact interface, by either strengthening the materials or lowering the adhesive force between the materials composing the sliding system. From these considerations, as the important material factors affecting adhesive wear, the mechanical property (strength) and the chemical property (compatibility) [2,5] between the materials composing the sliding system, can be extracted. Especially the compatibility between the materials is important,

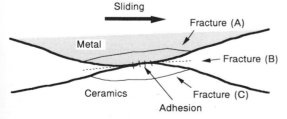

Fig. 2. Mechanism of adhesive wear.

because if we choose a material which tends to react and adhere at the interface, as the counter part in the sliding system, this leads to an increase [6] in both the resistance and the amount of wear, even if the materials have a high hardness with appropriate stiffness. Inversely, a sliding pair with a good compatibility exhibits good sliding properties with a smaller amount of wear, even if the strength and hardness of the materials composing the sliding pair are relatively low. Ag/Fe [6] and Cu/Al_2O_3 [7] are examples of such a kind of sliding pair. Concerning the efforts to understand the factors controlling the compatibility between materials, there are several interpretations. For example, one [6] is based upon the phase diagram of the binary system forming the sliding pair for a combination of pure metals and another is based upon the oxidizing activation [7] of metals in the sliding wear of some pure metals against Al_2O_3. Unfortunately, these are only specific solutions, and a general solution for the practical wear phenomena is not yet found.

The atmosphere, with factors such as the humidity [8] and the existence of oxygen [9], must be taken into account as a chemical factor affecting the wear. In some cases, it is effective to decrease the wear through forming a protective film consisting of oxides and/or hydrates, or, on the other hand, there are cases that it increases the wear by increasing the adhesive force at the contact interface, depending upon the combination of materials and/or sliding conditions.

Although there are some papers [1,10] describing that the incorporation of whiskers and/or particulates in the ceramics is effective for decreasing the wear of ceramics, the sliding pairs appearing in these papers are self-mated materials, i.e. combinations of the same ceramic composites and composites against ceramics. Thus, it was difficult to find papers reporting on the sliding wear between ceramic matrix composites and metallic materials. However, as mentioned previously, there is a high possibility that ceramics are chosen as the sliding counterpart of metals. In the following, the results of investigations on the sliding wear of ceramic matrix composites against metallic materials are described. From the tribological point of view, the expectations for ceramic matrix composites are not only improved mechanical properties, such as strength and toughness, compared to the mono- lithic ceramic but also improved compatibility with the sliding partner through changes of the chemical properties. The latter, the changes of the chemical properties, influences the wear properties significantly. For example, even in the case of macroscopic hybridization, such as using a sandwich structure [11], it was reported that an enhanced transition of severe to mild wear [12] and a sudden decrease in the higher wear rate in the initial stage of sliding during the course of wear of metallic materials, occurred through the formation of a protective layer on the sliding surface [13]. The occurrence of mild wear was confirmed by Yust and Allard [10] for self-mated sliding of Al_2O_3–SiC-whisker composite, and it is considered that an investigation of the effects of changes in chemical properties on the wear is important for future application of ceramic matrix composites.

The tribology of ceramic matrix composites in sliding with metallic materials is described and discussed in this chapter with special emphasis on the compatibility-related chemical properties. More precisely, the effect of the incorporation of SiC whiskers in both silicon nitride- and alumina-based ceramics on the sliding-wear properties for sliding against bearing steel is described. Also, the effect of structural anisotropy on the tribological properties of SiC-whisker reinforced silicon nitride ceramics, prepared using the hot-pressing method, was investigated. Finally, the tribological properties of various ceramic matrix composites in sliding with pure iron are discussed.

2. Apparatus and experimental procedure

A pin-on-disk type wear tester, whose appearance and the test principle are shown in fig. 3, was used in the experiment. The dimensions and geometry of the specimens are also shown in the figure. The test section of the apparatus was covered by a closed chamber, in which the atmosphere could be controlled. For the test in vacuum a pressure of the order of 10^{-6} Pa could be achieved by evacuation using a turbomolecular pump. The level of vacuum was checked using an ionization gauge and the remaining gases were detected by the mass-spectrum analyzers.

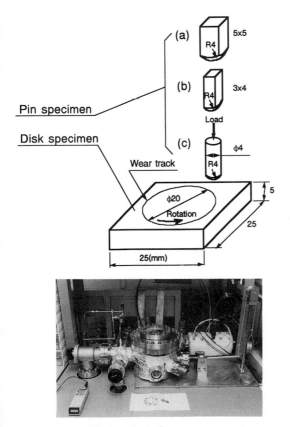

Fig. 3. Shape and geometry of the test pieces for both the pin and the disk, and general view of the pin-on-disk wear tester.

The load was applied to the pin specimen by a lever from the outside of the closed chamber via vacuum bellows. the applied load and the friction force could be detected by a strain gauge placed on a point on the lever (50 mm away from the end). The specimen was placed at the end of the lever. Each reading of the vacuum level, the friction force and the load was recorded using a personal computer after A/D conversion. The load was measured only during the calibration and the vacuum level was monitored once every 10 s. The friction force was recorded during 1 s (150 times) with 10 s intervals and the friction coefficient was calculated by dividing by the applied load.

The sliding surfaces of both the ceramic pin and the bearing-steel disk were polished to an average roughness, R_a, of 0.08 μm or less using diamond paste. The extent of the wear was detected by measuring the weight of the samples before and after testing on a 10^{-2} mg scale using an electronic scale and then calculating the difference in weight. To eliminate the effect of contamination and attachment of wear debris, the weight of the specimens was measured after drying prior to cleaning in an acetone bath with application of ultrasonic vibration. Both the pin and the disk specimens were treated carefully to avoid contamination from oil elements and were then placed in the tester, if it was necessary the atmosphere was changed before the tests.

3. Effect of SiC whisker addition on the tribological characteristics of silicon nitride ceramics when sliding against bearing steel

At the start of a series of investigations on the tribology of ceramic matrix composites, the effect of the amount of the phase incorporated in the composite materials, on the wear properties must be known [14]. As a typical example, the amount of SiC whiskers in Si_3N_4 matrix was studied. The ceramic samples used were hot-pressed SiC-whisker reinforced Si_3N_4 ceramics prepared in-house [15], where the amount of whiskers was varied in the range from 0 to 30 wt%. Table 1 shows the composition and some mechanical properties of the ceramic specimens used. The experimental conditions in the pin-on-disk tests are shown in table 2.

TABLE 1
Composition and mechanical properties of the ceramic specimens.

Matrix	SiC whiskers (wt%)	Sintering method	Density (g/cm³)	Flexural strength (MPa)	Fracture toughness (MPa m$^{1/2}$)	Hardness (Hv)
Si_3N_4	0	Hot pressing	3.24	1029	5.5	1540
	10		3.24	1025	6.7	1820
	20		3.22	989	6.9	1990
	30		3.23	940	7.7	1790
SiC (Ibiden SC850)	0	Normal sintering	3.11	850	2.5	2400

TABLE 2
Experimental conditions.

Sliding speed (mm/s)	100
Load (N)	28
Sliding distance (m)	1000
Atmosphere	dry air

The disk sample was of bearing steel (JIS SUJ2). The geometry of the pin sample, made of ceramics, is shown in fig. 3a.

The experimental results, i.e. the effect of the amount of whisker contents on the wear of the ceramic pin and bearing-steel disk, are shown in fig. 4. The wear of the ceramic pin decreased with increasing amount of SiC whiskers. The wear of the pin with 30 wt% SiC_w was smaller than that of pure SiC ceramics (100 wt% SiC). This tendency agreed well with the description in refs. [1,10] that the addition of whiskers suppressed the wear of ceramics. From the observation and analysis of the sliding surface of the ceramics was found that, although a large amount of Fe transfer and extensive adhesion occurred for the sample without containing whiskers, the percentage of this extensive adhesion decreased with increasing whisker content up to 30 wt%, and the adhered material became finer and flaky (fig. 5). Using an optical microscope, it was also observed that adhesion of Fe occurred selectively to the silicon nitride matrix, avoiding the silicon carbide whiskers (fig. 6). This observation was consistent with the phenomenon that massive adhesion of iron occurred to the silicon nitride ceramics in spite of the little adhesion to the silicon carbide.

According to the results described above, the reasons for the decreasing wear of silicon nitride due to the incorporation of silicon carbide whiskers, are not only the improvement of the mechanical properties but also the change in the compatibility with Fe, resulting in fracture of the junction at (B) in fig. 2 because of the decreasing adhesion.

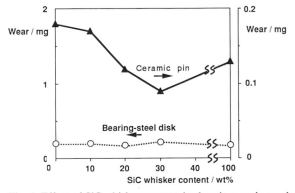

Fig. 4. Effect of SiC-whisker content in the pin sample on the amount of wear of both the ceramic pin made of SiC-whisker reinforced Si_3N_4 ceramics, and the disk made of bearing steel.

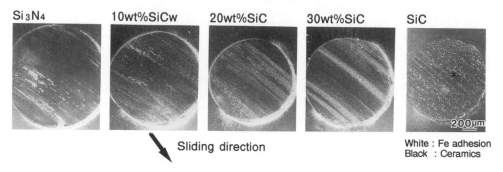

Fig. 5. SEM micrographs of worn surfaces of Si_3N_4, its composites and SiC, showing the Fe adhesion on the surface.

4. Tribological anisotropy of silicon nitride–silicon carbide whisker composites

It is known that a material having structural and mechanical anisotropy exhibits anisotropy also in the tribological properties. For example, in polymeric materials reinforced with unidirectionally oriented carbon fibres, the tribological properties varied depending on the sliding surface and direction, which can be perpendicular or parallel to the orientation of the fibres [16]. And in the case of the sliding surface with the fibre oriented parallelly, the properties are different again depending on the sliding direction being longitudinal or transversal to the fibres. In many cases composite materials show structural anisotropy, therefore it is important to realize that tribological anisotropy exists when composite materials are applied as sliding parts.

The test pieces prepared for the experiments were sintered bodies of Si_3N_4 and Si_3N_4–SiC-whiskers composites. They are all consolidated by hot-pressing; the method is uniaxial pressing using dies and constraining the outer sides using a rigid mold, which results in anisotropy of the sintered bodies. Actually, since anisotropy in the flexural strength [17] and the fracture toughness [18] was confirmed for

Black : Fe adhesion
White : SiC whisker
Gray : Si_3N_4 matrix

Fig. 6. Fe adhesion on the sliding surface of the Si_3N_4–SiC-whisker composite.

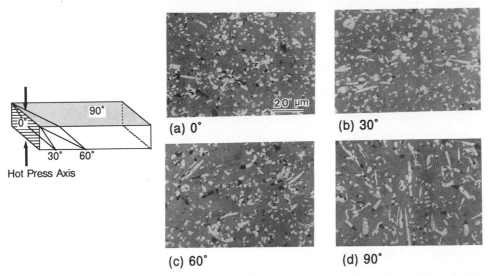

(a) 0°   (b) 30°

(c) 60°   (d) 90°

Fig. 7. Optical micrographs of the various surfaces (planes) shown in the schematic drawing on the left, having an angle with the hot-pressing axis of (a) 0°, (b) 30°, (c) 60°, and (d) 90°.

silicon nitride ceramics and Al_2O_3- or Si_3N_4-based SiC-whisker composites, it is reasonable to assume that their tribological properties also exhibit anisotropy.

The materials used were Si_3N_4 and Si_3N_4–20 wt% SiC-whisker composite, see table 1. Figure 7 shows optical micrographs of surfaces of the composite with angles of 0, 30, 60, and 90° with the hot-pressing direction, as schematically shown in the left part of the figure. The cross-sectional images of the SiC whiskers in the figure indicate that the maximum aspect ratio of the whiskers appears in the case of 90°, where the whiskers are aligned with the surface (plane). Pin test pieces were machined with the sliding surfaces identical to each of the surfaces shown in fig. 7 (see fig. 8). The disk sample which was used as the counterpart for sliding, was chromium-plated carbon steel and the thickness of the plating was about 90 μm. The test conditions adopted are summarized in table 3. It was confirmed

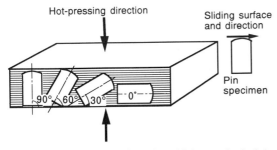

Fig. 8. Schematic illustration of machining method of the pin specimens having various angles with the hot-pressing direction.

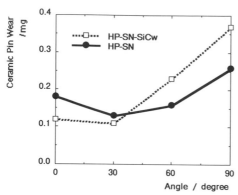

Fig. 9. Effect of anisotropy on the wear of hot-pressed ceramics.

that in each test sliding was performed between the ceramic pin and the chromium plate without touching the steel substrate.

Figure 9 shows the variation of the amount of wear of the ceramic pin with the angle between the sliding surface and the hot-pressing direction of the pin sample. From the figure it is apparent that hot-pressed silicon nitride has certainly anisotropy in its wear properties and improvement of the tribological properties due to incorporation of whiskers was brought about when the angle was lower than 60°. The angle dependence of the wear was higher for the whisker-reinforced materials than for silicon nitride ceramics without whiskers, i.e. the former showed very significant anisotropy. In the materials tested here, the wear became minimal for an angle of 30°.

5. Tribology of various ceramic matrix composites against pure iron

The composite ceramics used were SiC-whisker reinforced Al_2O_3, TiC-particulate dispersed Al_2O_3 and SiC-whisker reinforced Si_3N_4, where the former two were the commercial products designated WG300 (Greenleaf corp.) and A65 (Kyocera), respectively, and the last one was prepared in-house. For comparison, various commercial ceramics, Al_2O_3(A479/Kyocera), Si_3N_4(FX950/Toshiba tungalloy) and SiC(SC850/Ibiden), were also tested under the same conditions. In order to investigate the compatibility of the sliding pair, tests were conducted under high-vacuum conditions of 1.5×10^{-5} Pa, which eliminates the effects of the atmosphere on the wear. The other test conditions are the same as those summa-

TABLE 3
Pin-on-disk test conditions.

Sliding speed (mm/s)	62.8
Load (N)	8.3
Sliding distance (m)	314
Atmosphere	dry air

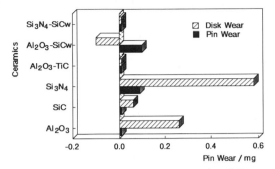

Fig. 10. Amount of wear in the Fe/ceramics sliding system.

rized in table 3. In these tests pure iron was used as pin sample (fig. 3c) and each ceramic material was used as disk sample.

The test results are shown in fig. 10. The weight change of the disk sample does not necessarily represent the amount of wear, mainly due to adhering of iron from the pin sample, like the case of WG300, which exhibited a weight increase. By comparing the amount of wear of both monolithic ceramics, Al_2O_3(A479) and Si_3N_4(FX950), with those of the composite ceramics, A65 and SiC-whisker reinforced Si_3N_4, an apparent decrease in wear by the incorporation of the second phase can be observed for each matrix. Although both the pin and the disk sample were worn extensively in the test of FX950, only little wear was observed for both the pin and the disk for the case of SiC-whisker reinforced Si_3N_4. This improvement in the tribological properties was brought about by the change in compatibility due to the incorporation of SiC whiskers in the Si_3N_4. As apparent from the wear-test results with SiC(SC850), there was good tribological compatibility between SiC and iron. It was also estimated that the incorporation of TiC particles in Al_2O_3 was effective to improve the tribological compatibility of Al_2O_3 and iron. In the tests of SiC-whisker reinforced Al_2O_3(WG300) significant wear of the iron pin as well as weight gain of the ceramic disk was observed. Since the compatibility between Al_2O_3 and iron is not so good, intensive adhesion occurred at the sliding interface. This kind of junction formed by adhesion, tends to break at the Al_2O_3 side of the interface, due to its lower shear strength. Through these processes, the wear of alumina became significant. Even in the case of incorporation of SiC whiskers in Al_2O_3, the compatibility of Al_2O_3 with iron was not improved progressively, and intensive adhesion occurred at the sliding interface. Since the shear strength of alumina was increased by the incorporation of SiC whiskers, the breaking of the joint interface (contact) was shifted to the iron side. So the wear of iron increased significantly. In Si_3N_4-based ceramics, the incorporation of SiC whiskers improved not only the mechanical properties of the matrix but also the compatibility of it with iron. The reduction of the adhesion by the change of the chemical properties at the sliding interface decreased the wear of both the ceramics and the iron because the breaking of the junction occurred exactly at the contact interface.

6. Future image of tribology of composite ceramics

Although it is apparent that ceramic matrix composites have various attractive characteristics for tribological applications, to get further spreading in their application in the near future, it is required to thoroughly understand the materials characteristics and also to develop the materials taking their wear mechanisms into account.

In this chapter, as example, the effect of structural anisotropy on the wear properties of ceramic matrix composites were described, which is important for actual application of composite materials. From the standpoint of investigation of the wear mechanism, adhesion was discussed in the sliding wear between iron and ceramic matrix composites. Since the tribological phenomena are very complicated, depending on various factors, the results described in the text might be only a minor example. Although it is desirable to collect and store data for future development of practical application(s) (methods) of ceramics, the application in the fields for which the characteristics of ceramics are really sought, has to be performed based on a deep understanding of their characteristics.

We would be very pleased if this chapter has a role in promoting the usage of ceramic matrix composites as tribological materials.

List of symbols

R_a average roughness (μm)

References

[1] C. Lhymn, Effect of normal load on the specific wear rate of fibrous composites, Wear 120 (1987) 1–27.
[2] E. Rabinowicz, The determination of the compatibility of metals through static friction tests, ASLE Trans. 14 (1971) 198–205.
[3] K.T. Faber and A.G. Evans, Crack deflection processes – I, Theory, Acta Metall. 31(4) (1983) 565–576.
[4] P.F. Becher, C.H. Hsueh, P. Angelini and T.N. Tiegs, Toughening behavior in whisker-reinforced ceramic matrix composites, J. Am. Ceram. Soc. 71(12) (1988) 1050–1061.
[5] F.P. Bowden and D. Tabor, The Friction and Lubrication of Solids (Clarendon Press, Oxford, 1954).
[6] T. Sasada, S. Norose and H. Mishina, Effect of mutual alloyability on metallic wear, in: Proc. 20th Japan Congress on Materials Research (1977) pp. 99–104.
[7] A. Enomoto, K. Hiratsuka and T. Sasada, Friction and wear between metal and oxide, in: Proc. 5th Int. Congr. on Tribology (EUROTRIB'89), Vol. 1 (1989) pp. 111–116.
[8] K. Fukuda, S. Norose and T. Sasada, Effect of humidity on sliding wear in Fe and Cu rubbing system, in: Proc. 28th Japan Congress on Materials Research (1985) pp. 89–92.
[9] G.J. Tennenhouse and F.D. Runkle, The effects of oxygen on the wear of Si_3N_4 against cast iron and steel, Wear 110 (1986) 75–81.
[10] C.S. Yust and L.F. Allard, Wear characteristics of an alumina–silicon carbide whisker composite at temperatures to 800°C in air, STLE Tribol. Trans. 32(3) (1989) 331–338.
[11] S. Norose, T. Sasada and K. Maruyama, Anti-wear property of two phasic metal composite, in: Proc. 27th Japan Congress on Materials Research (1984) pp. 119–124.
[12] J.F. Archard and W. Hirst, The wear of metals under unlubricated conditions, Proc. R. Soc. A 236 (1955) 397–410.
[13] J.K. Lancaster, The formation of surface films at the transition between mild and severe metallic wear, Proc. R. Soc. A 273 (1962) 466–483.

[14] K. Fukuda, Y. Sato, T. Sato and M. Ueki, Wear properties of SiC-whisker-reinforced ceramics against bearing steel, in: Proc. Conf. on Tribology of Composite Materials, 1–3 May, 1990, eds P.K. Rohatgi, C.S. Yust and P.J. Blau (ASM International, Oak Ridge, TN, 1990) pp. 323–328.

[15] Y. Sato, T. Sato and M. Ueki, Processing and mechanical properties of SiC-whisker reinforced Si_3N_4 composites, 92nd Annual Meeting, American Ceramic Society, April 22–26, 1990, Dallas, Texas, USA (1993).

[16] K. Friedrich, O. Jacobs, M. Cirino and G. Marom, Hybrid effects on sliding wear of polymer composites, in: Proc. Conf. on Tribology of Composite Materials, 1–3 May, 1990, eds P.K. Rohatgi, C.S. Yust and P.J. Blau (ASM International, Oak Ridge, TN, 1990) pp. 227–285.

[17] S.T. Buljan, J.G. Baldoni and M.L. Huckabee, Si_3N_4–SiC composites, Am. Ceram. Soc. Bull. 66(2) (1987) 347–352.

[18] P.F. Becher and G.C. Wei, Toughening behavior in SiC-whisker-reinforced alumina, J. Am. Ceram. Soc. 67(12) (1984) C-267–C-269.

Advances in Composite Tribology
edited by K. Friedrich

Chapter 8

Wear Mapping for Metal and Ceramic Matrix Composites

P.K. ROHATGI [a], Y. LIU [b] and S.C. LIM [b]

[a] *Materials Department, University of Wisconsin-Milwaukee, Milwaukee, WI 53201, USA*
[b] *Department of Mechanical and Production Engineering, National University of Singapore,*
10 Kent Ridge Crescent, Singapore 0511, Singapore

Contents

Abstract

Some issues concerning wear-map construction for metal matrix and ceramic matrix composites have been addressed in this paper. The physical processes by which composites wear during sliding are still imperfectly understood, and this renders the construction and calibration of wear-mechanism maps for these composites more difficult than for monolithic materials, such as steels. However, some possible variables and effects, such as bulk heating and flash heating, have been analyzed. Equations from earlier models for single-phase materials have been adjusted to incorporate the influence of particle dispersion and any film formation at the mating interface. Some maps, from different sources, have been presented in terms of the test variables, and possible shifts of the boundaries between the various competing wear mechanisms due to particle dispersion and changes in the sliding parameters have been discussed.

1. Introduction

There are many mechanisms of wear, including adhesive, abrasive, erosive, oxidative, delaminative and fatigue. Although different mechanisms may operate simultaneously during wear, it can be assumed that for any particular system one will generally dominate, implying that it may account for the greatest part of the wear phenomenon. However, the dominant mechanism may change when the sliding conditions change. A casual observer is struck by the complexity of this change and by the lack of an overall framework or pattern into which the individual wear mechanisms can be fitted. Investigations have so far tended to focus on the details of individual mechanisms, and not on the interrelationships between the competing mechanisms.

Some attempts have been made earlier [1–6] to present an overview of the wear behavior of materials, mostly steels, by producing diagrams relating the sliding conditions to different aspects of wear, such as the rate of wear, and the conditions leading to mild or severe forms of wear and to seizure. These included the work by Okoshi and Sakai [1], Welsh [2], Childs [3], Eyre [4], Marciniak and Otimianowski [5], and Egawa [6]. However, these diagrams only deal with a limited range of sliding conditions (with the exception of Childs' [3]), and the regimes of mild and severe wear described are phenomenological; strictly speaking, a general classification of wear based on the observable results of wear rather than based on the dominant wear mechanisms operative in that regime.

One way to more satisfactorily examine this overall behavior, as suggested by Tabor [7], might be to construct wear-mechanism maps (by analogy with deformation-mechanism maps [8]). These wear-mechanism maps should summarize data and models for wear, showing how the mechanisms operate and interact, allowing the dominant mechanism for any given set of conditions to be identified, and to indicate the conditions under which a particular mechanism may cease to be effective [9]. Such an approach has been adopted by Lim and Ashby [10], which resulted in a presentation of a wear-mechanism map summarizing the unlubricated wear behavior of steels over a very wide range of load and sliding velocities (fig. 1). There are in principle no obvious obstacles for applying this approach to other materials, or combinations of materials, assuming that all materials exhibit distinct wear behaviors characteristic of predominant modes. Adopting the same methodology, a wear-mechanism map for aluminum and aluminum alloys was recently proposed by Liu et al. [11] – a considerable improvement over the early empirical wear map for aluminum alloys [12].

Earlier, attempts have also been made to provide an overall framework for lubricated wear in the form of diagrams and maps. Notable examples include the map for steels under lubricated conditions as a function of the specific film thickness [13] and the transition diagram of the International Research Group on Wear of Engineering Materials (IRG-OECD) [14,15]. Beerbower's map [13] is based on a large body of experimental data from the technical literature, and is qualitative in nature; the IRG diagram, on the other hand, presents the critical load–velocity curves for failure in sliding, lubricated, concentrated contacts, de-

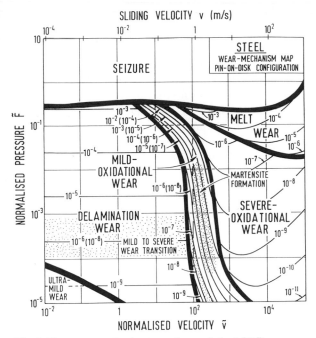

Fig. 1. The wear-mechanism map for steels (ref. [10]).

marcating three regimes of different tribological behavior. More recently, Akagaki and Kato [16–18], considering abrasion as the fundamental mechanism of wear of well-lubricated machine elements, proposed several wear-mode diagrams for the lubricated wear of steels, relating the dominant wear mechanism to the sliding velocity and the contact pressure. The effects of material hardness and lubricant additives on the wear-mode diagrams were also discussed.

This concept of wear-mechanism mapping is not restricted to metals: substantial progress has been achieved in the development of wear maps for ceramics. Two major research groups have proposed ceramic wear maps based largely on their own experimental data [19–25]. In these diagrams, regimes of different dominant wear mechanisms are demarcated and, in some, the sliding conditions leading to transitions between mild and severe wear are indicated. Development in mapping polymeric wear has, however, so far been slow, with the exception of the deformation map for the unlubricated sliding of PTFE on steel [26].

In view of the current paucity of quantitative wear-mechanism maps in the literature, the construction of new wear maps for other advanced and engineered materials used in sliding applications, as well as the continual refinement of existing ones, is, therefore, an area where increasing research efforts should be directed. Some issues related to the construction of wear-mechanism maps for a few metal matrix composites have been addressed recently [27]. In what follows, the basis and various considerations involved in the construction of wear-mechanisms maps for both metal matrix and ceramic matrix composites will be enumerated and elucidated. Special attention is given to the wear map of aluminum–graphite-particle composites, which show lower friction and seizing tendency compared to monolithic aluminum alloys.

2. Theoretical basis for the construction of wear maps for composites

Engineered composites, a new generation of tailor-made materials, have been of considerable interest in tribology research in recent years. The cost advantage of particle-dispersed composites and the improvements in engineering properties such as wear and seizure resistance, make them attractive candidates for various industrial applications [28–31], including bearings, electrical contacts, liners, brakes, and wear plates.

Aluminum-based solid-lubricant composites reduce material loss and friction under both dry and lubricated conditions, and these have been tried out as automotive components [29,30], including bearings, pistons, and liners. The abrasive-wear resistance of some aluminum-based ceramic-particle composites is as high as that of steel. Ceramics are particularly useful for high-temperature service conditions. However, extensive application of such materials has been limited by their brittleness. This disadvantage can be overcome by the incorporation of a second phase in the form of fibres, whiskers, or particles, giving rise to ceramic matrix composites. Ceramic-based composites are potentially suitable for applications such as bearings and liners for low heat rejection engines or gas turbines [20,32,33].

While a considerable wealth of empirical information is available on the wear and friction characteristics of composites, the capability to predict these properties by elucidating the operative mechanisms of wear is a relatively recent phenomenon. In particular, the kinetics of formation of a lubricating film, its stability and the chemistry of this film in metal matrix composites have only recently attracted some attention [28]. The phenomenon of film formation also appears in ceramic matrix composites, as reported recently [34].

2.1. Basic variables

According to Lim and Ashby [10] a wear-mechanism map should clearly delineate the wear rate and the regime of dominance of the wear mechanisms. The maps for steel were constructed by using two converging approaches:
  (1) by plotting experimental wear data and normalized test variables and regrouping them according to the experimentally observed mechanisms of wear, and
  (2) by employing model-based equations describing the wear rate caused by each mechanism, and by correlating these equations to the experiments.
In composite materials, since experimental information on the wear mechanisms and transitions between different mechanisms are limited and somewhat controversial, a true correlation of the model equations proposed here with experiments is difficult. The approach proposed here, therefore, does not purport to be

predictive, but merely attempts to provide one possible methodology for construct-ing wear maps and organizing different wear data.

Since the wear rate, W, of a sliding surface depends upon several mechanisms involving the thermal, mechanical and chemical properties of the materials in-volved, no single choice of variables is ideal to systematize the wear data. However, as proposed by Lim and Ashby [10], the variations and diversity in wear data from different sources, the different sample geometries, and the methods of testing of materials, including composites, may be best described in terms of the following dimensionless variables:

$$\bar{F} = (\text{normalized pressure}) = F/A_n H_0,$$

$$\bar{V} = (\text{normalized velocity}) = Vr_0/a,$$

$$\bar{W} = (\text{normalized wear rate}) = W/A_n,$$

where F is the applied load, A_n the nominal area of contact, H_0 the room-temper-ature hardness, V the sliding speed, r_0 the pin radius, a the thermal diffusivity $(K/C\rho)$, W the wear rate, K the thermal conductivity, C the specific heat, and ρ the density.

For two-phase materials, such as composites, the properties of the composites are different from those of the isotropic matrix alloys. Under certain simplifying assumptions (no interfacial chemical reactions, and particle interactions), the effects of second-phase dispersion on the properties can be accounted for by considering the weighted average values of the properties (rule of mixture). Thus, the effective density, ρ_e, and hardness, H_e, values for composites may be expressed by

$$\rho_e = \rho_m(1 - \phi) + \rho_p\phi \tag{1}$$

and

$$H_e = H_m(1 - \phi) + H_p\phi. \tag{2}$$

The thermal parameters of interest (K, C, ρ and a) can also be obtained for composites by using suitable relationships for two-phase systems, as discussed below. Here, ϕ denotes the second-phase volume fraction, and the subscripts e, m, and p refer to effective composite, matrix and particle, respectively.

2.2. Thermal effects

Since the process of dry sliding of surfaces may generate heat, the local temperature distribution is affected. The frictional heat generated at the mating interface may modify the chemical and metallurgical characteristics of both the surfaces, oxidize, deform plastically or even melt the surfaces. This may cause a change in the dominant wear mechanism. Two important thermal effects relating to the sliding surfaces are the bulk temperature and the flash temperature at the contacting surface asperities. In the following, we use the expressions for the bulk and flash heating temperatures given by Lim and Ashby [10], and, with suitable modifications, apply them to composites.

Since the thermal properties (K, C, a, and ρ) are involved in the definitions of the test parameters used for the construction of wear maps, the thermal properties of the two-phase materials must be obtained from the individual properties of the constituents.

Assuming a homogeneous distribution of equisized spherical particles, the following expression for the effective thermal conductivity may be used [27]:

$$(K_e)_b = \left[\frac{1 - 2\phi\left[1 - (K_p/K_m)\right]/\left[2 + (K_p/K_m)\right]}{1 - \phi\left[1 - (K_p/K_m)\right]/\left[2 + (K_p/K_m)\right]} \right] K_m \tag{3}$$

and

$$(C_e)_b = C_m(1 - \phi_w) + C_p\phi_w, \tag{4}$$

where ϕ_w is the weight fraction of the dispersoid and $(K_e)_b$ and $(C_e)_b$ are the effective values of the thermal conductivity and the specific heat of the composite, respectively.

2.2.1. Bulk temperature

In a pin-on-disk wear test, the heat generated at the sliding surfaces is partitioned between the pin and the disk. Based on Lim and Ashby [10], the basic equation for the bulk surface temperature is given as:

$$T_b = T_0 + \frac{\alpha\mu FVl_b}{A_n K_m}, \tag{5}$$

where T_0 is the ambient temperature, α is the heat distribution coefficient, l_b is the linear diffusion distance, and μ is the coefficient of friction. Representing the heat absorbing capacities of the pin and the disk by their respective heat flux, α can be written as follows [10]:

$$\alpha = \frac{1}{2 + l_b(\pi V/8ar_0)^{1/2}} \tag{6}$$

Equation (5) can be further simplified by introducing some new factors, such as:

$$\beta = l_b/r_0, \tag{7}$$

$$T^* = aH_0/K_m, \tag{8}$$

where r_0, a, and H_0 have been previously defined. For a composite pin, a, H_0, and K should be written in terms of $(a_e)_b$, $(H_e)_b$, and $(K_e)_b$. Substituting eqs. (6), (7), and (8) into eq. (5), one obtains the following expression for the bulk temperature, T_b:

$$T_b = T_0 + \frac{\mu\beta T^*}{2 + \beta(\pi\overline{V}/8)^{1/2}} \overline{FV}, \tag{9}$$

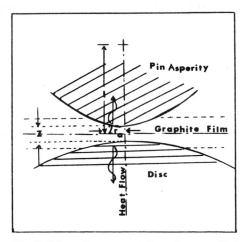

Fig. 2. Schematic representation of a magnified view of the region of contact between two asperities, showing an intervening graphite film between the Al–graphite pin and the steel disk (ref. [27]).

where

$$T^* = \frac{(a_e)_b (H_e)_b}{(K_e)_b} \quad \text{and} \quad (a_e)_b = \frac{(K_e)_b}{(C_e)_b (P_e)_b}.$$

2.2.2. Flash temperature

Figure 2 schematically represents the region of contact between two asperities, with an intervening film of a separate phase. For an Al–steel-disk system, a thin oxide layer may adhere to the bearing pin asperities; for Al–graphite composites, the film may consist predominantly of graphite [35]. Let Z be the thickness of this film between the contact points and let l_f be the total heat diffusion path to the bulk pin, as shown in fig. 2. The heat flow to the pin and the disk will be affected by the presence of this film. Since the thermal diffusivity of graphite is much less than that of Al alloys (see appendix II), the presence of the graphite film causes an obstruction of the heat flow. This is especially true since the graphite film will generally be oriented with its basal plane parallel to the tribosurface.

However, due to the reduced coefficient of friction in Al–graphite composites, the heat flux, $(\mu F V / A_n)$, generated at the interface during dry sliding is also smaller. Hence, relative to the base alloy, the flash temperature in composites such as Al–graphite, is expected to be determined by the competition between the reduced heat flux and the reduced thermal diffusivity of the composite.

Considering the formation of a graphite film at the mating interface, an expression can be derived for the interface or flash temperature when an Al–graphite pin is mated against a steel disk (cf. appendix I and II). The bulk and flash heating equations presented here are approximations, since they represent modifications of the corresponding equations for the monolithic matrix alloy. Since the transitions in wear mechanisms are also influenced by the bulk and flash heating,

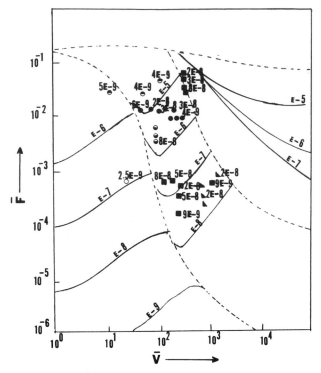

Fig. 3. Experimental wear data on Al–graphite composites superimposed on the map for Al alloys (ref. [27]).

the boundaries between the various interfacing wear mechanisms in the graphite-free base alloys are expected to shift accordingly. However, as shown in fig. 3, a comparison of the theoretical boundaries of the Al-alloy wear map and the Al–graphite-composite wear data [35–46] does not show any congruence between them. This reflects possible basic differences between the wear mechanisms of the matrix alloy and the graphite-dispersed composite when tested against steel under identical conditions, and also the differences in their wear behavior.

3. Wear maps of metal matrix composites

A wear map [11] was presented earlier for Al alloys (fig. 4), by using the empirical wear data and the physical modeling approach of Lim and Ashby [10]. The field boundaries between the various interfacing mechanisms of dry wear were constructed using the critical values of the experimental wear data, resulting in discontinuities observed in the slope of the wear curves. The wear mechanisms successfully modeled for Al alloys include oxidation-dominated wear, delamination wear, severe plastic deformation wear, and melt wear [11].

In the case of aluminum-alloy composites (MMCs), the particle volume fraction in the metal matrix plays an important role in determining its tribological behavior.

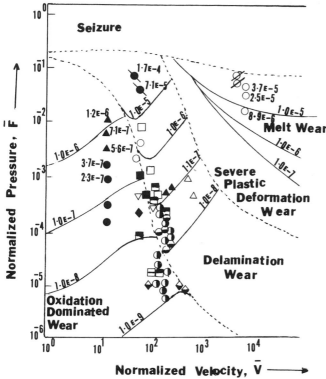

Fig. 4. A quantitative wear-mechanism map for Al alloys, showing experimental wear data and the regions of dominance of each wear mechanism (ref. [11]).

It appears that there is a critical volume fraction of particles which is required for achieving improvements in the wear resistance through a change in the wear mechanism. Gibson et al. [45] reported that only 2 wt% of graphite in Al–Si-alloy matrix composite improved the wear resistance. Rohatgi [28] reported that aluminum alloys do not seize even under boundary lubrication when above 2% graphite is present in their matrix. Similarly, Bruni and Iguera [29] concluded that only 3 wt% of graphite in Al–Cu-alloy MMCs increased the wear resistance significantly. This tends to indicate that about 3 wt% of graphite in Al-alloy MMCs may represent the critical volume fraction for improving the tribological properties of Al–graphite composites.

Above the critical value, the graphite volume fraction may significantly affect the wear rate and possibly the coefficient of friction of composites. Figure 5 shows the influence of the graphite volume fraction on the wear rates in both Al- and Cu-based composites. At a constant absolute sliding velocity and a given normalized wear rate, the normalized pressure, $\bar{F}$, for the same wear rate, increases with increasing graphite percentage; the value of $\bar{F}$ for the composite may increase by an order of magnitude compared to the corresponding value for the base alloy. Figure 6 shows the effect of graphite volume fraction on the velocity to achieve a

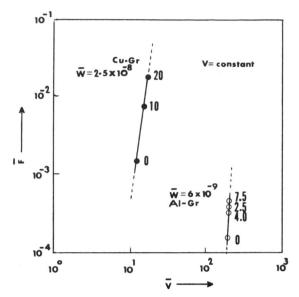

Fig. 5. Experimental normalized isowear-rate lines for Al–graphite and Cu–graphite composites plotted on a normalized $\bar{F}-\bar{V}$ diagram, showing the influence of graphite content and matrix material (ref. [27]).

given wear rate. Under a constant absolute (unnormalized) pressure, with increasing graphite content lower velocities result in the same wear rate. The experimental results show that the wear rate is a relatively weak function of the sliding velocity in Al–graphite composites over the range of sliding velocities ($\leqslant 8$ m/s) studied in this work.

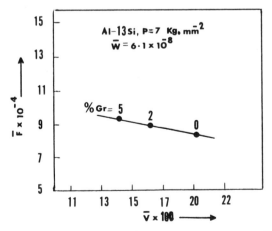

Fig. 6. Influence of the normalized sliding velocity, $\bar{V}$, on the normalized pressure, $\bar{F}$, (at constant absolute load F) for a constant normalized wear rate $\bar{W}$ in Al–Si–graphite composites with various graphite volume fractions (ref. [27]).

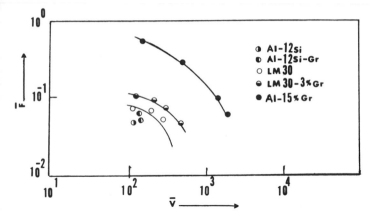

Fig. 7. Movement of seizure boundary in Al–Si–graphite composites in the $\bar{F}-\bar{V}$ field as a function of the graphite content.

The $\bar{F}-\bar{V}$ limit line also depends upon the volume fraction of graphite in the Al–graphite composites. Das et al. [37] have reported that only 3 wt% graphite in a heat-treated LM 30 alloy increased the $\bar{F}-\bar{V}$ limit by about half an order of magnitude; an increase has also been reported in other literature [36,37,43], as shown in fig. 7. It appears that wear data of aluminum composites cannot be superimposed on a well-developed wear map for similar Al matrix alloys [11]. This is not just a reflection of the different wear mechanisms that are operative in Al alloys and their composites, but also of the fact that the empirical wear data do not agree at constant $\bar{F}$ and $\bar{V}$. A forced superimposition of the wear data on Al–graphite composites, on the Al alloy wear map [11] shown in fig. 3, indicates that a quantitative wear map for composites cannot be derived from simple improvement and extention of the wear map of the corresponding monolithic Al alloy.

Figure 8 shows an empirical wear diagram in terms of the dimensionless test parameters $\bar{F}$ and $\bar{V}$, showing the experimental wear data from the literature on a wide variety of composites containing both soft and hard particles. As shown in this figure, composites containing hard, abrading particles like silicon carbide, alumina, zircon and silica, have been tested at relatively low normalized loads ($\bar{F} \leqslant 10^{-4}$), while composites containing soft particles such as graphite, have been tested at relatively high normalized loads and velocities ($\bar{F} \geqslant 10^{-4}$ and $\bar{V} \geqslant 10$), where there is a possibility of more effectively realizing their antifriction and antiwear properties.

As mentioned earlier, the wear mechanisms in composite materials are not well understood and this represents a prime difficulty in the construction and calibration of metal matrix composite wear maps.

However, some qualitative remarks may now be made concerning the relative positions of the various boundaries between the interfacing mechanisms of wear of Al alloys as a result of graphite additions. As pointed out in ref. [47], both the

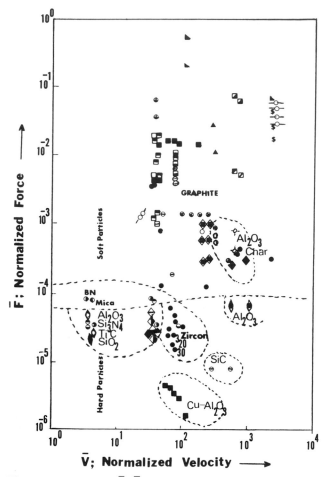

Fig. 8. An empirical $\bar{F}-\bar{V}$ diagram for various metal-matrix particulate composites, showing the experimental ranges of $\bar{F}-\bar{V}$ used by various authors (ref. [27]).

experimental evidence and the suggested models indicate that plastic deformation of the matrix alloy may become easier due to stress concentration at the edges of the pits left behind by graphite particles which are removed. This may suggest that the delamination-wear or the plasticity dominated wear regime may also extend to lower normalized velocities and other load regimes. Similarly, since graphite dispersions considerably enhance the seizure resistance of Al alloys, the seizure line for Al–graphite composites is expected to shift to higher normalized loads, as in fig. 7.

The seizure line for most metals can be adequately described by the modified Tabor equation [9], which for composites can be simply written as

$$\bar{F} = \frac{1}{\left(1 + \alpha_t \mu_e^2\right)^{1/2}} \left(\frac{(H_e)_T}{(H_e)_{RT}} \right), \tag{10}$$

where μ_e refers to the effective friction coefficient of the composite and $(H_e)_T$ and $(H_e)_{RT}$ refer to the effective hardness of the composite at a temperature T (due to frictional heating during sliding) and the effective hardness at room temperature, respectively. Lim and Ashby [10] refined the above equation for the seizure line by introducing the influence of strain rate hardening of a metal during sliding, although they indicate that the approximation $H_T = H_{RT}$ is adequate for most situations and is thus considered here. By ignoring the strain rate hardening effects, the seizure equation for Al–graphite composites can be approximated as

$$\bar{F} = \frac{1}{\left(1 + \alpha_t \mu_{eff}^2\right)^{1/2}},$$

where α_t is an adjustable constant. For steels and some other metals, a value of $\alpha_t = 12$ fits the experimental data well; for composites its value needs to be calibrated from experiments. These remarks show the possible effects that graphite (or other particle) dispersions may have on the process of dry sliding wear in Al-alloy composites. A more rigorous evaluation of these effects is necessary before a more quantitative mapping procedure for the wear mechanisms in composites can emerge. This evaluation further necessitates elucidation of wear transitions in composites from carefully designed experiments.

4. Wear maps of ceramic matrix composites

Quantitative wear maps of ceramics and ceramic matrix composites have recently been published by Lee et al. [20]. Wear maps of alumina and SiC-whisker reinforced alumina composites are shown in figs. 9a and b, respectively. The wear maps were constructed as a function of speed and normal load under dry sliding conditions. Under dry sliding conditions, the transition zones from mild to severe wear for both monolithic alumina and SiC-whisker reinforced alumina matrix composites tend to cross the maps diagonally from low sliding speed, low load to higher sliding speed and higher load [20].

Figures 10a and b show the wear mechanisms in maps for both monolithic alumina and alumina matrix composites. Figures 10a and b distinguish three levels of wear, indicating different wear mechanisms, ranging from mild wear (plastic deformation) to severe wear to ultra-severe wear. In comparison to monolithic alumina, the isowear lines of the composite move to the lower-end regimes of the higher sliding speeds and loads, indicating that the reinforced composite shows considerably greater wear resistance than monolithic alumina.

5. Concluding remarks

Although the first wear-mechanism map for carbon steels was published in 1987, the concept of wear mapping is still new. The data available are not adequate for the construction of more meaningful and qualitatively correct wear-mechanism maps for composites, like the map by Lim and Ashby; development of

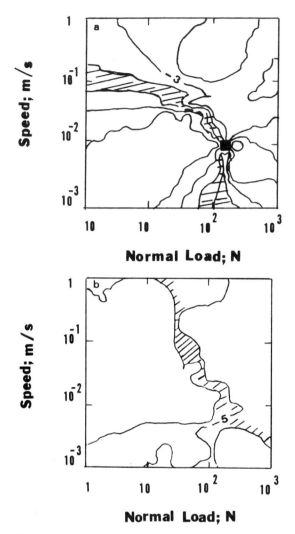

Fig. 9. Contour wear maps in dry air for (a) alumina and (b) SiC$_w$ reinforced alumina. (Curves of constant wear volume per unit sliding distance are labeled in units of (mm^3/m) (ref. [20]).)

significant qualitative maps will require a continuing and extensive effort. Some issues concerning wear-map construction for metal matrix and ceramic matrix composites have been addressed in this chapter. The physical processes by which composites wear during sliding are still imperfectly understood, and this renders the construction and calibration of wear-mechanism maps for these composites more difficult than for monolithic materials, such as steels. However, some possible variables and effects, such as bulk heating and flash heating, have been analyzed. Equations from earlier models for single-phase materials have been adjusted to incorporate the influence of particle dispersions and any film forma-

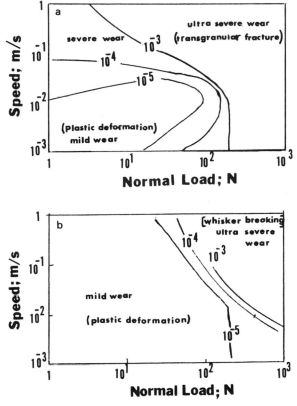

Fig. 10. (a) Wear-mechanism map of alumina in dry air (ref. [20]). (b) Wear-mechanism map of SiC-whisker reinforced alumina in dry air (ref. [20]).

tion at the mating interface. Some maps, from different sources, have been presented in terms of test variables, and possible shifts of the boundaries between the various competing wear mechanisms due to particle dispersion and changes in sliding parameters have been discussed.

Acknowledgements

The authors acknowledge the support of ONR and valuable comments by S. Bhattacharyya, S. Ray, and J. Achary. The word processing by Karen Steldt is greatly appreciated.

Appendix I. Relative heat fluxes in the pin and disk

As an index representing the heat absorbing capacity of the sliding pin and steel disk pair, a new thermal heat diffusivity ($KC\rho$), and not the conventional thermal diffusivity $(K/C\rho)$, is used in this analysis. The heat flux, q, generated at the interface is partitioned as shown below.

The heat flux into the pin, q_{pin}:

$$q_{pin} = \frac{q[(K_e)_b(C_e)_b(\rho_e)_b]^{1/2}}{[(K_e)_b(C_e)_b(\rho_e)_b]^{1/2} + [K_D C_D \rho_D]^{1/2}}, \tag{A.1}$$

the heat flux into the steel disk, q_{disk}:

$$q_{disk} = \frac{q[K_D C_D \rho_D]^{1/2}}{[(K_e)_b(C_e)_b(\rho_e)_b]^{1/2} + [K_D C_D \rho_D]^{1/2}}, \tag{A.2}$$

and q, the total frictional heat generated at the interface, is

$$q = \frac{\mu FV}{A_n}, \tag{A.3}$$

where μ is the friction coefficient of the composite; K_D, C_D, and ρ_D are the thermal conductivity, specific heat, and density of the steel disk, respectively. $(K_e)_b$, $(C_e)_b$, and $(\rho_e)_b$ are the corresponding properties of the composite pin.

Representing the square roots of the heat diffusivities of the pin and disk by D_B and D_D, respectively, eqs. (A.1) and (A.2) become

$$q_{pin} = \frac{qD_B}{D_B + D_D}, \tag{A.4}$$

$$q_{disc} = \frac{qD_D}{D_B + D_D}. \tag{A.5}$$

If l_b denotes the heat diffusion distance in the pin, then

$$q_{pin} \simeq \left(\frac{T_b - T_0}{l_b}\right)(K_e)_b = \frac{qD_B}{D_B + D_D}.$$

Appendix II. The effect of a graphite film formed on the tribosurface of the composite

Since the experiments show that a graphite film adheres to both the pin and the mating steel disk surface, the effect of the film on the heat flow can be considered by associating half the film thickness $(Z/2)$ with the pin and the remaining half with the steel disk. In the one-dimensional heat-flow approximation, the graphite layer of thickness $Z/2$ contributes an additional series resistance to the flow of heat over the distance l_f (fig. 2) in the pin. Hence, on adding up series resistances, we get

$$\frac{(l_f - Z/2)}{(K_e)_b} + \frac{(Z/2)}{K_p} = \frac{l_f}{(K_e)'_b} \tag{A.6}$$

or

$$(K_e)_b' = \frac{K_p(K_e)_b}{K_p(1 - Z/2\beta r_a) + (Z/2\beta r_a)(K_e)_b}. \tag{A.7}$$

The temperature increment from injecting an energy density αq into an area πr_a^2 (the area of an asperity contact) is given by Carslaw and Jaeger (and cited by Lim and Ashby [10]) as:

$$T = T_b + \frac{\alpha q r_a}{\sqrt{\pi}(K_e)_b'} \tan^{-1}\left(\frac{4(a_e)_b t^{1/2}}{r_a^2}\right), \tag{A.8}$$

where $t = 2r_a/V$ is the time of heat injection into an asperity and α is the fraction of the heat injected into the pin.

Expressing the above equation in terms of normalized variables, we obtain the interface or flash temperature, T_f:

$$T_f = T_b + \frac{\alpha \mu \overline{FV} T_e^*}{\sqrt{\pi}} \frac{r_a}{r_0} \frac{A_n}{A_r} \tan^{-1}\left(\frac{8r_0}{\overline{V} r_a}\right)^{1/2}, \tag{A.9}$$

where

$$T_e^* = \frac{(a_e)_b(H_e)_b}{(K_e)_b'} \tag{A.10}$$

and $(K_e)_b'$ is given by eq. (A.7).

Experimental evidence [47] also indicates that plastic deformation at the contacting asperities occurs in composites like Al–graphite. Hence, the ratio of real to nominal areas of contact is taken as $\overline{F}$, i.e.

$$\frac{A_r}{A_n} = \overline{F} = N\frac{\pi r_a^2}{\pi r_0^2}, \qquad \frac{r_a}{r_0} = \left(\frac{\overline{F}}{N}\right)^{1/2}, \tag{A.11}$$

where N is the number of contacting asperities per unit area, and is given by Lim and Ashby [10] as:

$$N = \left(\frac{r_0}{r_a}\right)^2 \overline{F}(1 - \overline{F}) + 1. \tag{A.12}$$

Substituting eq. (A.11) in eq. (A.9) for T_f, we obtain

$$T_f = T_b + \frac{\alpha \mu FV T_e^*}{\sqrt{\pi}} \left(\frac{\overline{F}}{N}\right)^{1/2} \left(\frac{r_0}{r_a}\right)^2 \frac{1}{N} \tan^{-1}\left(\frac{8r_0}{\overline{V} r_a}\right)^{1/2}, \tag{A.13}$$

where r_a (asperity radius) is typically taken as $r_a \simeq 10^{-5}$ [10] and N is given by eq. (A.12) above. The equation for the flash temperature takes the influence of the graphite film on the heat flow effects into account.

TABLE 1
Material properties used in normalization of test values [27].

Material	Thermal conductivity (kcal/cm °C s)	Specific heat (cal/gm °C)	Density (gm/cm^3)	Thermal diffusivity * (cm^2/s)
Al	0.560	0.21	2.705	0.98
Cu	0.927	0.09	8.890	1.15
Al–13Si	0.289	0.23	2.657	0.47
Al–4Cu	0.289	0.22	2.80	0.47
Graphite	0.057	0.165	2.25	0.15
SiC$_p$	0.031	0.201	3.20	0.05
Steel	0.230	0.14	7.86	0.21

* Calculated values.

Sample values for typical metals and alloys, as well as particles, to use in these calculations are shown in table 1.

References

[1] M. Okoshi and H. Sakai, Trans. JSME 7 (1941) 29.
[2] N.C. Welsh, Philos. Trans. R. Soc. A 257 (1965) 31.
[3] T.H.C. Childs, Tribol. Int. 13(6) (1980) 285.
[4] T.S. Eyre, Powder Metall. 24(2) (1981) 57.
[5] J. Marciniak and T. Otimianowski, in: Proc. 3rd Int. Tribology Congress, Eurotrib 81, Warsaw, Poland, 21–24 September, 1981, Vol. 1A (The Polish Tribology Council, 1981) p. 241.
[6] K. Egawa, J. JSLE Int. Ed. 3 (1982) 27.
[7] D. Tabor, in: Proc. Int. Conf. Tribology in the 80's, NASA Lewis Research Center, Cleveland, Ohio (NASA, 1983) p. 1.
[8] H.J. Frost and M.F. Ashby, Deformation-Mechanism Maps. The Plasticity and Creep of Metals and Ceramics (Pergamon Press, Oxford, 1982).
[9] D. Tabor, Friction, lubrication and wear, fifty years on, in: Proc. Int. Conf. on Tribology, London (Inst. Mech. Eng., London, 1987) p. 157.
[10] S.C. Lim and M.F. Ashby, Acta Metall. 35 (1987) 1.
[11] Y. Liu, R. Asthana and P.K. Rohatgi, J. Mater. Sci. 26 (1991) 99.
[12] R. Antonious and C. Subramaniam, Scripta Metall. 22 (1988) 809.
[13] A. Beerbower, US Army Contract No. DAHC19-69-C-0033 (Office of the Chief of Research and Development, June, 1972).
[14] G. Salomon, Wear 36 (1976) 1.
[15] A.W.J. de Gee, A. Begelinger and G. Salomon, in: Proc. 11th Leeds–Lyon Symp. on Tribology, Leeds (Butterworths, London, 1985) p. 105.
[16] T. Akagaki and K. Kato, Wear 129 (1989) 303.
[17] T. Akagaki and K. Kato, Wear 141 (1990) 1.
[18] T. Akagaki and K. Kato, Wear 143 (1991) 119.
[19] S.M. Hsu, Y.S. Wang and R.G. Munro, Wear 134 (1989) 1.
[20] S.W. Lee, S.M. Hsu and R.G. Munro, in: Proc. Conf. Tribology of Composite Materials, Oak Ridge, TN, May 1–3, 1990, eds P.K. Rohatgi, C.S. Yust and P.J Blau (ASM International, Oak Ridge, TN, 1990) p. 35.
[21] Y.S. Wang, S.M. Hsu and R.G. Munro, Lubr. Eng. 47 (1991) 63.
[22] X. Dong, S. Jahanmir and S.M. Hsu, J. Am. Ceramic Soc. 99 (1991) 999.

[23] K. Kato, Wear 136 (1990) 117.
[24] K. Adachi and K. Hokkirigawa, in: Proc. Int. Conf. on Wear of Materials, Orlando, FL, April 7–11, 1991 (ASME, New York, 1991) p. 333.
[25] K. Hokkirigawa, in: Proc. Int. Conf. on Wear of Materials, Orlando, FL, April 7–11, 1991 (ASME, New York, 1991) p. 353.
[26] B.J. Briscoe and P.D. Evans, in: Proc. Int. Conf. on Wear of Materials, Denver, CO, April 9–13, 1989 (ASME, New York, 1989) p. 449.
[27] P.K. Rohatgi, Y. Liu and R. Asthana, in: Proc. Conf. on Tribology of Composite Materials, Oak Ridge, TN, May 1–3, 1990, eds P.K. Rohatgi, C.S. Yust and P.J Blau (ASM International, Oak Ridge, TN, 1990) p. 69.
[28] P.K. Rohatgi, Metals Handbook, Vol. 15, 9th Ed. (1989) 840.
[29] L. Bruni and S.P. Iguera, Automotive Eng. 3 (1978) 29.
[30] S.V. Prasad and P.K. Rohatgi, J. Metals 33 (1987) 22.
[31] M. Suwa, J. Jpn. Inst. Metals 40 (1976) 1074.
[32] C. DellaCorte, S.C. Farmer and P.O. Book, in: Proc. Conf. on Tribology of Composite Materials, Oak Ridge, TN, May 1–3, 1990, eds P.K. Rohatgi, C.S. Yust and P.J Blau (ASM International, Oak Ridge, TN, 1990) p. 345.
[33] C.S. Yust, in: Proc. Conf. on Tribology of Composite Materials, Oak Ridge, TN, May 1–3, 1990, eds P.K. Rohatgi, C.S. Yust and P.J Blau (ASM International, Oak Ridge, TN, 1990) p. 25.
[34] V.K. Sarin and M. Ruhle, Composites 18 (1987) 129.
[35] P.K. Rohatgi, Y. Liu, M. Yin and T.L. Barr, Mater. Sci. & Eng. A 123 (1990) 213.
[36] P.R. Gibson, A.J. Clegg and A.A. Das, Mater. Sci. & Technol. 1 (1985) 559.
[37] S. Das, S.V. Prasad and T.R. Ramachandran, Wear 133 (1989) 173.
[38] M. Suwa, Hitachi Graphite-Dispersed Cast Alloy-Gradia, Hitachi Report (1986).
[39] W.K. Chou and C.H. Hong, J. Korean Inst. Met. 17 (1979) 474.
[40] S.K. Biswas and B.N. Pramila Bai, Wear 68 (1981) 347.
[41] J.M. Carstevens, Wear 49 (1978) 169.
[42] J.M. Carstevens, Wear 50 (1978) 371.
[43] M. Muran and M. Srnanek, Metall. Mater. (Kovove Materialy) 23 (1985) 107.
[44] E. Yuasa, T. Morooka and F. Hayama, J. Jpn. Inst. Metals 49(11) (1985) 981.
[45] P.R. Gibson, A.J. Clegg and A.A. Das, Wear 95 (1984) 193.
[46] Y.T. Suya, J. Jpn. Inst. Composites 11 (1985) 127.
[47] P.K. Rohatgi, Y. Liu and T.L. Barr, in: Proc. Conf. on Tribology of Composite Materials, Oak Ridge, TN, May 1–3, 1990, eds P.K. Rohatgi, C.S. Yust and P.J Blau (ASM International, Oak Ridge, TN, 1990) p. 113.

Advances in Composite Tribology
edited by K. Friedrich

Chapter 9

Performance of Metal Matrix Composites in Various Tribological Conditions

S. WILSON and A. BALL

Department of Materials Engineering, University of Cape Town, Rondebosch 7700, Republic of South Africa

Contents

Abstract

    The abrasive, sliding, cavitation and particle erosive wear resistances of two aluminium alloys (Al6061 and Al2014), reinforced with 10, 15 and 20 vol% alumina particulates, and the Al6061 alloy reinforced with 20 vol% silicon carbide particles, have been measured. The modes by which these materials are worn by deformation and fracture processes have been examined by scanning electron microscopy. An attempt has been made to reconcile the steady-state wear rates of the alloys containing the different amounts of particles, with their observed wear modes and bulk mechanical properties of the composites.

The resistance to abrasion is dominated by the composite's ability to resist indentation by hard particles. Thus there is a correlation between abrasion resistance and macrohardness. Abrasion with fine grit particles leads to a reduced load per particle and a correspondingly significant reduction in abrasive wear.

The resistance to sliding wear is improved by factors of up to three orders of magnitude and this is attributed to the increased resistance to surface shear provided by the particulates themselves and their constraining effects on the matrix. The particulates become load bearing and protect the matrix by reducing the metal to counterface adhesion and wear. However, the counterface wear increases due to this interaction with the hard reinforcements.

In contrast to the results for abrasive and sliding wear, the composites showed increasingly inferior erosion resistance with increasing volume fractions of particulates. This effect was especially evident for particle erosion. This depreciating effect can be correlated with the composite's increasing inability to accommodate the increments of strain which accompany the erosive processes. Thus microtoughness is the major requirement for erosion resistance and not hardness or strength. The presence of hard brittle particulates which constrain the matrix give the composites inferior erosion properties as compared to those of the base alloys.

In all of the wear tests the Al6061 alloy containing 20 vol% silicon carbide gave better performances than the same alloy containing 20 vol% alumina. This effect is considered to be the result of the increased integrity of both the silicon carbide particles themselves and their interfaces with the metal matrix.

1. Introduction

Aluminium alloys are used extensively for the manufacture of low-density, high-strength components for high-speed reciprocating mass components and structural members in the aerospace and general transport sectors. Recently, the dominance of aluminium alloys has been challenged by the development of advanced polymeric composite materials having stiffness, strength and density characteristics ideally suited for high-performance energy-efficient applications. In response to this competition there have been a number of developments in the aluminium industry; the major ones being the high-modulus aluminium–lithium alloys, the aluminium–polymeric (aramid fibre) composite laminates and aluminium alloys reinforced with continuous and discontinuous ceramic reinforcement (metal matrix composites) [1].

Heat treatable aluminium alloys reinforced with discontinuous ceramic fibres and particulates are the most commonly available and easily produced metal matrix composites to date. These materials were developed originally for their greater stiffness and strength compared to ordinary monolithic alloys, and have been used in a limited but growing number of applications. In the early 1980s, a motor company initiated the use of alumina fibres to reinforce the piston ring area of their diesel engine pistons with successful results [2]. Other engine components being considered for manufacture from aluminium composites are drive shafts, connecting rods and rocker arms [3]. The thermal expansion properties of silicon carbide reinforced aluminium alloys can be tailored to levels well below those of conventional aluminium alloys [4]. The use of reinforcing ceramics with high thermal conductivity and specific heat, such as silicon carbide, makes these materials excellent heat sinks and ideal for housing electronic instruments. Recent research has focussed on the wear behaviour of aluminium matrix composites [5–23]. As metal matrix composites are considered as candidate materials in applications such as engines and aerospace structures, the need to characterise their behaviour in a range of different wear environments has become critically important.

Much of the research carried out on the wear behaviour of aluminium alloys reinforced with hard ceramic particulates, has concerned their abrasive and sliding wear resistances with little having been reported on their performance in other tribological conditions, e.g. cavitation and erosion by solid particles. Aluminium matrix composites have been found to outperform their matrix alloy counterparts in a wide variety of abrasive wear conditions. Banerji et al. [5] ascribed a simple rule of mixtures behaviour to an aluminium alloy (Al–11.8Si–4Mg) containing up to 35 vol% of zircon particulates (100 μm mean diameter) in abrasion tests against 80 mesh aluminium oxide grit impregnated cloth. Worn surfaces of each composite surface revealed particulate fracture but no evidence of pull out or debonding. Surappa et al. [6] reported abrasive wear rates for cast Al–Si eutectic alloys containing 5 wt% alumina particulates (100 μm size), that were superior to hypereutectic Al–Si alloys, indicating that alumina reinforcement could be used as a suitable replacement for silicon in aluminium alloys for wear resistant applications. Wang and Hutchings [7] conducted two-body abrasive wear tests on an Al6061 alloy reinforced with up to 30 vol% discontinuous alumina fibres. The wear resistance of the composites was found to increase with added reinforcement, up to a volume fraction of approximately 20% fibres, when abraded against large silicon carbide abrasive particles (240 mesh – 60 μm mean diameter). An alloy containing additional reinforcement (30 vol%) displayed a subsequent decrease in abrasion resistance. When the composites were abraded against fine silicon carbide abrasive particles (600 mesh – 20 μm mean diameter) a linear increase in wear resistance with increasing fibre volume fractions was obtained. Under these low contact stress conditions, wear rates of up to six times below that of the matrix alloy were displayed. The transition from high contact stress (large abrasive grit sizes) to low contact stress conditions (small abrasive grit sizes) was accompanied by a change in wear mode from extensive fibre fracture and debonding to plastic deformation and minimal fibre damage, respectively. Using parameters such as fibre tensile strength, fibre dimensions and calculated contact pressures present at abrasive grit contact points, a model was developed to determine the critical abrasive particle size required to initiate fibre fracture in the aluminium matrix. A critical abrasive grit diameter of 30–35 μm was obtained from the model, which correlated reasonably well with the experimentally observed value of 20–28 μm.

Prasad et al. [8] conducted low-stress abrasion tests using a rubber wheel abrasion apparatus (ASTM G65) with quartz (50–70 mesh) as abrasive on Al–11.8Si–4.0Mg zircon particulate composites. Wear rates five times below that of the unreinforced matrix alloy were obtained with an alloy containing 35 vol% particulates, showing a marked deviation from the rule of mixtures behaviour observed under high-stress two-body abrasive conditions by Banerji et al. [5]. The surfaces of the composites abraded under low-stress conditions displayed minimal damage (microfracture) of the zircon particulates, which tended to stand proud of the surrounding matrix alloy.

Investigations into the abrasive wear behaviour of silicon carbide particulate and whisker reinforced Al7091 alloys by Wang and Rack [9] found similar improvements in wear resistance over that of the unreinforced matrix alloy, with the greatest improvements found for abrasion against finer abrasive particles. They proposed that the ratio of the average abrasive penetration depth to the size of the reinforcing particulates was the critical parameter controlling the relative abrasive wear resistance of the silicon carbide reinforced composites. Thus it becomes apparent that by reducing asperity sizes and contact stresses, the probability of catastrophic failure of reinforcing particulates in an aluminium matrix is diminished. The reinforcement becomes load bearing and material loss occurs through plastic deformation and ductile removal of the matrix alloy accompanied by microfracture of the reinforcing particulates.

Tribological environments in which the contacting asperities are no longer able to indent the surface to an extent that substantial material removal occurs, can be referred to as sliding or adhesive wear conditions [10]. The expected behaviour of aluminium composites in sliding wear situations can be partially inferred from their performance under low-stress abrasion; the markedly reduced contact stresses and resulting minimal damage to the load bearing reinforcement should result in very low material removal rates. Nevertheless, variables such as sliding speed, contact temperature, chemical environment and the magnitude of frictional stresses developed at the sliding wear interface can become rate controlling factors [11].

A common feature of investigations which involve sliding contact between aluminium matrix composites and steel surfaces, is a lowering of the amount of adhesive transfer of aluminium to the steel, resulting in reduced wear rates. Hosking et al. [12] reported wear rates of silicon carbide and alumina reinforced Al2024 and Al2014 alloys that were in some cases two orders of magnitude below those of their unreinforced matrix alloys. While the wear mechanism for the unreinforced alloys was purely adhesive, that for the composites was of a mixed oxidative–abrasive nature, with the steel ball against which the composites were loaded undergoing significant abrasive wear. The lowest composite wear rates were associated with alloys containing the highest volume fractions of ceramic particulates, they also displayed smaller coefficients of sliding friction. Alloys reinforced with silicon carbide particulates showed superior wear resistances; this was attributed to the greater hardness of silicon carbide compared to the softer alumina.

Another factor which may be playing a role is the greater thermal conductivity of the silicon carbide, over that of the alumina, which allows for increased transport of thermal energy away from the sliding interface. Furthermore, silicon carbide has a lower friction coefficient against steel than alumina [13]. The strong surface bonding between iron and alumina results in increased adhesive transfer of steel to the alumina reinforcement and greater counterface wear. Wear rates of the steel ball were not recorded by Hosking et al. [12], so it is difficult to assess the extent of abrasive damage imparted by the ceramic particulates, however, it can be inferred that the reduced coefficients of sliding friction for greater volume fractions of reinforcement are indicative of lower steel removal rates. In this regard, the importance of counterface wear is dependent on the type of engineering application in which the steel and composite come into contact.

Other investigations involving aluminium matrix composites sliding against hardened steels have reported similar features to those of Hosking et al. [12]. Rana and Stefanescu [14] recorded reductions in the sliding friction coefficients with increasing volume fractions of silicon carbide particulates in an Al–1.5% Mg alloy sliding against steel. Composites with smaller average particulate sizes displayed lower frictional forces in comparison to alloys containing larger particulate sizes with the same volume fraction. The reduction in friction coefficient was attributed to the greater surface area of silicon carbide on the composite surface in contact with the steel and less cutting action afforded by the lower contact stresses. Yang and Chung [15] recorded decreased weight losses in sliding wear of an Al–Si eutectic alloy reinforced with bauxite particulates in contact with a steel counterface. The lowest wear rates were associated with alloys containing the highest bauxite content. The addition of graphite particles to the bauxite reinforced aluminium alloys decreased their sliding wear rates even further, producing a layer of graphite solid lubricant on the steel surface. Surappa et al. [6] determined the sliding behaviour of Al–Si alloys reinforced with up to 5 wt% alumina particulates, which showed reductions in the amount of adhesive transfer to a hardened steel counterface with increase in reinforcement content. Similar reductions in composite wear rate with increasing volume fractions of alumina reinforcing particulates have been reported by Anand and Kishore [16], who also observed significant removal and transfer of steel counterface material to the composite sliding surface.

The surface microstructures of aluminium matrix alloys which have been in sliding contact with hardened steel surfaces have been found to consist of a transfer layer mixture of oxides, steel debris and fractured reinforcing particulates. You et al. [17] characterised the nature and formation of the transfer layer in an Al2124 aluminium alloy reinforced with 20 vol% silicon carbide particulates sliding against a hardened 1045 steel. In the early stages of wear, the load bearing silicon carbide particulates remove counterface steel material in an abrasive "micromachining" operation. The steel debris generated by this process is compacted between the reinforcing particulates, followed by mechanical mixing and plastic deformation, giving rise to a "steady-state" transfer layer consisting of aluminium, iron and silicon carbide. A substantial degree of subsurface deformation was also found up to depths of approximately 50 μm in the composite, indicating that large shear stresses are developed at the sliding contact interface. Similar transfer layer characteristics have also been reported by other researchers [18–20].

The addition of ceramic particulates to aluminium matrices not only improves their sliding wear performance, but also has a marked effect on their bulk mechanical properties. In particular, alloys reinforced with increasing volume fractions of ceramic reinforcement show significant reductions in ductility. Recent investigations have shown that the sliding wear resistance of composites having low strains to failure are reduced to levels close to those of their unreinforced states. Pan et al. [21] observed severe particle "pull-out" or pitting accompanied by large wear rates in an A356 Al–Si casting alloy reinforced with 15 vol% silicon carbide particulates, in rolling contact with nodular cast iron. The composite had an elongation to failure of 0.3% and a yield strength of 317 MPa. In comparison, an Al2014 alloy containing 15 vol% alumina reinforcement with a ductility of 2.3% and a yield strength of 476 MPa, had a wear rate approximately half that of the A356 composite, despite having a higher sliding friction coefficient. The poor wear behaviour of the silicon carbide reinforced A356 alloy was attributed to its reduced fatigue resistance, being unable to withstand the cyclic loading stresses associated with rolling wear. Alpas and Embury [22] reported similar wear resistances for both an Al2014 20 wt% SiC composite and a monolithic Al2024 alloy in dry sliding contact under equivalent loading conditions against a steel counterface. They inferred that the low ductility and fracture toughness of the reinforced alloy, irrespective of the higher hardness, were responsible for controlling its wear resistance. While the monolithic alloy exhibited extensive strain accommodation beneath its sliding surface, the reinforced alloy had a very large subsurface strain gradient associated with cracking and delamination at silicon carbide particulates. The continued exposure of reinforcing particulates by delamination of the surface transfer layer was suggested as being the major reason for increased abrasion of the steel counterface. The addition of a lubricating oil to the sliding wear interface of both the composite and the unreinforced matrix alloy, was found to decrease the subsurface strains and damage accumulation to such an extent that the composite showed wear rates which were an order of magnitude below those of the monolithic alloy.

The diminished sliding wear performance of aluminium matrix composites with low strain energies to failure is indicative of the important control that a material's ability to accommodate strain, both elastically and plastically, has over its wear resistance. The magnitude of shear stresses imparted to the composite in sliding wear situations, is sufficient to initiate subsurface failure only in composites having very low strain energies to fracture. If the magnitude of the strain energies imparted to composite surfaces are dramatically increased however, their ability to accommodate strain becomes rate determining. This is indeed the situation when aluminium matrix composites are eroded by gas-borne particles travelling at high velocities. Goretta et al. [23] studied the erosion by alumina particles of cast Al2014 aluminium alloys reinforced with 20 vol% silicon carbide and 20 vol% alumina inclusions, respectively. The steady-state erosion rates of the alumina and silicon carbide reinforced alloys were similar for a range of erodent sizes and impact velocities, but markedly greater than those of the unreinforced matrix alloy. Lack of ductility was ascribed as being the major reason for reduced erosion resistance in the composites, with properties such as hardness and elastic modulus being of little influence. Similar findings have been made by Hutchings and Wang [24], who studied the solid particle erosion resistance of Al6061 aluminium reinforced with up to 30 vol% discontinuous alumina fibres. The extent to which fibres had fractured during particle impact was described as the dominant erosion rate controlling factor. The solid particle erosion resistance of an Al6061 aluminium alloy, reinforced with 20 vol% silicon carbide particulates, was determined by Wilson and Ball [18] using a silicon carbide erodent. The steady-state erosion rates of the composite were approximately twice those of the unreinforced matrix alloy for a range of erodent impact angles. The eroded surfaces of the composite material revealed a wear mode displaying extensive constrained and localised plastic deformation accompanied by widespread reinforcement fracture.

Another common form of erosive wear is cavitation erosion, which often affects engineering components used in liquid environments. Cavitation is the repeated formation and collapse of bubble cavities arising from high-frequency pressure disturbances in a liquid medium. Erosion damage occurs as a result of the collapse of bubble clouds close to the material surface. Shock-impact and tensile load cycles are developed in the surface regions of the material, which may deform in an elastic or plastic manner, depending on its mechanical properties. Heathcock et al. [25] studied the behaviour of a wide range of materials in cavitation erosion conditions. They established that improved performances were attained by materials with one or more of the following properties: a high elastic resilience, a high resistance to the accumulation of fatigue damage under repeated shock loading conditions, a tough microstructure which is resistant to the propagation of microcracks.

Aluminium alloys generally have low resistances to cavitation erosion [26], due to their inability to absorb impact energy in an elastic manner. An Al6061 alloy reinforced with 20 vol% silicon carbide particulates displayed an erosion resistance which was slightly above that of its unreinforced state [18]; the presence of high-modulus silicon carbide reinforcement possibly improving the elastic resilience of the alloy to a certain degree. Filler materials may also provide regions of elastic modulus mismatch in a matrix and can act as nuclei for rupture [25], giving higher wear rates. However, the excellent bond between ceramics, such as silicon carbide and alumina, and aluminium [27,28], lowers the possibility of interfacial rupture. The influence of reinforcing particulate fracture toughness and composite failure strain on cavitation erosion resistance still need to be determined however.

This chapter focusses on the wear behaviour of two age hardening commercial aluminium alloys, each reinforced with different volume fractions of ceramic particulates. While most of the research done to date generally tends to concentrate on the behaviour of aluminium matrix composites in one type of wear situation, our approach is based on an attempt to characterise the responses of each reinforced and unreinforced alloy in a variety of tribological environments.

This should provide an insight into the various modes of wear and the controlling mechanisms. The chapter initially deals with the tensile behaviour of each material. Uniaxial tensile tests were conducted on the reinforced and unreinforced alloys in order to determine how the addition of ceramic particulates influences the stress–strain response of the matrix aluminium alloys. The failure modes and strengthening mechanisms of each material are characterised and discussed in terms of the properties of the ceramic reinforcements and matrix alloys used. The tensile characteristics of each composite are then used to facilitate interpretation of the often complex deformation processes that occur in the tribological situations.

The wear resistance of each material is determined for four different tribological environments, namely two-body abrasion, reciprocating sliding wear, solid particle erosion and cavitation erosion. The two-body abrasion resistances of each material are determined and compared, against fine-mesh (low contact stress) and coarse-mesh (high contact stress) abrasive papers. A reciprocating sliding wear testing arrangement is used to determine the behaviour of each reinforced and unreinforced alloy in sliding contact against a hardened stainless steel counterface. The erosion rates and wear modes are then determined for each material in cavitation erosion conditions in an aqueous environment and solid particle impact conditions by silicon carbide particles. The results obtained are then discussed in terms of the bulk and surface properties of each material, taking into account the different deformation modes, strain rates and contact stresses associated with each form of wear.

2. Materials

The materials investigated consisted of the aluminium alloys Al6061 (1.0 Mg, 0.6 Si, 0.2 Cu, 0.2 Cr in wt%), a commercial structural extruding alloy and Al2014 (4.4 Cu, 0.8 Si, 0.4 Mg, 0.8 Mn in wt%), an alloy used widely in aerospace structures, each reinforced with 10, 15 and 20 vol% alumina particulates. The Al6061 alloy was also reinforced with 20 vol% SiC particulates. Each composite and unreinforced alloy was received in the extruded state having been solution treated and artificially aged to peak hardness (T6). Optical micrographs of each of the composites are shown in fig. 1. The distribution of reinforcing particulates is uniform in the alloys containing 15 and 20% volume fractions with some clustering of particulates being evident. The two 10% volume fraction composites show widespread particulate alignment in the extrusion direction, leaving large particulate-free zones. The average particulate sizes for each of the composites are listed in table 1 together with hardness readings, calculated interparticulate spacing data and Young's modulus values. The elastic constant values were determined using the in-plane resonance spectra of thin disks of each material [29]. The largest particulates are found in the 20% Al_2O_3 reinforced Al6061 alloy (average 32.2 μm in diameter) and 20% Al_2O_3 reinforced Al2014 alloy (average 27.2 μm), while those measured for the other Al_2O_3 and SiC reinforced composites were in the size range 11–17 μm. The macrohardness values (HV_{20kg}) of each of the reinforced

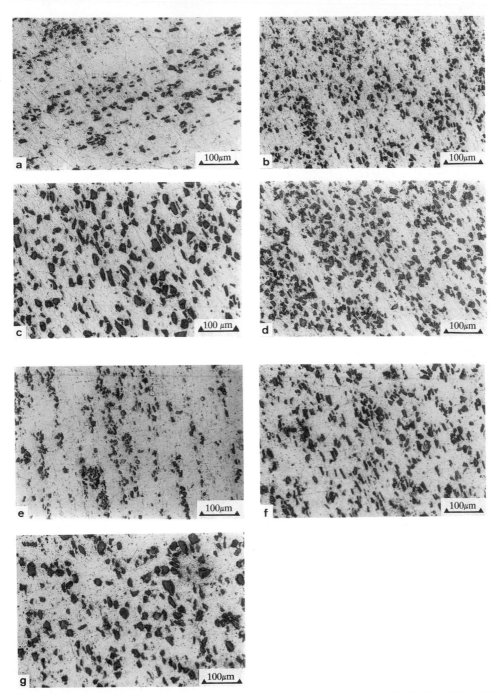

Fig. 1. (a)–(d) Optical micrographs of the Al6061 aluminium alloy reinforced with 10, 15 and 20 vol% alumina particulates and 20 vol% silicon carbide particulates, respectively. (e)–(g) Optical micrographs of the Al2014 aluminium alloy reinforced with 10, 15 and 20 vol% alumina particulates, respectively.

TABLE 1
Average particulate diameters, inter-particulate spacing (calculated), hardness and Young's modulus values for each of the reinforced and unreinforced aluminium alloys.

Alloy	Volume fraction	Particulate diameter (μm)	Inter-particle spacing (μm)	Hardness (HV$_{20kg}$)	Young's modulus (GPa)
Al6061	0	–	–	123.8	69.6
	10% Al$_2$O$_3$	14.1	84.5	130.4	78.5
	15% Al$_2$O$_3$	15.1	40.3	135.3	85.1
	20% Al$_2$O$_3$	32.2	121.6	151.7	91.1
	20% SiC	17.0	45.2	159.6	95.7
Al2014	0	–	–	151.4	74.7
	10% Al$_2$O$_3$	11.4	68.1	162.5	80.4
	15% Al$_2$O$_3$	13.6	51.3	173.8	93.6
	20% Al$_2$O$_3$	27.2	72.5	196.9	97.0

and unreinforced alloys are shown together with their respective interparticulate matrix alloy microhardness values (HV$_{15g}$) in fig. 2. The composites and unreinforced alloys which have the same alloy composition, appear to show no significant difference in matrix microhardness, indicating that the matrix alloy in each

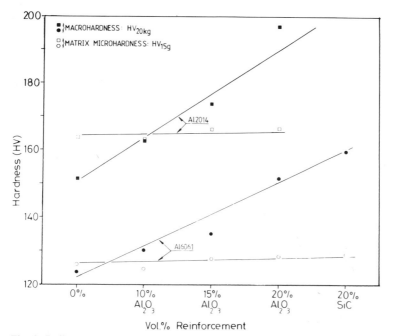

Fig. 2. Bulk macrohardnesses (HV$_{20kg}$) and matrix alloy microhardness values (HV$_{15g}$) measured for each reinforced and unreinforced aluminium alloy.

composite has been artificially aged to the same strength. The bulk macrohardness values of each material increase linearly with reinforcement content.

Sample preparation for each of the tribological and tensile tests was carried out by cutting specimens from extruded bar stock using an abrasive cut-off wheel, followed by machining to the required dimensions. The surfaces were then diamond polished to reveal microstructural details. The composites containing alumina particulates required special attention during mechanical polishing due to the friable nature of the alumina. An alumina suspension (1.0 μm) was used as a polishing agent to facilitate the removal of particulate debris from the matrix.

3. Experimental

3.1. Tensile testing

The stress–strain behaviour of each composite and unreinforced alloy was determined by conducting uniaxial tensile tests to failure on tensile specimens machined from each material. Gauge lengths of each specimen were mechanically polished to remove machining marks, followed by an electropolishing procedure in an 80:20 methanol/nitric acid mixture at $-17°C$ and 20 V for 4–5 s. The electropolishing procedure was used to remove all traces of particulates which had fractured during mechanical polishing, as well as to obtain a finish in the matrix alloy suitable for microstructural observation. Testing was carried out at a strain rate of 2.5×10^{-4} s^{-1}, and scanning electron microscopy was employed to establish the post-fracture deformation characteristics and fracture mode for each specimen.

3.2. Abrasion testing

Dry abrasion testing was performed using a modified Rockwell belt sanding machine [30]. A continuous bonded abrasive belt is run horizontally at a constant velocity, against which a specimen is loaded perpendicularly on its cross-section face (10×10 mm). The specimen is made to traverse normal to the direction of the belt movement so that it always abrades against unworn particles. The total abrasion distance traversed by each of the composites and unreinforced alloys was about 20 m, with the mass losses being determined at specific intervals. The wear rates were calculated from the mass losses averaged over the total distance abraded by each specimen.

The conditions employed for testing are shown in table 2. The wear rate of each material was determined using four different grit sizes of alumina bonded abrasive belt. Scanning electron microscopy was employed to characterise the average abrasive grit size for the different grades of abrasive belt. The worn surfaces of the composites and matrix alloys were also examined in the scanning electron microscope, in order to establish their respective wear modes.

TABLE 2
Abrasive belt grit sizes and conditions employed for abrasion testing of each reinforced and unreinforced alloy.

Nominal contact pressure	0.1 MPa
Applied load	10 N
Abrasive type	Aluminium oxide
Abrasive belt speed	0.34 m s^{-1}
Abrasive grit mesh nos.	220 320 400 600
Abrasive grit size (μm) (average)	56 44 34 16

3.3. Reciprocating sliding wear tests

Reciprocating sliding wear tests were conducted on the composites and matrix alloys using 431 stainless steel as the counterface material. Each counterface was heat treated to a hardness in the range 441–452 HV and surface ground to an average roughness of $R_a = 0.3$ μm. The rig used consists of a reciprocating base arrangement to which the counterface specimen is attached [31]. The material to be tested is loaded against the counterface and distilled water is used as a coolant and debris remover.

The wear tests were conducted using a counterface reciprocating rate of 31.5 cps, which corresponds to an average sliding speed of 1.9 m s^{-1} and a maximum speed of 3 m s^{-1} during each cycle. A load of 20 N (0.2 MPa nominal pressure) was applied to each sliding specimen for the duration of each test, using a load cell arrangement as monitor. Mass loss/gain readings of both the sliding specimens and counterfaces were recorded at specific intervals of sliding distance. The wear rates were then calculated from cumulative mass loss measurements averaged over the sliding distance in the steady-state wear regime. Scanning electron microscopy was employed to establish the wear mode for the counterfaces and sliding specimens. The composite sliding specimens were sectioned perpendicularly and at a taper angle (5°) to their worn surfaces in order to reveal subsurface deformation. EDS X-ray mapping was used to determine the extent of material transfer between sliding specimen and counterface.

3.4. Cavitation erosion testing

Coupon specimens of each of the composites and matrix alloys were polished to a 1 μm surface finish and eroded in a vibratory cavitation erosion rig [32]. The cavitation apparatus consists of an ultrasonic drill, which generates high-frequency oscillations in the 18.5 to 22 kHz range. The rig has been designed so that samples can be clamped beneath the drill tip, followed by immersion in a bath containing the cavitating liquid. A separating distance between the drill tip and specimen surface of 0.35 mm was used throughout the testing routine and the vibratory amplitude of the drill tip was calibrated at between 75 and 100 μm.

Distilled water was used as the cavitating liquid, which was kept at a temperature of 25°C by a thermostat/cooling coil arrangement. Each specimen was eroded and weighed initially at time intervals of two minutes in order to establish their respective erosion incubation periods. Thereafter longer erosion time intervals were used in the steady-state erosion region, up to a total cumulative period of one hour. The cavitation erosion rates were determined from the steady-state regions of cumulative mass loss and erosion time plots. The changes in surface topography of the eroded specimens were monitored using scanning electron microscopy. Subsurface damage in each composite was inspected through the use of taper sections of eroded surfaces.

3.5. Solid particle erosion tests

A conventional air blast type erosion rig [33] was used to erode coupon specimens of each composite and matrix alloy. Erodent particles are fed via a vibratory hopper into an airstream at controlled pressure, in which they are accelerated towards the target sample. Silicon carbide particles (120 grit, approximately 100 μm in diameter) were used as the erodent. An average particle velocity of 60 m s^{-1} was used and was calculated using the double rotating disk method of Ruff and Ives [34].

Each specimen was eroded to steady-state conditions, with mass loss/gain measurements being made after impact by 5 g of erodent in the incubation region and 10 g in the steady-state regime. The erosion rates were calculated from the steady-state regions on plots of the cumulative mass loss as a function of the total erodent mass. The erosion behaviour was determined for each specimen inclined at angles of 30° and 90° to the erodent stream. The modes of material removal for all the composites and unreinforced alloys were characterised using scanning electron microscopy. The extent of subsurface damage in each material was examined by taper sectioning of eroded sample surfaces.

4. Results

4.1. Tensile test results

The stress–strain curves obtained from uniaxial tensile tests of the 6061 matrix alloy and various composites are shown in fig. 3a. The 20% SiC reinforced alloy displays the greatest ultimate tensile strength, 423 MPa, with a failure strain of 2.8%. The 20% Al$_2$O$_3$ reinforced alloy is the strongest of the alumina reinforced Al6061 matrix composites, at 390 MPa having an elongation to failure of 2.5%, which is considerably below those of the 10% and 15% Al$_2$O$_3$ reinforced alloys. The ultimate tensile strength of the control alloy Al6061 is similar to that of the 20% Al$_2$O$_3$ composite at 391 MPa and exceeds the strength of the 10% (370 MPa) and 15% Al$_2$O$_3$ (380 MPa) composites. Of note is the sharper yield characteristics obtained for the unreinforced alloy, which work-hardens to some 8% strain, whereas the yield behaviour of the composites appear less defined, indicating that microyielding is followed by a period of rapid work hardening.

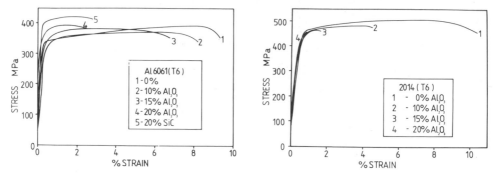

Fig. 3. (a) Stress–strain curves corresponding to the Al6061 matrix alloy and its various reinforced states. (b) Stress–strain curves corresponding to the Al2014 matrix alloy and its various reinforced states.

The stress–strain curves corresponding to the Al2014 matrix alloy and various reinforced states are shown in fig. 3b. The elongation to failure of each composite is strongly affected by the volume fraction of alumina reinforcement, with the 20% Al$_2$O$_3$ material failing at 0.9% strain. The tensile strength of each composite is slightly reduced with greater amounts of reinforcing particulates. The Al2014 reinforced and unreinforced alloys are noticeably stronger than the Al6061 matrix alloy and its composites. Note that the strains to failure of the Al2014 composites are below those of their Al6061 matrix alloy counterparts.

Examination of post-fractured tensile specimen gauge lengths for each of the composites revealed a number of characteristic features, the most prominent being that of reinforcing particulate fracture in the high-strain region immediately below each fracture surface. Fractured particulates were also found in regions of the gauge length situated well away from the fracture surface, particularly for composites containing 15% and 20% reinforcement. However, the populations of these fractured particulates were considerably below those found in the immediate fracture-surface region. The alloys containing 10% Al$_2$O$_3$ reinforcement exhibited very little particulate fracture elsewhere in the gauge length.

The electron micrograph shown in fig. 4, shows the tensile specimen gauge length of the 20% Al$_2$O$_3$ Al2014 composite after it has been tested to tensile failure. The specimen gauge length was electropolished prior to tensile straining. An area approximately 1 mm below the fracture surface is displayed, where a large proportion of particulates have fractured in an orientation perpendicular to the tensile direction. There is some evidence of void nucleation occurring in the matrix alloy and adjacent to particulate interfaces. The bottom area of the micrograph shows crack development localised around a cluster of particulates; a common feature found in gauge lengths of each composite, particularly those containing 10% Al$_2$O$_3$ reinforcement, where inhomogeneous distribution of particulates was prevalent.

The tensile specimen gauge length of the lower-strength Al6061 alloy, reinforced with 15% Al$_2$O$_3$ particulates, is shown in the electron micrograph in fig. 5.

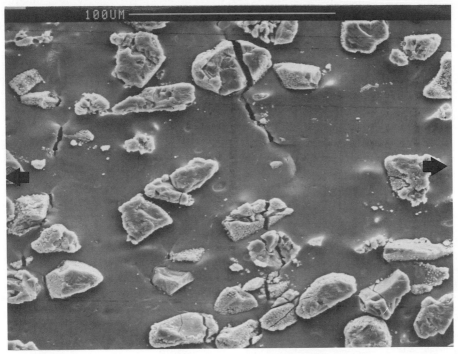

Fig. 4. Electron micrograph of the 20% Al$_2$O$_3$–Al2014 composite gauge length after tensile failure. The area shown is approximately 1 mm below the specimen fracture surface – arrows indicate the tensile direction.

The specimen has been tensile tested to failure, and the area shown is a region of particulate damage just below the fracture surface, where a large amount of localised slip is associated with the formation of a cavity in the matrix alloy in the immediate vicinity of the cracked particulates. All the composites containing 15% and 20% reinforcement showed little or no evidence of plastic slip in the polished interparticulate matrix alloy, except when associated with localised intensive shear and cavity formation after particulate fracture.

An electron micrograph of the fracture surface of the Al6061 alloy reinforced with 15% Al$_2$O$_3$ is displayed in fig. 6. Several fractured reinforcing particulates are observed, each partially or completely surrounded by a ridge of plastically deformed matrix alloy, described as "tear ridges" by Davidson [35], with isolated patches of small dimples (2–3 μm) in the surrounding matrix. Similar features to these were found on the fracture surfaces of all the composites, with the Al2014 reinforced alloys displaying significantly less ductility associated with tear ridge formation and larger populations of very fine dimples (0.5–1.0 μm). A typical example of the fracture-surface morphology of the less ductile 20% Al$_2$O$_3$ reinforced Al2014 alloy is shown in fig. 7. Decohesion of matrix alloy from particulate interfaces was not a common observation in any of the post-fractured composite tensile specimens.

4.2. Abrasive wear results

The abrasive wear rates for each of the Al6061 and Al2014 matrix alloys and their composites are shown in figs. 8 and 9, respectively. The Al6061 matrix alloy displays the largest wear loss during abrasion against all sizes of abrasive grit. In general, the Al6061 matrix composites show greater wear rates than their Al2014 matrix alloy counterparts, with the exception of the SiC reinforced Al6061 alloy, which is the most abrasion resistant of all the composites. It is interesting to note that the 20% SiC composite has a hardness below that of all the Al2014 matrix alloy composites, despite being more abrasion resistant. Likewise, of all the alumina reinforced alloys, the abrasion resistance of the Al6061 alloy reinforced with 20% Al_2O_3 is only exceeded by that of the 20% Al_2O_3 reinforced Al2014 alloy, despite having a hardness lower than all the Al2014 composites.

There is a definite trend towards improved wear resistance of reinforced alloys compared to those of their respective matrix alloys, when they are subjected to the low-stress wear conditions associated with the use of finer abrasive grit sizes. This pattern is more appropriately displayed in figs. 10 and 11, which show values of composite abrasion resistance calculated relative to those of the Al6061 and

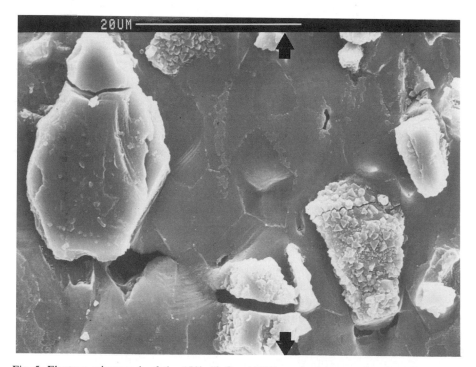

Fig. 5. Electron micrograph of the 15% Al_2O_3–Al6061 composite gauge length after tensile failure. The area shown is approximately 0.5 mm below the specimen fracture surface – arrows indicate the tensile direction.

Fig. 6. Electron micrograph showing the tensile overload fracture surface of the Al6061 alloy reinforced with 15% Al_2O_3 particulates.

Al2014 matrix alloys, respectively, for each abrasive grit size. The SiC reinforced Al6061 alloy exhibits the greatest increase in relative abrasion resistance when compared to the behaviour of the other composites during abrasion against finer abrasive grit sizes. An abrasion resistance 5.7 times that of the Al6061 matrix alloy is obtained when the SiC reinforced alloy is worn against the finest abrasive grit size. In contrast, the alumina reinforced composites of both matrix alloys display significantly lower relative abrasion resistances as those found for the SiC composite.

Composites containing higher volume fractions of reinforcement show proportionately greater improvements in relative abrasion resistance when subjected to low-stress wear against finer abrasives. This tendency is evident in the behaviour of both the Al2014 and Al6061 matrix composites. The wear resistance of each composite, with the exception of the SiC reinforced Al6061 alloy, approaches that of its respective matrix alloy upon being abraded by the coarsest abrasive grit.

The scanning electron micrographs in figs. 12a and b show typical representations of the plastic cutting wear mode found in the unreinforced matrix alloys. The two micrographs display the Al6061 alloy after abrasion by the largest and smallest abrasive grits, respectively, with the only difference being in the depth and width of wear grooves on each surface. Electron micrographs of abraded surfaces of the Al6061 alloy reinforced with 20% SiC are shown in figs. 13a and b. The composite

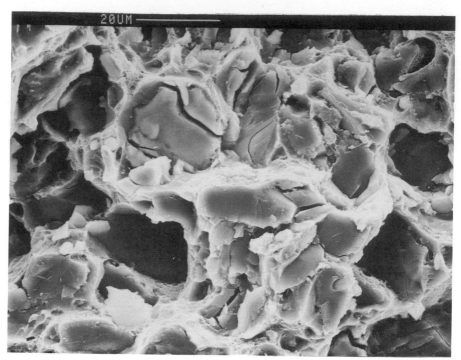

Fig. 7. Electron micrograph showing the tensile overload fracture surface of the Al2014 alloy reinforced with 20% Al_2O_3 particulates.

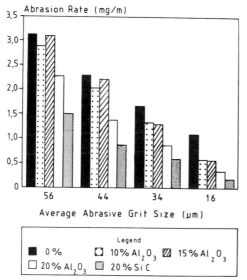

Fig. 8. Abrasive wear rates of the Al6061 matrix alloy and respective composites, determined as a function of the abrasive grit size.

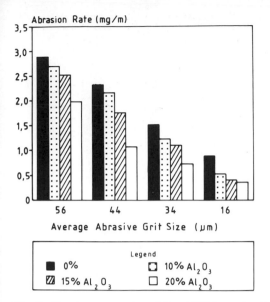

Fig. 9. Abrasive wear rates of the Al2014 matrix alloy and respective composites, determined as a function of the abrasive grit size.

surface after being subjected to high-stress abrasion by the coarsest abrasive grit is shown in fig. 13a, where extensive chipping of the matrix alloy and particulate fracture is evident. In contrast, the composite worn under low-stress conditions using the finest abrasive grit is shown in fig. 13b. The worn surface exhibits little

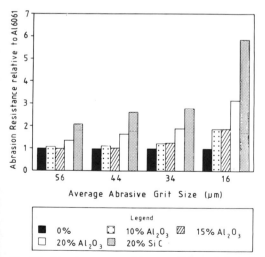

Fig. 10. Abrasive wear resistances of the reinforced Al6061 alloys, determined relative to the unreinforced matrix alloy and as a function of the abrasive grit size.

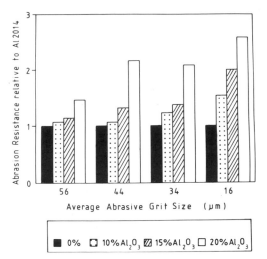

Fig. 11. Abrasive wear resistances of the reinforced Al2104 alloys, determined relative to the unreinforced matrix alloy and as a function of the abrasive grit size.

particulate damage and the wear mode is characteristically attritive in nature. Similar transitions in wear mode were noted for the 15% and 20% alumina reinforced composites of both matrix alloys, but to a lesser extent for the 10% alumina reinforced materials, where the wear mode was predominantly plastic cutting, with some particulate fracture and chip formation at clustered regions of reinforcement.

An electron micrograph of the abraded surface of the 15% Al_2O_3 reinforced Al2014 alloy is shown in fig. 14a. The composite has been abraded with the coarsest abrasive grit and the morphology of the reinforcing particulates is revealed by removing the surrounding plastically deformed matrix alloy by electropolishing for 1–2 s in a methanol/nitric acid mixture. Extensive fracture of the alumina reinforcement is evident, indicating that the stress imparted to the composite surface during abrasion is high enough to damage the reinforcing particulates. Conversely, the nature and extent of particulate damage is markedly decreased upon abrasion by the finest grit abrasive, as shown in fig. 14b for the same composite as fig. 14a. Particulate damage is minimal, showing the upper surface of each particulate having been removed by a polishing mechanism. This abrasion mode was found to be characteristic in each composite, especially in those containing larger volume fractions of reinforcement. Similar particulate damage characteristics were found in the SiC reinforced Al6061 alloy and the other alumina reinforced composites when abraded with the finer grit abrasive belts.

4.3. Reciprocating sliding wear results

The reciprocating sliding wear rates obtained for each of the Al6061 and Al2014 matrix alloy composites and the stainless steel counterfaces are represented in figs.

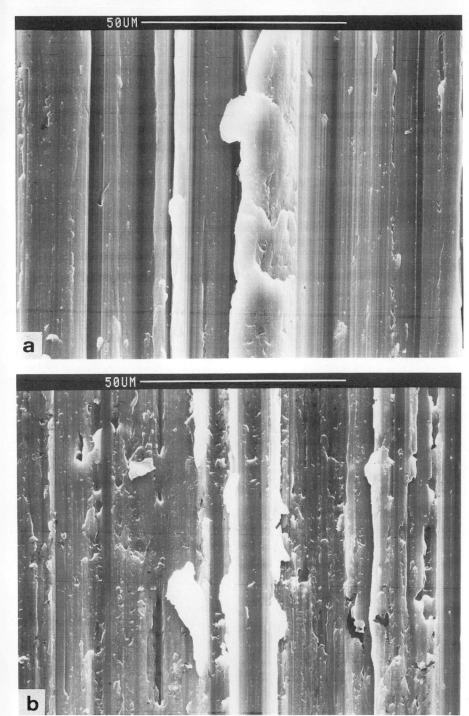

Fig. 12. (a), (b) Electron micrographs of the Al6061 matrix alloy after being worn by the coarsest (56 μm) and finest (16 μm) abrasive grit size, respectively.

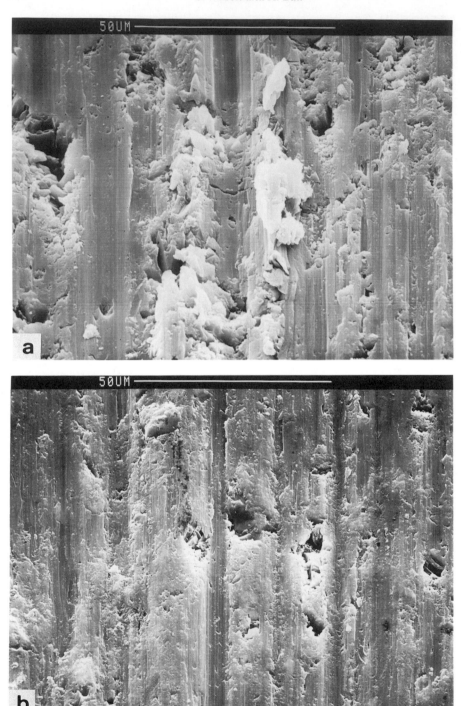

Fig. 13. (a), (b) Electron micrographs of the SiC reinforced Al6061 alloy after being worn by the coarsest (56 μm) and finest (16 μm) abrasive grit size, respectively.

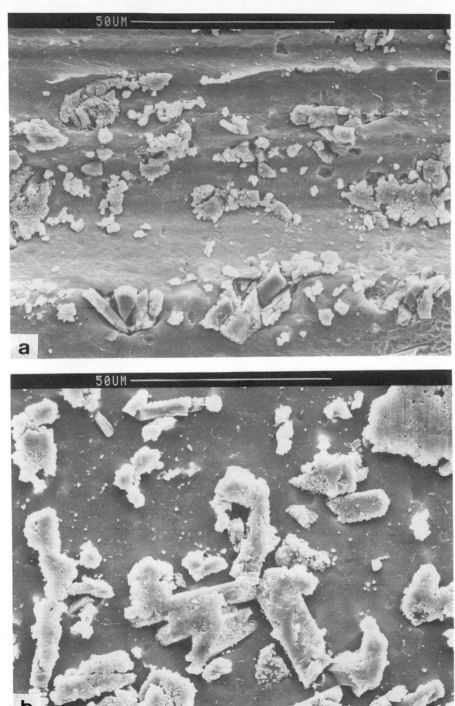

Fig. 14. (a), (b) Subsurface damage in the 15 vol% Al_2O_3–Al2014 composite after abrasion by the coarsest (56 μm) and finest (16 μm) abrasive grit, respectively. The worn surfaces were electropolished for 1–2 s to reveal particulate damage.

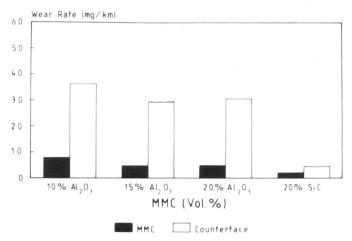

Fig. 15. Reciprocating sliding steady-state wear rates for each of the Al6061 composites and respective stainless steel counterfaces.

15 and 16, respectively. These wear rates were calculated from steady-state regimes of material loss over a minimum sliding distance of 24 km. Each composite displayed an initial run-in period during which the wear sample achieved full geometrical contact with the steel counterface. In some cases the run-in period was almost immediate for the low volume fraction composites, whereas that for the higher volume fractions of reinforcement took up to 30 km (20% SiC Al6061). A characteristic feature of the wear behaviour of all the reinforced alloys tested is the loss of material from both the counterface and the composite sliding sample. In contrast, the unreinforced aluminium alloys show wear losses that exceed those of the composites by almost three orders of magnitude, with no removal of

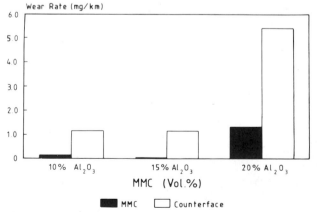

Fig. 16. Reciprocating sliding steady-state wear rates for each of the Al2014 composites and respective stainless steel counterfaces.

counterface material occurring. The unreinforced aluminium sliding samples had steady-state wear rates of 272 and 28 mg km^{-1} for the Al6061 and Al2014 matrix alloys, respectively. The 10% Al$_2$O$_3$ reinforced Al6061 and Al2014 sliding samples showed wear rates of 0.78 and 0.15 mg km^{-1} and had counterface wear rates of 3.67 and 1.13 mg km^{-1}, respectively.

The Al6061 matrix composites and associated counterfaces display a slight trend towards lower wear rates with increasing volume fraction of reinforcement. The wear rate of the 20% SiC reinforced material is approximately half that of the 20% Al$_2$O$_3$ reinforced Al6061 alloy and displays a markedly lower counterface wear rate. When the wear behaviours of the Al6061 composites are compared to those of the Al2014 composites shown in fig. 16, the Al2014 composites reinforced with 10% Al$_2$O$_3$ and 15% Al$_2$O$_3$ particulates exhibit lower sample and counterface material removal rates, with the exception of the Al6061 alloy reinforced with 20% SiC; this material has the lowest counterface wear rate of all the composites. The wear rate of the 20% Al$_2$O$_3$ reinforced Al2014 alloy is noticeably higher than those of all of the composites and exhibits the greatest counterface material removal rate.

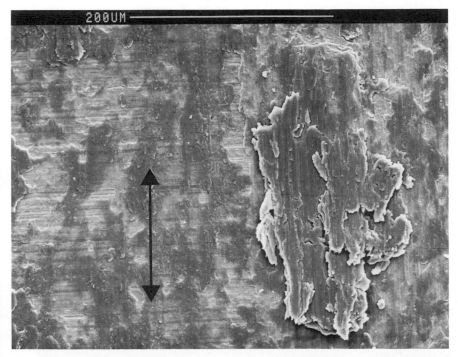

Fig. 17. Electron micrograph of stainless steel counterface after being worn to steady state against the Al6061 matrix alloy. Aluminium has transferred to the counterface surface during the sliding process, with the generation of flake-like debris. Arrow indicates sliding direction.

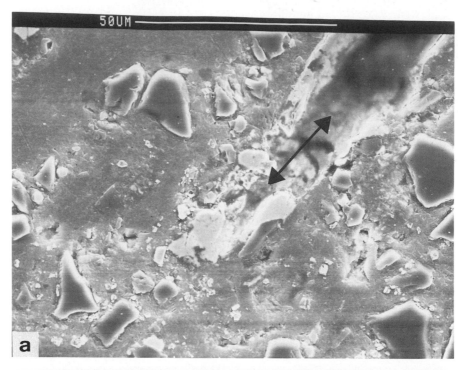

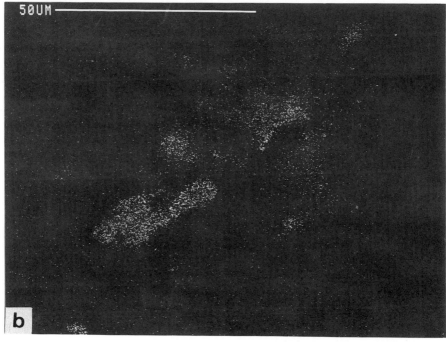

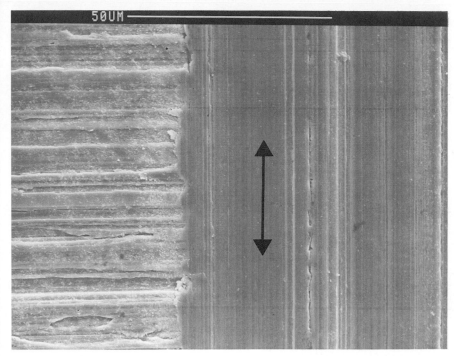

Fig. 19. Electron micrograph showing the abrasive wear mode of a stainless steel counterface after reciprocating sliding against the 15% Al_2O_3–Al6061 composite. Arrow indicates sliding direction.

The surface of a counterface worn by the unreinforced Al6061 alloy is shown in fig. 17. Aluminium transfer to the counterface is evident, with a large flake-like debris particle about to decohere. The Al2014 alloy showed a similar transfer mode, but there was less evidence of large flake debris formation.

An electron micrograph of the 20% SiC reinforced Al6061 alloy after 60 km of sliding is shown in fig. 18a. The worn surface of the composite has been taper sectioned and polished to reveal the extent of subsurface damage. The micrograph shows the base of a wear groove where fractured particulate debris is evident together with stainless steel transfer from the counterface. The counterface material is lighter in contrast compared to the surrounding matrix alloy and is concentrated in the area near the base of the wear groove. An Fe X-ray map of the same area as in fig. 18a is displayed in fig. 18b, showing the distribution of the counterface fragments in the subsurface wear zone. Similar features were observed in taper sections of each worn composite. The electron micrograph shown in fig. 19

Fig. 18. (a) Taper section of the 20% SiC–Al6061 composite surface after being worn to steady state during reciprocating sliding against a stainless steel counterface. Stainless steel debris, transferred during the sliding process, is visible as lighter inclusions in the subsurface region of a wear groove. Arrow indicates sliding direction. (b) Fe X-ray map of the region shown in fig. 18a, displaying the distribution of steel debris in the subsurface wear zone.

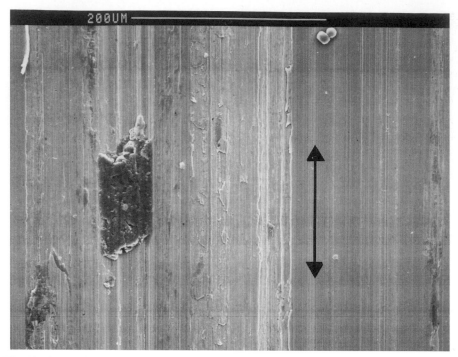

Fig. 20. Counterface surface after being worn by the 20% Al_2O_3–Al2014 composite, showing debris transfer to the stainless steel surface. Arrow indicates sliding direction.

shows a counterface having been worn by the 15% Al_2O_3 reinforced Al6061 alloy. The original grinding marks on the counterface have been completely worn away and there is no evidence of aluminium transfer from the sliding sample to the steel surface. Counterfaces worn by the alumina reinforced alloys did show some evidence of transfer of sliding-sample material. This was particularly so for the 10% Al_2O_3 Al6061 composite and the 20% Al_2O_3 reinforced Al2014 alloy, and fig. 20 is an example of this phenomenon.

Backscattered electron micrographs of the worn surfaces of the 10% Al_2O_3 and 20% Al_2O_3 reinforced Al2014 alloy are shown in figs. 21 and 22, respectively. The worn surface of the 10% Al_2O_3 composite is typical of that found for the reinforced Al6061 alloys and the Al2014 alloy containing 15% Al_2O_3. The composite matrix alloy, together with oxide products and particulate debris, has smeared plastically over the contact surface during the sliding operation. The worn surface of the 20% Al_2O_3 composite, shown in fig. 22, is notably more fractured in appearance; patches of smooth deformed matrix material, similar in morphology to that observed in fig. 21, are interspersed with areas of extensive chipping and particulate exposure.

Electron micrographs of the 10% Al_2O_3 and 20% Al_2O_3 reinforced Al2014 sliding specimens are again shown in section in figs. 23 and 24, respectively. Each composite has been sectioned perpendicular to the worn surface and parallel to

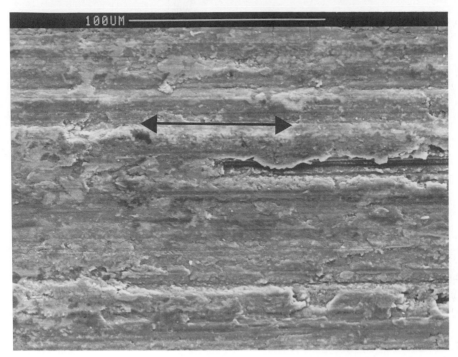

Fig. 21. Steady-state reciprocating sliding worn surface of the 10% Al_2O_3–Al2014 composite. Arrow indicates sliding direction.

the direction of sliding, so as to reveal the extent of plastic strain accumulation in the subsurface microstructures. The 10% Al_2O_3 alloy has been severely strained with deformation extending some 50 μm into the subsurface zone. In contrast, the 20% Al_2O_3 composite has undergone very little subsurface plastic strain, there is evidence also of subsurface microcracking and delamination in the immediate vicinity of the worn surface.

4.4. Cavitation erosion results

Plots of the cumulative mass losses for each of the Al6061 composites and monolithic alloy as a function of cavitation erosion time are shown in fig. 25. Each composite and unreinforced alloy displays an incubation period of some four minutes before appreciable material losses occur. The mass losses extend into a steady-state regime, with the alumina reinforced composites displaying higher erosion rates. The erosion rate of the unreinforced alloy is below that of the alumina reinforced composites and is marginally above that of the SiC reinforced alloy.

The cavitation erosion mass losses for each of the Al2014 composites and the monolith are shown in fig. 26. All the Al2014 materials display equal incubation erosion periods of some six minutes; this is slightly greater than those obtained for

Fig. 22. Typical steady-state reciprocating sliding worn surface of the 20% Al_2O_3–Al2014 composite is shown. Arrow indicates sliding direction.

the Al6061 materials. Overall, the Al2014 composites and the unreinforced alloy display erosion rates below those of their Al6061 counterparts. A definite trend in erosion loss, based on the volume fraction of the reinforcement, is evident, with the 20% Al_2O_3 reinforced alloy showing the highest and the unreinforced alloy the lowest erosion rates in the steady-state regime.

Electron micrographs of the surface of the SiC reinforced Al6061 alloy, after exposure to cavitation erosion for two, five and sixty minutes are displayed in figs. 27a to c, respectively. The same area of the composite is shown for the two and five minute exposures, displaying an overall increase in plastic deformation of the matrix alloy with cavity formation at particulate interfaces and within the matrix itself. Particulates become more exposed as matrix alloy is eroded away, but their integrity does not appear to be affected by the erosion process. Massive plastic deformation of the composite surface is evident after one hour of cavitation and particulates are difficult to discern from the surrounding heavily deformed alloy. An examination of the subsurface microstructure of this composite, after one hour exposure, reveals no particulate damage, as shown in the optical micrograph taper section in fig. 28, where the matrix alloy has been eroded away from around the reinforcing particulates. Likewise, the composite which displayed the greatest wear rate of the Al2014 matrix materials, namely that reinforced with 20 vol% alumina

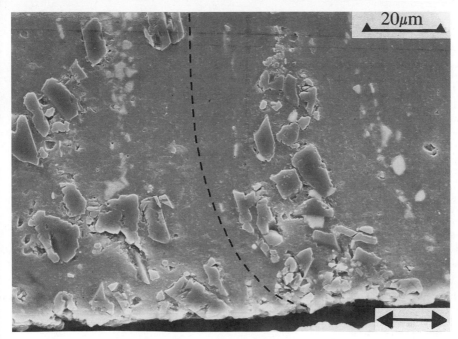

Fig. 23. Section taken perpendicular to the worn surface, showing subsurface plastic strain in the 10% Al$_2$O$_3$–Al2014 composite after reciprocating sliding against a stainless steel counterface to steady state. Arrow indicates sliding direction at the wear interface.

particulates, shows no evidence of reinforcement damage in the immediate subsurface wear zone (fig. 29).

4.5. Solid particle erosion

Plots of the cumulative mass loss in the solid particle erosion tests, conducted at an erosive impact angle of 90° to the sample surface for the Al6061 and Al2014 unreinforced and reinforced alloys, are shown in figs. 30a and b, respectively. Steady-state erosion conditions are achieved almost immediately for the harder 20% Al$_2$O$_3$ and 20% SiC reinforced Al6061 alloys and the 20% Al$_2$O$_3$ reinforced Al2014 alloy. The erosion behaviour for the other Al6061 composites and unreinforced alloy is characterised by an initial mass-gain "incubation period", with steady-state behaviour noticeable after erosion by 35 g of erodent in the case of the unreinforced alloy. The unreinforced Al2014 alloy reaches steady state after erosion by 25 g of erodent and is noticeably less erosion resistant than the Al6061 alloy. The Al2014 composites show a definite trend towards greater erosion rates with increasing volume fraction of reinforcement, unlike the behaviour of the Al6061 composites, where size and type of reinforcement also appear to influence the erosion rates.

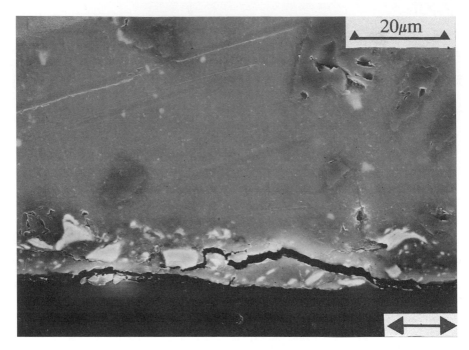

Fig. 24. Subsurface damage in the 20% Al_2O_3–Al2014 composite after reciprocating sliding to steady state. The micrograph shows a section taken perpendicular to the composite's worn surface where microcracking and delamination are visible at the wear interface. Arrow indicates sliding direction.

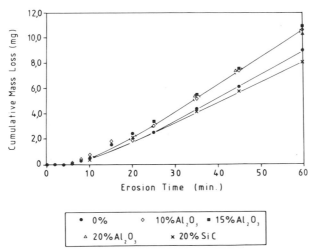

Fig. 25. Cumulative mass losses for the Al6061 matrix alloy and respective composites as a function of exposure time to cavitation erosion.

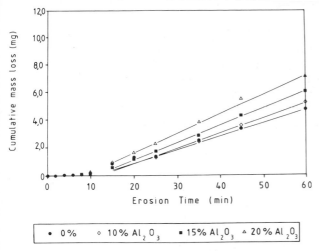

Fig. 26. Cumulative mass losses for the Al2014 matrix alloy and respective composites as a function of exposure time to cavitation erosion.

Erosion at a 30° impact angle is for all the composites characterised by the immediate onset of steady-state conditions. Figure 31 shows the steady-state erosion rates obtained at 90° and 30° impact angles for each composite and the unreinforced alloy. The erosion rates were determined from the slopes of linear regions (steady state) in the cumulative erosion mass loss curves of the materials. The erosion rates at the 30° impact angle are higher for each composite and the unreinforced alloy. The erosion rates for the 10% Al_2O_3 and unreinforced Al6061 at 30° are higher than those of their Al2014 counterparts with equivalent volume fraction, which, in contrast, show the greater wear rates at 90° impact conditions. The 20% Al_2O_3 reinforced Al6061 alloy exhibits similar erosion rates as the Al2014 alloy reinforced with 20% Al_2O_3 at both impact angles. The 20% SiC reinforced Al6061 alloy displays an erosion rate between that of the 15 and 20% Al_2O_3 reinforced Al6061 alloys.

Electron micrographs of the unreinforced Al6061 alloy and its 20% SiC reinforced counterpart, showing typical steady-state eroded surfaces at impact conditions of 90° and 30°, are displayed in figs. 32a to d. The composite surface eroded at 90° impact shows heavy plastic deformation, with some fragmentation of the deformed metal being evident. The unreinforced alloy displays a similar morphology, but is more ductile in appearance with less evidence of fragmentation of the deformed metal. A transition to micromachining or ploughing of material is observed at the 30° impact angle. The eroded surface of the composite exhibits less plastic shear and is highly fractured, in contrast to the unreinforced alloy.

The subsurface taper section optical micrographs in figs. 33a and b show the extent of microstructural deformation in the 20% SiC reinforced Al6061 alloy at erosion angles of 90° and 30°, respectively. The erosion at 90° shows a large amount of subsurface plastic deformation, cavity formation and particulate fracture compared to erosion at 30°.

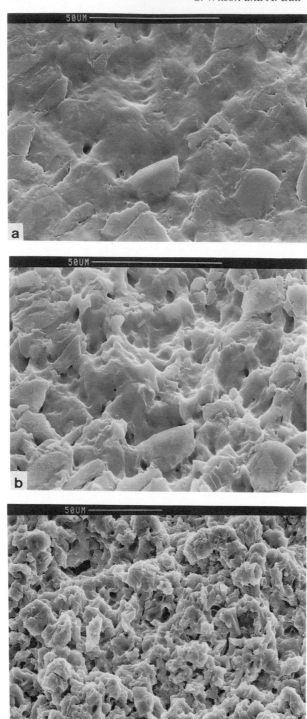

Fig. 28. Optical micrograph of a taper section of a cavitation eroded surface of the 20% SiC reinforced Al6061 alloy.

5. Discussion

5.1. Tensile properties

The tensile behaviours of monolithic aluminium alloys are characterised by a period of work hardening in the post-yield strain region, followed by plastic instability and then rupture. The fracture surface morphology is typically "dimpled", resulting from a process of void initiation and growth from nucleation points, such as intermetallics and other microdefects [36]. In each of the composites under consideration, the work hardening and fracture properties of the respective ductile matrices reinforced with ceramic particulates were altered significantly. Our observations of post-fracture tensile specimen gauge lengths of each composite indicate that the plastic slip processes in the interparticulate matrix alloy have been restricted. The absence of slip in the "necking" region of tensile specimens was particularly marked in the composites containing higher volume fractions of reinforcement. Regions of matrix alloy which did exhibit some evidence of plastic strain accommodation, were those devoid of particulate rein-

Fig. 27. (a)–(c) Electron micrographs of the SiC reinforced Al6061 Alloy after exposure to cavitation erosive conditions for two, five and sixty minutes, respectively. The same region of the composite is shown in (a) and (b).

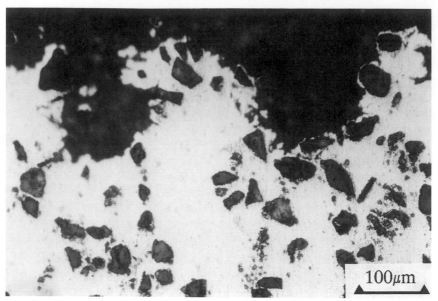

29. Optical micrograph of a taper section of a cavitation eroded surface of the 20% Al_2O_3 reinforced Al2014 alloy.

forcement; this was a common feature found in the 10 vol% Al_2O_3 reinforced matrices, which lacked homogeneous particulate distribution. All the composites, particularly those with the lower-strength Al6061 as matrix, did show highly localised strain accommodation by intensive shear in regions of the matrix next to where reinforcing particulates, or clusters of particulates, had fractured. In this regard it appears that the reinforcing particulates play a major part in preventing slip in the surrounding matrix alloy. The almost total absence of any matrix debonding at particulate interfaces is indicative of excellent interface integrity and bond strength, lending credibility to the proposal that the particulate interfaces constrain the matrix alloy from undergoing plastic deformation.

Several recent experimental and numerical investigations into the deformation characteristics of discontinuously reinforced aluminium alloys [37–45] have indicated that triaxial stresses are developed within the interparticulate matrix alloy during tensile straining. The composite matrix is constrained from plastic flow by the reinforcement and the hydrostatic nature of the stresses developed are sufficient to inhibit any yielding phenomena and encourage void growth [39,41]. In composites containing higher volume fractions of reinforcement, the levels of triaxial stress are sufficient to initiate fracture in the ceramic reinforcement [38,40,45]. Likewise, localised regions with high particulate density in the matrix alloy (clusters) will also fail by particulate fracture. The release of strain energy during tensile overload occurs when the ceramic particulates fracture, resulting in localised intensive shear accompanied by rapid strain hardening and tearing of the matrix alloy adjacent to the cracked particulates. Eventual failure of the composite

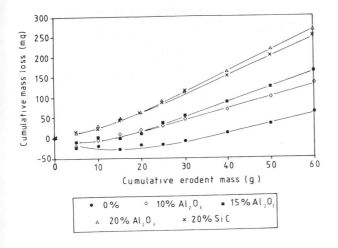

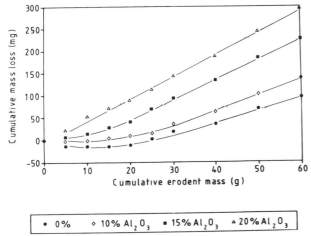

Fig. 30. (a) Solid particle erosion: cumulative mass losses for the Al6061 matrix alloy and composites for erosion at a 90° impact angle. (b) Solid particle erosion: cumulative mass losses for the Al2014 matrix alloy and composites for erosion at a 90° impact angle.

occurs through linkage of torn matrix "cavities" situated around broken particulates and clusters of particulates. Further support for this failure mode is observed in the tensile overload fracture-surface morphologies of the composites under investigation. Each reinforced alloy exhibited particulate fracture in the plane normal to the tensile axis. The fracture surfaces of the alloys containing 15 and 20 vol% reinforcement, showed widespread particulate cleavage, associated with extensive matrix tearing forming "tear ridges" [35] surrounding each broken particulate. The composites containing 10 vol% of reinforcement had fracture surfaces showing particulate failure and matrix tearing, together with widespread dimple formation, associated with the release of strain energy by void nucleation

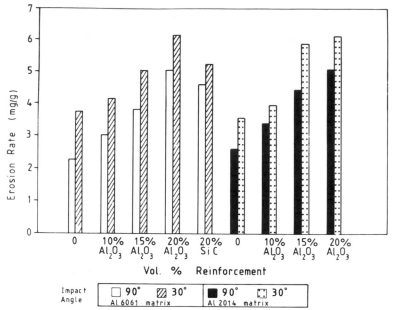

Fig. 31. Steady-state solid particle erosion rates for the two matrix alloy composite systems at erodent impact angles of 90° and 30°.

and growth in the less constrained regions of the matrix which contain low densities of reinforcing particles.

The stress–strain response of each of the reinforced alloys is determined predominantly by the volume fraction of ceramic particulates present in the ductile matrices. With larger volume fractions of high-modulus particulates there is a buildup of constraint within the matrix, resulting in a more efficient transfer of load to the reinforcement. Strain energy is thus accumulated, and liberated, in a more elastic fashion by the added ceramic reinforcement, resulting in lower plastic strains during tensile overload.

The strength of each base matrix alloy appears to have an influence on the stress–strain response of each reinforced system. The high strength and ductility of the Al2014 alloy is compromised with the addition of alumina reinforcement. Thus inferring that the magnitude of the stresses developed in the early stages of strain hardening in the interparticulate matrix alloy of each composite are of the same order of magnitude as those required to initiate fracture in the alumina reinforcement. The weaker and less ductile Al6061 matrix alloy undergoes moderate to significant strengthening effects with the addition of alumina particulates, showing proportionately higher plastic extensions in comparison to the Al2014 alloys reinforced with the same volume fraction of particulates. It is evident that the Al6061 composites require additional straining and strain hardening in order to generate matrix triaxial stresses sufficient to initiate fracture in the alumina reinforcement. The plausibility of this argument is improved when considering the

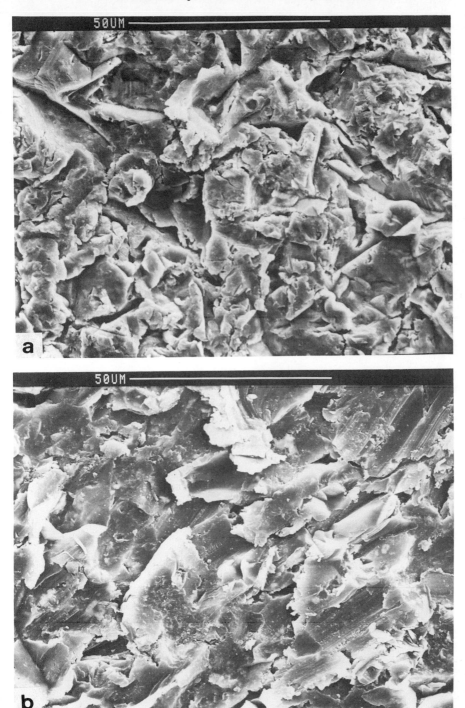

Fig. 32. (a), (b) Steady-state solid particle eroded surfaces of the unreinforced Al6061 alloy at erodent impact angles of 90° and 30°, respectively. (c), (d) Steady-state solid particle eroded surfaces of the SiC reinforced Al6061 alloy at erodent impact angles of 90° and 30°, respectively.

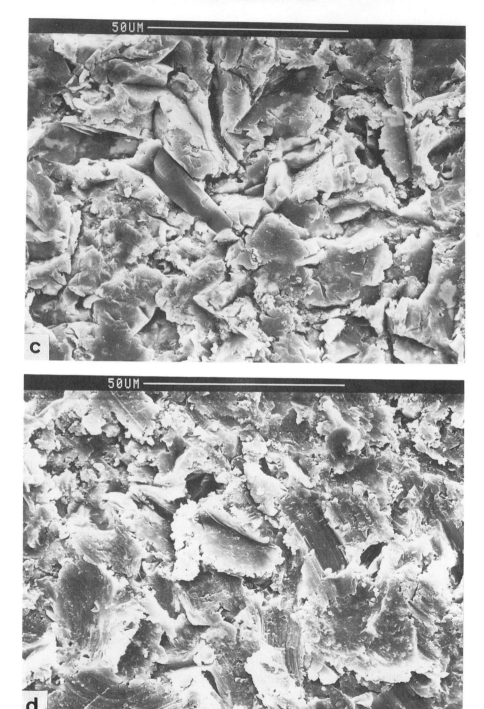

Fig. 32 (continued).

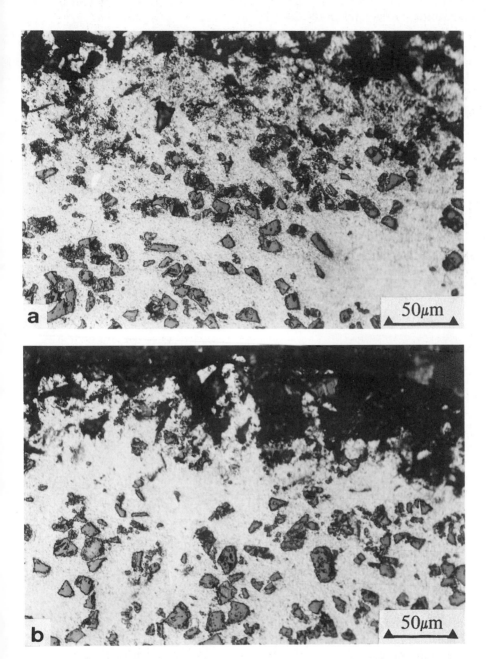

Fig. 33. (a), (b) Subsurface taper section optical micrographs showing the extent of microstructural damage in the 20% SiC reinforced Al6061 alloy after solid particle erosion at erodent impact angles of 90° and 30°, respectively.

behaviour of the Al6061 alloy reinforced by the harder, tougher and stiffer ceramic silicon carbide [46,47]. The extra strain energy required to initiate fracture in the silicon carbide particulates in the 20 vol% SiC reinforced Al6061 alloy is achieved by the additional work hardening and consequent strength increment of the SiC composite as compared to the alumina reinforced matrices.

The rationale behind characterising the mechanical properties of the reinforced and unreinforced aluminium alloys was to facilitate the interpretation of their behaviour in various tribological situations. While the stresses imparted to a material as it rubs, abrades or impacts against another are often more complex in nature than those developed in a simple uniaxial tensile test, it is hoped, nevertheless, that the current assessment of the mechanical response will contribute towards predicting and interpreting the tribological behaviour. The majority of tribological environments involve situations where a hard body indents, impinges or slides against another, softer surface. In this regard, the hardness, work hardening ability and strain to fracture of each material involved in the mechanical interaction are important factors which determine the degree and depth to which strain accumulates below each surface prior to microfracture and wear loss. Furthermore, the magnitude of the stresses as well as the rate at which strain energy is imparted and dissipated between two bodies as they interact, also determines the respective levels of subsurface strain accumulation and fracture.

5.2. Abrasion

Each of the unreinforced aluminium matrix alloys displays abrasive wear modes that are typical of those found for ductile metals; extensive plastic deformation and cutting of the aluminium is evident, together with lower wear rates for the harder Al2014 aluminium alloy. All the reinforced aluminium alloys display two-body abrasion resistances that are greater than those of their respective matrix alloys, in both high and low contact stress abrading conditions. The interaction between the reinforcing particulates on the surface of each composite and abrasive particles results in the generation of stresses at asperity contact points which are sufficient to initiate damage in the ceramic reinforcement. The extent of particulate damage depends on the magnitude of these contact stresses as well as the toughness and hardness of both the reinforcing ceramics and abrasive particles. When a composite is abraded against the coarsest abrasive, the reinforcing particulates fracture as a result of the large stresses present at asperity contact points. The surface morphology of each worn composite correlates well with observations made by other researchers using coarse abrasives against a variety of aluminium matrix composites [5–7,9]; the common characteristic features of high contact stress abrasion involve widespread particulate or fibre fracture, accompanied by extensive plastic deformation and cutting of the matrix alloy.

The use of finer abrasives is associated with a larger number of asperity contact points per unit area and thus lower average contact stresses. The abrasive particles are unable to indent the composite surfaces and damage to the ceramic reinforcement is restricted to microfracture, with very little evidence of catastrophic failure.

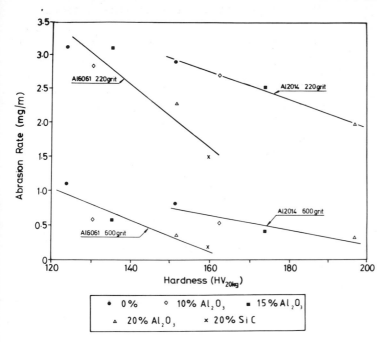

Fig. 34. Abrasion rates of each reinforced and unreinforced alloy, plotted as a function of their respective macrohardness (HV_{20kg}) values. The wear rates are those for abrasion against the coarsest (220 grit) and finest (600 grit) abrasives.

Thus the indentation stresses during abrasion by fine abrasives are so low that little plastic deformation is caused on each pass. The onset of the transition from general reinforcement fracture and fragmentation to low-stress abrasion, where the integrity of the ceramic inclusions is maintained, is associated with the use of abrasive grit particles that are in the size range of approximately 34–16 μm. Similar values were obtained by Wang and Hutchings [7] for alumina fibre reinforced Al6061 alloys, where abrasive particle sizes of between 20–28 μm no longer fractured the reinforcing fibres. A material's resistance to indentation appears to be the dominant factor in determining the abrasion resistance under both high and low contact stress conditions. This is further illustrated by fig. 34, which shows a plot of the abrasion rates of each reinforced and the unreinforced alloy as a function of their respective bulk hardness values. The wear rates for abrasion against both the coarsest and the finest abrasives are shown and they exhibit a decrease in abrasion rate with increase in each composite's bulk hardness, under both low and high average contact stress conditions.

The importance of the type of reinforcement used in each composite in determining the wear behaviour, is illustrated by comparing the alumina reinforced alloys to the Al6061 alloy reinforced with 20 vol% of silicon carbide particulates. Each of the alloys reinforced with alumina displays wear rates well above those of the alloy containing silicon carbide, in both high and low contact stress environments. A possible explanation for the excellent behaviour of this composite may partly be the mechanical properties of the silicon carbide particulates themselves. Previous research has established that silicon carbide is stronger, tougher and harder than alumina [46,47]. Our own tensile testing investigations have established that silicon carbide reinforcement was able to withstand greater stresses than alumina reinforcement before eventual catastrophic failure. In this regard, it is most likely that the extent of plastic strain accumulation and strain hardening in the silicon carbide reinforced alloy during abrasion, is much greater than that found in the alumina reinforced alloys, where particle fracture is a precursor to microfracture of the composite and wear loss.

Another factor which plays an important role in the wear behaviour of each reinforced alloy, is the hardness of the abrasive particles against which the composite is loaded. The aluminium oxide abrasive particles are themselves abraded by the alumina and silicon carbide reinforcing particulates present in the aluminium alloys. A situation is reached where aluminium oxide abrasive particles and ceramic reinforcing particulates are involved in a process of mutual microfracture as they move across each other. The silicon carbide reinforcement present in the 20 vol% Al6061 alloy, is able to indent the softer abrading particles, resulting in very low wear rates for this composite. By taking into account the excellent abrasion resistance of the silicon carbide reinforced alloy, when it is abraded against an abrasive which is softer than silicon carbide, it can be inferred that the use of softer abrasives (e.g. silica), will result in even better abrasion resistances for all the composites.

The influence that the reinforcing particulate size has on the abrasion resistance is best illustrated by focussing on the behaviour of the softer alumina reinforced Al6061 alloys. The 20 vol% composite has a wear resistance which out-performs that of the harder Al2014 alloy reinforced with 15 vol% alumina, in both high and low-stress abrasion. A possible explanation for such behaviour may be associated with the large average diameter of this composite's reinforcing particles, which are approximately twice the size of the alumina particulates in the 10 and 15% reinforced Al6061 alloys and the 10 and 15% reinforced Al2014 alloys. Thus, an abrading particle has to indent the surface of the 20% Al6061 alloy to a proportionately greater depth than that of the other composites in order to effect the same degree of damage to the larger particulates. Bigger reinforcing particulates would also require greater amounts of kinetic energy to be transferred during abrasion, in order to initiate their fracture.

The 10 and 15 vol% alumina reinforced Al6061 alloys have similar particulate sizes and are able to accommodate strain energy in a more plastic fashion compared to the harder 20 vol% alumina reinforced Al6061 alloy. The indentation depths in these two composites are more likely to exceed the diameters of their respective reinforcing particulates under high contact stress abrasion, increasing the likelihood of particulate fracture and greater material removal rates. The relatively small difference in bulk hardness between these two composites indicates that the depth of subsurface deformation during high contact stress abrasion is similar. The material removal rates can then be rationalised in terms of the frequency with which ceramic particulates fracture, which in the case of the 15 vol% composite would be much higher than for the 10 vol% composite. The increased frequency of particulate fracture in the 15 vol% composite give rise to its greater wear rate in high contact stress abrasion. The transition to low contact stress abrasion results in the 15 vol% reinforced alloy having a marginally lower wear rate than the 10 vol% composite since the indentation depths are reduced.

The importance of the ratio of abrasive particle penetration depth to the size of reinforcing particulates was also highlighted by Wang and Rack [9]. They established that the transition from low contact stress abrasion to high stress cutting and fracture of reinforcing particulates arose once the abrasive particle penetration depth approached the average diameter of the reinforcing particulates. Further increases in abrasive particle penetration depth, through the use of coarser abrasives, would have very little effect on the abrasion resistance. However, their investigations involved the use of silicon carbide reinforced aluminium alloys which were abraded against silicon carbide abrasives. This raises the question as to whether alumina reinforcing particulates would fracture according to the same relative penetration-depth criteria as found for silicon carbide reinforced alloys. Alumina is both softer and less tough than the silicon carbide reinforcement and would be expected to fracture at a lower relative penetration depth. Nevertheless, our results show a similar lack of change in wear resistance for the softer 10 and 15 vol% alumina reinforced Al6061 alloys and the 10 vol% alumina reinforced Al2014 alloy during abrasion against abrasive particles bigger than an average diameter of 44 μm (figs. 10 and 11).

5.3. Reciprocating sliding wear

A common feature of the interaction between aluminium and steel, as they come into sliding contact, is the rapid formation of a transfer layer of aluminium on the steel surface, due to the high adhesive bonding force between iron and aluminium [48]. During steady-state wear a situation develops where the aluminium alloy eventually ends up sliding against its own transfer layer, with friction coefficients approaching those measured for the aluminium alloy sliding against itself. Thus, the rate of material removal from an aluminium alloy surface is primarily dependent on its resistance to the large shear stresses developed in its microstructure, caused by the frictional welding occurring at the sliding interface.

The steady-state sliding wear behaviour of the two unreinforced aluminium alloys is characterised by high wear rates and transfer of the aluminium to the stainless steel counterfaces. Of the two alloys, the harder Al2014 monolith displayed a wear rate that was approximately ten times below that of the Al6061 alloy. Our examination of the counterface which had been worn to steady state by the Al6061 alloy, displayed a large amount of transferred aluminium and a preponderance of big flakes of the alloy adhering to its surface. In comparison, the counterface worn by the Al2014 alloy showed markedly less transfer. The generation of large debris flakes by the Al6061 alloy during sliding contact is an indication of its lower strength, compared to that of the stronger Al2014 alloy.

The sliding wear rates of the aluminium alloys reinforced with ceramic particulates are almost three orders of magnitude below those of their respective unreinforced alloys. Similar reductions in aluminium transfer to steel counterfaces have been reported by several other investigators [6,12–15,22]. The incorporation of particulate reinforcement into each alloy has resulted in a lowering of the adhesive transfer rate of aluminium to the stainless steel surfaces. By increasing the volume fraction of ceramic particulates in each alloy the surface area of matrix alloy that comes into contact with the counterface is reduced as the particulates themselves become load bearing, resulting in a reduction in the rate of aluminium to steel transfer. Our tensile test investigations have recorded that plastic flow of the matrix alloy of each composite is inhibited by the constraints imposed by the particulate interfaces during tensile straining. Similar effects can be expected at the sliding wear interface, where plastic flow of the interparticulate matrix alloy, arising from frictional shear stresses generated at the composite sliding surface, would be prevented depending on the level of plastic constraint in the matrix. Thus, the probability of shear failure in the aluminium matrix is diminished by the reduced ability of the composite to accommodate strain by plastic slip, resulting in less transfer to the steel counterface. In this regard, the Al6061 composites and the 10 and 15 vol% alumina reinforced Al2014 alloys display steady-state wear rates which decrease as the volume fraction of reinforcement is increased. The 10 vol% Al6061 alloy shows the greatest wear rate, which is in accordance with its proportionately higher surface area of matrix alloy in contact with the counterface, and increased susceptibility to shear failure, afforded by the lack of significant matrix constraint by the reinforcing particulates. This is further corroborated by the existence of patchy areas of transferred aluminium alloy on the counterface surface against which the composite was worn. The sliding wear rates of the Al2014 alloys reinforced with 10 and 15 vol% alumina are significantly below those of their Al6061 counterparts, due to the higher hardness and shear strength of the Al2014 alloy.

Each of the stainless steel counterfaces displayed steady-state wear rates during sliding contact with the reinforced alloys. The wear mode for all the counterfaces was distinctively abrasive, indicating that the ceramic reinforcing particulates are responsible for removal of the steel. During sliding contact with the counterface, the exposed reinforcing particulates in each alloy abrade and cause the transfer of steel debris to the surface of the composites (fig. 18). The large adhesive bonding forces between aluminium and steel result in the accumulation of counterface debris in the interparticulate matrix regions of the composites. It is also highly probable that some of the steel debris embedded between particulates is transferred back to the counterface. Should such a process occur, it would be dependent on the probability of steel debris coming into adhesive contact with the counterface, as well as on the strength of the aluminium matrix to which it has bonded. A weaker interparticulate matrix alloy, such as Al6061, would allow for increased rates of back transfer of steel debris and matrix alloy to the counterface. The abrasive action of the freshly exposed reinforcing particulates in the composite will then generate more counterface debris as well as remove any adhering aluminium and steel debris from the counterface surface, which is compacted into the composite interparticulate matrix regions. This process of transfer and back transfer of steel debris and matrix aluminium would also result in their surfaces becoming heavily oxidised, hence the high proportion of oxide products on the worn surfaces of the composites (fig. 21).

The counterfaces worn by the 10 and 15 vol% alumina reinforced Al2014 matrix alloys, display lower wear rates compared to the alumina reinforced Al6061 alloys. This is probably due to the greater shear strength of the Al2014 matrix, which would reduce chances of decohesion, followed by transfer of aluminium matrix and steel debris between the two sliding surfaces. The Al6061 alloys reinforced with alumina, display counterface wear rates which show a slight decrease with increase in volume fraction of reinforcement. The decrease in wear rate can be attributed to the higher number of reinforcing particulates per unit area in contact with the counterface and the resultant lower contact stresses. However, once a significant transfer layer of steel debris, aluminium and oxide products has been formed, the contact stresses would be markedly reduced. Of greater influence, perhaps, is the ease with which interparticulate matrix alloy and steel debris is able to be transferred to the counterface, thereby exposing the ceramic particulates to the counterface and initiating further abrasion.

The lowest counterface wear rate is displayed by the silicon carbide reinforced Al6061 alloy. Silicon carbide is harder, tougher and shows a much lower affinity towards adhesive bonding with steels than alumina. In contrast, alumina forms a strong adhesive bond with steels during sliding contact [13]. This effect has also been reported for sliding contact between ceramic matrix composites and steels, where steel was found to transfer preferentially to alumina surfaces as opposed to carbide or nitride reinforcements [49,50].

The shear stresses developed in the immediate wear zone beneath composite surfaces in sliding contact with their respective counterfaces, results in the formation of shear strains, which are then accommodated either plastically or elastically. The magnitude of the strain accumulation depends on each composite's ability to plastically deform. Factors which influence this ability are the strength of the matrix alloy as well as the degree to which it has been constrained by the reinforcing particulates. The compaction of steel and oxide debris into the interparticulate matrix regions of each composite would also generate added straining effects in the composite microstructure. By referring to these considerations, the low wear resistance of the Al2014 alloy reinforced with 20 vol% alumina can be explained. This composite displays a strain to failure of 0.9% and is weaker than the unreinforced matrix alloy. The extent to which the composite accumulates strain in a plastic manner is shown in the subsurface section in fig. 24. The composite displays a large strain gradient, with deformation concentrated primarily in the immediate wear zone up to some 10 μm below the worn surface. This is in contrast to the smaller subsurface strain gradient displayed by the more ductile 10 vol% alumina reinforced Al2014 alloy (fig. 23), which shows plastic accumulation of strain to some 50 μm beneath its worn surface.

The low strength and inability of the 20 vol% composite to accommodate strain results in rapid failure and delamination of surface material and an overall increase in the rate of transfer of aluminium and steel debris to the counterface. Nevertheless, the wear rate of this composite is still some three orders of magnitude below that of its matrix alloy. Subsurface cracking is evident in the immediate wear zone of the composite (fig. 24), together with significant transfer of aluminium debris to the counterface, as is shown in fig. 20. These observations agree with those reported by Alpas and Embury [22], who obtained a similar high wear rate for an Al2014 alloy reinforced with 20% silicon carbide. This composite also displayed a very low strain to failure and a wear mode involving extensive delamination and cracking of the surface transfer layer, resulting in exposure of reinforcing particulates and significant damage to the counterface.

5.4. Cavitation erosion

The two unreinforced aluminium alloys each display cavitation erosion characteristics similar to those of other age-hardening alloys previously investigated by Vaidya and Preece [26]. They established that aluminium alloys generally show greater erosion resistances when their strengths are increased, either by age hardening or solute content additions. This is verified by the erosion data for the two monoliths in figs. 25 and 26, where the harder and stronger Al2014 alloy shows a longer incubation period of exposure to cavitation erosion before mass losses occur, accompanied by a lower steady-state erosion rate in comparison to the behaviour of the softer Al6061 alloy.

The erosion behaviours of the composite materials are similar to those of their respective matrix alloys. Apart from slight variations in steady-state erosion rate, their incubation periods show no deviation from those of the monolithic alloys, indicating that the dominant mode of material removal occurs through strain accumulation and rupture in the matrix alloy. The steady-state erosion rate of each material is displayed in fig. 35, where they are plotted as a function of their respective work to fracture ($E_{fracture}$) values. The work to fracture value of each composite and monolith is calculated from the area beneath its respective stress–strain curve (fig. 3), and is a measure of the strain energy required for tensile fracture. The erosion rates of the alumina reinforced alloys do show a reasonable correlation with $E_{fracture}$ values, indicating that the dominant mode of failure occurs through strain accumulation and fracture, as opposed to an indentation wear mechanism, where the high bulk hardnesses would contribute to greater erosion resistances. The silicon carbide reinforced Al6061 composite has an erosion resistance well below that of its matrix alloy and alumina reinforced counterparts. In this regard, the micrograph sequence in fig. 27 shows the extent and nature of cavitation damage in the silicon carbide reinforced Al6061 alloy. The strain energy transmitted to the composite surface during cavitation has resulted in extensive plastic deformation of the aluminium matrix, accompanied by rupture. The erosion process has not affected the silicon carbide particulates in any way, apart from exposing them through removal of surrounding matrix alloy. The steady-state eroded surface of the composite is similar in morphology to that of the

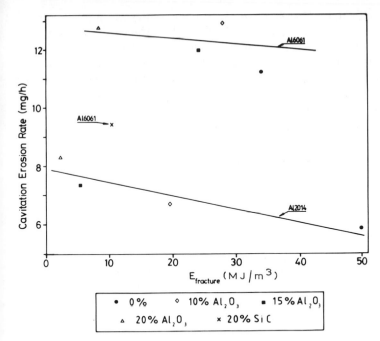

Fig. 35. Cavitation erosion rates of each reinforced and unreinforced alloy, plotted as a function of their respective work to fracture (E_{fracture}) values.

unreinforced alloys, showing massive plastic deformation and cavity formation. Subsurface damage is limited to removal of the aluminium matrix from around the reinforcing particulates (figs. 28 and 29). Both subsurface taper sections reveal very little evidence of damage to the silicon carbide or alumina reinforcing particulates. While particulate integrity is maintained, there remains a possibility that microfracture of the softer and less tough alumina reinforcement surfaces is occurring. This may account for the difference in erosion rates between the alumina and silicon carbide reinforced Al6061 alloys; the tougher silicon carbide particulates being more resistant to microfracture and displaying a slightly lower erosion rate. Surface microfracture may also account for the slight increases in erosion rate of the Al2014 alloy with greater volume fractions of alumina.

Other factors which may be playing a role in determining the erosion behaviour, are the effect of modulus mismatch between particulate and matrix interfaces as well as particulate morphologies. The difference in modulus between aluminium and silicon carbide is greater than that found between aluminium and alumina. As a result of this there would be an expected increase in the probability of matrix rupture at the silicon carbide particulate interfaces, and higher erosion rates. However, the silicon carbide reinforced alloy has a much lower erosion rate, indicating that the matrix–particulate interface strengths are probably very high, thereby preventing void initiation and rupture at interfaces. Evidence from the micrographs in fig. 27 suggests that the matrix alloy remains bonded to the ceramic reinforcement during cavitation, with a few inclusions acting as sites for matrix rupture. Similar characteristics were observed for the alumina reinforced alloys, and evidence from the tensile test investigations suggests that alumina and silicon carbide form very strong bonds with aluminium. Thus, although interfacial rupture does occur on a limited scale, the bulk of the nuclei for cavity formation and rupture would be preferentially generated in the matrix, as opposed to the stronger interfaces. The alumina particulate reinforcements are also more angular in morphology when compared to the silicon carbide particulates. The increased angularity would increase the probability of added stress concentration effects and concentration of deformation at alumina–aluminium interfaces, giving rise to increased matrix and particulate failure and higher erosion rates.

Heathcock et al. [25] established that materials with a greater elastic resilience displayed better resistances to cavitation erosion. The addition of high-modulus ceramics to aluminium alloys results in an overall increase in their ability to absorb and dissipate impact energy in an elastic manner. Thus, an increase in erosion resistance with greater volume fractions of high-modulus reinforcement would be expected, if the predictions of Heathcock et al. are valid. However, the erosion data for each of the alumina reinforced alloys contradict these considerations, with composites having higher elastic moduli showing greater wear losses. We can infer from these observations that the depth to which strain accumulation occurs in the matrix alloy of each composite during cavitation is very small, due to the fact that there is negligible transfer of strain energy to the high-modulus reinforcement from the matrix. Thus, deformation in these composites is restricted to the immediate subsurface regions during cavitation. This is corroborated by the lack of any visible subsurface bulk deformation and particulate fracture in the composites (figs. 28 and 29). The extent to which strain accumulation and hardening occurs in similar age-hardening alloys during cavitation has also been demonstrated to be restricted to a subsurface depth of about 20 μm, when they are in the peak aged condition [26].

5.5. Solid particle erosion

The two unreinforced matrix alloys display erosive wear modes that are characteristic of those found for ductile metals. The steady-state eroded surface of each alloy shows extensive plastic deformation and indentation at erodent impingement angles that are normal to the target surface (fig. 32a). The energy transmitted to an aluminium surface during normal impact by the silicon carbide erodent is accommodated primarily through strain hardening and lateral displacement of the surface metal. In contrast, during erosion at a lower impact angle of 30° the energy transferred to the metal surface by each erodent particle is largely consumed through continuous removal of the aluminium metal by a cutting and ploughing mechanism. This results in a markedly higher material removal rate, accompanied by less subsurface deformation.

The onset of steady-state erosion in each matrix alloy is preceded by an initial incubation period of mass gain, where erodent particles become embedded in the softer aluminium matrix. This effect is most marked at 90° erosion, where there is almost total conversion of impact energy into plastic deformation of the aluminium. The harder Al2014 alloy shows a smaller incubation region, accompanied by a slightly greater erosion rate when compared to the softer Al6061 alloy. The depth to which erodent particles indent the surface of the harder alloy at impact, is smaller than that found for the Al6061 alloy, with there being less chance of erodent particles embedding into its surface. In addition, the energy transmitted to the Al2014 surface as an erodent particle strikes is concentrated into a smaller impact site when compared to the softer Al6061 alloy. As a result of this, the impact stresses are proportionately greater for the harder alloy, thus increasing the rate of strain hardening and rupture of the metal's surface. Silicon carbide erodent fragments, which have embedded into the Al6061 alloy and increased its hardness, will also be responsible for added erosion resistance over that of the Al2014 alloy. When the impact energies are largely consumed by cutting and ploughing of each alloy's surface at a 30° erosion angle, the softer Al6061 displays a greater wear rate than the Al2014 alloy, erodent particles being able to displace and remove greater volumes of the softer Al6061 alloy.

Solid particle erosion of each of the reinforced aluminium alloys is characterised by increases in erosion rate over their respective matrix alloys. Similar findings have been reported by Goretta et al. [23] and Hutchings and Wang [24]. The eroded surface of each composite shows similar features to those found in each of the matrix alloys at the two impact angles. However, there is also evidence of extensive reinforcing particulate fracture and the deformed metal is less ductile in appearance, indicating constrained plastic flow, particularly in alloys with higher reinforcement contents (figs. 32c and d). Subsurface taper sections (figs. 33a and b) reveal extensive particulate fracture and fragmentation. The depth of damage accumulation is greatest in the composites subjected to erosion at 90°, where the impact energies are almost completely absorbed through straining of the subsurface material.

The erosion rate of the composites appear to be influenced by their respective ductilities, with the least ductile materials exhibiting the greatest mass losses and smallest incubation periods. This argument has also been proposed by Goretta et al. [23] to explain the greater erosion losses in particulate reinforced Al2014 alloys. While ductility remains an important consideration in analysing erosion behaviour, a more useful parameter is the strain energy required to initiate failure in a material. This is determined from the area beneath a material's tensile load–extension curve and is an indication of its toughness or work to fracture. A particle striking the surface of a material transmits a certain amount of kinetic energy into its microstructure. If the impact energy exceeds that required to initiate failure in the material, then subsurface rupture will occur. Alloys or composites which have low fracture energies will then become more susceptible to erosive wear loss. In this regard, Srinivasan et al. [51] investigated the erosion behaviour of a series of Al–4% Cu metal matrix composites containing up to 30 vol% of alumina fibres. The erosion rates of each of these materials were found to increase significantly with added reinforcement contents. A good correlation between erosion resistance

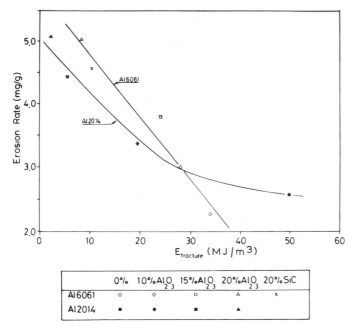

Fig. 36. Solid particle erosion resistance of each of the composites, at 90° incidence, plotted as a function of their E_{fracture} parameters.

and a work to fracture parameter (maximum tensile strength × strain to failure), calculated from tensile test curves, was obtained for each composite. Alloys which had large reinforcement volume fractions displayed lower erosion resistances and work to fracture values in comparison to composites with less reinforcement.

The areas beneath the stress–strain curves of each of the Al6061 and Al2014 reinforced and unreinforced alloys were determined so as to give the parameter, E_{fracture}, that is proportional to the work to fracture of each material. The E_{fracture} values are then plotted as a function of the erosive wear rates of the composites and the unreinforced alloy at a 90° erosive impact angle, as shown in fig. 36. Both the Al2014 and Al6061 composite materials display reasonable correlation between their erosion rates and E_{fracture} values. The two curves do not show ideal linear behaviour, which is most probably due to the different strain accumulating characteristics of each of the alloys and composites, the Al6061 composites being able to accumulate plastic strain to a greater extent before fracture, and thus displaying higher fracture energies than the Al2014 composites. Composites with small fracture energies are prone to rapid material losses. The unreinforced alloys have the greatest fracture energy values and hence show minimal wear losses.

6. Conclusions

An attempt has been made to characterise the wear behaviour of two age-hardening commercial aluminium alloys, each reinforced with different volume fractions of hard ceramic particles, in a variety of tribological situations. The tensile characteristics of each composite were also investigated in order to facilitate interpretation of their various tribological responses. The results of this investigation can be summarised as follows.

The strength of each reinforced aluminium alloy was found to be dependent on the degree to which plastic slip processes had been inhibited by the constraining effects of the ceramic particulates, the strength of the matrix alloy itself and the ability of the reinforcing ceramic to resist fracture. Composites having the weaker Al6061 alloy as matrix, required additional straining and work hardening in order to generate sufficient levels of triaxial stress so as to initiate fracture in the ceramic inclusions. In contrast, the stress levels generated in the stronger Al2014 matrix alloy were sufficient to initiate fracture in the alumina reinforcing particulates after very little work hardening. As a result of this the reinforced Al2014 alloys displayed a deterioration in strength and strain to failure with increase in reinforcement content.

The abrasive wear rate of each composite was found to be lower than that of its respective matrix alloy, with greatest improvements occurring during low contact stress conditions. All the reinforced and unreinforced alloy's abrasion resistances were found to be related to their respective macrohardness values, which, in turn, are dependent on the strength of the matrix alloy, the level to which the matrix alloy has been plastically constrained by the presence of the reinforcement, as well as the toughness of the ceramic particulates themselves. The silicon carbide reinforced Al6061 alloy had the greatest wear resistance of all the composites. This was attributed to the higher hardness and toughness of the silicon carbide particulates compared to that of the alumina reinforcement, as well as the relatively soft alumina abrasive against which it was worn.

The adhesive transfer and wear loss of aluminium to steel surfaces during reciprocating sliding wear conditions were found to be reduced by almost three orders of magnitude by the addition of ceramic reinforcement to each of the aluminium alloys. Alloys containing higher volume fractions of particulates displayed lower wear losses. Steel counterfaces were also found to undergo wear losses due to the abrasive action of the ceramic particles, with the silicon carbide reinforced Al6061 imparting the least counterface damage and removal. The sliding wear behaviour of each composite was characterised by the formation of a transfer layer on the surface in contact with the steel counterface. The stability of this transfer layer was found to be dependent on the strength and ductility of each composite. By plastically constraining the matrix alloy, ceramic reinforcing particulates prevent shear flow of the aluminium matrix at the sliding interface. The particulates also become load bearing while in contact with the steel counterface, thereby protecting the matrix alloy from shear forces created by contact. In this regard, the Al2014 alloy reinforced with 20 vol% alumina displayed wear losses that were significantly greater than any of the other composites. This was attributed to its poor strength and low strain to failure, resulting in an inability to accommodate shear strains developed at the sliding interface, with subsequent rapid deterioration of the transfer layer and increased abrasion of the counterface.

The cavitation erosion resistances of the composites were similar to those of their respective matrix aluminium alloys. The incubation periods showed no deviation with the addition of ceramic reinforcement, indicating that the major material removal mode was through strain accumulation and rupture in the matrix alloy. The alumina reinforced alloys displayed a slight trend in increasing erosion rate with higher volume fractions of particulates and lower work to fracture (E_{fracture}) values. The possibility that the alumina reinforcement may be responsible for higher erosion rates by concentrating deformation at the particulate interfaces and initiating their microfracture is proposed. In contrast, the silicon carbide reinforced alloy had a slightly lower erosion rate compared to its matrix Al6061 alloy, attributable to the greater integrity of the silicon carbide particles, being tougher, less angular in morphology and possibly having a stronger interfacial bond with the matrix compared to the alumina reinforcement.

A reasonable correlation was obtained between the solid particle erosion resistances of each of the composite materials and their respective work to fracture (E_{fracture}) values calculated from tensile test data. The addition of ceramic reinforcement to each alloy resulted in a deterioration in their erosion resistances, indicating that the mass losses were not dependent on the resistance to indentation, but rather on each material's ability to accumulate strain, followed by shear failure of the matrix. The eroded surfaces displayed increasing amounts of plastic constraint and microfracture of the matrix alloy with the addition of higher volume fractions of reinforcement.

Acknowledgements

The authors would like to thank Dr. T. Hurd and Mr. B.I. Dennis of Hulett Aluminium Ltd. at Pietermaritzburg for supply of material, financial support and helpful discussions. The help of Prof. J.F.W. Bell with elastic constant measurements is gratefully acknowledged.

List of symbols

HV_{20kg} Vickers indentation hardness (20 kg load) [kg mm^{-2}]
HV_{15g} Vickers indentation microhardness (15 g load) [kg mm^{-2}]
R_a Average roughness [μm]
E_{fracture} Work to fracture [MJ m^{-3}]

References

[1] A.K. Vasudevan and R.D. Doherty, eds, Aluminium Alloys – Contemporary Research and Applications, Treatise on Material Science and Technology, Vol. 31 (Academic Press, San Diego, CA, 1989).
[2] W.C. Harrigan, in: Metal Matrix Composites, Processing and Interfaces, Treatise on Materials Science and Technology, eds R.K. Everett and R.J. Arsenault (Academic Press, San Diego, CA, 1991) ch. 1, pp. 1–15.
[3] M. Hunt, Mater. Eng. January (1989) 37–40.

[4] T.A. Hahn, in: Metal Matrix Composites, Mechanisms and Properties, Treatise on Materials Science and Technology, eds R.K. Everett and R.J. Arsenault (Academic Press, San Diego, CA, 1991) ch. 11, pp. 329–354.

[5] A. Banerji, S.V. Prasad, M.K. Surappa and P.K. Rohatgi, Wear 82 (1982) 141–151.

[6] M.K. Surappa, S.V. Prasad and P.K. Rohatgi, Wear 77 (1982) 295–302.

[7] A. Wang and I.M. Hutchings, Mater. Sci. & Technol. 5 (1989) 71–76.

[8] S.V. Prasad, P.K. Rohatgi and T.H. Kosel, Mater. Sci. & Eng. 80 (1986) 213–220.

[9] A. Wang and H.G. Rack, Wear 146 (1991) 337–348.

[10] J. Larsen-Basse, Scripta Metall. 24 (1990) 821–826.

[11] K.-H. Zum Gahr, ed., Microstructure and Wear of Materials, Tribology Series, Vol. 10 (Elsevier, Amsterdam, 1987) p. 352.

[12] F.M. Hosking, F.F. Portillo, R. Wunderlin and R. Mehrabian, J. Mater. Sci. 17 (1982) 472–498.

[13] D.H. Buckley and K. Miyoshi, in: Structural Ceramics, Treatise on Materials Science and Technology, Vol. 29, ed. J.B. Wachtman Jr (Academic Press, San Diego, CA, 1989) ch. 7, pp. 293–363.

[14] F. Rana and D.M. Stefanescu, Metall. Trans. A 20 (1989) 1564–1566.

[15] J. Yang and D.D. Chung, Wear 135 (1989) 53–75.

[16] K. Anand and Kishore, Wear 85 (1983) 163–169.

[17] C.P. You, W.T. Donlon and J.M. Boileau, in: Proc. Conf. on Tribology of Composite Materials, Oak Ridge, TN, May 1–3, 1990, eds P.K. Rohatgi, C.S. Yust and P.J Blau (ASM International, Oak Ridge, TN, 1990) pp. 157–167.

[18] S. Wilson and A. Ball, in: Proc. Conf. on Tribology of Composite Materials, Oak Ridge, TN, May 1–3, 1990, eds P.K. Rohatgi, C.S. Yust and P.J Blau (ASM International, Oak Ridge, TN, 1990) pp. 103–112.

[19] M.L. Shaw, P.H.S. Tsang and S.K. Rhee, in: Wear of Materials, eds K.C. Ludema and R.G. Bayer (ASME, Orlando, FL, 1991) p. 167–175.

[20] A.T. Alpas and J.D. Embury, Scr. Metall. Mater. 24 (1990) 931–935.

[21] Y. Pan, M.E. Fine and H.S. Cheng, in: Proc. Conf. on Tribology of Composite Materials, Oak Ridge, TN, May 1–3, 1990, eds P.K. Rohatgi, C.S. Yust and P.J. Blau (ASM International, Oak Ridge, TN, 1990) pp. 93–101.

[22] A.T. Alpas and J.D. Embury, in: Wear of Materials, eds K.C. Ludema and R.G. Bayer (ASME, Orlando, FL, 1991) pp. 159–166.

[23] K.C. Goretta, W. Wu, J.L. Routbort and P.K. Rohatgi, in: Proc. Conf. on Tribology of Composite Materials, Oak Ridge, TN, May 1–3, 1990, eds P.K. Rohatgi, C.S. Yust and P.J. Blau (ASM International, Oak Ridge, TN, 1990) pp. 147–155.

[24] I.M. Hutchings and A. Wang, in: Proc. Conf. on New Materials and their Applications, Warwick, England, April 1990 (Institute of Physics Conf. Series, 1990) pp. 111–120.

[25] C.J. Heathcock, B.E. Protheroe and A. Ball, in: Proc. 5th Int. Conf. on Erosion by Solid and Liquid Impact, Cambridge, England (1979) pp. 219–224.

[26] S. Vaidya and C.M. Preece, Metall. Trans. A 9 (1978) 299–307.

[27] Y. Flom and R.J. Arsenault, Mater. Sci. & Eng. 77 (1986) 191–197.

[28] D.L. Davidson, Metall. Trans. A 18 (1987) 2115–2128.

[29] J.F.W. Bell and J.C.K. Sharp, Rev. Int. Hautes Temp. Refract. 12 (1975) 40–43.

[30] C. Allen, A. Ball and B.E. Protheroe, Wear 74 (1981–82) 287–299.

[31] U.F.B. Kienle, M.Sc. Thesis (University of Cape Town, 1988).

[32] C.J. Heathcock, Ph.D. Thesis (University of Cape Town, 1980).

[33] R.C. Pennefather, M.Sc. Thesis (University of Cape Town, 1986).

[34] A.W. Ruff and L.K. Ives, Wear 35 (1975) 195.

[35] D.L. Davidson, Metall. Trans. A 22 (1991) 113–123.

[36] P.F. Thomason, Ductile Fracture of Metals (Pergamon Press, Oxford, 1990) p. 19.

[37] T. Christman, A. Needleman and S. Suresh, Acta Metall. Mater. 37 (1989) 3029–3050.

[38] D.J. Lloyd, Acta Metall. Mater. 39 (1991) 59–71.

[39] J. Llorca, A. Needleman and S. Suresh, Acta Metall. Mater. 39 (1991) 2317–2335.

[40] Z. Wang and R.J. Zhang, Metall. Trans. A 22 (1991) 1585–1593.

[41] V. Tvergaard, Acta Metall. Mater. 39 (1991) 419–426.

[42] C.P. You, A.W. Thompson and I.M. Bernstein, Scripta Metall. 21 (1987) 181–185.

[43] C.P. You, M. Dollar, A.W. Thompson and I.M. Bernstein, Metall. Trans. A 22 (1991) 2445–2450.

[44] T.P. Johnson, J.W. Brooks and M.H. Loretto, Scripta Metall. Mater. 25 (1991) 785–789.

[45] P. Mummery and B. Derby, Mater. Sci. & Eng. A 135 (1991) 221–224.

[46] M. Srinivisan, in: Structural Ceramics, Treatise on Materials Science and Technology, Vol. 29, ed. J.B. Wachtman Jr (Academic Press, San Diego, CA, 1989) ch. 3, pp. 99–148.

[47] R.A. Vaughan and A. Ball, in: Wear of Materials, 1991, eds K.C. Ludema and R.G. Bayer (ASME, Orlando, FL, 1991) pp. 71–75.

[48] D.H. Buckley, ed., Surface Effects in Adhesion, Friction, Wear and Lubrication, Tribology Series, Vol. 5 (Elsevier, Amsterdam, 1981) p. 267.

[49] H. Liu, M.E. Fine and H.S. Cheng, in: Proc. Conf. on Tribology of Composite Materials, Oak Ridge, TN, May 1–3, 1990, eds P.K. Rohatgi, C.S. Yust and P.J. Blau (ASM International, Oak Ridge, TN, 1990) pp. 329–336.

[50] K. Fikuda, Y. Sato, T. Sato and M. Veki, in: Proc. Conf. on Tribology of Composite Materials, Oak Ridge, TN, May 1–3, 1990, eds P.K. Rohatgi, C.S. Yust and P.J. Blau (ASM International, Oak Ridge, TN, 1990) pp. 323–328.

[51] S. Srinivasan, R.O. Scattergood and R. Warren, Metall. Trans. A 19 (1988) 1785–1793.

Advances in Composite Tribology
edited by K. Friedrich

Chapter 10

Tribological Properties of Unidirectionally Oriented Carbon Fibre Reinforced Glass Matrix Composites

ZAIPING LU *

Department of Material Science and Engineering, Beijing University, of Aeronautics and Astronautics, Beijing 100083, China

Contents

Abstract

A fundamental study on the dry sliding and abrasive wear behaviour of a unidirectional carbon fibre reinforced glass matrix composite was made at ambient temperature. The wear rate and friction coefficients against different metals and SiC papers were experimentally determined; in addition, scratch tests by the use of a diamond indentor were performed. The resulting wear mechanisms could be studied by scanning electron microscopy.

* Present address: Institute for Composite Materials, University of Kaiserslautern, W-6750 Kaiserslautern, Germany.

Three principle sliding directions relative to the dominant fibre orientation in the composite were selected. When sliding took place against smooth, hard metals, the highest wear resistance and the lowest friction coefficient were observed in the antiparallel direction. For both the composite and the counterparts, the wear rates decreased as the hardness of counterpart material increased. Two types of coloured wear debris were observed; they played an important role in the wear mechanisms during sliding. The type of contact configuration with smooth metals affected the wear rates especially for the P-direction.

The effects of the sliding direction and the grain size of the abrasive paper on the wear characteristics of the composite was also studied. In contrast to sliding against smooth metals, the wear rate in the AP-direction of the composite was higher than in the other two directions. The dominating wear mechanisms were in-plane bending of the fibres, which caused fibre fracture, and matrix ploughing and cutting. In addition, it was found that the wear rate increased with increasing abrasive grain size. For most cases, the wear rates of the composite were higher than those of the unfilled glass. By the use of single scratch tests, the abrasive mechanisms could be better explained.

1. Introduction

The use of glass as a structural material has always been limited by its susceptibility to brittle failure due to small cracks or flaws acting as stress concentrators. One approach to reduce the detrimental effects of these flaws is to reinforce the glass with fibres so as to produce a glass matrix composite. Glass and glass–ceramic matrix composites reinforced with fibres have several potentially attractive properties [1–3], including high specific stiffness and strength, good corrosion resistance, and retention of these properties to high temperature. These make them suitable for applications requiring chemical inertness, toughness and stability at high temperature. By reinforcing glass with carbon fibres it has been possible to develop composites whose friction coefficient and mechanical properties compare favourably with those of resin matrix composites, and whose environmental stability is comparable to those of glasses and ceramics.

Much research and development has been carried out on these materials especially with respect to their mechanical behaviour [4–7]. Prewo and co-workers have done extensive research [3–5,8,9] on carbon and silicon carbide fibre reinforced glass and glass–ceramics composites with regard to their manufacturing, mechanical and chemical behaviour. The work of Chou and his colleagues [10] on borosilicate glass composites with different fibres was focussed on analysis and modelling of the thermomechanical properties and the fatigue characteristics. Hegeler and Brückner [1,2,6,11] investigated a new technique for manufacturing and the reinforcement mechanisms of fibre reinforced glass composites.

However, very little attention has been given to studies on sliding and abrasive wear of glass matrix composites reinforced with carbon fibres. Minford and Prewo [12] illustrated the friction and wear properties of graphite fibre reinforced glass matrix composites with a symmetric 0°/90° cross-ply configuration when sliding against metal. In this case, the wear rate is dependent on the fibre orientation. They observed that the lowest values for both the coefficient of friction and the wear rate were obtained when the fibres in the centre layer were oriented parallel to the direction of sliding.

Several authors [13–15] have examined the friction and wear of structural ceramics, when rubbed against flat counterparts, using the pin-on-disc technique or a roller-on-beam tribometer; in these cases, different wear mechanisms have been identified (adhesion, delamination, oxidation, abrasion, melting, etc.). Some authors (for example Torrance [16]) used a scratch technique to model the abrasion and the friction generated by the interaction of rough surfaces in sliding contact. Suh and Sin [17] considered the contribution of the wear particles as a component of the frictional force. The third-body approach [18,19] tried to rationalize the many aspects which are common to different types of material in different types of rubbing contact. Extensive research has been carried out on the consequences and influencing parameters in the abrasive wear of metals [20,21]. In this case, wear debris formation occurs by ploughing, cutting and cracking mechanisms induced by the hard asperities of the counterface. Various models have been used to describe abrasive wear of these isotropic materials. Friedrich [22] has classified the dominant micro-wear mechanisms which occur during abrasive wear of different polymer composites.

This chapter describes a fundamental approach to the investigation of the dry sliding, the abrasive and the scratch wear behaviour of a unidirectional carbon fibre reinforced glass matrix composite. The wear rates and friction coefficients of the composite against different metals and SiC papers are experimentally determined. In addition, a microscopic study of the resulting wear mechanisms helps to find out under which external conditions carbon fibre/glass matrix composites are useful candidates for wear resistant components needed in special technical applications.

2. Glass matrix composites

The concept of glass matrix composites has been considered since the 1960s and work on them started in the early 1970s. Several glass matrices reinforced with carbon fibres [23,24] showed significant improvements in mechanical properties compared with those of the unreinforced matrix. The object of incorporating fibres into glass and glass–ceramics is, essentially, to change the failure mode from the catastrophic brittle characteristic of the unreinforced matrix to a controlled failure mechanism associated with energy absorbing processes such as interface debonding, fibre breakage, fibre peeling and fibre pull-out in the composite.

Interlaminar fracture toughness measurements indicated that the critical strain energy release rate (G_{IC}) of the glass increased from 0.02 to 0.1 kJ/m^2 after reinforcement with carbon fibre (T300). Prewo et al. [3] compared the retention of the flexural strength in argon of both borosilicate glass matrix and two polymer matrix (polyether sulfone and polysulfone) composites, all reinforced with the same amount of HM carbon fibres (shown in fig. 1). While in this case the polymer

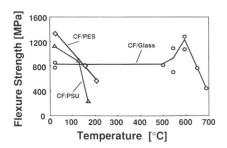

Fig. 1. Flexural strength comparison for composites reinforced with HM unidirectional carbon fibres and tested at room temperature in inert argon [3].

matrix systems were stronger at room temperature, at elevated temperatures the borosilicate system was far superior, with eventually strength loss occurring above 600°C due to softening of the glass matrix.

Because of the fact that the glass matrix can be readily deformed and can even flow in its low-viscosity state at elevated temperatures, the processes for making glass matrix composites in current use are: (1) hot pressing of infiltrated uni-tape and fabric lay-ups, (2) hot matrix transfer into woven preforms and (3) hot injection molding of chopped fibre compounds or preforms.

The composite samples used here for tribological characterization consisted, of unidirectional carbon fibres (T800 or M55) and a borosilicate glass matrix (Duran®), which consists approximately of 80.7 wt% SiO_2, 12.8 wt% B_2O_3, 3.5 wt% Na_2O and 2.4 wt% Al_2O_3, with traces of oxides such as K_2O. The unidirectional composite was produced as a 3 mm thick plate by hot press densification at Schott Glaswerke in Germany. Details about the fabrication procedure can be found elsewhere [1,2]. The process for making a CF/glass composite was started by winding slurry-impregnated yarn onto a mandrel to form monolayer tapes, as illustrated in fig. 2a. These tapes were cut up to make plies, which were stacked and densified to form the final composite in a hot-pressing operation. A typical glass matrix composite micrograph is shown in fig. 2b. The properties of the composite (with carbon fibre T800) and its constituent materials are shown in table 1. It can be seen, that an excellent improvement in Young's modulus and the tensile strength of the glass is achieved with carbon fibre reinforcement. The composite plates were carefully cut with a diamond saw. This was followed by a polishing procedure using fine grinding paper (grain 800); no cracks were visible on the specimens, having a 3×4 mm^2 cross section, which were used as the sliding surface in the tribological experiments.

3. Sliding wear with metal counterparts

3.1. Experimental procedure

Three principal sliding directions with respect to the fibre orientation could be distinguished. Experiments were performed either with the fibre orientation paral-

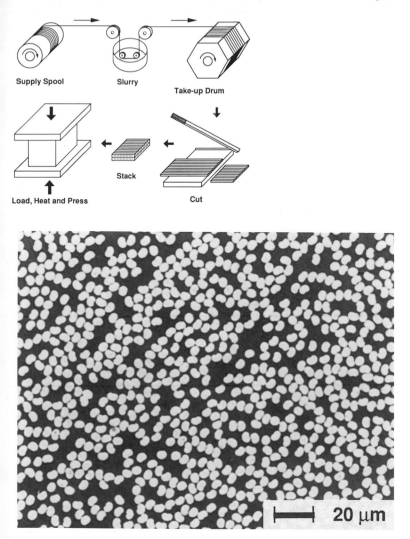

Fig. 2. (a) Schematic processing of the carbon fibre reinforced glass matrix composites. (b) Micrograph of a typical carbon fibre–glass matrix composite.

lel (P), antiparallel (AP) or normal (N) to the counterface (fig. 3). Testing of dry sliding wear took place under ambient temperature using devices with two differ-ent configurations: (a) a block-on-ring and (b) a pin-on-disc apparatus. Their experimental layout is depicted schematically in figs. 4a and b. The reason for this choice was to be able to find out how much the wear of a particular type of composite is affected by the test configuration. In addition, the block-on-ring facility available had the advantage that more than one sample could be worn at a time. The disadvantage that no friction coefficients could be measured with this device, was compensated by the pin-on-disc machine, which allowed to test single

TABLE 1
Properties of unidirectional continuous carbon fibre reinforced glass matrix composite and constituent materials tested.

Materials	Density (g cm^{-3})	Elastic modulus (GPa)	Tensile strength (MPa)	Tensile elongation (%)	Fibre content (vol%)
Glass matrix (DURAN)	2.22	60	10–100[a]	0.1–0.2	–
Carbon fibre (CF; Toray T-800)	1.81	300	5600	1.9	–
CF-Duran	2.05	160	1200	0.9	40

[a] Dependent on surface quality.

specimens under continuous monitoring of the friction coefficient and the wear rate.

As the counterparts in the block-on-ring device different metallic rings of 60 mm diameter were used; their properties are listed in table 2. The profile of the roughness grooves running parallel to the sliding direction was produced by pressing grinding paper with a grain size of 7 μm (grain 800) on the rotating rings prior to the actual wear measurements. The roughness was measured according to the DIN 4768 standard with a surface measuring instrument, Perthometer S5P. Further details about the typical preparation and testing procedures with this device are comprehensively described in previous papers [25,26]. The apparatus was adjusted to run at different sliding speeds (v). The variation of the normal contact pressure (p) was chosen between 0.3 and 3.0 MPa in such a way that different pressure–velocity (pv) levels from 0.1 to 4 MPa m s^{-1} were maintained. After the running-in period, when the ring had rubbed its circular profile into the specimen, the mass loss Δm was determined for each p and v combination by the use of more than two specimens, each tested at intervals of 4 h and over at least three mass loss measurements. For the actual experiments, i.e. after the running-in period, the specimens were slid against unused "tracks" on the metallic rings, so

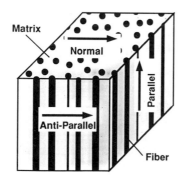

Fig. 3. Relative sliding direction with respect to the fibre orientation of a unidirectional carbon fibre–glass matrix composite.

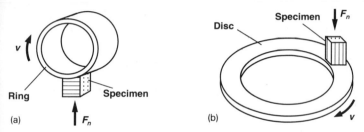

Fig. 4. Schematic sketches of the sliding-experiment arrangements: (a) block-on-ring, (b) pin-on-disc.

that the initial roughness was the same for each test. The different measurements allowed to distinguish the period in which steady-state conditions predominated the wear of the tribo-system.

On the pin-on-disc testing machine, glass-composite pins were run against stationary metallic discs (table 2). Continuous monitoring of the torque allowed to determine the coefficient of friction (μ), and on-line measurements of the reduction of the length of the glass composite pin allowed to obtain the specific wear rate. In this way it was possible to distinguish the changes in the wear rate of the system between the initiation period and the steady-state, multipass sliding condition, in which the wear rate of the system approached an equilibrium level. A comparison of the different sliding direction and testing conditions and their influence on the wear of the different partners in the tribo-system (composite pins and steel counterparts) was made only for the steady-state condition, which in terms of the wear life of the system has to be considered as the dominant period here. Therefore, for calculations of the specific wear rates, only the testing time and the weight losses during the steady-state period were used. Prior to the actual wear testing, a preparation phase was performed, in which the pins were "pre-worn" in the pin-on-disc device using grinding paper (grain 800). With this method the roughness of the specimens before testing was adjusted to be always the same, and so was the apparent area of contact with the disc because of better parallelism between the two contacting surfaces. For each sliding direction three specimens were used. Each specimen was run on a fresh disc. Further design details of this apparatus were published in refs. [27,28]. Before the wear tests, the surfaces of both the specimens and the counterparts were cleaned carefully with acetone, in order to eliminate impurities and get reproducible results.

TABLE 2

Properties of counterpart materials as ring and disc.

Conterpart material	Density (g cm^{-3})	Tensile strength (MPa)	Hardness (HB)	Roughness (R_a) (μm)
Steel (100Cr6)	7.8	2000	620 ± 10	0.06–0.2
Cast iron	7.1	250	210 ± 20	0.1 –0.2
Al Alloy	2.7	195–245	$90– 125$	0.15–0.2

Before and after each run, the mass losses of the specimens and the counter-parts were measured on a Mettler analytical balance (sensitivity: 0.01 mg) under the same environmental conditions as the test was conducted.

In the present paper the wear results are given in terms of the specific wear rate $\dot{w}_s$, calculated according to:

$$\dot{w}_s = \frac{\Delta m}{\Delta t} \frac{1}{v \rho F_n},$$

where Δm is the mass loss during the test time Δt (as taken from quasi-steady state wear conditions), v the sliding velocity, ρ the density of the specimen and F_n the normal load. This results in dimensions of $mm^3/(N\ m)$, which describes the physical nature of the specific wear rate as a volume loss of material at a certain input of energy. The inverse of the specific wear rate, $\dot{w}_s^{-1}$ is referred to as the wear resistance.

In the pin-on-disc device Δm was determined as:

$$\Delta m = \Delta h A \rho,$$

where Δh is the decrease in length of specimen during the test time Δt and A is the apparent contact area of the specimen. A torque load cell continuously monitored the frictional force between the specimen contact surface and the counterface in the pin-on-disc device. Applying the Amonton–Coulomb law of dry friction, the coefficient of friction (μ) was calculated as:

$$\mu = F_f/F_n,$$

where F_f is the measured frictional force.

In order to investigate the wear mechanisms, most of the worn surfaces were examined using a scanning electron microscope (SEM). For some samples energy dispersion X-ray analysis (EDS) was also carried out. The wear samples were coated with gold prior to inspection. The debris produced during the sliding test was collected, and characterized with an analytical chemistry method. Profilograms across the wear track on the ring and disc were taken using a surface measuring instrument.

3.2. Results of sliding wear tests with the block-on-ring device

The wear rates of the glass composite with carbon fibre reinforcement as a function of the sliding direction are given in fig. 5. These results stemmed from tests performed at a constant pv-level of 1 MPa m s^{-1}, but obtained with different sliding speeds and apparent contact pressures. Under the external conditions chosen here, it can be seen from fig. 3 that there was no remarkable dependency of the specific wear rates on the sliding speed or apparent contact pressure, at a given sliding direction. Only a clear anisotropy in the wear behaviour due to the different fibre orientations could be detected. The AP-direction generated the lowest, whereas the P-direction resulted in the highest wear rates. This was also true for the counterpart material (steel in this case).

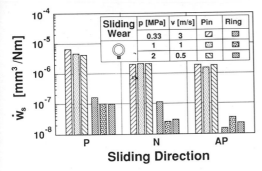

Fig. 5. Average specific wear rates of both the glass composite and the steel ring as a function of the sliding direction under $pv = 1$ MPa m s^{-1}.

The influences of the testing conditions were further investigated. The wear rates are plotted in fig. 6 as a function of the pv-factors applied. In this case, a clear response of the specific wear rates to the pv-factors seemed to occur only above about 2 MPa m s^{-1} and in particular for the composite in the parallel direction. Therefore, below this level, the specific wear rate can be nearly considered as a material related property, which mainly depends on the direction of the fibre orientation, but being almost unaffected by the mechanical testing conditions. Within this range, the specific wear rate ($\dot{w}_s$) can therefore be used to calculate (for design purposes) the length wear rate of the material ($\dot{w}_t$ in e.g. μm/h) according to the relationship [25,27,29]:

$$\dot{w}_t = \dot{w}_s\, pv.$$

From fig. 7, a three-dimensional diagram, it can further be seen, that the influences of the sliding speeds and apparent contact pressures on the specific wear rates of CF/glass composites were different. A clear response of the specific wear rate to the sliding speed was found in the parallel direction. On the other hand, there were no pronounced differences in the specific wear rate for the

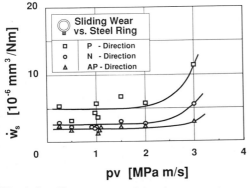

Fig. 6. Specific wear rates of the glass composite as a function of the pv-factors applied.

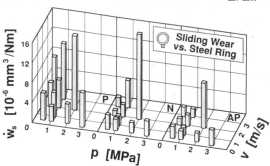

Fig. 7. Interrelationship between the specific wear rate, the apparent contact pressure and the sliding velocity for the glass composite.

AP-direction, at least in the range of sliding speeds employed in these experiments with a p of 2 MPa. An increase of the apparent contact pressure was effective in increasing the wear rate of the composite in the P-direction, i.e. the same trend as observed for changes in sliding speed existed, but this effect seemed not to be obvious in the AP- and N-direction. Further, it was noticed that the wear rates showed a more or less significant scatter ($\pm 30\%$), although they were obtained on identical specimens. This may be explained by different wear mechanisms, which can be investigated by using a scanning electron microscope (SEM) attached to an energy-dispersive X-ray spectrometer [30].

A typically worn surface of the specimen tested in the normal (N) direction is shown in fig. 8. The characteristic feature is an accumulation of brown wear debris in the form of a patchwork layer in some regions of the worn surface. This wear debris layer could not be removed by blowing with a compressed air stream, i.e. it stuck on the surface very strongly. Energy-dispersive X-ray spectrometry of the worn surface gave evidence for compositional differences between region (A) with the brown debris layer and region (B) without such a layer (fig. 9). Region A showed a great amount of iron, while silicon and oxygen were mainly detected in region B. The two different regions resulted also in the fact that two kinds of debris were produced during sliding wear. One of them was of a brown colour and the other one was black. The debris was characterized with an analytical chemistry method. Evidence of oxidized iron was found in both cases. It was proved that the brown debris contained a lot of iron (III)-oxide (Fe_2O_3), while the black debris consisted only partially out of iron (II)-oxide (FeO). The compacted layer of this wear debris played an important part as a third body [18] in the wear process. If a large debris layer is deposited on the surface of the specimen, this can result in a decrease in the removal of the bulk material; on the other hand, the wear of the counterpart is increased in this case, and a lot of brown debris is produced during the sliding motion. The change of the thickness and size of the debris layer can cause large variations in wear rate of both the composite specimen and the steel counterpart. This may explain the large scatter of the wear rates, especially at lower pressure levels (about $\pm 30\%$, to be added to the average values given in figs. 6 and 7).

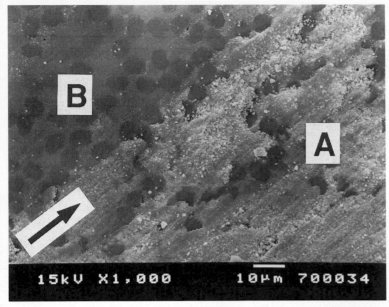

Fig. 8. SEM micrograph of the worn surface of the glass composite specimen tested at $p = 1$ MPa and $v = 1$ m s^{-1} in the normal direction (arrow indicates sliding direction).

A similar formation of a transfer film was also observed when sliding took place in the P-direction, as illustrated in fig. 10. Under this particular condition, a great deal of rupturing and debonding of carbon fibres occurred during sliding against steel. This resulted in the larger wear rates measured in the parallel direction. The worn surfaces of the composite in the antiparallel (AP) direction have essentially similar features at different pv-factors. It is, however, rather difficult to find large-scale rupturing and debonding of carbon fibres, and in addition, only a thin debris layer could be seen (fig. 11). This led to a smaller wear rate of the AP-specimens. Regarding the wear of the steel counterpart, this appeared at lower rates, and smaller scatter in wear data was found.

The effects on the wear performance of the glass matrix composite of varying the counterpart material were investigated under the condition $p = 1$ MPa and

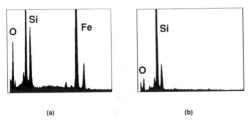

(a) (b)

Fig. 9. Energy-dispersive X-ray spectrum of the worn surface shown in fig. 8: (a) region A with debris layer; (b) region B without debris layer.

378 Z. Lu

Fig. 10. SEM micrograph of the worn surface of the glass composite specimen tested at $p = 1$ MPa and $v = 1$ m s^{-1} in the parallel direction (arrow indicates sliding direction).

Fig. 11. SEM micrograph of the worn surface of the glass composite specimen tested at $p = 1$ MPa and $v = 1$ m s^{-1} in the antiparallel direction (arrow indicates sliding direction).

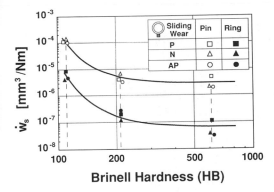

Fig. 12. Specific wear rates of both the glass composite specimens and the counterparts as a function of the hardness of the counterpart materials, tested at $p = 1$ MPa and $v = 1$ m s^{-1}.

$v = 1$ m s^{-1}. The average specific wear rates of both the composite specimens and the counterparts as a function of hardness of the counterpart materials are shown for a pv-factor of 1 MPa m s^{-1} in fig. 12. In this case, for both the composite and the counterparts, the wear rates decreased as the hardness of the counterpart material increased. However, it must be noted at this point that this general statement is somewhat misleading because the counterparts were not only different in hardness but also in the material itself. Therefore, the increased wear in the case of the Al alloy was probably not only because of the lower bulk hardness of the aluminium, but also due to an existing Al$_2$O$_3$-layer on the surface. The latter has a higher hardness, but when being removed it can act, even under steady-state conditions, as a third-body abrasive, thus causing a considerably higher amount of wear than expected from just the lower hardness of the bulk Al alloy. The wear rates of the composite for all sliding directions were between one or two orders of magnitude higher than those of the corresponding counterparts. Similar as in the case of wear against steel rings, the wear rates of the composite in the AP-direction against these three types of counterparts were lower than those in the two other directions. In addition, it became obvious that the wear rates against aluminium rings were by more than one order of magnitude higher than those against the other metals. The surface profilometer tracks of aluminium rings after wear testing gave the impression as if these surfaces had been machined with a hard tool.

   In order to investigate the effect of an oxide film on the wear behaviour of the composite, wear tests were performed against (a) an as-received original surface of cast iron (with a roughness of $R_a = 0.9$ μm) and (b) a freshly polished one. In the first case, the composition of the iron surface was mainly oxide. Figure 13 shows the results when compared to the wear data obtained with the smooth surface of the freshly prepared rings, i.e. those polished with grinding paper (grain 800) down to a roughness of $R_a = 0.2$ μm. It is clearly seen that in all three sliding directions the wear rates of both specimens and rings were lower in the case of the original, oxidized surface than when measured with the freshly produced smooth surface. It

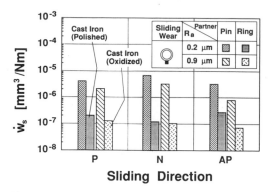

Fig. 13. Specific wear rates of both the glass composite and the counterpart rings made of cast iron with different roughness, as a function of the sliding direction, tested at $p = 1$ MPa and $v = 1$ m s^{-1}.

is therefore assumed that the wear mechanisms are influenced by the oxide on the surface.

3.3. Results of sliding wear tests on the pin-on-disc facility

The friction and wear performance of the glass composite was also investigated on a pin-on-disc apparatus with the different metal counterparts under a pv-condition of 1 MPa m s^{-1}. The wear rates of the CF/glass composite in the three directions are shown in fig. 14. Similar as in the case of wear tests on rings, the wear rate against aluminium discs was by more than one order of magnitude higher than against the other metals, and the lower wear rate was found for the AP-orientation against both cast iron and steel discs.

Figure 15 clearly illustrates how the fibre orientation in the glass matrix composite can reduce the friction coefficient of the material. Also here, similar to findings by Shim et al. [31], the AP-orientation leads to lower values of μ than the

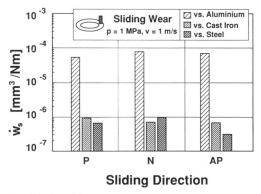

Fig. 14. Specific wear rates of the glass composite as a function of the sliding direction and the counterpart material, tested on the pin-on-disc device at $p = 1$ MPa and $v = 1$ m s^{-1}.

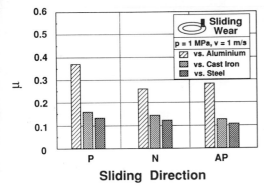

Fig. 15. Friction coefficient of the glass composite as a function of the sliding direction and the counterpart material, tested on the pin-on-disc device at $p = 1$ MPa and $v = 1$ m s^{-1}.

P- or N-orientation (in case of steel and cast iron counterparts). With regard to the aluminium counterpart, the coefficients of friction were clearly higher than for the other two metallic partners. Typical wear surfaces of specimens tested in the N-orientation on the pin-on-disc apparatus show a variety of different wear mechanisms (fig. 16). The worn surfaces of the composite against discs of steel or cast iron appeared at some sites as being smoothly polished, with some ploughing features in the matrix between the fibres (fig. 16a). Other spots showed entrapped

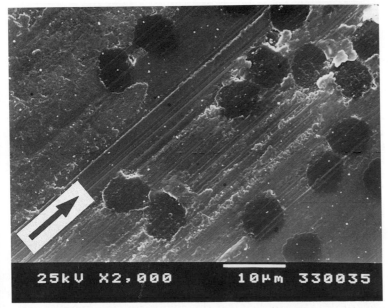

Fig. 16. SEM micrograph of the worn surface of the glass composite tested on the pin-on-disc device at $p = 1$ MPa and $v = 1$ m s^{-1} in the normal direction (arrow indicates sliding direction); (a) tested on steel, (b) tested on cast iron and (c) tested on aluminium.

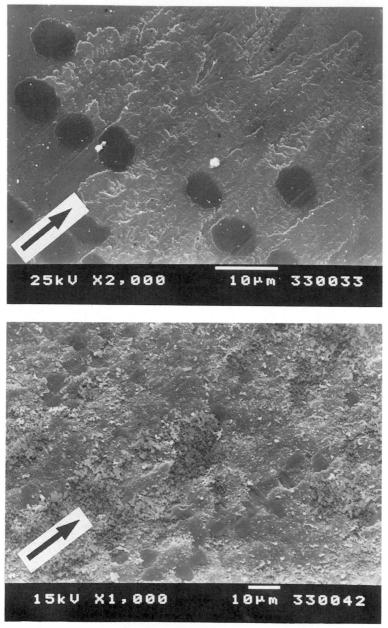

Fig. 16 (continued).

wear debris in the form of densely packed patches attached to the smoothly polished surface underneath (fig. 16b). When the composite was, however, worn against the aluminium counterpart (fig. 16c), the surfaces showed clear evidence of a fine, powder-like structure as a result of micro-abrasive mechanisms (cutting,

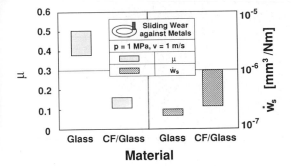

Fig. 17. Comparison of the tribological properties of unfilled glass and its composite with carbon fibre, tested on the pin-on-disc device against cast iron and steel counterparts at $p = 1$ MPa and $v = 1$ m s^{-1}.

cracking and ploughing) of both the aluminium and the components of the composite (glass and carbon fibre). In contrast with polymer or ductile metal matrix composites [32,33], there was no fibre bending and fracturing in the subsurface of the CF/glass composite due to the rigid nature of the glass matrix, for all directions at a moderate pv-value.

3.4. Discussion of other material effects

Carbon fibres can act effectively as an internal lubricant, due to the fact that loose graphite is produced under wear loading [34,35]. The wear rates and friction coefficients of unfilled glass and the CF/glass composite in all directions under the condition $p = 1$ MPa and $v = 1$ m/s, against cast iron and hard steel counterparts are shown in fig. 17. It can be seen that carbon fibres in the glass matrix improve obviously the friction coefficient of the material. For the specimen on a disc of cast iron or steel, the friction coefficient (μ) of glass decreased drastically from about 0.4 to 0.1 when carbon fibres with $V_f = 40\%$ are incorporated. However, an improvement of the resistance to sliding wear of the glass with carbon fibre reinforcement could not be detected. The opposite, an increase in wear against all metals applied, was detected. It should be noted though, that the wear test of unfilled glass against aluminium could not be carried out due to brittle fracture of the glass specimens under the given conditions. The worn surface of the unfilled glass against hard steel or cast iron appeared at some sites in the form of entrapped wear debris, building up densely packed patches on the bulk glass surface (fig. 18). This protects the further wearing of the glass, but increases the friction resistance, thus leading to more material removal from the counterparts. In addition, it became obvious that for the unfilled glass, the value of μ against cast iron was by more than 20% lower than that against steel. This can be partly attributed to the presence of the oxide during the wear test [19] and graphite in the cast iron, which is supposed to have some lubricating effect on the system as well [36].

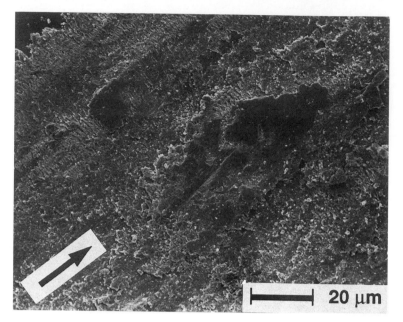

Fig. 18. SEM micrograph of the worn surface of the glass tested on the pin-on-disc device at $p = 1$ MPa and $v = 1$ m s^{-1} against a steel counterpart (arrow indicates sliding direction).

Regarding a comparison with the block-on-ring data, the cylindrical surface of the ring results in locally different fibre orientations, for instance, when sliding in the parallel direction. This may lead to different points of stress concentration in the contact area, which in turn can lead to a slightly higher amount of wear, under the same external conditions, than measured in the case of the pin-on-disc configuration. This is in fact seen very clearly for the P-orientation, but also evidenced for the two other directions, for which the block-on-ring data were by factors of two to five higher than those achieved on the pin-on-disc machine (fig. 19).

Different types of fibres result in different composite properties, including their tribological behaviour. To demonstrate this effect, the wear data of the composite with T800 carbon fibres (cf. table 1) were compared with the wear behaviour of a glass composite containing M55 carbon fibres (as a non-optimized, preliminary laboratory product). The M55 carbon fibre has a higher E-modulus of 540 GPa but lower fracture strain of 0.7. The specific wear rates of the glass composites with carbon fibre of T800 and M55 under the pv conditions 0.1 and 0.25 MPa m s^{-1} are shown in fig. 20. Under the external conditions chosen here, there was a remarkable difference in the specific wear rates of the two composites. The composite with T800 generated the lower, whereas the composite with M55 resulted in the higher wear rates. This was also true for the steel counterpart (fig. 21). It can be further seen from fig. 20 that a much greater response of the specific wear rates to pv-factors seemed to occur for the composite with fibre M55.

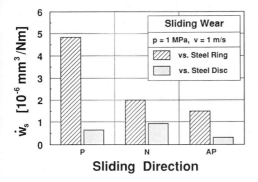

Fig. 19. Specific wear rates of the glass composite as a function of the sliding direction, tested on a steel ring and disc under $pv = 1$ MPa m s^{-1}.

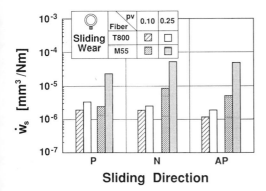

Fig. 20. Specific wear rates of the glass composites with carbon fibre of T800 and M55 under the pv conditions 0.1 and 0.25 MPa m s^{-1}.

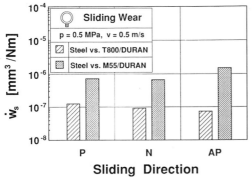

Fig. 21. Comparison of the specific wear rates of the steel counterpart against glass composites reinforced with different carbon fibres.

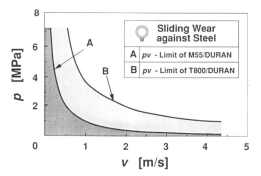

Fig. 22. pv-range for wear tests of the glass composites reinforced with different carbon fibres against steel using the block-on-ring apparatus.

In a block-on-ring configuration, macroscopic failure of the CF/glass composite specimens could occur if the pv-value applied was higher than a critical value. Below the critical value, specimens could bear the wear load without macro-fracture. In fact, the situation of the specimens in this case can be compared with a complex loaded cantilever beam. If the normal load or sliding speed is very high, then the higher frictional force can cause the specimen to break at the bottom constrained by the holding device of the block-on-ring configuration. For the glass composite with T800 fibres the critical pv-value was about 4 MPa m s^{-1}, but only 1 MPa m s^{-1} for the glass composite with M55 fibres. Figure 22 shows schematically the pv-range of wear tests of different carbon fibre reinforced glass composites against hard steel using the block-on-ring apparatus.

It is remarkable that the pv-range of the T800/glass composite was significantly larger than that of the M55/glass composite. One reason was that in this non-optimized laboratory product (M55/glass) there were more pores (fig. 23); this caused a lower flexural strength, in contrast to the T800/glass composite, in which pores or unfilled space between fibres and matrix could not be observed (fig. 2b).

Pores and a poor fibre–matrix interfacial bond quality result in poor resistance against shear-splitting parallel to the fibres. That is why the failure of the glass composites during wear testing above the critical pv-level took place always in the form of interfacial shear, even for tests in the N-direction, which also showed the fracture mode interfacial debonding and fragmentation of the matrix around fibres. These schematic failure features of the glass composite after wear by use of a block-on-ring apparatus for all three fibre-orientations are shown in fig. 24. It should be noted that the critical pv-value was not identical for different wear directions of the same CF/glass composite. From our experiments, the load capacity in the P-direction appeared to be higher than in other directions, although the wear rate in this case was greater than in the N- or AP-direction.

It is worthwhile to compare the results observed here with those for a similar material tested by Minford and Prewo [12]. They investigated the friction and wear properties of graphite fibre reinforced/glass matrix composites with a symmetric

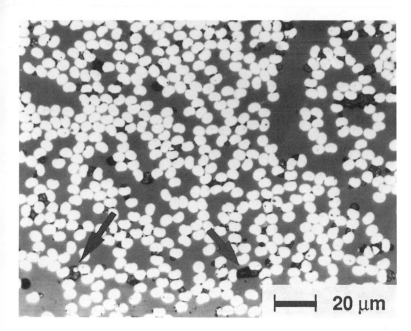

Fig. 23. Micrograph of the glass matrix composite with reinforcement of M55 carbon fibre (arrow indicates the pores).

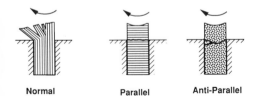

Normal   Parallel   Anti-Parallel

Fig. 24. Schematic of fracture features in the glass composite after wear test on the block-on-ring apparatus.

TABLE 3

Comparison of our results with the data of Minford and Prewo [12]. The μ- and $\dot{w}_s$-data of our CF/glass results are illustrated in more detail in figs. 14 and 15, respectively.

Composite pin against steel disc		From literature [12]		From our test	
Sliding direction		P	AP	P	AP
$\dot{w}_s$	CF/glass	$2\text{--}3\times10^{-6}$	$1\text{--}1.5\times10^{-5}$	6.6×10^{-7}	3.2×10^{-7}
(mm^3/Nm)	Steel	7.5×10^{-7}	$1.5\text{--}8\times10^{-6}$	$5\text{--}9\times10^{-8}$	$2\text{--}4\times10^{-8}$
μ		$0.2\text{--}0.3$	0.6	0.13	0.10
Condition		$pv_{max} = 0.385$ MPa m s^{-1}		$pv = 1$ MPa m s^{-1}	

0°/90° cross-ply configuration, fabricated at United Technologies Research Centre, with regard to sliding against a steel disc (with 60 HBC). Although a lower pv-factor was applied by them, both the wear rate and the friction coefficient were higher than those of the composite used here, regardless of the sliding direction (table 3); and also the wear rates of the steel counterpart were obviously higher than those observed here. This proved that the material under consideration here is at least of a similar quality with respect to its wear performance, as was known from the material published in the literature [12].

4. Abrasion by hard abrasive particles

4.1. Testing procedure

A pin-on-flat type apparatus was used for abrasive wear tests, as schematically shown in fig. 25. A more detailed description of the apparatus is given in ref. [37]. Silicon carbide (SiC) abrasive papers, waterproofed and of three different grain sizes, were used and adhered to the mill table. According to the manufacturer, the average particle diameters (D) were 18, 52 and 75 μm, respectively. The specimens were slid against the SiC paper at a constant velocity (v) of 0.005 m s^{-1} and at an apparent normal pressure (p) of 2.2 MPa over a sliding distance (L) of 0.5 m. In each of the three principal directions relative to the fibre orientation, four runs were made for each specimen. Each run was on a fresh track of SiC paper (single-pass conditions).

4.2. Results and discussion

Figure 26 shows the effect of the sliding direction on the specific wear rate of the composite slid on two kinds of abrasive paper (grain size (D) of 18 and 75 μm respectively). Opposite to the results obtained under sliding wear against smooth metallic counterparts, the greatest wear rates were measured here for the AP-direction. The orientation effect was, however, not very pronounced when the grain size of the SiC particles was small. In order to demonstrate how different types of wear mechanisms (e.g. adhesion relative to abrasion) affect the wear rate of the composite at different fibre orientations, several results of sliding and abrasive wear were plotted in the form of one diagram (fig. 27). It becomes evident that the

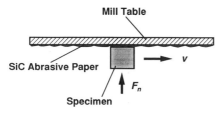

Fig. 25. Schematic sketches of the abrasive-experiment arrangement.

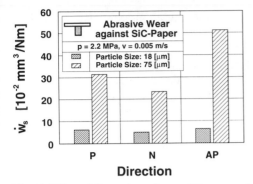

Fig. 26. Effect of fibre orientation at different particle sizes at $p = 2.2$ MPa and $v = 0.005$ m s^{-1} on the abrasive wear performance of the glass composite.

abrasive wear conditions lead to 10^5 higher wear rates, although the pv-factor was much smaller than that at sliding wear. In addition, the wear rate in the AP-direction was more sensitive to the change in wear mechanism from adhesive to abrasive. This means that the different dominating wear mechanisms could result in different tendencies in the wear anisotropy of the glass composite. In fact, when comparing the wear-surface photographs obtained for the composite under sliding wear conditions (figs. 8, 10, 11 and 16) with those of the samples worn against abrasive papers (figs. 28 to 30), one finds enormous contributions of micro-cracking and micro-cutting effects in the second group. In particular, it was found for the N-orientation case (figs. 28a,b), that the fibre ends were cut or fractured by the abrasive particles in a slicing manner. In the case against SiC paper with $D = 75$ μm (fig. 28a), fibres were broken at deeper locations than the worn surface. Broken fibre pieces, which were about 50 μm in average length, were observed on the counterpart surface of SiC paper; therefore, the dominant mechanism for the removal of fibre material could be considered as micro-fracture. In contrast, fibres

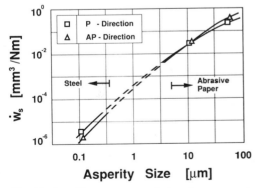

Fig. 27. Specific wear rates of the glass composite with carbon fibre reinforcement in the P- and AP-direction as a function of the "asperity size" of the counterparts (i.e. surface roughness of steel and SiC particle diameter), tested at $p = 1$–2.2 MPa and $v = 0.005$–1 m s^{-1}.

Fig. 28. SEM micrograph of the worn surface of the glass composite tested in the N-direction on different SiC paper at $p = 2.2$ MPa and $v = 0.005$ m s^{-1} (arrow indicates sliding direction); (a) tested on particle size of 75 μm, (b) tested on particle size of 18 μm.

were broken almost in the same plane in which matrix damage occurred if the glass composite was worn against SiC paper with $D = 18$ μm (fig. 28b). On the surface of the SiC counterpart, the cut fibre pieces were not longer than 10 μm. In both

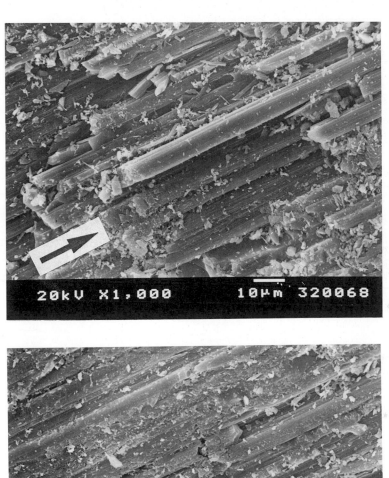

Fig. 29. SEM micrograph of the worn surface of the glass composite tested in the P-direction on different SiC paper at $p = 2.2$ MPa and $v = 0.005$ m s^{-1} (arrow indicates sliding direction); (a) tested on particle size of 75 μm, (b) tested on particle size of 18 μm.

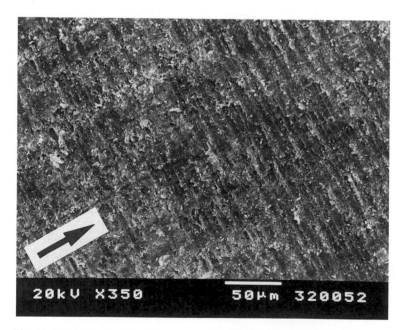

Fig. 30. SEM micrograph of the worn surface of the glass composite tested in the AP-direction on different SiC paper at $p = 2.2$ MPa and $v = 0.005$ m s^{-1} (arrow indicates sliding direction); (a) tested on particle size of 75 μm, (b) tested on particle size of 18 μm.

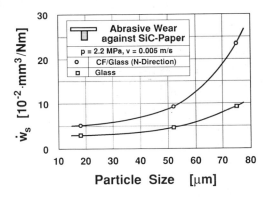

Fig. 31. Specific wear rates of both the unfilled glass and its composite in the N-direction as a function of the abrasive particle size of the SiC paper counterparts at $p = 2.2$ MPa and $v = 0.005$ m s^{-1}.

cases, the dominant mechanism of matrix removal was ploughing and micro-cutting. In the P-orientation against the SiC paper with $D = 75$ μm (fig. 29a), fibre fracture occurred in the contact region as well as in subsurface regions; their contributions to the wear of the composite were enhanced by the ploughing, cutting and cracking effects on the removal of the matrix between the exposed fibres. The ploughing and cutting wear mechanisms remained almost the same for the smaller abrasive grain size ($D = 18$ μm), while fibre fracture events were drastically reduced in this case (fig. 29b). In the AP-orientation (figs. 30a,b), fibre or fibre bundles were broken in the contact surface due to a bending action imposed by passing abrasive particles. This resulted in the removal of whole fibre packages about 50 μm in width in the case of the 75 μm SiC paper, as can be estimated from the width of the wear tracks which remained on the worn surface of the composite (fig. 30a). In the centre part of this figure, it can be seen that ploughing by SiC particles occurred to a depth of several fibre layers. This was, of course, less pronounced in the case of the smaller SiC particles ($D = 18$ μm), for which the width of the wear tracks was only about 10 μm, and the depth amounted to one half or one complete fibre diameter (fig. 30b).

It should be noted that, similar to the sliding wear conditions mentioned in section 3.4, a decrease of the abrasive wear resistance of the glass occurred when it was reinforced with carbon fibres. As can be seen from fig. 31, the increasing tendency in the wear rate (if the glass was reinforced with carbon fibres) was more significant with increasing abrasive particle size. The main reason for this could be attributed to different wear mechanisms. The appearance on the worn surface of unfilled glass against SiC paper with grain size of 75 μm (fig. 32a) showed mainly continuous, deeper grooves impressed in the glass surface with apparent lateral cracking of the glass into pieces between these grooves and larger adjacent fragments. The glass surface, worn against SiC paper with grain size of 18 μm (fig. 32b), showed, on the other hand, mainly a wear pattern which looked rather smooth and which seemed to exist of deformed, smeared out glass patches of compacted wear debris, partly back-transferred to the worn surface underneath. In

Fig. 32. SEM micrograph of the worn surface of unfilled glass tested on different SiC paper at $p = 2.2$ MPa and $v = 0.005$ m s^{-1} (arrow indicates sliding direction); (a) tested on particle size of 75 μm, (b) tested on particle size of 18 μm.

comparison with the CF/glass composite against SiC paper of the same grain size, unfilled glass exhibited much smaller pieces of debris. This seemed to be due to the fact that in case of the glass against SiC paper no interfacial debonding and fibre cutting effects could happen. The latter resulted, on the other hand, in the case of the CF/glass composite to an easier formation of a large amount of wear debris and therefore higher wear rates.

By comparison with metallic materials [38], which have higher hardness and fracture toughness, and with polymeric materials [22,39], which have lower hardness but higher fracture toughness, the effect of the grain size on the wear of glass is very much greater. For polymeric materials under similar conditions as the case tested here, the differences in wear rates between $D = 70$ µm and $D = 15$ µm [39] were of about half those measured for glass between $D = 75$ µm and $D = 18$ µm. It is suggested that when the material has higher hardness [40] but lower toughness, the mechanism of material removal can change from a plastic deformation mode to a fracture mode at an increase of the abrasive grain size, thus leading to greater differences in wear rate.

5. Scratching by a diamond indentor

5.1. Testing procedure

In order to better understand the wear mechanisms of the CF/glass composite against abrasive particles, scratch tests were carried out using an apparatus which has been described fully elsewhere [41,42]. The apparatus, schematically shown in fig. 33, consists of a balanced beam, of which at one end a diamond pyramid-shaped indentor is loaded normally against the composite surfaces, which had been previously polished. A normal load of between 0.5 and 3.0 N was applied directly to the cone, which was driven across the surface at a velocity of 1.8×10^{-6} m/s by means of a motor. The horizontal frictional force during scratching was measured continuously by strain gauges.

5.2. Results and discussion

Typical friction curves of the CF/glass composite for two principle fibre directions (P,N) are shown in fig. 34. Substantial structure-related changes in the frictional force were obtained. Due to the different fibre orientation the appear-

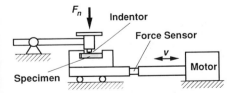

Fig. 33. Schematic illustration of the scratch testing device.

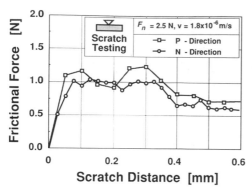

Fig. 34. Frictional force of the glass composite during scratch testing at $F_n = 2.5$ N and $v = 1.8 \times 10^{-6}$ m s^{-1}.

ance of the scratches in the CF/glass composite was also quite different. The observations on the scratch morphology after scratching under a normal load of $F_n = 2.5$ N, as shown in fig. 35, indicated that the scratch-damaged region when testing parallel to the fibre orientation was much greater than that under normal fibre orientation. In the case of parallel fibre direction, the fibres and glass matrix between the fibres were "extruded" along the scratch and pushed away from the composite as large-size fragments; the frictional force, shown in fig. 34, had a higher value (more than 1 N). Under normal fibre orientation, on the other hand, the scratch-damaged region appeared smaller, with matrix cracks and depression of fibres. Very little material removal could be seen; the frictional force showed a slightly smaller value than that of the P-direction. The lower-level regions of the frictional forces under the two testing directions reflected testing conditions at which scratching occurred only in the glass matrix between the fibres (as seen from comparsions of the real scratches on low-magnification micrographs, e.g. fig. 35c, with the force–distance curves). If the diamond indentor scratched only over the unfilled glass, the scratch-damaged region seemed small, and plastic ploughing with micro-cracking was dominant (fig. 36).

Figure 37 shows the frictional force as a function of the normal load during scratching of the CF/glass composite in the anti parallel fibre orientation. At a lower normal load of 1.0 N, the frictional force curve was rather smooth. From the SEM observation, the scratch was virtually featureless, i.e. no cracks could be found; the remaining track of the indentor was only a shallow, plastic groove. At higher normal forces, micro-chipping and -cracking occurred along the scratches, and, in particular, debonding of fibres during the chipping effect of the indentor could be found (fig. 38a). At a normal load of 3.5 N, the friction curve increased drastically, i.e. very high frictional forces were measured at some locations. The reason for this can be found from fig. 38b. Due to the increase in penetration depth of the diamond indentor, the scratch depth was detected to be more than one fibre layer; hence fibre debonding took place in a larger region, thus causing a higher frictional force and a greater amount of wear.

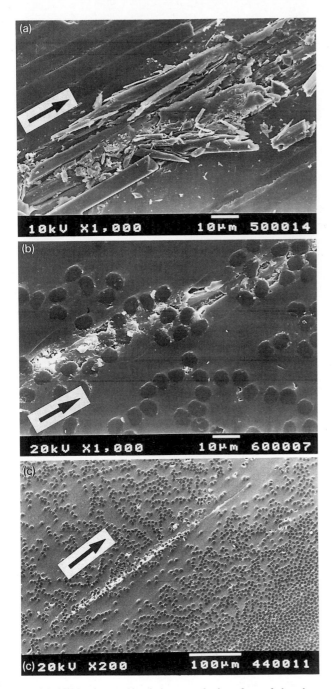

Fig. 35. SEM micrograph of the scratched surface of the glass composite tested at $F_n = 2.5$ N and $v = 1.8 \times 10^{-6}$ m s^{-1} (arrow indicates scratching direction): (a) in the P-direction, (b) and (c) in the N-direction.

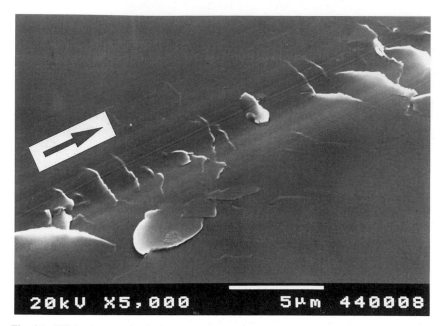

Fig. 36. SEM micrograph of the scratched surface of the unfilled glass tested at $F_n = 2.5$ N and $v = 1.8 \times 10^{-6}$ m s^{-1} (arrow indicates scratching direction).

Together, the wear and friction phenomena described above make it clear that scratching normal to the fibre direction results in the best wear characteristics. This finding is in agreement with the abrasive test results against SiC paper. In addition, it can be understood from the scratch tests why the wear rate in the AP-direction against SiC particles had a higher value, in particular, when the grain size was large and therefore the point load by each grain was high.

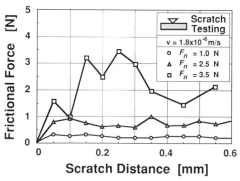

Fig. 37. Frictional force of the glass composite during scratch testing in the AP-direction under different normal loads at $v = 1.8 \times 10^{-6}$ m s^{-1}.

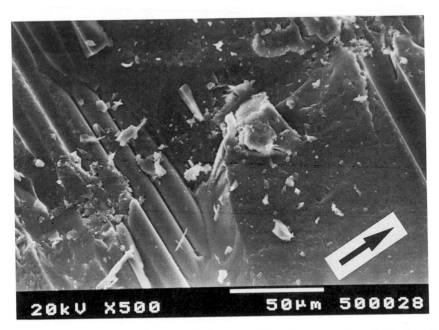

Fig. 38. SEM micrograph of the scratched surface of the glass composite tested in the AP-direction at $v = 1.8 \times 10^{-6}$ m s^{-1} under the normal loads (a): 2.5 N and (b): 3.5 N (arrow indicates scratching direction).

6. Conclusions

Based on the wear and friction results obtained with the carbon fibre/glass matrix composite in this study, the following conclusions can be drawn:

(1) The effects of the sliding direction on the wear behaviour of the unidirectionally reinforced composite can be different depending on the type of wear action applied.

(2) When sliding took place against smooth hard metals under ambient temperature, the highest wear resistance and the lowest friction coefficient were observed in the antiparallel direction.

(3) For both the composite and the counterparts, the wear rates decreased as the hardness of the counterpart material increased. This statement can, however, not be seen as a general one, because the types of the metallic counterparts were also different.

(4) The wear rates of the composite were about one or two orders of magnitude higher than those of the corresponding counterparts, for all sliding directions.

(5) Two types of coloured wear debris were observed, and they played an important role in the wear mechanisms during sliding.

(6) The type of contact configuration with smooth metals affects the wear rates, especially for the P-direction.

(7) Addition of carbon fibre to the glass reduced the friction coefficient drastically.

(8) The wear characteristics of the composite changed when the specimens were run against SiC papers of various grain size.

(9) Opposite to the sliding wear situation, the abrasive wear rate in the AP-direction of the composite was higher than in the other two directions. It was found that the dominating wear mechanism was in-plane bending of fibres, which caused fibre fracture and removal of whole fibre packages from the contact surface.

(10) By the use of a scratch test the abrasive mechanisms could be better understood.

In summary, one can learn from this study that for the use of this material under sliding wear applications against metallic counterparts the following conditions should be maintained in order to achieve the best tribological benefits:

(a) Sliding should take place in the anti parallel fibre direction,

(b) the counterpart should be either cast iron or hard steel, and

(c) loading conditions should be up to a maximum of $pv = 3$ MPa m s^{-1} (with v between 0.3 to 3 m s^{-1}), so that a constant wear factor (identical to the specific wear rate) of about 2×10^{-6} mm^3/N m can be assumed for further design calculations.

Under abrasive wear conditions, i.e. when very hard particles scratch along the surface of this composite, it is more favourable to position the fibres under normal orientation.

Acknowledgements

I am very grateful to Prof. Dr. K. Friedrich for his guidance in this field of study and, especially, for his valuable discussions on this work as well as his critical reading of the manuscript. Further thanks are due to Schott Glaswerke, Mainz, in particular, Dr. W. Pannhorst and Mr. J. Heinz, for the supply of the CF/glass composites and for fruitful discussions. The author gratefully acknowledges Mr. H. Nöcker for his assistance in the scratch studies at the Institute for Materials Science, Ruhr University Bochum. In addition, the help of Mr. S. Bleser during the performing of the experiments is gratefully mentioned. Finally, I wish to thank the Deutscher Akademischer Austauschdienst (DAAD) for my fellowship.

List of symbols

A	apparent contact area (mm^2)
D	average size of the counterpart asperities (μm)
E	elastic tensile modulus (MPa)
F_f	frictional force (N)
F_n	normal load (N)
G_{IC}	critical strain energy release rate (kJ/m^2)
L	sliding distance (m)
p	apparent contact pressure (MPa)
pv	product of apparent contact pressure and sliding velocity (MPa m s^{-1})
pv_{max}	maximal value of pv
R_a	average roughness (μm)
ρ	density of material tested (g cm^{-3})
V_f	fibre volume fraction (%)
v	sliding velocity (m s^{-1})
$\dot{w}_s$	specific wear rate (mm^3 N^{-1}m^{-1})
$\dot{w}_s^{-1}$	wear resistance
$\dot{w}_t$	time related (length-) wear rate (μm h^{-1})
μ	coefficient of friction
Δh	reduction in length during wear (μm)
Δm	mass loss during wear (g)
Δt	testing time (h)

Abbreviations

Al_2O_3	aluminium oxide
AP	antiparallel
B_2O_3	boron oxide
CF	carbon fibre
Duran®	a type of borosilicate glass
EDS	energy dispersion X-ray analysis
Fe	iron
Fe_2O_3	iron (III) oxide

FeO	iron (II) oxide
HM	higher modulus carbon fibre
K_2O	potassium oxide
N	normal
Na_2O	sodium oxide
O	oxygen
P	parallel
PES	polyether sulfone
PSU	polysulfone
SEM	scanning electron microscope
Si	silicon
SiC	silicon carbide
SiO_2	silicon oxide

References

[1] M. Spallek and W. Pannhorst, Hochtemperaturverbundwerkstoffe, SCHOTT Information 50 (1989) 16.
[2] H. Hegeler and R. Brückner, Fibre reinforced glasses, J. Mater. Sci. 24 (1989) 1191.
[3] K.M. Prewo, J.J. Brennam and G.K. Layden, Fiber reinforced glasses and glass-ceramics for high performance applications, Amer. Ceram. Soc. Bull. 65 (1986) 305.
[4] K.M. Prewo and J.A. Batt, The oxidative stability of carbon fibre reinforced glass-matrix composites, J. Mater. Sci. 23 (1988) 523.
[5] N. Takeda, O. Chen, T. Kishi and K. Prewo, Matrix microcracking damage characterization of discontinuous carbon fiber reinforced glass-matrix composites, in: Proc. ICCM/8, Honolulu, July 15–19 (1991).
[6] H. Hegeler and R. Brückner, Fibre-reinforced glasses: influence of thermal expansion of the glass matrix on strength and fracture toughness of the composite, J. Mater. Sci. 25 (1990) 4836.
[7] S.M. Bleay and V.D. Scott, Microstructure and micromechanics of the interface in carbon fibre reinforced Pyrex glass, J. Mater. Sci. 26 (1991) 3544.
[8] K.M. Prewo, B. Johnson and S. Starrett, Silicon carbide fibre-reinforced glass-ceramic composite tensile behaviour at elevated temperature, J. Mater. Sci. 24 (1989) 1373.
[9] K.M. Prewo and J.F. Bacon, Glass matrix composites – I, graphite fiber reinforced glass, in: Proc. 2nd Int. Conf. on Composites, ed. B. Noton (AIME, 1978) pp. 64–74.
[10] Composites Update, Researchers Meld Work in Ceramics and Textiles, Newsletter of the University of Delaware's Center for Composite Materials, Fall (1991) p. 5.
[11] H. Hegeler and R. Brückner, Mechanical properties of carbon fibre-reinforced glasses, J. Mater. Sci. 27 (1992) 1901.
[12] E. Minford and K. Prewo, Friction and wear of graphite-fiber-reinforced glass matrix composites, Wear 102 (1985) 253.
[13] T.E. Fischer and H. Tomizawa, Interaction of tribochemistry and microfracture in the friction and wear of silicon nitride, Wear 105 (1985) 29.
[14] O.O. Ajayi and K.C. Ludema, Mechanism of transfer film formation during repeat pass sliding of ceramic materials, Wear 140 (1990) 191.
[15] J. Denape and J. Lamon, Sliding friction of ceramics: mechanical action of the wear debris, J. Mater. Sci. 25 (1990) 3592.
[16] A.A. Torrance, A three-dimensional cutting criterion for abrasion, Wear 123 (1988) 87.
[17] N.P. Suh and H.-C. Sin, The genesis of friction, Wear 69 (1981) 91.
[18] M. Godet, The third-body approach: a mechanical view of wear, Wear 100 (1984) 437.
[19] Y. Berthier, Experimental evidence for friction and wear modeling, Wear 139 (1990) 77.

[20] M.M. Khruschov, Principles of abrasive wear, Wear 8 (1974) 69.

[21] W.M. Garrison, Abrasive wear resistance, the effects of ploughing and the removal of ploughed material, Wear 114 (1987) 239.

[22] K. Friedrich, Wear of reinforced polymers by different abrasive counterparts, in: Friction and Wear of Polymer Composites, Composite Materials Series, Vol. 1, ed. K. Friedrich (Elsevier, Amsterdam, 1986) pp. 233–287.

[23] R.A.J. Sambell, D.H. Bowen and D.C. Phillips, Carbon fiber composites with ceramic and glass matrices, Part 1, Discontinuous fibers, J. Mater. Sci. 7 (1972) 663.

[24] R.A.J. Sambell, J.A. Batt, D.C. Phillips and D.H. Bowen, Carbon fiber composites with ceramic and glass matrices, Part 2, Continuous fibers, J. Mater. Sci. 7 (1972) 676.

[25] H. Voss and K. Friedrich, On the wear behaviour of short-fibre-reinforced PEEK composites, Wear 116 (1987) 1.

[26] K. Friedrich, J. Karger-Kocsis and Z. Lu, Effects of steel counterface roughness and temperature on the friction and wear of PE(E)K composites under dry sliding conditions, Wear 148 (1991) 235.

[27] K. Friedrich and J.S. Wu, Polymer composites with high wear resistance, in: Encyclopedia of Composite Materials, Vol. 4, ed. S.M. Lee (VCH Publ., New York, 1991) pp. 255–279.

[28] K. Friedrich, Z. Lu and R. Scherer, Wear and friction of composite materials, in: Composite Materials, eds A.T. Di Benetto, L. Nicolais and R. Watanabe (Elsevier, Amsterdam, 1992) pp. 67–85.

[29] K. Friedrich, Sliding wear performance of different polyimide formulations, Tribol. Int. 22 (1989) 25.

[30] Z. Lu, K. Friedrich, W. Pannhorst and M. Spallek, Sliding wear behaviour of a unidirectional carbon-fiber-reinforced glass matrix composite against steel, in: Proc. DGM-Symposium Verstärkung keramischer Werkstoffe, Hamburg, Germany, 8–9 October, pp. 261–274 (1991).

[31] H.H. Shim, O.K. Kwon and J.R. Youn, Effects of structure and humidity on friction and wear properties of carbon fiber reinforced epoxy composites, in: Proc. SPE ANTEC '91 (1991) p. 1997.

[32] H. Voss and K. Friedrich, Wear performance of a bulk liquid crystal polymer and its short-fiber composites, Tribol. Int. 19 (1986) 145.

[33] H. Nayeb-Hashemi, J.T. Blucher and J. Mirageas, Friction and wear behavior of aluminium–graphite composites as a function of interface and fiber direction, Wear 150 (1991) 21.

[34] O. Jacobs, Scanning electron microscopy observation of the mechanical decomposition of carbon fibres under wear loading, J. Mater. Sci. Lett. 10 (1991) 838.

[35] H.H. Shim, O.K. Kwon and J.R. Youn, Friction and wear behavior of graphite fiber-reinforced composites, Polym. Composites 11 (1990) 337.

[36] J. Sugushita and S. Fujiyoshi, The effect of cast iron graphites on friction and wear performance, I: Graphite film formation on grey cast iron surfaces, Wear 66 (1981) 209.

[37] H. Voss, Aufbau, Bruchverhalten und Verschleisseigenschaften kurzfaserverstärkter Hochleistungs-thermoplaste, Fortschr.-Ber., VDI Zeitschr. Reihe 5, Nr. 116 (VDI Verlag, Düsseldorf, 1987).

[38] R.C.D. Richardson, The wear of metals by relatively soft abrasives, Wear 11 (1968) 245.

[39] M. Cirino, R.B. Pipes and K. Friedrich, The abrasive wear behaviour of continuous fibre polymer composites, J. Mater. Sci. 22 (1987) 2481.

[40] M.A. Moore and F.S. King, Abrasive wear of brittle solids, Wear 60 (1980) 123.

[41] H. Nöcker, Ch.Y. Gan and E. Hornbogen, Reibung- und Verschleißmessung an Al–Si-Legierungen, in: Berichte der Metallographie-Tagung Garmisch-Partenkirchen, 28–30 September (1988).

[42] H. Nöcker and E. Hornbogen, Friction and wear measurements with a new metallographic scratching method, Pract. Met. 26 (1989) 455.

Advances in Composite Tribology
edited by K. Friedrich
© *1993 Elsevier Science Publishers B.V. All rights reserved*

Chapter 11

Friction and Wear Characteristics of Advanced Ceramic Composite Materials

BRAHAM PRAKASH

Industrial Tribology Machine Dynamics & Maintenance Engineering Centre (ITMMEC),
Indian Institute of Technology, Delhi Hauz Khas, New Delhi-110016, India

Contents

Abstract

Advanced ceramics both of oxide and non-oxide types have been the subject of many tribological studies in the last few decades. These monolithic ceramics, such as alumina, zirconia, silicon nitride, sialons, silicon carbide, titanium nitride and titanium carbide, etc., are characterized by good resistance to wear/corrosion/erosion, good high-temperature strength and good dimensional stability, which makes them attractive for use in tribological applications involving high temperatures and reactive environments. However, the major obstacles in attempts to the use of monolithic ceramics in actual engineering applications are their low mechanical reliability, due to their inherent brittleness, and difficulties involved in producing finished complex machine parts. To overcome the problem of low reliability, ceramic matrix has been reinforced with whiskers, fibres and refractory particles, leading to the development of ceramic composites with enhanced mechanical strength and fracture toughness.

The emphasis of tribological investigations in the past has been mainly on monolithic ceramics. Only a few studies concerning the friction and wear behaviour of ceramic composite materials have been carried out. The aim of this article is to review the results of the friction and wear studies conducted so far on different ceramic composite materials. The test techniques adopted for the tribological evaluation and some tribological applications have also been discussed. Further, some remarks concerning future directions for tribological research on ceramic composite materials have been included.

1. Introduction

Advanced ceramics are gradually emerging as potential tribological materials. Both oxide and non-oxide ceramics are in active use in some tribological applications, because of several advantages over conventional materials. These advanced ceramic materials, also known as high-tech or fine ceramics, include alumina (Al_2O_3), zirconia (ZrO_2) based ceramics, sialons or silicon oxynitrides ($Si_{6-z}Al_zO_zN_{8-z}$) and titanium carbide (TiC), etc. Their low coefficients of friction, good resistance to wear/corrosion/erosion and high-temperature strength make them attractive tribological materials, particularly in applications where machine components are exposed to high temperatures and reactive environments. Additionally these ceramic materials are characterized by good dimensional stability, close tolerances, good surface finish, good thermal-shock resistance and low weight. In spite of these unique properties, the introduction of ceramic materials in actual engineering applications has been rather slow. This is primarily due to the low mechanical reliability of ceramic materials, owing to their inherent brittleness, flaw sensitivity and difficulties in manufacturing complex engineering components from ceramic materials. Recent technological advances in processing and manufacturing techniques of ceramics have, however, led to the development of some ceramic components and their introduction in actual use. These application examples include ceramic turbocharger rotors, ball bearings and ceramic cutting tool bits. Further efforts have gone into developing ceramic composite materials by dispersing whiskers, fibres or particles in the ceramic matrix. These ceramic composite materials are characterized by superior mechanical strength and fracture toughness, thereby enhancing the mechanical reliability of the ceramic materials. Several mechanisms, such as load transfer, prestressing, microcracking, phase transformation, crack impediment or deflection, fibre pull-out and crack bridging contribute to toughening of ceramic composite materials [1]. In addition to enhancements in the mechanical strength, ceramic composites can also be developed using suitable fillers such as solid lubricants, or any such reinforcing agents

TABLE 1

Ceramic composite materials of current interest (●, industrial applications; ○, research work) (from ref. [1]).

Second phase	Matrix					
	Al_2O_3	ZrO_2	Glass	Si_3N_4	SiC	C
Carbide, nitride or boride particles	●	○	○	●	○	
ZrO_2 particles	●	●	○	○		
Si_3N_4 whiskers				○	○	
SiC whiskers	●	○	○	○	○	
SiC fibres	○		●	○	○	
C fibres	○		●	○	○	●

which lead to improvements in hardness, wear resistance and reduction in the friction coefficient. Some typical ceramic composite systems of current interest and which are the subject of industrial research and development are listed in table 1.

The tribological research efforts in the past have been mostly on monolithic ceramics. However, the development of ceramic composite materials with enhanced mechanical properties has also given an impetus to friction and wear studies on some ceramic composite systems.

2. Friction and wear behaviour of ceramic composite materials

As mentioned earlier, the reinforcement of ceramic matrix with whiskers, fibres or particulates results not only in enhanced fracture toughness but it also gives improvements in the hardness and other properties of the ceramic composites. Thus, ceramic composites are also expected to offer greatly improved friction and wear characteristics vis-à-vis monolithic ceramics. Ceramic composites have been developed using metals, refractories or other fillers such as whiskers/fibres or particulates. However, the emphasis in this chapter is mainly on ceramic composite materials obtained through reinforcement of ceramic materials with ceramic whisker/fibres, particulates and some transformable phases. Further, ceramic composites using lubricant fillers have also been covered, in view of their influence on the tribological properties. Tribological studies conducted on ceramic composite materials include friction and wear under sliding (adhesive) and grooving (abrasive) conditions. Of course, several studies on ceramic composites have been conducted concerning their erosive-wear properties, but the scope of this review is limited to sliding and abrasive wear of these materials. The processing techniques employed for producing ceramic composites have a great influence on their friction and wear behaviour and these techniques have been briefly mentioned while discussing the friction and wear characteristics.

3. Sliding friction and wear characteristics of ceramic composite materials

The sliding friction and wear properties of several ceramic composite materials have been studied. The composites are based on a matrix of Al_2O_3, Si_3N_4, SiC and TiN. Reinforcements of these ceramics have been done using SiC whiskers, ZrO_2, TiC, TiN and BN particulates.

The friction and wear studies have been carried out using self-mated ceramic–ceramic or metal–ceramic tribological pairs. Also, tribological tests have been done under unidirectional or reciprocating sliding conditions at temperatures ranging from room temperature up to 1000°C. The friction and wear studies have mostly been conducted without lubrication, barring few exceptions. In some cases the operating environment has also been varied. In view of the diverse test conditions, geometry and test techniques adopted, the friction and wear characteristics of each ceramic composite material have been discussed separately.

4. Whisker reinforced ceramic composites

Ceramic composite materials obtained through reinforcements with whiskers are characterized by extremely good mechanical properties coupled with good hardness and fracture toughness. The toughening of the ceramic matrix with whisker reinforcement takes place mainly through crack bridging, crack deflection and whisker pull-out mechanisms [1].

Friction and wear studies have been conducted on ceramic composites employing alumina (Al_2O_3), zirconia toughened alumina (ZTA) and silicon nitride matrices reinforced with silicon carbide (SiC) whiskers.

4.1. SiC-whisker reinforced alumina

The friction and wear behaviour of SiC-whisker reinforced ceramic composites has been studied by Yust et al. from room temperature up to 800°C in dry nitrogen and air environments on self-mated tribological pairs of such ceramic composite material using a pin-on-disc machine [2,3] with environmental control.

The Al_2O_3–SiC_w composite was processed by hot-pressing mixtures of alumina powder (Baikowski Cb-10) and 20 vol% SiC whiskers (ARCO SC-9), using magnesia and yttria (0.5% each) as densification aids. The resulting matrix had a grain size of 4 μm and a whisker aspect ratio of 50 with a nominal diameter of 0.6 μm. Also, the composite material had a preferential alignment of the longitudinal axis of the whiskers in the plane perpendicular to the hot-pressing direction, but within that plane the whiskers were randomly oriented.

There was a significant improvement in the mechanical properties of the composite material. The flexural strength was 650 MPa as against 300 MPa for monolithic Al_2O_3. Likewise, the fracture toughness was 8.3 MPa $m^{1/2}$ compared to 4.5 MPa $m^{1/2}$ for monolithic Al_2O_3. The test specimens for the friction and wear tests were in the form of flat discs and 6.35 mm diameter pins with a spherical tip of 6.35 mm radius, both polished to a surface finish with an R_a of 1 μm or less. The disc surfaces were prepared perpendicular to the hot-pressing direction, which yielded wear surfaces with a large proportion of SiC whiskers lying almost in the wear plane. The whisker orientation in the pin specimens was not controlled.

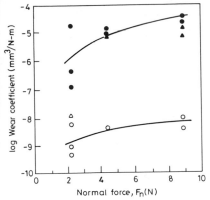

Fig. 1. Wear coefficients of SiC-whisker reinforced Al_2O_3 as function of applied force. Disc-wear coefficients were determined for those experiments which yielded measurable wear. The elevated temperature values correspond to severe wear while room temperature values represent mild wear. The curves represent least-squares fits of functions of the form $y = m \ln(x) + b$ (○ – pin, 20°C; ● – pin, 425°C; △ – disc, 20°C; ▲ – disc, 425°C). From ref. [2].

The wear coefficients obtained from tests in dry nitrogen, are plotted in fig. 1 as function of the normal load. The wear coefficients for this composite are 10^{-8} to 10^{-9} mm^3 N^{-1} m^{-1} and 10^{-4} to 10^{-5} mm^3 N^{-1} m^{-1} at 20 and 425°C, respectively. The corresponding values for commercial alumina without whisker reinforcement are 1×10^{-4} and 1×10^{-3} mm^3 N^{-1} m^{-1} at 20 and 425°C, respectively. It can thus be seen that very significant enhancement in the wear resistance of Al_2O_3 ceramic occurs as a result of SiC-whisker reinforcement. The wear reduction is about four orders of magnitude at 20°C and about two orders of magnitude at 425°C. It is also observed that at both temperatures the wear coefficients do not vary much with the change of normal load, from 2.2 to 8.9 N.

The friction coefficients for the Al_2O_3–SiC$_w$ composite at 20°C are about 0.6 to 0.7, except at the highest load, 8.9 N, at which the friction coefficients observed are 0.17 and 0.31 for the one hour and the two hour duration test, respectively. This indicates a reduction in friction at higher loads. The highest coefficient of friction was observed at 425°C and at the lowest load, 2.23 N, and was 1.64. The trend of lower friction coefficients at higher loads was observed at 425°C also.

Examination of the worn surfaces from the tests conducted at 20°C indicated a layer of additional material of about 0.2 μm thickness formed on the wear track of the disc. This layer is continuous but irregular in shape, and is typically a dry, mild form of wear. On the other hand, the worn surfaces from the tests carried out at 425°C are very irregular, and show deeply grooved and flattened regions extending in the sliding direction, and these are caused by the occurrence of severe wear. Further examination of the severely worn surface at higher magnification revealed that the material removal occurred through the formation of surface cracks. The transition from mild to severe wear takes place with the formation of surface cracks which intersect with whiskers and are then deflected, thereby inhibiting further crack growth.

TABLE 2
Summary of the friction and wear results [a] on SiC-whisker reinforced alumina (from ref. [3]).

Ambient temperature (°C)	Stress [b] (MPa)	Sliding velocity (m/s)	Sliding distance [c] (km)	Relative humidity (%)	Wear coefficient (mm³/(N m)) Pin	Disk [d]	Friction coefficient
20	~ 80	1.0	2		9.6×10^{-6}	HL	
20	~ 80	1.0	1.4		4.2×10^{-6}	HL	
20	~ 80	1.0	1	42	4.9×10^{-6}	HL	
20	~ 80	1.0	1	37	1.3×10^{-5}	HL	
20	30	0.1	8	38	4.8×10^{-9}	VSL	0.4
20	30	0.1	8	50	1.1×10^{-8}	VSL	0.5–0.7
20	50	0.1	8	45	1.0×10^{-8}	VSL	0.4–0.5
20	50	0.1	12	42	9.3×10^{-9}	VSL	0.5
20	30	0.5	8	35	1.5×10^{-8}	VSL	0.5–0.6
20	30	0.5	8	55	1.8×10^{-8}	VSL	
20	50	0.5	6	55	9.6×10^{-7}	VSL	0.4–0.5
					8.5×10^{-9}	HL	0.5–0.7
20	50	0.5	6	50	7.7×10^{-7}	VSL	
					7.2×10^{-9}	HL	
20	75	0.1	8	80	1.3×10^{-8}	VSL	0.2–0.4
20	75	0.1	7		1.2×10^{-8}		
20	75	0.5	5		1.8×10^{-8}		
20	75	0.5	10	55	1.4×10^{-8}	VSL	0.5
400	~ 15	0.1	2	39	3.5×10^{-5}	6.8×10^{-5}	0.7–0.8
400	~ 15	0.1	1	42	1.6×10^{-5}		
400	~ 15	0.5	2	39	8.9×10^{-6}	1.61×10^{-5}	0.6
800	30	0.1	2	50	3.0×10^{-8}	NML	0.5–0.6
800	30	0.5	4	43	1.6×10^{-9}	NML	0.5–0.6
800	150	0.5	2	35	7.7×10^{-8}	NML	0.5

[a] All tests in air of indicated humidity.
[b] Based on the apparent contact area.
[c] Includes the initial run-in period.
[d] HL = heavy loss, NML = no measurable loss, VSL = very small loss.

Yust and Allard further studied the friction and wear of Al_2O_3–SiC_w composite material up to 800°C [3] and the results are summarized in table 2. The results of this study also indicated the occurrence of mild wear at room temperature (20°C) up to a sliding velocity of 0.5 m/s. The wear coefficients during this stage of wear were about 10^{-8} mm³/N m. Transition from mild wear to severe wear occurred at 0.5 m/s at 50 MPa and after a sliding distance of 6 to 8 km. This transition was characterized by a two orders of magnitude increase in the wear coefficient. Severe wear occurred soon after the start of sliding at a sliding velocity of 1 m/s. This has been attributed to the combined effects of a very high stress imposed on the contact area at the start of sliding and the generation of high temperatures in the sliding interface. Wear tests conducted at 400°C indicated the occurrence of severe wear in all tests within the first kilometre of sliding.

The wear behaviour of the Al_2O_3–SiC_w composite at 800°C was in marked contrast with that at 400°C. Mostly mild wear occurred in all tests, with wear rates in the range 10^{-8} to 10^{-9} mm^3/N m. The friction coefficients, which were measured periodically throughout the tests at 20°C, were between 0.2 and 0.7. In the elevated-temperature tests these measurements were made only in the beginning, because of the problem of thermal drift. The friction coefficients measured at 400°C were between 0.6 and 0.8 and those at 800°C were between 0.5 and 0.6.

In mild wear, fine wear debris was observed and the composite disc material showed worn surfaces with an adhered debris layer, with the appearance of compressed and sheared fine particles. The initiation of fracture in the composite occurs at the SiC-whisker–Al$_2$O$_3$-matrix interface and is promoted by tensile forces generated in the surface beneath the sliding contact. Fracture at the interface may further be caused by cyclic tensile stresses as a result of repeated sliding, by the fatigue mechanism. This leads to removal of a relatively large volume of material and higher wear coefficients. At 400°C the transition to severe wear occurs soon after the start of sliding, and the interfacial debris formed is composed of very fine particles of 10 to 50 nm size. The continuing formation and shearing may be responsible for the reduction in size of the wear particles. The occurrence of mild wear at 800°C is attributed to oxidation of SiC and the formation of a mixed aluminium and silicon oxide layer on the worn surface, which may inhibit surface fracture and modify the stress state in the composite's surface. The protective role of the surface layer was confirmed by the fact that removal of this layer revealed the undamaged composite-material surface.

4.2. SiC-whisker reinforced zirconia toughened alumina

Addition of SiC whiskers to zirconia toughened alumina resulted in better properties, as a result of the multiple toughening produced by both the zirconia particles (mainly transformation toughening) and the silicon carbide whiskers (mainly crack bridging, crack deflection and whisker pull-out) [1,4,5]. Bohmer and Almond studied the mechanical and wear properties of SiC-whisker reinforced zirconia toughened alumina [5].

The composite-material samples of zirconia toughened alumina (ZTA), containing about 17 wt% of commercial SiC whiskers, were produced through axial hot-pressing at 1600°C under a pressure of 30 MPa. The hot-pressing produced directionality in the whisker orientation, and the whiskers were preferentially oriented in the plane perpendicular to the hot-pressing axis. Reinforcement of ZTA matrix with SiC whiskers resulted in considerable improvements in the mechanical properties of the ceramic composite material (table 3). The fracture toughness of the ceramic composite was enhanced to 5.6 MPa m$^{1/2}$ and the flexural strength to 980 MPa. This enhancement is attributed to the combined effect of two toughening mechanisms, i.e. dispersion toughening by the ZrO$_2$ particles and reinforcement by the SiC whiskers. The ZrO$_2$ particles were more effective in enhancing the strength than the fracture toughness, whereas the SiC whiskers enhanced the fracture toughness but decreased the strength. The hardness, stiffness and creep resistance of the SiC reinforced ZTA composite also improved.

TABLE 3
Influence of SiC-whisker reinforcement on the physical and mechanical properties of zirconia toughened alumina (ZTA) (from ref. [5]).

Property	Unreinforced ZTA	SiC-whisker reinforced ZTA
Density (mg/m^3)	4.16	3.95
Modulus of elasticity (GPa)	325	342
Shear modulus (GPa)		128
Poisson's ratio		0.28
Vickers hardness (HV30)	1700	1905
Fracture toughness (MPa m$^{1/2}$)		5.6
Flexural strength (MPa)		980

The sliding friction and wear properties of self-mated tribological pairs of SiC-whisker reinforced zirconia toughened alumina (ZTA) composite were studied under dry sliding conditions on a pin-on-disc machine. The pin and the disc test surfaces were polished to an average surface roughness, R_a, of 0.3 μm, and the wear test surfaces were perpendicular to the hot-pressing direction. The friction and wear results of this study are given in fig. 2. It may be observed that the wear resistance of ZTA–SiC$_w$ composite considerably improved vis-à-vis both pure alumina and zirconia toughened alumina. The wear resistances of both pure

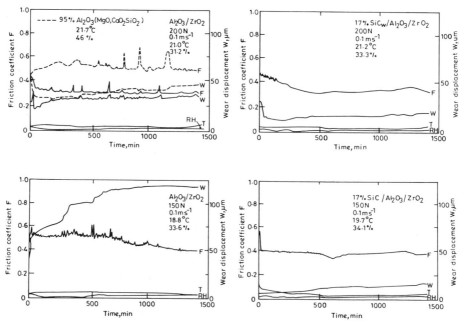

Fig. 2. Friction coefficients and wear of Al$_2$O$_3$–ZrO$_2$ ceramics with and without SiC-whiskers reinforcement as function of sliding distance (T: test temperature; RH: relative humidity). From ref. [5].

alumina and unreinforced ZTA were almost similar. Also, no change in the friction coefficient of the ceramic composite was observed by the addition of SiC whiskers.

4.3. SiC-whisker reinforced silicon nitride

The tribological properties of SiC-whisker reinforced silicon nitride composite have been investigated by Ishigaki et al. [6]. SiC-whisker containing Si_3N_4 composites were developed with the aim of utilizing the good tribological properties of SiC and the strength properties of Si_3N_4. SiC possesses superior friction and wear properties but its fracture toughness is low, whereas hot-pressed Si_3N_4 is characterized by a higher fracture toughness. The Si_3N_4–SiC_w composite material is thus expected to have enhanced mechanical and tribological properties.

The composite-material samples were prepared by hot-pressing a mixture of SiC whiskers (Tokai Carbon Ltd.) and silicon nitride powder (Hermann C Starck's H1) at a temperature of 1800°C and 40 MPa pressure for 60 min in a graphite mould. The densification aids were 7.5 mol% each of Y_2O_3 and La_2O_3. Si_3N_4 composite samples containing 10, 20 and 30 wt% SiC whiskers were prepared. The tribological-test specimens were polished using a #200 diamond whirl.

Microscopic examination revealed that the SiC whiskers in the Si_3N_4–SiC_w composite were oriented in the direction perpendicular to the direction of application of pressure during hot-pressing. This was also confirmed through electrical-resistance measurements. The electrical resistance between surfaces perpendicular to the direction of pressure during hot-pressing was higher than that between surfaces parallel to the direction of pressure application. It was further observed that the electrical resistance increased with increasing SiC-whisker content.

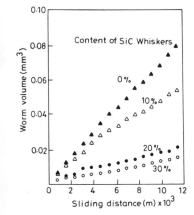

Fig. 3. Variation of wear volume as function of sliding distance for different concentrations of SiC-whiskers in silicon nitride matrix while rubbing against monolithic silicon carbide (rotational speed: 200 rpm). From ref. [6].

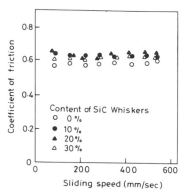

Fig. 4. Variation of coefficient of friction as function of sliding distance for different concentration of SiC-whiskers in silicone nitride matrix while rubbing against monolithic silicon carbide (load: 5 N; RH: 50 ± 5). From ref. [6].

Friction and wear studies were carried out on a ring-on-block tribometer at sliding velocities of 10 to 100 mm/s. Further, a pin-on-flat instrument was used to study the friction characteristics at very low sliding velocities of 10 mm/min. In the block-on-ring tribometer the friction and wear characteristics of SiC-whisker reinforced Si_3N_4 ceramic composite were studied by rubbing the composite material against monolithic silicon carbide and silicon nitride ceramics. In Si_3N_4–SiC_w composite sliding against monolithic SiC there was a marked decrease in wear with increasing SiC-whisker content (fig. 3), whereas no change in the friction coefficient occurred (fig. 4). Similarly, the wear of Si_3N_4–SiC_w composite when rubbed against monolithic Si_3N_4 also decreased with higher concentration of SiC whiskers in the Si_3N_4 matrix. Again, no significant change in the coefficient of friction occurred due to a higher SiC-whisker content (fig. 5).

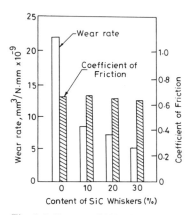

Fig. 5. Influence of SiC-whisker content in silicon nitride matrix on wear rate and coefficient of friction of the composite while rubbing against monolithic silicon nitride. From ref. [6].

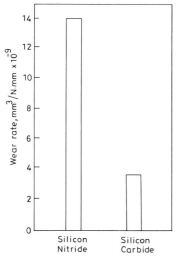

Fig. 6. Comparative wear rates of monolithic silicon nitride and silicon carbide while rubbing against silicon carbide (speed: 200 rpm; load: 5 N; RH: 50 ± 5). From ref. [6].

The improvement in wear resistance as the result of SiC-whisker reinforcement of Si_3N_4 matrix is attributed to the inherently higher wear resistance of monolithic SiC over Si_3N_4 (fig. 6) and the increased hardness of the composite due to the addition of SiC whiskers (fig. 7). Further, the studies revealed that the orientation of the SiC whiskers in the Si_3N_4 ceramic matrix considerably influences the wear characteristics of the composite material (fig. 8). It was observed that the wear resistance of the ceramic composite is superior when the SiC whiskers are oriented in the direction perpendicular to the direction of application of pressure during

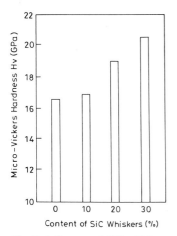

Fig. 7. Influence of SiC-whisker content in silicon nitride matrix on microhardness of the composite material. From ref. [6].

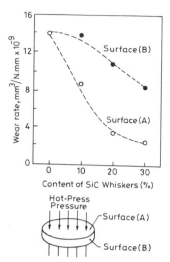

Fig. 8. Influence of the orientation of SiC-whiskers in silicon nitride matrix on wear rate of the composite materials. From ref. [6].

hot-pressing. The wear resistance of the ceramic composite decreased when the SiC whiskers in the composite are oriented parallel to the direction of pressure during hot-pressing.

5. Particulate reinforced and other ceramic composites

The ceramic composite materials of which the friction and wear characteristics have been studied, include those reinforced with zirconia (ZrO_2) particles for transformation toughening and those reinforced with carbides, nitrides and borides, etc. In some other ceramic composites use of solid lubricant fillers, such as graphite, has been made. Important ceramic composite systems include Al_2O_3–ZrO_2, Si_3N_4–BN, Si_3N_4–TiC, Si_3N_4–TiN, SiC–TiC, B_4C–TiB_2–W_2B_5, Al_2O_3–TiN, Al_2O_3–graphite and Si_3N_4–graphite.

5.1. Al_2O_3–ZrO_2 ceramic composites

Trabelsi et al. studied the relationship between the mechanical properties and the wear resistance of Al_2O_3–ZrO_2 ceramic composites [7]. Alumina is characterized by a high modulus of elasticity and a good wear resistance owing to its high hardness. On the other hand, partially stabilized zirconia has good mechanical properties such as flexural strength and fracture toughness, but poor wear resistance due to its low hardness. The dispersion of ZrO_2 particles was thus aimed at obtaining good mechanical and tribological properties of the ceramic composite materials.

Alumina–zirconia composites containing up to 45 vol% of ZrO_2, were prepared from raw powders of Al_2O_3, ZrO_2 and Y_2O_3 by hot-pressing at 1500°C, and in

TABLE 4
Physical, microstructural and hardness characteristics of zirconia toughened alumina (ZTA) (from ref. [7]) *.

Composition	Vol% ZrO$_2$	p (g/cm^3)	Tr (%)	D Al$_2$O$_3$	D ZrO$_2$	HV (GPa)
Al$_2$O$_3$ (RC172)	0	3.98		2.5		20.5
A5Z0Y	5	4.22	95.7	1.56	0.63	19.3
A10Z0Y	10	4.22	82.4	1.16	0.62	18.4
A15Z0Y	15	4.29	23.4	0.96	0.62	17.0
A20Z0Y	20	4.35	6.5	1.30	0.90	14.0
A20Z1Y	20	4.38	30.3	1.00	0.69	16.7
A20Z2Y	20	4.40	97.4	0.94	0.54	18.4
A20Z3Y	20	4.39	100	0.91	0.64	18.7
A45Z3Y	45	4.91	99.6	0.83	0.65	17.0
TZ3Y	100	6.03	100		0.70	15.1

* p: density (Archimede's method), D Al$_2$O$_3$, ZrO$_2$: mean grain size (SEM observations), HV: Vickers hardness (10 N).

TABLE 5
Room-temperature mechanical properties of zirconia toughened alumina (ZTA) (from ref. [7]).

Composition	σ (MPa)	K_{IC}^* (MPa m$^{1/2}$)	E (GPa)
RC172	610	4.5	392
A5Z	670	8.1	379
A10Z	795	10	370
A15Z	750	6.5	340
A20Z	625	6	330
A20Z1Y	825	8.2	350
A20Z2Y	1080	10.1 (9+)	349
A20Z3Y	700	8.5 (7.6+)	345
A45Z3Y	1700	13.5 (9+)	296
TZ3Y		20.1 (11.25+)	

* + (annealed: 950°C to 1000°C, 15 min).

some cases 0.3 mol% of Y$_2$O$_3$ was used for stabilization purposes. Test specimens for mechanical and wear studies were machined from hot-pressed cylinders. The characteristics of zirconia toughened alumina (ZTA) composite are given in tables

TABLE 6
Temperature dependence of K_{IC} in some ZTA composites (from ref. [7]).

Composition	RT	600°C	800°C	1000°C	1200°C
RC172	4.5		4.1	3.6	3
A10Z0Y	10	7.5	6.5	6.0	5
A45Z3Y	13.5	5.2	4.75	3.1	4
TZ3Y	20.1	3.8	3.7	4.25	4.1

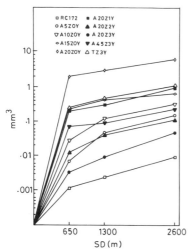

Fig. 9. Wear volume of different ZTA composites as function of sliding distance while rubbing against steel ring in air. From ref. [7].

4–6. It can be noticed that all the mechanical properties, except the hardness, were enhanced with the addition of ZrO_2 to Al_2O_3.

Wear studies on zirconia toughened alumina containing different concentrations of ZrO_2 were conducted on a block-on-ring tribometer using a ceramic composite block–steel ring tribological pair under dry and water-lubricated conditions.

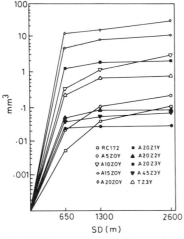

Fig. 10. Wear volume of different ZTA composites as function of sliding distance while rubbing against steel ring in water environment. From ref. [7].

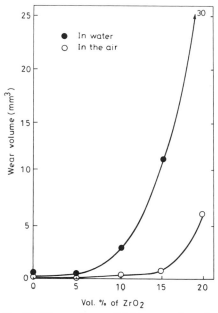

Fig. 11. Wear volume of ZTA composites after 2600 m of sliding in air and water environments. From ref. [7].

The wear for all ZTA composites increased with increasing sliding distance both in dry air and water environments (figs. 9, 10). The wear resistance of ZTA was generally higher in air than in a water environment, except for some composites for which the wear is nearly the same in both environments (fig. 11). The influence of the zirconia content on the mechanical and the wear properties revealed that addition of ZrO_2 to Al_2O_3 matrix reduced the wear resistance (except for some composites in a water environment), but enhances the fracture toughness of ZTA composites (fig. 12).

5.2. Silicon nitride and silicon carbide based ceramic composites

Prakash et al. studied the friction and wear behaviour of hot-pressed Si_3N_4, reinforced with BN (10 vol%) and TiC (10 and 15 vol%), composites when rubbed against bearing steel under reciprocating sliding conditions in air at room temperature [8]. The studies were conducted on a SRV Optimol test machine using a bearing-steel ball–ceramic composite disc tribological pair. The wear resistance of Si_3N_4 ceramic composite significantly improved with the addition of 15 vol% TiC. Also, the wear rate of this composite remained almost constant with increasing sliding distance (fig. 13). The wear rates of mating bearing steel balls were also decreased while rubbing against this ceramic composite material (fig. 14). TiC when sintered with Si_3N_4 formed TiN, as revealed by XRD analysis of the sintered samples. In most cases about 80% of TiC was converted to TiN, which is

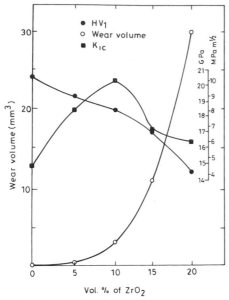

Fig. 12. Effect of ZrO$_2$ content on wear and fracture properties of ZTA. From ref. [7].

comparatively more wear resistant. The friction coefficient did not show any significant change due to the presence of the different kinds of particulate reinforcement in the silicon nitride matrix under high-frequency reciprocating sliding conditions.

Woydt et al. studied the friction and wear behaviour of the self-mated tribological pairs of Si$_3$N$_4$–TiN and SiC–TiC ceramic composites at room temperature, 400°C and 800°C [9].

The Si$_3$N$_4$–TiN ceramic composite contained 30 wt% of TiN, whereas SiC–TiC contained 50 wt% TiC. The mechanical and other properties of the ceramic composites, along with those of monolithic Si$_3$N$_4$ and sintered silicon carbide,

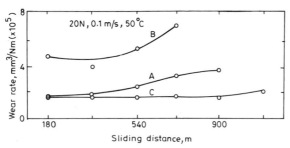

Fig. 13. Wear rate of silicon nitride composites as function of sliding distance in bearing steel ball/ceramic composite disc tribo pair. (A–BN 10 vol%; B–TiC 10 vol%; C–TiC 15 vol%). From ref. [8].

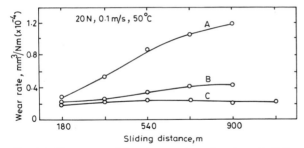

Fig. 14. Wear rate of bearing steel as function of sliding distance in bearing steel ball/ceramic composite disc tribo pair (A–BN 10 vol%; B–TiC 10 vol%; C–TiC 15 vol%). From ref. [8].

which were studied for comparison, are listed in table 7. The friction and wear characteristics of these ceramic composites and monolithic ceramics were studied on a specially built, high-temperature tribometer employing a self-mated disc-on-disc test configuration under dry sliding conditions.

The wear volume of Si_3N_4–TiN ceramic composite material was similar to that of monolithic Si_3N_4 at room temperature (fig. 15), but at 800°C the wear volume of the composite was about one to two orders of magnitude lower than that of Si_3N_4 ceramic (fig. 16), when studied as a function of the sliding velocity. The total wear coefficients indicated a similar trend (fig. 17). Further, it was revealed that reinforcement of Si_3N_4 matrix with TiN did not have any beneficial effect in reducing the friction. On the contrary, the friction coefficients of the Si_3N_4–TiN composite self-mated tribological pair increased for low sliding velocities and at 800°C for medium sliding velocities (fig. 18).

For self-mated tribological pairs, the wear volume of SiC–TiC composite was lower than that of monolithic SiC for all conditions, except at 400°C for a sliding velocity of 3 m/s, for which the wear volume of the rotating disc made of the composite is higher (figs. 19, 20).

The total wear coefficients indicated the same trend (fig. 21). Also, the friction coefficient of the SiC–TiC ceramic composite was almost half that of monolithic SiC at room temperature, but it increased considerably at 400°C (fig. 22).

Analysis of the worn surfaces of the Si_3N_4–TiN and SiC–TiC ceramic composites indicated that the friction and wear process in these ceramic composite materials is controlled by tribochemical-reaction layers, which may probably be composed of SiN_xO_y and TiO_2. For the SiC–TiC composite the concentration of TiO_2 on the wear track was higher than that for the Si_3N_4–TiN composite, because of the higher concentration of TiC (50 wt%) in SiC–TiC as compared to that of TiN (30 wt%) in S_3N_4–TiN. The modification of the friction and wear in SiC–TiC might be the result of TiO_2 and SiC_xO_y layers and the adsorption of water molecules. The higher toughness of the SiC–TiC composite could also lead to lower wear.

B. Prakash

TABLE 7
Properties of Si_3N_4, SiC, Si_3N_4–TiN and SiC–TiC ceramic materials (from ref. [9]).

Material	Young's modulus E (GPa)	Weibull modulus m	Bending strength σ (MPa) 22°C	Toughness K_{IC} (MPa m$^{1/2}$)	Thermal conductivity λ (W/m K)	Thermal expansion α RT–1000°C (10^{-6}/K)	Hardness HV 0.2 * 22°C	Roughness C.L.A. (μm) disc
Si_3N_4 (7 variations)	280–320	9–20	698–1050	5.6–8.2	20–46	3.1–3.7	1470–1740	0.028–0.091
70 Si_3N_4 30 TiN	295	–	735	8.2–9.2	45	5.7	1620	0.038
SSiC	410	–	410	3.2	110	4.0	2840	0.03
50 SiC 50 TiC (HP)	456	20	600	5.4	–	5.7	1700–3100	0.130

* DIN 50133.

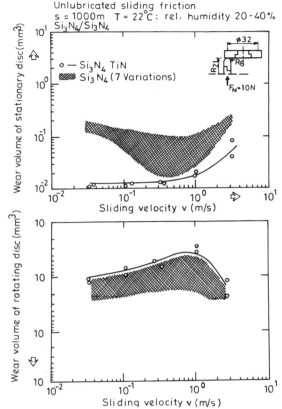

Fig. 15. Wear volume of Si_3N_4–TiN composite and Si_3N_4 monolithic ceramic at room temperature. From ref. [9].

5.3. B_4C–TiB_2–W_2B_5 ceramic composite

The wear of hot isostatically pressed (HIP) B_4C–TiB_2–W_2B_5 ceramic composite has been studied by Lörcher et al. while cutting Al–Si alloy, C15 steel and clinker discs [10]. The influence of the mechanical properties on the wear has also been studied and the wear results have been compared with those of hot-pressed SiC–TiC composite and slip-cast ZrO_2 ceramic.

The wear studies were conducted on a specially designed pin-on-disc machine. The pin samples ($93 \times 4 \times 11$ mm) with one face lapped and polished and having a cutting angle of 5°, were made from B_4C–TiB_2–W_2B_5, SiC–TiC and ZrO_2 ceramic material. The discs were made from AlSi17 alloy, C15 steel and clinker plates. The characteristics of the pin and the disc materials are summarized in table 8.

The tribological tests were carried out under dry conditions in ambient air with increasing loads of 20, 30 and 50 N up to 300 s duration.

When a B_4C–TiB_2–W_2B_5 pin was slid against an Al–Si alloy disc large material transfer took place, leading to the formation of a built up edge on the composite

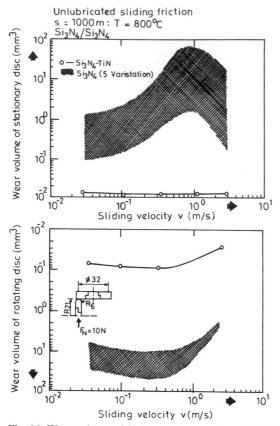

Fig. 16. Wear volume of Si_3N_4–TiN composite and Si_3N_4 monolithic ceramic at 800°C. From ref. [9].

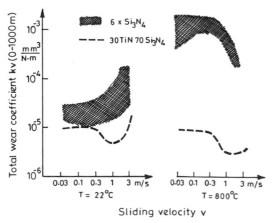

Fig. 17. Total wear coefficient of Si_3N_4–TiN/Si_3N_4–TiN and Si_3N_4/Si_3N_4 at 22°C and 800°C. From ref. [9].

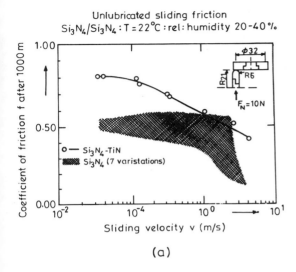

(a)

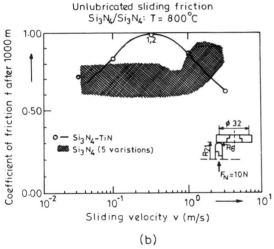

(b)

Fig. 18. Friction coefficient of Si_3N_4-TiN/Si_3N_4-TiN and Si_3N_4/Si_3N_4 at (a) 22°C and (b) 800°C. From ref. [9].

pin surface. This built up edge had a layered structure and indicated the occurrence of adhesive wear. The wear tracks on the Al–Si alloy were very rough, with depths between 5 and 9 mm. For a $B_4C-TiB_2-W_2B_5$ pin sliding against a steel disc relatively less material transfer occurred, and the operating mechanisms of wear were adhesion and abrasion. Also, the wear tracks on the steel discs were relatively less deep (2–4 mm).

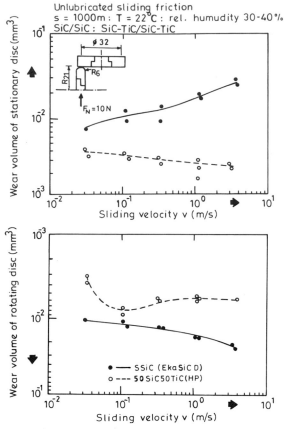

Fig. 19. Wear volume of SiC–TiC composite and SiC monolithic ceramic at room temperature. From ref. [9].

In both cases the damage to the pin was caused by chipping of large material fragments behind the cutting edge of the pins, although relatively smaller when sliding against a steel disc. Chipping of material from the pin surface is brought about by the different thermal expansion coefficients of B_4C and the reinforcing particles of TiB_2 or W_2B_5 causing a mismatch of the internal thermal stresses. The superposing of the sliding shear stresses creates a complicated stress system which leads to chipping of material fragments from the pin surfaces. For a B_4C–TiB_2–W_2B_5 ceramic composite pin sliding against a clinker disc no material transfer to the disc surface occurred, and the worn surface indicated wear as a result of abrasion and surface fatigue. Also, the damage to the pin was lower than that encountered in sliding against AlSi alloy and C15 steel. The damage to the ceramic composite pin decreased in the order Al–Si alloy, C15 steel, clinker disc.

The wear volumes for B_4C–TiB_2–W_2B_5, SiC–TiC and ZrO_2 ceramic material have been plotted against the fracture toughness (fig. 23) and the hardness (fig. 24) in sliding against C15 steel, Al–Si alloy and clinker discs at different loads.

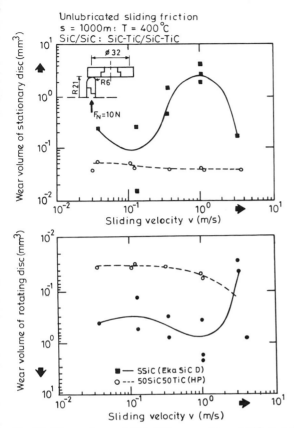

Fig. 20. Wear volume of SiC–TiC composite and SiC monolithic ceramic at 400°C. From ref. [9].

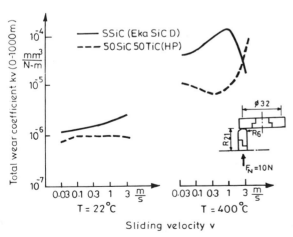

Fig. 21. Total wear coefficients of SiC–TiC/SiC–TiC and SiC/TiC at 22°C and 400°C. From ref. [9].

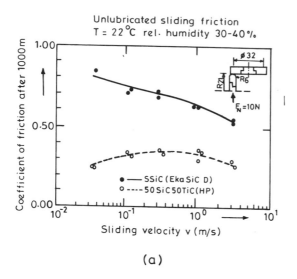

(a)

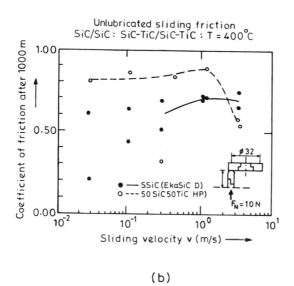

(b)

Fig. 22. Friction coefficient of SiC–TiC/SiC–TiC and SiC/SiC at (a) 22°C and (b) 400°C. From ref. [9].

Relatively lower wear in the ceramic composite materials occurred with increasing fracture toughness at low loads in sliding against C15 and clinker materials, i.e. in situations where no adhesive wear occurred. When adhesive wear occurred, no clear load dependency was observed. In all the ceramic materials studied higher wear was noticed for increasing hardness, independent of the disc material. Further, if an Al–Si alloy disc was cut by the ceramic material, the wear did not

TABLE 8

Characteristics of $B_4C–TiB_2–W_2B_5$ composite and other materials studied (from ref. [10]).

Material	H (GPa)	K_{IC} (MPa m$^{1/2}$)	G_B (MPa)	E (GPa)
Pin samples				
Hot isostatically pressed $B_4C–TiB_2–W_2B_5$	29–25	3.6–5.2	880–960	400–436
Hot-pressed SiC–TiC	24	5.1–5.8	620–730	441–462
Slip-cast ZrO_2	11	7	400	218
Disc samples				
Al–Si alloy (17 wt% Si)	0.9			
Steel C15 (0.15 wt% C, 0.25 wt% Si, 0.4 wt% Mn)	2			
Clinker tiles (with burned glassy surfaces)	6			

show any dependence on the hardness in the case of SiC–TiC and ZrO_2. However, the wear of $B_4C–TiB_2–W_2B_5$ increased at higher loads.

5.4. Alumina–titanium nitride ceramic composite

The tribological properties of alumina–titanium nitride ($Al_2O_3–TiN$) ceramic composite have been studied by Prakash et al. [11].

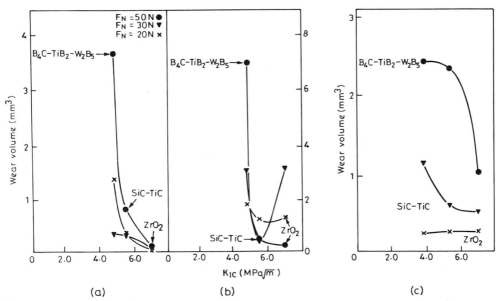

Fig. 23. Relationship between the wear volume and fracture toughness for different ceramic pins in contact with (a) C15 steel, (b) AlSi17 and (c) Clinker as disc material. From ref. [10].

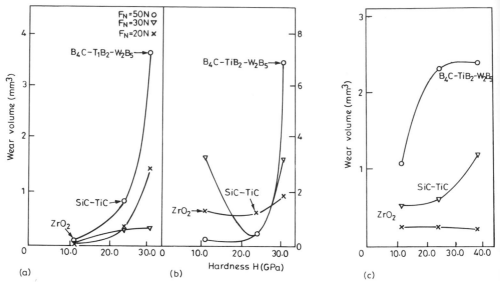

Fig. 24. Relationship between wear volume and hardness of different pins cutting (a) C15 steel (b) AlSi17 and (c) Clinker discs. From ref. [10].

The samples were prepared from an intimately mixed powder of Al_2O_3–TiN, which was obtained by an in situ reaction between TiO_2 and AlN at 1600°C in nitrogen. This mixture contained about 40 mol% TiN. The powder was ball-milled in acetone using alumina balls and then hot-pressed at 1450°C temperature and 30 MPa pressure. The processing details of this ceramic composite have earlier been reported by Mukerji and Biswas [12]. Some of the important properties of Al_2O_3–TiN composite are given in table 9.

The tribological studies were conducted on a reciprocating friction and wear test (SRV Optimol) machine using a bearing-steel ball–ceramic composite disc tribological pair. The friction and wear characteristics of both the steel balls and

TABLE 9
Properties of hot-pressed Al_2O_3–TiN ceramic composite (from ref. [11]).

Property	Value
Flexural strength (MPa)	430–4
(4 pt., 40–20 mm span)	
K_{IC} (indentation) (MPa m$^{1/2}$)	4.7–0.5
Young's modulus (ultrasonic) (GPa)	353
Poisson's ratio (ultrasonic)	0.2
Coeff. of thermal expansion (°C^{-1})	8.78×10^{-6}
(RT–1000°C)	
Hardness (kg/mm^2)	1800
(VMH 1 kg)	
DC resistivity (Ω)	57

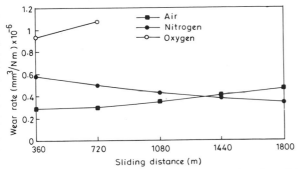

Fig. 25. Wear rate of Al_2O_3–TiN composite as function of sliding distance in different environments while rubbing against bearing steel ball. From ref. [11].

Al_2O_3–TiN composite were studied in air, nitrogen and oxygen environments up to a temperature of 250°C.

The wear rate of the Al_2O_3–TiN composite increased with increasing sliding distance in air and oxygen environments, but decreased in a nitrogen atmosphere (fig. 25). The maximum wear rate of the ceramic composite occurred in an O_2 environment. Also, the wear rate of the steel balls increased with increasing sliding distance in air and oxygen environments, but remained almost constant in nitrogen (fig. 26).

The wear rates of both alumina–titanium nitride composite and bearing steel were lower in a nitrogen environment than in air. No appreciable change in the wear rate of the ceramic composite was observed when the reciprocating sliding velocity was varied from 0.05 to 0.15 m/s. The wear of both the ceramic composite and the steel increased marginally when the temperature was raised up to 250°C

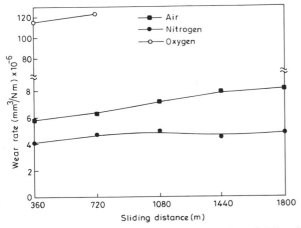

Fig. 26. Wear rate of bearing steel ball as function of sliding distance while rubbing against Al_2O_3–TiN composite disc. From ref. [11].

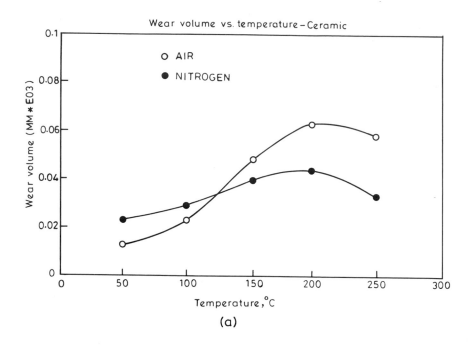

(a)

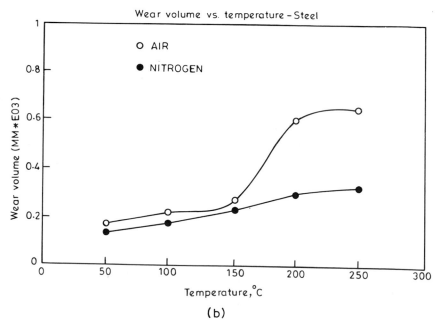

(b)

Fig. 27. Wear rate of (a) Al_2O_3–TiN disc and (b) bearing steel ball as function of temperature in different environments. From ref. [11].

TABLE 10
Properties and compositions of alumina and silicon nitride ceramics impregnated with graphite (from ref. [13]).

	α-Alumina	Silicon nitride
Impurities (wt%)	0.03% Na_2O, 0.01% K_2O 0.02% TiO_2, <0.1% SiO_2	0.4–0.6% Mg, 0.18–0.3% Al, 0.16–0.35% Fe, 0.05% Mn, 0.006–0.03% Ca
Grain size (μm)	3	0.1
Porosity (%)	0.2	≈ 0
Knoop hardness (GPa)	15.2 (at 1000 g)	17.9 (at 500 g)
Elastic modulus (GPa)	345	310
Compressive strength (MPa)	2071	3000
Thermal conductivity (W/m K)	29.4	32
Fracture toughness (MPa m$^{1/2}$)	3.8	5.4
Thermal expansion coefficient (°C^{-1})	6.7×10^{-6}	3.5×10^{-6}

and the wear of both the Al_2O_3–TiN composite and the steel were lower in N_2 than in air (fig. 27). The higher wear in air and oxygen environments is attributed to oxidation since TiN oxidizes above 800°C and forms TiO_2. In nitrogen environment the wear is low since the N_2 protects the surfaces from oxidation.

The friction coefficient for the Al_2O_3–TiN composite and bearing steel tribological pair was about 0.48 and remained the same in air, nitrogen and oxygen environments. Also, almost no change in the friction coefficient occurred when the temperature was raised to 250°C.

5.5. *Solid lubricant filled ceramic composites*

The impregnation of ceramic materials with solid lubricants aims at reducing the friction coefficient of monolithic ceramics and also to avoid the use of liquid lubricants, which create problems at elevated temperatures. The influence of graphite solid lubricant fillers in Si_3N_4 and Al_2O_3 on the friction and wear characteristics of the composite has been studied by Gangopadhyay and Jahanmir [13]. In this study the Si_3N_4–graphite and Al_2O_3–graphite composites were prepared from commercially available high-purity alumina, hot isostatically pressed silicon nitride and nickel chloride ($NiCl_2$) intercalated graphite powder. The properties and compositions of Si_3N_4 and Al_2O_3 ceramics impregnated with graphite are listed in table 10. The $NiCl_2$ intercalated graphite, with an average particle size of about 100 μm in the planar direction, contained 81.5% C, 5.8% Ni, 12.2% Cl and less than 0.5% ash. Graphite pins were made by pressing $NiCl_2$ intercalated powder into a tool-steel die at a pressure of 1.5 GPa.

The Al_2O_3–graphite and Si_3N_4–graphite composites were prepared by drilling a series of 1 mm deep holes of different diameters by ultrasonic machining in the surfaces of cylindrical pins (6.25 mm diameter and 9 mm long) of the two ceramics. These holes in the ceramic pins were drilled in one direction along the diameter of

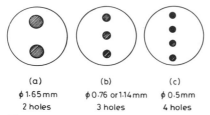

(a)   (b)   (c)
φ 1·65mm   φ 0·76 or 1·14mm   φ 0·5mm
2 holes   3 holes   4 holes

Fig. 28. Schematic of pin specimen geometries used for composites. From ref. [13].

the pins, as shown in fig. 28. These Al_2O_3 and Si_3N_4 pins with holes drilled in their surfaces were inserted in a tool-steel die and $NiCl_2$ intercalated graphite powder was pressed in the holes at 0.42 GPa pressure, and the extra graphite material was removed using 600 grit abrasive paper.

The tribological studies were conducted using a pin-on-ring configuration, employing a ceramic pin (filled with graphite)–steel ring tribological pair, in which the ceramic composite pin was held perpendicular to the axis of the steel ring. The material of the steel ring was AISI 52100 grade steel, with a hardness of 58 RC and a surface roughness of 0.23 μm rms. The ceramic composite pins were further polished using 1 μm diamond paste before the tests. In the friction and wear tests the pins were oriented in such a way that the contact with the steel ring counterface was lined up with rows of graphite regions. For comparison, friction and wear tests were also conducted using Al_2O_3, Si_3N_4 and graphite pins rubbing against the steel ring.

The coefficients of friction for the composite materials, monolithic ceramics and graphite when rubbed against steel, have been plotted as a function of the sliding distance in fig. 29. The friction coefficient of Si_3N_4 reduced from 0.43 to 0.17 when filled with graphite, but no significant friction reduction occurred in Al_2O_3 due to

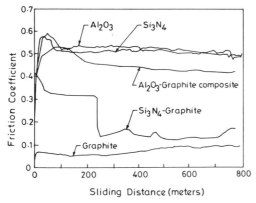

Fig. 29. Coefficient of friction traces as function of sliding distance of alumina, alumina–graphite composite, silicon nitride, silicon nitride–graphite composite and graphite sliding against steel ring. From ref. [13].

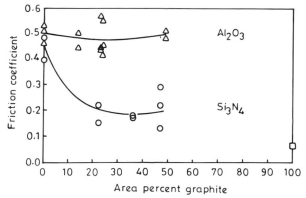

Fig. 30. The effect of graphite addition on the friction coefficient of alumina and silicon nitride. From ref. [13].

graphite addition. Further studies indicated that no significant change in the friction coefficients occurred with increasing graphite content of the surface beyond a particular concentration, as shown in fig. 30. In this figure the friction coefficients are plotted against the area percentage of graphite which was calculated from the amount of graphite regions in contact with the steel counterface.

The wear rate of the Si_3N_4–graphite composite did not change up to 36 area percent graphite and increased beyond this graphite content (fig. 31). The wear rate of Al_2O_3–graphite was lower than that of the Si_3N_4–graphite composite and it did not change for 24 to 30 area percent graphite.

From these observations it was deduced that the application of solid lubricant is controlled by the wear of the ceramic material. The higher wear rate of Si_3N_4 than of Al_2O_3 makes more solid lubricant available on the wear track to form a self-lubricating film on the ceramic composite and steel counterface surfaces. On the other hand, the lower wear rate of Al_2O_3 prevents this and, hence, no friction

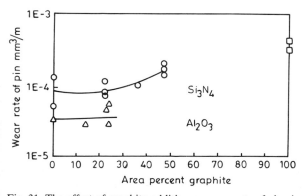

Fig. 31. The effect of graphite addition on wear rate of alumina and silicon nitride. From ref. [13].

reduction is observed. Additionally, the difference in surface chemistry of Al_2O_3 and Si_3N_4 could also influence the formation of transfer films on rubbing surfaces.

6. Abrasive-wear characteristics of ceramic composite materials

Only a few studies have been reported in the literature concerning the abrasive-wear characteristics of ceramic composite materials. These studies pertain to whisker reinforced and other ceramic composite materials.

The abrasive-wear behaviour of SiC-whisker reinforced ZTA (zirconia toughened alumina) has been studied by Bohmer and Almond [5]. The sliding-wear characteristics of whisker reinforced ZTA containing different amounts of SiC whiskers, has been discussed earlier in this chapter.

The abrasive-wear studies were conducted on a pin-on-disc type machine using aqueous slurries of different abrasives. The abrasives used were alumina, silicon carbide and boron carbide, with a nominal particle size of 29 μm. The abrasive–wear resistances of ZTA ceramic composites containing varying amounts of SiC whiskers were compared with that of commercial grade alumina. The abrasive-wear behaviour obtained using SiC abrasive was similar to that obtained with alumina and boron carbide. The results indicated that the abrasive-wear resistance of Al_2O_3 improved both with SiC-whisker and ZrO_2-particle reinforcement (fig. 32). This enhancement is attributed to an increased hardness and fracture toughness of the ceramic composite. The material removal occurred as a result of indentation cracking.

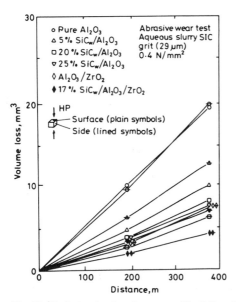

Fig. 32. Variation in abrasive wear with sliding distance for different ceramic-based materials. From ref. [5].

In another study, Holz et al. studied the abrasive wear of several ceramic-matrix composite materials [14]. Abrasive wear of various commercial as well as experimental ceramic and ceramic composite materials was obtained through experiments conducted on a pin-on-disc machine. The pin samples of 3×4.5 mm size were made from ceramic materials and the wear test surfaces (3×4.5 mm) were polished with 15 μm diamond grit. The experiments were carried out at a constant speed of 300 mm/min and a pressure of 2.2 MPa. For all ceramic materials abrasive wear was obtained through abrading against 600 grit (15 μm) and 180 grit (70 μm) SiC paper. For the various ceramic materials studied, their abrasive wear and some of their important properties are listed in table 11 and a comparison of the abrasive wear of some ceramic materials is given in fig. 33. It can be seen that pressure-infiltrated Si in reaction-bonded silicon nitride (RBSN) greatly enhances the abrasive-wear resistance. However, the incorporation of SiC platelets (> 100 μm) reduces the wear resistance of RBSN. Likewise, the abrasive-wear resistance of Al_2O_3 also reduces with the addition of SiC platelets. Further, the addition of SiC whiskers to mullite was found to be very effective in enhancing the abrasive-wear resistance of mullite composites.

7. Friction and wear test techniques

It is now well understood that the friction and wear characteristics of materials are not intrinsic properties but are influenced by various factors related to the tribological system. Some of the important variables which significantly influence the tribological process are:
- operating temperature,
- applied load (direction, amplitude),
- type of motion (rolling/sliding, unidirectional/reciprocating),
- sliding velocity,
- contact configuration,
- contact area,
- sliding duration,
- surface roughness related parameters,
- mode of lubrication,
- type of lubricant,
- operating environment,
- vibration.

In view of the complex nature of the tribological process and the large number of influencing variables, friction and wear evaluation is still considered an art rather than a science. Unlike other mechanical properties, the friction and wear of materials are highly system dependent and there is no simple measurement available which can qualify a material as suitable for a given tribological application. For meaningful tribological evaluation of materials it is essential to select the most appropriate tribological test rig (tribometer) and to choose test parameters which to an extent correspond to the application for which the test material pair is intended.

TABLE 11

Abrasive wear of various ceramic materials tested with the pin-on-disc technique on 600 and 800 grid SiC paper (relevant mechanical data are also given) (From ref. [14]).

Num-ber	Material	Wear, w (μm/m)		Hardness HV30 (GPa)	Toughness K_{IC} (MPa m$^{1/2}$)	Strength [a] (MPa)	Young's modulus (GPa)	Density (g/cm^3)	Remarks
		600	180						
1	RBSN	36	147	4.5	2.5 [b]	200	175	2.58	ACG [c]
2	RBSN + 10 vol% SiC$_{pl}$	57	346	4.5	2.5 [b]	130	–	2.45	ACG, SiC platelets, American Matrix
3	RBSN + 20 vol% SiC$_{pl}$	74	630	4.1	2.0 [b]	65	–	2.41	ACG, SiC platelets, American Matrix
4	RBSN	43	177	5.1	2.7 [d]	227	–	2.83	Annawerk
5	RBSN + Al (99.9%)	15.0	40	11.0	5.0 [b]	478	–	3.08	No. 4 gas-pressure infiltrated at ACG
6	RBSN + Al–Si–Mg alloy	16.7	42	11.5	4.5 [b]	463	–	2.96	
7	RBSN + Si (99.99%)	7.1	26	15.3	3.9 [b]	427	–	2.85	
8	χ-sialon	5.9	14.5	18.0	4.2 [e]	430	325	3.29	MPI [e], 76Si$_3$N$_4$, 15 AlN, 9 Y$_2$O$_3$
9	HP β-sialon	3.8	2.5	15.3	4.1 [e]	410	310 ^{g}	3.16	MPI [f], Si$_{5.3}$Al$_{0.7}$O$_{0.7}$N$_{7.3}$
10	HP Si$_3$N$_4$ + TiC/TiN	13.6	23	16	7.5 [g]	900 ^{g}	300 ^{g}	3.16	Krupp Widia, Widianit, N 1000 [h]
11	HP Si$_3$N$_4$ + TiC/TiN	9.2	20	14	7.0 [g]	1000 ^{g}	280 ^{g}	3.26	Krupp Widia, Widianit, N 2000 [i]
12	Mullite	28	69	8.3	2.3 [d]	200	–	3.14	ACG, HP at 1600°C
13	Mullite + 20 vol% ZrO$_2$	20	39	10.4	3.5 [d]	260	–	3.58	ACG, HP at 1610°C
14	Mullite + 22 vol% ZrO$_2$	20	40	10.8	5.0 [j]	260	–		MPI, reaction sintered
15	HP Mullite + 10 vol% ZrO$_2$ + 20 vol% SiC$_w$	6.4	14.6	12.9	5.4 [d]	580	240	3.42	MPI, HP, SiC whiskers, Tatco

No.	Material								Manufacturer
16	Al_2O_3	6.0	32	14.5	4[g]	330[g]	380[g]	3.76	Feldmuhle, V 38, 96%
17	Al_2O_3	9.8	40	15.1	4[g]	330[g]	380[g]	3.79	Hoechst Ceram Tec, 98%
18	Al_2O_3	7.8	50	16.8	4[g]	325[g]	380[g]	3.94	Friedrichsfeld, 99.8%
19	$Al_2O_3 + ZrO_2$	7.3	34	17.2	5.8[g]	500[g]	380[g]		Feldmühle, BN 70
20	$Al_2O_3 + ZrO_2$	5.9	12.3	17[g]	5.1[g]	800[g]	410[g]	4.16	Krupp Widia, Widalox U
21	HP $Al_2O_3 + ZrO_2$ + TiC/TiN	2.3	8.8	19.3[g]	4.5[g]	620[g]	400[g]	4.25	Krupp Widia, Widalox H
22	Al_2O_3 + 10 vol% SiC_{pl}	8.2	51	12.2	4.1[d]	200	–	3.70	ACG SiC platelets,
23	Al_2O_3 + 15 vol% SiC_{pl}	11.3	56	13.9	5.2[d]	215	–	3.67	ACG American Matrix
24	HP Al_2O_3 + 15 vol% SiC_{pl}	4.4	15	18.2	5.7[d]	290	–	3.75	ACG
25	Al_2O_3 + Al	17	44	10.2	4.5[d]	250	–	3.47	Melt-oxidation derived
26	Mg-PSZ	16	29	11.4	8.1[g]	520[g]	210[g]	5.78	Feldmuhle, ZN 40
27	Mg-PSZ	8.2	32	10.7	8.9[g]	525[g]	210[g]	5.75	Friedrichsfeld
28	HIP 12 Ce-TZP	17.5	32	9.2	9.5[b]	480	205	6.26	ACG
29	HIP 3 Y-TZP	15.5	18	14.0	5.4[b]	1560	210	6.10	ACG
30	HIP 3 Y-TZP + 20 vol% Al_2O_3	10.2	16	14.4	4.2[b]	1280	258	5.51	ACG
31	HIP 3 Y-TZP-Duplex	13.8	18	13.4	6.9[d]	558	247	5.38	ACG

[a] 4-point bending.
[b] Chevron notch.
[c] Prepared at Advanced Ceramics Group TUHH.
[d] Indentation-strength method.
[e] Indentation crack length method.
[f] Max-Planck-Institute of Metals Research Stuttgart FRG.
[g] Data supplied by manufacturer.
[h] Contains about 2 wt% Al_2O_3 and 2 wt% MgO.
[i] Contains about 6 wt% Y_2O_3 and 35% Al_2O_3.
[j] SENB method.

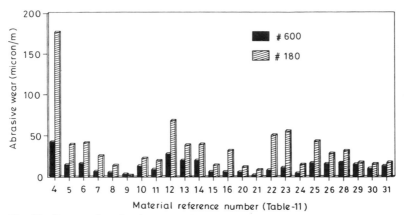

Fig. 33. Comparative abrasive wear of various ceramic materials obtained from pin-on-disc machine tests. From ref. [14].

Various types of tribological test rigs have been developed over the years, ranging from simple laboratory material-screening machines to full-scale simulation test rigs. A compilation of the different friction and wear test devices was brought out by the American Society of Lubrication Engineers (ASLE) in 1976 [15], and several other tribometers have since been reported in the literature. During the discussions on the friction and wear characteristics of ceramic composite materials some of these test rigs have already been mentioned. Some of the more recent test rigs, which enable friction and wear evaluation at high temperatures and in a controlled atmosphere, are briefly described here.

A high-temperature, controllable-environment tribometer as illustrated in fig. 34, has been developed within the framework of EURAM (European Research on Advanced Materials) [16]. Both a ball-on-disc and a disc-on-disc contact configurations can be employed for friction and wear evaluation of materials. A ball and disc tribological pair minimizes the interaction with the wear debris, whereas a disc-on-disc configuration allows for maximum interaction of the wear debris with the tribological process. The sliding velocity can be varied in the range 0.01 to 1.2 m/s. Loads of 5 to 10 N can be applied in the ball-on-disc configuration and 200 to 10 000 N in the disc-to-disc configuration. This tribometer is built to study friction and wear up to a temperature of 1000°C in a controlled atmosphere.

Another high-temperature and controllable-environment tribometer, which employs a pin-on-disc tribological pair, has been developed at VAMAS (Versailles Project on Advanced Materials and Standards), and is shown in fig. 35 [17]. This tribometer also can be adapted to study friction and wear using a disc-on-disc contact configuration. Tribological tests can be conducted up to a temperature of 1000°C in vacuum and 1400°C in a controlled atmosphere. Both tribometers described above provide unidirectional sliding conditions only.

The SRV Optimol tester shown in fig. 36, is a reciprocating friction and wear test machine in which point-, line- and area-contact geometrics can be studied up

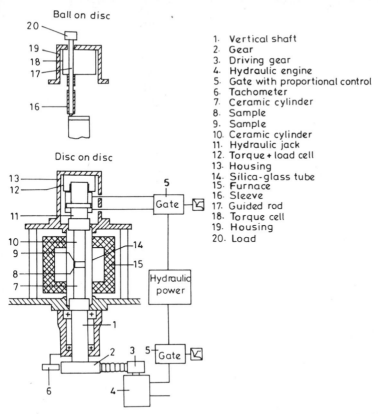

Ball on disc

Disc on disc

1. Vertical shaft
2. Gear
3. Driving gear
4. Hydraulic engine
5. Gate with proportional control
6. Tachometer
7. Ceramic cylinder
8. Sample
9. Sample
10. Ceramic cylinder
11. Hydraulic jack
12. Torque + load cell
13. Housing
14. Silica-glass tube
15. Furnace
16. Sleeve
17. Guided rod
18. Torque cell
19. Housing
20. Load

Fig. 34. Schematic of high-temperature tribometer developed through EURAM programme. From ref. [16].

to a temperature of 900°C. This machine can also be effectively used to study the fretting-wear process since the machine can be run at low reciprocating strokes and high frequencies [18].

Sliney and Dellacorte have recently reported the development of a new friction and wear test machine, on which studies can be carried out under both unidirectional sliding and oscillating sliding conditions up to a temperature of 1200°C in various atmospheres [19]. In addition to the above-mentioned tribometers, several other tribological test machines, which are suitable for friction and wear evaluation of ceramic-based materials, have also been reported [20–22].

In friction and wear testing, apart from the selection of an appropriate tribometer, due consideration must be given to the selection of the test parameters, the design of the experiments and the test procedures in order to obtain meaningful and reliable data. Various useful publications are available in the literature, concerning these aspects [23–26]. Only a few aspects, which are more pertinent to friction and wear tests on ceramic-based materials, are briefly discussed here.

The load and speed are the most important parameters in any friction and wear test, and these should correspond to some extent to the intended application for

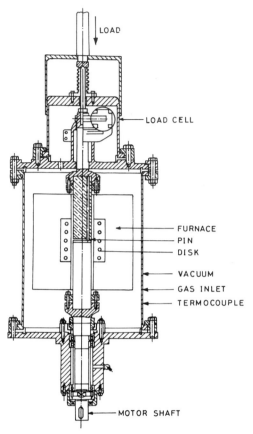

LOAD

LOAD CELL

FURNACE
PIN
DISK

VACUUM
GAS INLET
TERMOCOUPLE

MOTOR SHAFT

Fig. 35. A pin-on-disc tribometer for high temperature and controlled atmosphere developed under VAMAS programme. From ref. [17].

which the materials are being evaluated. The typical load and speed ranges for various moving machine assemblies are shown in fig. 37 [27]. The contact configuration is also extremely important in evaluation of ceramic materials, Here, the contact geometry of the tribological test specimens along with the design of the tribometer will determine the interaction of the wear debris generated during the tests, with the tribological process. For ceramic materials the wear debris generated during sliding is very hard and is likely to have greater influence on the wear processes than that for metallic materials. Likewise, the influence of vibration, both due to internal as well as external sources, may have significant influence on the wear of ceramic-based materials in unlubricated sliding conditions [28,29]. In metallic materials sliding normally takes place under lubricated conditions, where the damping capacity of the lubricant film tends to prevent the transmission of vibration across the sliding contacts. Also, the operating temperature and environment significantly influence the friction and wear in ceramic materials. As such, these parameters must be carefully selected.

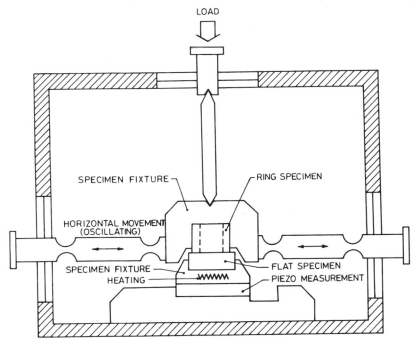

Fig. 36. Schematic of SRV optimal tester.

Due to the use of different tribometers and the many test variables which influence the tribological process, large variations in friction and wear results have been observed. These problems can, however, be overcome if the experimental conditions (particularly the cleaning procedure, etc.) are well defined, as proved through a VAMAS round-robin exercise on Al_2O_3/steel and Al_2O_3/Al_2O_3 tribological pairs [30].

The attempts in friction and wear evaluation tests should be aimed at reproducing the same wear mechanism as is prevalent in practice. This may, however, be difficult for ceramic materials since information is scarce at present. Of course, the increasing emphasis on tribological studies on ceramic-based materials and their gradual introduction in actual engineering applications will help in assimilating knowledge for devising better evaluation techniques and procedures.

8. Tribological applications

Ceramic materials are characterized by their unique physical, mechanical and thermal properties. Some of the tribologically relevant properties include high hardness, good resistance to abrasion/erosion, resistance to wear at elevated temperatures, resistance to chemical attack, high-temperature strength, high rigidity, thermal stability, low coefficient of thermal expansion, thermal-shock resistance, dimensional stability and low density.

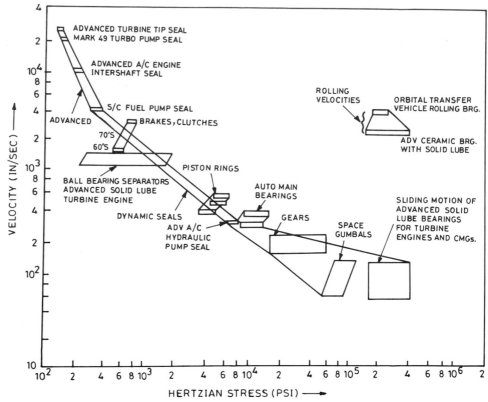

Fig. 37. Load-speed map of various commonly used bearing elements (1000 psi = 6.8948 MPa; in/sec = 2.54×10⁻² m/s). From ref. [27].

In view of these properties, ceramic-based materials are finding increasing usage as tribological materials, particularly in applications involving high temperatures and reactive environments. Some of the typical applications of ceramics as tribological materials are briefly summarized in the following sections.

8.1. Metal cutting and forming applications

One important area of application of ceramic-based materials is that of metal cutting and forming, where both monolithic ceramics and ceramic composites have been effectively used in view of their high hardness and wear resistance at elevated temperatures. In metal cutting the tool-tip temperature can be as high as 1000°C. Monolithic ceramics, such as sialon, Si_3N_4, TiB_2, PSZ, etc., and ceramic composites, such as SiC-whisker reinforced Al_2O_3, B_4C, Al_2O_3–TiC and Si_3N_4-based materials, have been used as cutting tool inserts. The wear resistance of SiC-whisker reinforced Al_2O_3 and Si_3N_4-based ceramic composite materials is almost twice that of Al_2O_3, with Si_3N_4 composite being more suitable at high speeds, as can be

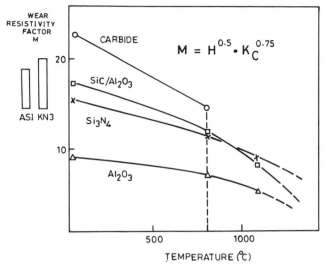

Fig. 38. Wear resistivity factor M for cutting tool materials at elevated temperatures (ASI–SiC-whisker reinforced Al_2O_3; KN_3–Si_3N_4 composite). From ref. [31].

seen from fig. 38 [31]. For metal forming applications PSZ, Si_3N_4 and sialon ceramic materials have been used for extrusion, drawing, bending and tube expanding operations [32].

Ceramic bearings, both complete ceramic and partial ceramic (steel races and ceramic balls/rollers), have been developed for applications involving high temperatures, high abrasion, high erosion and corrosive environments. Further, the low density/weight of ceramic materials has led to the development of bearings having DN (diameter × rpm) values of up to two million [32]. Si_3N_4 ceramic bearings, both in complete and partial ceramic designs, are commercially available [33].

8.2. Heat engine applications

Typical applications of ceramic-based materials in heat engines include ceramic rocker arm pads, cam noses, Si_3N_4 turbocharger rotors and ceramic antifriction bearings [34,35]. The most significant potential of ceramic-based materials is expected in adiabatic and minimum-friction engines. These applications include cylinder liners, valves, valve seat inserts, valve guides, cam shaft lobes, antifriction bearings, turbo-rotors, piston rings, etc. Some ceramic-based materials which can meet the requirements for these components are summarized in table 12 [35].

8.3. Chemical processing industry and other applications

Ceramic-based materials have been used in various applications in chemical processing and other applications where elevated temperatures, abrasion, erosion

TABLE 12
Summary of high-technology ceramics for adiabatic engines (from ref. [35]).

Adiabatic components	Desired characteristics							High-technology ceramics *
	Low friction	Light weight	Insulation	Wear resistance	Heat resistance	Corrosion resistance	Expansion coefficient	
Piston		•	•			•	•	Si_3N_4, PSZ, TTA
Piston ring					•	•	•	SSN, PSZ, Coating
Cylinder liner	•			•		•		Si_3N_4, PSZ, Coating
Prechamber		•	•	•	•	•		PSZ, Si_3N_4
Valve			•	•	•	•		SSN, PSZ, Composite
Valve seat insert			•	•	•	•		PSZ, SSN
Valve guides	•		•		•	•		PSZ, SSN, SiC
Exhaust/intake ports			•		•	•		ZrO_2, Si_3N_4, TiO_2 Al_2O_3
Manifolds			•		•	•		ZrO_2, Si_3N_4, TiO_2 Al_2O_3
Tappets		•		•		•		PSZ, SiC, Si_3N_4
Mechanical seals	•			•		•		SiC, Si_3N_4, PSZ
Turbocharger								
Turbine rotor		•	•		•	•	•	Si_3N_4, SiC
Turbine housing			•			•	•	LAS
Heat shield			•		•	•	•	ZrO_2, LAS
Ceramic bearings		•		•	•	•	•	SSN

* SSN: sintered silicon nitride, PSZ: partially stabilized zirconia, LAS: lithium alumina silicate, TTA: transformation toughened alumina.

and corrosive environmental conditions are commonly encountered. PSZ, Al_2O_3 and SiC ceramic balls are used in check valves in the chemical process industry. Nozzles for rockets, spray dryers and sand-blasting machines have been made from SiC, mainly to provide resistance against erosion/abrasion [32]. WC sleeve and Al_2O_3 bearings are used in ash scrappers of coal-fired boilers. SiC and Si_3N_4 sleeve bearings have been successfully used for river-water pumps to provide protection against corrosion and abrasion due to muddy water [22].

The above-mentioned applications primarily indicate the use of monolithic ceramics and only in few cases ceramic composite materials have been employed. This is due to the fact that ceramic composites have been developed only recently. However, in view of the improvements in mechanical reliability and the enhancement of several other properties, including the friction and wear characteristics, through reinforcement of the ceramic matrix, ceramic composite materials have great potential as tribological materials in various applications, such as metal cutting and forming, aerospace and military applications, and adiabatic and minimum-friction engines.

9. Concluding remarks and future research directions

The friction and wear characteristics of various ceramic composite materials, both in sliding as well as abrasion conditions, have been discussed in this chapter. It can be noticed that reinforcement of the ceramic matrix with whiskers, fibres or particulates improves the wear resistance of the composite significantly, but not much change occurs in the friction coefficients, except in the cases where solid lubricant fillers have been used. Also, the friction coefficients observed are rather high. It is further seen that the environment also has significant influence on the friction and wear of ceramic composite materials. Most of the friction and wear studies on ceramic composite materials have been conducted on self-mated tribological pairs for unidirectional sliding without lubrication. Fretting and abrasive wear of ceramic composites have also received very little attention and these wear mechanisms are not yet very clear. In view of these observations it is felt that future research should focus on following aspects:

– Lubricated friction and wear studies, with a view to understand the boundary-lubrication mechanism of ceramic composites and to identify the most compatible lubricant–additive combination.

– The influence of solid lubricants on the friction and wear behaviour, with a view to eventually develop self-lubricating ceramic composites incorporating solid lubricants.

– Abrasive-wear studies on important ceramic composite systems, both under two-body and three-body abrasion conditions.

– Understanding the mechanisms of wear under conditions of dry sliding, abrasion and fretting.

– The influence of processing techniques, flaws and other material properties on the tribological performance.

– The influence of the environment on the friction and wear behaviour.
– The development of procedures for running-in of ceramic composites and understanding their seizure or scuffing behaviour.

Acknowledgements

I would like to gratefully acknowledge the researchers who have immensely contributed to this field and without reference to their work this article would not have been possible. I am particularly indebted to Profs. G. Petzow, K. Friedrich, Drs. M. Woydt, A. Skopp, K.H-. Habig, D. Holz, R. Janssen, N. Claussen, R. Lörcher, R. Telle, M. Bohmer, Germany, Drs. H. Ishigaki, R. Nagata, M. Iwasa, N. Tamari, I. Kondo, Japan, Drs. P. Boch, J. Glandus, F. Platon, R. Trabelsi, D. Treheux, G. Orange, F. Fantozzi, P. Homerin, F. Thevnot, France, Drs. S. Jahanmir, M.N. Gardos, C.S. Yust, J.M. Leitnaker, C.E. Devore, L.F. Allard, A. Gangopadhyay, R. Kamo, W. Bryzik, U.S.A., Drs. C. Palmonari, L. Esposito, E.H. Toscano, Italy, Drs. T.B. Troczynski, D. Ghosh, S. DasGupta, J.K. Jacobs, Canada, Dr. E.A. Almond, U.K., Drs. J. Mukerji, S.K. Biswas, P.K. Das, M.F. Wani, S. Kalia, India, for kindly acceptance the use of some of their earlier published results in this article.

References

[1] B. Cales, Ceramic matrix composites, in: 2nd European Symp. on Engineering Ceramics, ed. F.L. Riley (Elsevier Applied Science Publishers, Barking, UK, 1987) pp. 171–202.

[2] C.S. Yust, J.M. Leitnaker and C.E. Devore, Wear of alumina–silicon carbide whisker composite, Wear 122 (1988) 151–164.

[3] C.S. Yust and L.F. Allard, Wear characteristics of an alumina–silicon carbide whisker composite at temperatures to 800°C in air, STLE Tribol. Trans. 32(3) (1989) 331–338.

[4] N. Claussen and G. Petzow, Whisker reinforced oxide ceramics, J. Physique C 1 (1986) 693–702.

[5] M. Bohmer and E.A. Almond, Mechanical properties and wear resistance of a whisker-reinforced zirconia-toughened alumina, Mater. Sci. & Eng. A 105/106 (1988) 105–116.

[6] H. Ishigaki, R. Nagata, M. Iwasa, N. Tamari and F. Kondo, Tribological properties of SiC whisker containing silicon nitride composites, ASME Trans. J. Tribol. 10(3) (1988) 434–438.

[7] R. Trabelsi, D. Treheux, G. Orange, G. Fantozzi, P. Homerin and F. Thevenot, Relationship between mechanical properties and wear resistance of alumina–zirconia ceramic composites, STLE Tribol. Trans. 32(1) (1989) 77–84.

[8] B. Prakash, M.F. Wani, J. Mukerji and P.K. Das, Tribological studies of hot pressed silicon nitride and its composites, in: Proc. Japan Int. Tribology Conference, Nagoya (1990) pp. 1383–1388.

[9] M. Woydt, A. Skopp and K.-H. Habig, Dry friction and wear of self-mated sliding couples of SiC–TiC and Si_3N_4–TiN, in: Proc. ASME Conf. on Wear of Materials (1991) pp. 393–404.

[10] R. Lörcher, R. Telle and G. Petzow, Influence of mechanical and tribochemical properties on the wear behavior of B_4C–TiB_2–W_2B_5 composites, Mater. Sci. & Eng. A 105/106 (1988) 117–123.

[11] B. Prakash, J. Mukerji and S. Kalia, Tribological properties of Al_2O_3–TiN composite, to be published.

[12] J. Mukerji and S.K. Biswas, Synthesis, properties and oxidation of Al_2O_3–TiN composites. Am. Ceram. Soc. 73(1) (1990) 142–145.

[13] A. Gangopadhyay and S. Jahanmir, Friction and wear characteristics of silicon nitride–graphite and alumina–graphite composites, STLE Tribol. Trans. 34(2) (1991) 257–265.

[14] D. Holz, R. Janssen, K. Friedrich and N. Claussen, Abrasive wear of ceramic-matrix composites, J. Eur. Ceram. Soc. 5 (1989) 229–232.

[15] Friction and Wear Devices, American Society of Lubrication Engineers (now Society of Tribologists and Lubrication Engineering) (1976).

[16] P. Boch, J. Gilandus and S. Platon, Pre-standardization studies as means of helping the development of engineering ceramics, 2nd Eur. Symp. on Engineering Ceramics, London (1987) pp. 7–30.

[17] C. Palmonari, L. Esposito and E.H. Toscano, Tribological tests in ceramic materials development, a pin on disc tribometer for high temperature and controlled atmosphere, in: Proc. Int. Conf. on Evolution of Advanced Materials, Italy (ASM Europe, 1989) pp. 633–638.

[18] SRV Material Test System, Product bulletin (Optimol Instruments GmbH., Germany).

[19] H.E. Sliney and C. Dellacorte, A new test machine for measuring friction and wear in controlled atmosphere to 1200°C, STLE J. Lubr. Eng. (1991) 314–319.

[20] J. Lamon, J. Danape and D. Broussaud, Evaluation of ceramics for wear components in automotive engines, in: Proc. Int. Symp. on Ceramic Components for Engines, Hakone, Japan (1983) pp. 146–158.

[21] M. Woydt and K.-H. Habig, High temperature tribology of ceramics, Tribol. Int. 2(2) (1989) 75–88.

[22] S. Asanabe, Applications of ceramics for tribological components, Tribol. Int. 20(6) (1987) 355–364.

[23] W.O. Winer, The evaluation of materials for tribological applications, Mater. Res. Soc. Symp. Proc. 140 (1989) 387–395.

[24] T.S. Eyre, The testing and evaluation of materials in tribology, IXth Nat. Conf. of Industrial Tribology, Bangalore, India, Key note Papers (1991) 49–91.

[25] R.C. Erickson and W.A. Glaeser, Considerations for the design of accelerated wear experiments, ASM Metals Congress on Process Control and Reliability, Ontario, Canada (1985) Paper 8514–004.

[26] ASTM, Selection and Use of Wear Tests for Ceramics: ASTM STP 946 and ASTM STP 1010.

[27] M.N. Gardos, On ceramic tribology, STLE J. Lubr. Eng. (1988) 410–507.

[28] M.G. Gee and E.A. Almond, Effects from vibrations in wear testing of ceramics, Mater. Sci. & Technol. 4 (1988) 655–662.

[29] M.G. Gee and E.A. Almond, Effects of test variables in wear testing of ceramics, Mater. Sci. & Technol. 4 (1988) 877–884.

[30] H. Czichos, S. Becker and J. Lexow, Multilaboratory tribo-testing results from the Versailles advanced Materials and Standards programme on wear test methods, Wear 114 (1987) 109–130.

[31] T.B. Troczynski, D. Gosh, S. Das Gupta and J.K. Jacobs, Advanced ceramic materials for metal cutting, in: Proc. Int. Symp. on Advanced Structural Materials, Montreal, Canada (1988) pp. 157–168.

[32] S. Jahanmir, Tribological applications for advanced ceramics, Mater. Res. Soc. Symp. Proc. 140 (1989) 285–291.

[33] Koyo Seiko, Fine ceramic bearings, Cat. No. 118E-3 (Koyo Seiko Co., Japan).

[34] H. Kita, M. Arita and M. Kohzaki, Current status of development of ceramic tribo-elements in Japanese automotive industries, programme for the International Forum on Tribology of Advanced Ceramics, Tokyo, Japan (1990) p. 8.

[35] R. Kamo and W. Bryzik, Ceramics for adiabatic turbocompound diesel engines, in: Proc. Int. Symp. on Ceramic Components for Engine, Japan (1983) pp. 59–99.

Advances in Composite Tribology
edited by K. Friedrich

Chapter 12

Erosion of Metal Matrix Composites

IRWIN G. GREENFIELD

Department of Mechanical Engineering, University of Delaware, Newark, DE 19716, USA

Contents

Abstract

Erosive wear of metal matrix composites is dependent on the properties of each of the constituents of the target, the properties and kinetic energy of the erodent, the size of the microstructure relative to the impact-affected zones, and the angle of impact. With additional dependent variables associated with the above factors, the mechanisms of erosion of metal matrix composites hinge on the material systems used and the erosion environment. An understanding of the rules governing metal matrix composite erosion can be developed by considering elements of the process, and where possible, contrasting these to some of the better understood erosion mechanisms of homogeneous materials.

A simple reinforcement configuration in a composite with continuous, parallel fibres is suited for an initial analysis. For single or multiple impacts perpendicular to the direction of the fibres, the resulting crater's periphery is elliptical with the smaller axis corresponding to the direction of the fibres, i.e. the path of the greatest deformation constraint. If the impact particle trajectory is parallel to the fibres, the deformation penetration into the target surface is impeded more than lateral flow. Since the anisotropy of this type of composite is greater than that of a metal target, the flow asymmetry is more severe. For repeated impacts at one location, the kinetic energy lost in the formation of the crater by plastic deformation is affected by fibre reinforcement. As a result of local fibre fragmentation and work hardening during the repeated impacts the energy lost by plastic deformation approaches the value for the matrix for all fibre volume fractions and orientations.

Particular erosion mechanisms depend on the relative size of an impact-produced crater and the damaged region in the composite's surface. The probability of contact of either phase or both simultaneously depends on the geometric features, spacing of phases and relative size of the microstructure of the composite. When the impact crater is small with respect to the dimensions of the constituents in the target face, the initial erosion depends on a combination of the erosion rates of the constituents, however, the surface does not remain plane, and the modified impact angle perturbs the erosion conditions. If the deformation crater is much larger than the size and spacing of the constituents in the microstructure, the combined phases contribute to the erosion properties, and these are independent of the location of the impact. The process, however, is dependent on the angle of the stream of erosive particles.

Erodent particles are not rigid and consequently deform and fragment. Short-fibre α-Al_2O_3 in a 6061 aluminum alloy, eroded by either SiC or Al_2O_3, deteriorates at a higher rate with the harder and tougher SiC. Other investigated systems are considered in this review. Because of the many variables associated with erosion of MMCs, improved definition of experimental techniques and analytical characterization of composites are necessary.

1. Introduction

Erosive wear of ductile and brittle materials has been categorized, modelled, and in some instances specific physical mechanisms are described in detail; see the review by Ruff and Wiederhorn [1]. Experimentally determinable material parameters, such as hardness, energy of plastic deformation, toughness, changes in the affected microstructure, deformation, and fracture mechanisms, have been correlated with the rate of erosion wear. Each of these material parameters are non-linear with respect to the deformation generated in the multiple impact process. Moreover, the determination of the above parameters is generally based on bulk properties, for which the material is assumed to be a continuum and the size and contribution of the microstructure not of significance. For erosion, models based on bulk measured properties have described many aspects of multiple, small, hard particle impact induced material separation for different material types [1]. Usually, the non-uniformities of the microstructure are averaged, and models based on the assumption of homogeneity of the target material are frequently adequate.

A distinctive feature between the erosion of brittle and ductile targets is the dependence of the erosion rate on the incident angle. For a unidirectional stream of particles, the maximum erosion rate for brittle materials is perpendicular to the target surface, and for ductile materials the maximum erosion rate occurs at a stream angle between 18° and 20°. See ref. [1] for a general review of erosion. Because many aspects of the process are complex, both empirical and phenomenological descriptions are found. A critical analysis and discussion of a many erosion mechanisms of ductile metals was presented by Bahadur [2]; in summary, the models discussed were cutting [3], cutting plus deformation [4,5] dynamic indentation [6], delamination [7], low cycle fatigue [8], and shear deformation induced by localized heating [8–11].

The primary mechanisms of deformation and material removal are different with brittle targets; collisions by softer particles create regions with circumferential cracks about the areas of impact [12]; however, striking by harder particles creates zones of plastic deformation adjacent to the impact interface and the development of tensile residual stress fields. These stress distributions can be relieved by the creation of radial, conical, and lateral cracks [13,14]. In a brittle target the nucleated cracks generally grow radially [15] outward and finally toward the surface, creating disc-like fragments.

Metal matrix composites composed of a ductile matrix and hard, brittle reinforcing phases, generally display greater sliding wear resistance than the same alloy without the reinforcements [16]. Since many mechanical properties are enhanced by adding a hard wear-resistant phase to an alloy, a decrease in the erosion rate was expected with composites of the alloy; however, the reverse effect is found [17,18]. This chapter presents some experimental observation, modelling, and limitations to the physical representation of erosion of metal matrix composites.

2. Particle impact and erosion

Although erosion by a mass of solid-particle collisions with a target can affect large areas, the erosion process is composed of a multiple of individual impacts, each altering a local region. This altered zone is dependent on the size, shape, material, and kinetic energy of the colliding particle, the time-dependent pressure distribution, the dynamic triaxial stresses, the residual stresses, and the interaction of fracture modes of the target and the particle [14,19,20]. For a ductile, isotropic target material, where the effect of only one isolated particle is considered, the form and size of the crater depend on the incident angle [21], α, of the colliding particle. At 90°, the shape of the indentation crater for a spherical impacting surface is nearly a spherical cap (dependent on the elastic and plastic recovery of the deformed volume) [22], or nearly a mirror image of the impact region of an irregular particle [19]. For an impact of a rigid particle on a plastically deformable work-hardening target material, the rearrangement of material and the stress distributions can be determined [23]. For spherical particles striking the target at angles less than 90°, the imprint on the surface has an elliptical form [6], frequently with a lip of target material at the trailing end of the crater [24,25]. Irregular particles with protrusions, such as edges and corners, tend to produce more prominent lips than generated by smooth particles. The mechanisms leading to the rapid deformation are not only a function of α, but also of the target material properties associated with plastic deformation, the fracture mechanisms, and thermo-physical properties such as the melting temperature, enthalpy of melting, and thermal conductivity [9].

I.G. Greenfield

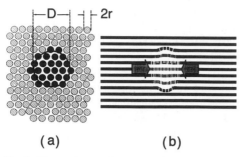

(a)   **(b)**

Fig. 1. Crater shape of a perpendicular impact on a surface of a parallel fibre composite: (a) circular outline; fibres in the direction of the impact, (b) elliptical outline with fragmented fibres; fibres perpendicular to the direction of the impact impede the flow (arrows).

The calculated stress distribution of a single impact on a elastic transverse-isotopic composite shows the influence of the imposed anisotropy by a geometrically oriented reinforcing phase [14]. Furthermore, with metal matrix composites that deform plastically, an anisotropic distortion of the crater takes place. In a composite with parallel continuous fibres normal to the target surface that is struck by a spherical particle (in the direction parallel to the fibres), the indentation is circular, see fig. 1a. Here D, the diameter of the impact impression, is somewhat larger than $2r$, the average diameter of the fibres; consequently, several fibres are contacted (indicated by filled circles). Because of the symmetry of the properties, the radial elastic and plastic deformation are the same in the target surface, and the crater has a circular border. Some of the fibres are not fully contacted during the impact in this illustration, and, thus, the outline of the crater deviates somewhat from a circle. Also, the penetration of the projectile into the composite is impeded in the stronger direction by the longitudinal fibres. When the impact is perpendicular to the fibre direction, fig. 1b, an elliptical crater is formed; the constraint in the fibre direction is depicted by the arrows. Similar effects can also be observed with impacts of anisotropic single crystals [26,27], because material flow is dependent on direction. In a composite that contains fibres, the plastic deformation of the matrix can bend some fibres to a curvature greater than the critical curvature for fracture; these fibres will fragment [28]. This type of multiple breaking of fibres in the elliptical crater is illustrated in fig. 1b. In a geometrically complex system of a composite composed of a more random array of short-fibre δ-Al$_2$O$_3$ (Saffil) reinforced Al–4 wt% Cu, erosion with angular Al$_2$O$_3$ particles (Norton Alundum, E-17) produced irregular craters. The bent fibres resulting from the non-uniform plastic flow of the metal matrix at the periphery of the impact crater, are fragmented [18]. The way the surface layers of composites fail at an impact is also dependent on the fibre arrangement: crack propagation in high-stress regions may be blocked by the constraints imposed by the high-strength fibres [14]. Similar anisotropic features observed in single impacts on a transverse-isotopic composite also appear after multiple impacts on a target. The SEM photomicrographs in fig. 2 illustrate the fragmentation of some reinforcing fibres that occurred after repeated impact with a spherical steel surface of aluminum matrix

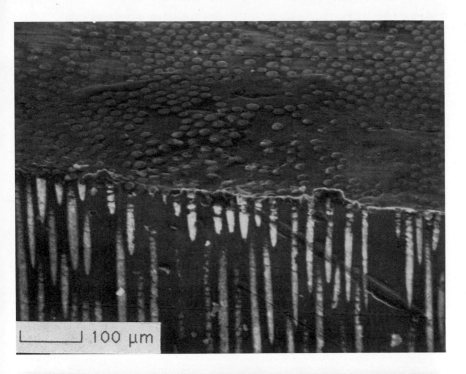

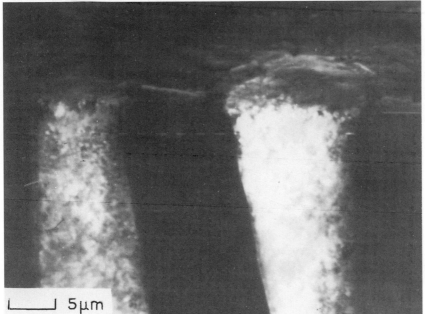

Fig. 2. Impact surface and cross section of a bisected scrater after ten repeated perpendicular impacts in the fibre direction; α-Al_2O_3/Al, $V_f = 0.35$: (upper) depression on the surface and fractured fibres below the crater, (lower) details of peened appearance of crushed fibres near the surface of the crater and fragmented fibres beneath the crater.

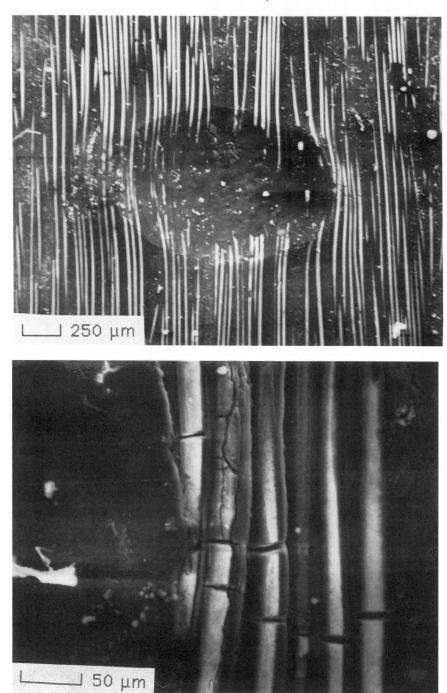

Fig. 3. Crater after 5000 impacts perpendicular to the fibre direction; α-Al_2O_3/Al, $V_f = 0.35$: (upper) elliptical shape because of constraint on flow in the fibre direction, and fibres at the crater base are heavily fragmented; (lower) detail of fractured fibres near the major axis where bending is maximum.

composite $V_{0.35}^{90}$. The volume fraction of fibres for a given orientation is indicated by V_f^ϕ, where V_f represents the volume fraction of the reinforcing fibres and the superscript ϕ is the angle between the unidirectional fibres and the surface of the target. These SEMs are shown with the surface of the specimen at an angle of about 30° with the electron beam, exposing a cut, polished, and etched cross section bisecting the crater on the target surface. For example, fig. 2a is a circular indentation resulting from ten impacts on the $V_{0.35}^{90}$ composite. The lower half of the photomicrograph is the bisecting cross section and the upper half is the target surface with the indentation. In addition to the plastic deformation of the ductile matrix, the fibres are also displaced. Since the fibres are brittle, they were slightly bent, heavily crushed, appearing in a peened shape near the surface, and fragmented by the impact. Concerning the peened appearance, transverse and shear fracture are visible at a higher magnification, see fig. 2b. For impacts on the $V_{0.35}^0$ specimen, the crater formed is elliptical due to the anisotropy of the composite. See fig. 3a. Fragmentation from bending of fibres near and on the major axis of the asymmetrical crater is shown in fig. 3b. These observations show that in a continuous fibre composite the amount of flow of a ductile matrix is dependent on the fibre orientation and the fractures of the fibres are dependent on the local strain in the region of the impact.

For repeated impacts at the same target location, changes of the microstructure from elastic, anelastic, and plastic responses of target material [29–31] depend on several non-linear processes, such as strain hardening, accumulated adiabatic shear [9], local melting [32], and decreasing strain rates with increasing strain of penetration of the impacting particle [27]. The combination of plastic deformation of the matrix and the fragmentation of the brittle reinforcing phase causes a thin surface layer of pulverized brittle material and matrix to be produced [33]. Even in a ductile target the confined flow associated with this mode of progressive deformation from repeated impacts leads to a residual stress field [22]. In a composite, the residual stress distribution is constrained and shaped by the geometric arrangement of the higher-modulus reinforcing phase [14].

Because of the simultaneous changes in mechanical properties from work hardening during the duration of the impact contact, less kinetic energy is expended by non-elastic deformation of the target. The increased availability of the target's elasticity boosts the rebound kinetic energy of the projectile [27]. The steady decrease of energy transferred to the target, evaluated by measuring the increase of the coefficient of restitution during led, repeated impacts, is shown in fig. 4. The coefficient of restitution, e, is the ratio of the square root of the rebound kinetic energy (the instant the projectile separates from the base of the crater) to the impact kinetic energy (at the initial contact with the target surface). That is, $e = (KE_R/KE_I)^{1/2} = v_R/v_I$, where v_R/v_I is the ratio of the rebound velocity to the impact velocity. At repeated low-velocity impact with a steel, spherical surface, where $v_I = 0.22$ m/s, for unidirectional MMC (0.35 or 0.55 volume fraction of α-Al_2O_3 fibres in an aluminum matrix) e increases to a steady-state value in the initial two to seven impacts because of the rapid progressive strain hardening of the impact region in the target. This trend is shown

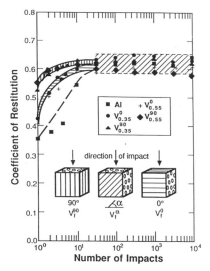

Fig. 4. The change in e as a function of repeated impacts and fibre orientation.

in fig. 4. For the aluminum matrix only, 20 to 30 impacts are necessary to reach the same steady-state value [28]. For a single impact on only the matrix metal, the largest portion of the energy is lost in plastic deformation of the target area, and a small fraction of energy is available for the elastic rebound; consequently, e is small. With $V_{0.55}^{90}$ (fibres perpendicular to the surface), the major portion of the elastic energy is returned to the erodent projectile. When the fibres are parallel to the surface ($V_{0.55}^{0}$), the plastic deformation in the first few impacts is greater than for $V_{0.55}^{90}$, but is less than for only the matrix. Nevertheless, in all cases, as the number of impacts increases the coefficients of restitution of the composites and of the neat matrix reach a steady-state range of e. The steady-state magnitude of e shows that the rebound energy is primarily dependent on the saturated value of the strain hardening of the matrix. The fragmented portions of fibres contribute insignificantly to the energy of the rebound at this maximum strain hardened condition. Thus, when the reinforcing fibres are shattered by erodents, the strained-hardened metal-matrix properties dominate the elastic rebound.

    For arrangements other than continuous parallel fibres, the local stress distributions and surface fracture patterns are more complex. Research on erosion of metallurgically produced two-phase materials and particulate composites have many similarities. For example, the rate of erosion is observed to be greater in alloys containing additional non-coherent phases with hardnesses different from the ductile matrix metal [29,34]. This effect is produced in alloys in which the microstructure has been altered and hardened by thermal treatment [35]. Similarly, regardless of the shape and arrangement of the of reinforcement phase in composites – unidirectional fibres, discontinuous fibres or particles – the rate of erosion is greater than if only the ductile matrix was eroded [18,36]. Moreover, it has been observed that particulate and fibre composites with the same reinforcing material (2014 aluminum with Al_2O_3 particles and Al–4 wt% Cu with Al_2O_3 fibres) have similar erosion rates [37]. In spite of efforts to derive universal physical or empirical representations, the models have only given some general relationships for particular groups of experiments for select combinations of materials.

Hovis et al. [38] modelled several multiphase systems and demonstrated that Kruschov's Rule (inverse rule of mixtures) for several of them (Al–12 wt% Si and Al–20 wt% Si) fits the experimental results well when the erodent particles are small with respect to the size of the microstructure. They showed that

$$\frac{1}{W} = \frac{m_1}{W_1} + \frac{m_2}{W_2}.$$

The average erosion rate in loss of mass of target sample per mass of erodent particles impacted on the target of a multiple component microstructure is W, where m_i and W_i are the mass fractions and the erosion rates (mass of component/mass of erodent of the ith component). For a variety of useful metal matrix composites, additional factors such as initial geometric arrangements of the reinforcing phases and kinetics of the development of a fragmented surface-layer, are essential for modelling.

An aggregate of ductile and brittle phases exhibits mechanical properties unique for the combination, depending on the properties of each phase, the geometric arrangements, and the relative sizes of the constituents. In the interaction of the target material with the erodent the mechanical properties, shape, and size of the erodent must be taken into account also. The erosion rate, W, will depend on the relative number of strikes on the ductile phase and on the brittle phase. Thus, W is a function of the impact diameters and the microstructural spacings. Thus, $W = f(D_{IM}, D_{IB}, s, r)$, if the impact-affected areas are described by parameters such as the average contact diameter of the impacts on the ductile phase (D_{IM}), the average contact diameter of the impacts on the brittle phase (D_{IB}), the spacing between the brittle phase regions (s), and the average diameter of the brittle phase regions ($2r$). In addition, the impact diameters depend on the character of the incoming particles and the mechanical reactions of the phases to the impacts.

Consider the relative-size effects for a particular geometric array in a fibre composite consisting of a brittle phase in an ideal hexagonal arrangement of equal-diameter parallel fibres intersecting the target surface at right angles. The distance relationship between the fibres [39] for two geometrical arrangements and a fibre volume fraction V_f is given by $s = 2r[\sqrt{g/V_f} - 1]$, where g is a numerical factor related to the distribution geometry; for a hexagonal array $g = \pi/2\sqrt{3}$, see fig. 5. Assume that a sphereical particle striking a target at 90° to the target, forms a spherical cap-shaped crater in the matrix with diameter D_{IM} in the plane of the target. In the hard phase, if the size of the impact is small with respect to its area, the stress distribution is described by Hertz contact loading [13] with a Hertz contact diameter D_{IB}.

For a general consideration, take both D_{IM} and D_{IB} about equal to an average

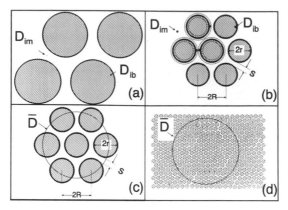

Fig. 5. Relative size of the impact diameter: (a) D_{IM} and $D_{IB} \lll$ microstructure, (b) D_{IM} and D_{IB} or $\bar{D} <$ microstructure, (c) D_{IM} and D_{IB} or $\bar{D} \approx$ microstructure, and (d) $\bar{D} \ggg$ microstructure.

diameter $\bar{D}$. If $R_{I/\mu}$ is the ratio of the impact diameter to a parameter that is characteristic of the size of the microstructure, then when $R_{I/\mu}$ is very small $\bar{D} \lll s$ and $\bar{D} \lll 2r$. See fig. 5a. The contribution of each of the phases in the target to the erosion is proportional to the erosion rate of each component, W_i, times the volume fraction of that phase. The probability of impact with the fibres and with the matrix is p_f and p_M, respectively, where $p_f + p_M \simeq 1$. The probabilities p_f and p_M for unidirectional, parallel fibre composites are merely proportional to the cross section of the microstructure on the target surface. Multiple surface impacts produce erosion almost exclusively in one or the other phase, with a minor effect from the interfacial region for a very small $R_{I/\mu}$. At the other extreme, when $R_{I/\mu}$ is large, as shown in fig. 5d, the microstructure is small with respect to the impacted region; then for an aggregate containing brittle reinforcements and matrix both are affected simultaneously.

Between the extremes, when $\bar{D}$ has a size that it intersects regularly interfacial boundaries but is also frequently contained in the exposed individual phases, the additional mechanisms involving edge fracture can significantly contribute to degradation of the surface. Figure 5b shows the condition where $\bar{D}$ is large enough to strike near the edge of the fibres. Because of the lack of sufficient transverse constraint at the periphery of the fibre, fracture of the weakly supported edges of the fibre can occur. In contrast, if the major contact of a particle is in the adjacent matrix, the flow of the plastically deformed metal is limited by the stiff fibre and the matrix will strain harden with less impacts. Also some matrix material can be extruded and form lip-like structures that would be separated by action of further impacts.

Figure 5c shows a surface-contact region with a larger size of the impact area, such that several fibres and some matrix are simultaneously involved. In addition to the mechanisms involved in the example in fig. 5b, some fibres experience a somewhat uniform compressive stress that may result in near-surface fibre crushing [28].

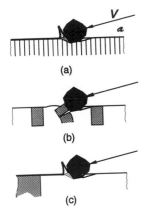

Fig. 6. Relative-size effect of particle to microstructure on deformation processes at impact angle α.

In figs. 5b and 5c, the microstructure and each impact diameter can be considered in the same size range; however, as the size of the erodent particle increases with respect to the microstructure, the impacted region transforms from either the fibre or the matrix to the combination of the fibre, matrix, and interface. These conditions exist when $R_{I/\mu}$ is in the range near unity. If the probability of the impact area overlapping the interface between the fibre and the matrix is p_{fM}, then $p_f + p_M + p_{fM} = 1$. Erosion processes in the regimes given in fig. 5, beside depending on the properties of the reinforcement and the matrix, depend on the geometric arrangement of the reinforcements, the distance between reinforcing phases, the size of impact area, D_{IM}, D_{IB}, and V_f.

By varying the angle of incidence, α, additional complexities are introduced because the impact particle has a cutting component. Figure 6 is an illustration with α about 16°. As discussed above, the mechanisms involved in the erosion process are dependent on relative-size factors of the impacting particle to the microstructure, and an analysis based on that of Finnie [3], in which a particle of mass m strikes a fibre reinforced composite surface with velocity v at an angle α, must be developed.

Erosion as a function of α and velocity, v, of short-fibre δ-Al$_2$O$_3$–aluminum–4 wt% copper matrix composites, $V_f = 0$, 0.05, 0.20, and 0.30, shows that range of angles with the maximum erosion rate broadened from about 20° for only the matrix to an angular spread of 20° to 50° for a V_f of 0.30 [18]. Since the specimens in the experiments were pressed into the final ingot form, a planar-radial symmetry with most of the fibres rotated away from the pressing direction is produced. This orientation distribution is not the ideal arrangement of parallel fibres described above. Comparison of erosion-rate (target mass loss per mass of erodent particles that collided with target) determinations in an aluminum alloy 6061 composite showed an increase in erosion rate as the reinforcing-phase volume fraction was increased [16]. Additional variables peculiar to composite materials must also be considered. The distribution, shape, size, orientation and relative target surface area of the reinforcing phase with respect to the matrix phase will have influence on the erosion rate [38].

Furthermore, the impacting particles are not rigid and their mechanical and physical properties for certain particle–target combinations influence the erosion process. SiC or SiO_2 particles at a velocity of about 65 m/s were used to compare the erosion effect on a Saffil reinforced aluminum alloy 6061 [17]. The less angular and softer SiO_2 erodent resulted in a lower erosion rate; actually, the erosion rate by the SiO_2 particles is about 50% the erosion rate for the SiC particles. Note that the kinetic energy at impact of the SiC is about 46% higher than with the SiO_2 particles. Since the toughness of SiC is greater than that of SiO_2 there is probably less fragmentation of the SiC at the higher-energy impacts. These two reasons are probably large contributors to the different erosion rates. For both erodents the erosion rates were lower at normal incidence than in the incident angular range 20° to 30°. This observed trend in incident angle versus erosion rate is indicative of ductile metal erosion. However, for the erodent SiO_2 the erosion rate was less than that of the matrix material for incident angles of 20°; hence, this experiment suggests that for certain material–erodent combinations, at low volume fraction of reinforcing phase, the erosion rate may be decreased [17]. Characteristics of the erodent particle such as elastic constant, hardness, fracture toughness, shape [37], size, mass, energy transferred to the impacted surface, and energy of cleavage, in addition to the angle of impingement, are fundamental in the erosion of the target material [5,16,40].

3. Summary

Since the erosion dependent variables have large values and are usually difficult to control and define precisely for metal matrix composites, usually general trends are reported in the scientific literature. However, some models for specific geometries have been discussed [40], and in this chapter some of the factors fundamental to erosion of metal matrix composites were appraised. The relative size of the impact area to the microstructure is an important parameter to consider, since the significant contributions to the erosion mechanisms change from small $R_{I/\mu}$ to large $R_{I/\mu}$. In the first case, an erodent particle will strike only one phase on impact of the target, and the erosion of the aggregate will be related to some combination, depending on the geometrical arrangement of the matrix and reinforcement, of the erosion rates of the components. Only a small influence from the interface regions is expected as a contribution to the process. On the other hand, when $R_{I/\mu}$ is large, many of the reinforcing structures are involved in each impact. Thus, a general deformation of the microstructure will take place. The near-surface portion of the brittle phase, constrained only by the highly deformable matrix, will fragment and chip at edges forming a surface layer of hard broken segments in a ductile matrix. The material removal after additional impacts occurs from this layer, which is similar to a particulate composite structure.

When the region impacted by one erodent particle has about the same size as the microstructure, $R_{I/\mu}$ is near to unity, a combination of edge fracture, brittle fracture, and ductile deformation is active. In this critical range, the erosion takes place at the transition between the mechanisms that occur at the extremes as described above.

In most reported experiments, addition of a stronger, more brittle phase in any geometry to a metal matrix increases the erosion rate. An exception was found with a more fragile erodent striking a surface of a composite that contained a low volume fraction of reinforcing material. Thus, in general, the fragmentation of the brittle phase and its impedance to plastic strain of the metal matrix leads to a more efficient removal of material.

A science of erosion of metal matrix composites has yet to be developed, because of lack of a universal and fundamental approach to this complex phenomenon. Not only were the composite microstructures in the erosion experiments reported different, but also the presentation of the results tended to reflect the way presentations have been made for monolithic erosion experiments. The major addition to the data is the effect of the volume fraction of the reinforcing phase. Characteristic information of the composite, such as the geometric arrangement of the hard phase on the target surface, orientation distributions of reinforcing materials (when appropriate), the homogeneity of the distribution, and the relative probability of each phase being hit by a erodent particle at the time of the exposure, is needed to develop consistent models for erosion. Experimental observations of the effect of cutting, crushing, flow, and fracture behavior of the phases during the initiation and the progress of the erosion process will give much of the fundamental information needed to assist in modelling the process. Finally, the mechanism of material removal from the surface containing the particular agglomeration of components that has been developed, must be investigated. The physical and mechanical constraints are more severe than encountered in the challenge with simple, ductile metals.

List of symbols

α	incident angle of the colliding particle(s) [°]
D	diameter of the impact impression [μm]
D_{IB}	average contact diameter of the impacts on the brittle phase [μm]
D_{IM}	average contact diameter of the impacts on the ductile phase [μm]
$\overline{D}$	average diameter of impact on the composite surface [μm]
e	coefficient of restitution
g	numerical factor at the perpendicular cross section of a regular geometric distribution of parallel fibres
KE_I	kinetic energy of impact particle [J]
KE_R	kinetic energy of rebound particle [J]
m	mass of a particle that strikes a fibre reinforced composite surface [g]
m_i	of mass fraction of ith component in the composite [g/g]
p_f	the probability of impact with the fibres
p_M	the probability of impact with the matrix
r	average radius of the fibres or the particulate brittle phase [μm]

$R_{I/\mu}$ ratio of the impact diameter to a parameter that is characteristic of the size of the microstructure of the composite

s the spacing between the brittle phase regions [μm]

V_f volume fraction of fibres

V_f^ϕ volume fraction of fibres of a given orientation

v_I velocity of impact particle [m/s]

v_R velocity of rebound particle [m/s]

W average erosion rate of target's mass loss per mass of erodent particles that impact the target of a multiple component microstructure [g/g]

W_i mass erosion rate of ith component per mass of erodent particles that impact the ith component [g/g]

Subscripts

B   brittle phase

f   reinforcing fibres

I   impact particle

M   metal matrix

μ characteristic size of microstructure of composite

R   rebound particle

Superscript

ϕ angle between the unidirectional fibres and the surface of the target [°]

References

[1] A.W. Ruff and S.M. Wiederhorn, in: Erosion by Solid Particle Impact, Treatise on Materials Science and Technology, Vol. 16, ed. C.M. Preece (Academic Press, New York, 1979) pp. 69–126.

[2] S. Bahadur, in: The Structure of Erosive Wear Models, eds K.C. Ludema and R.G. Bayer, ASTM STP 1105 (ASTM, Philadelphia, PA, 1991) pp. 33–50.

[3] I. Finnie, The mechanism of erosion of ductile metals, in: Proc. 3rd US National Congress of Applied Mechanics (1958) pp. 527–532.

[4] J.G.A. Bitter, Wear 6 (1963) 5–21; 169–190.

[5] J.H. Neilson and A. Gilchrist, Wear 11 (1968) 111–122.

[6] G.L. Sheldon and A. Kanhere, Wear 21 (1972) 195–209.

[7] N.P. Suh, Tribophysics (Prentice-Hall, Englewood Cliffs, NJ, 1986) pp. 305–346.

[8] I.M. Hutchings, Wear 70 (1981) 269–281.

[9] P.G. Shewmon, Wear 68 (1981) 253–258.

[10] G. Beckmann and J. Gotzmann, Wear 73 (1981) 325–353.

[11] G. Sundararajan and P.G. Shewmon, Wear 84 (1983) 237–258.

[12] F.P. Bowden and J.E. Field, Proc. R. Soc. London A 286 (1964) 331.

[13] A.G. Evans, M.E. Gulden and M. Rosenblatt, Proc. R. Soc. London A 361 (1978) 343–365.

[14] L.B. Greszczuk, Foreign Object Impact Damage to Composites, ASTM STP 568 (ASTM, Philadelphia, PA, 1975) pp. 183–211.

[15] F.C. Frank and B.R. Lawn, Proc. R. Soc. London A 299 (1967) 291–306.

[16] I.M. Hutchings, 2nd Eur. Conf. on Advanced Materials and Processes, Euromat '91, Cambridge, England, July 22–24 (1991).

[17] I.M. Hutchings and A. Wang, Inst. Phys. Conf. Ser. 111 (1990) pp. 91–100.

[18] S. Srinivasan, R.O. Scattergood and R. Warren, Metall. Trans. A 19 (1988) 1785–1793.

[19] P.E. Engel, Tribology Series, Vol. 2 (Elsevier, New York, 1978).
[20] I.M. Hutchings, R.E. Winter and J.E. Field, Proc. R. Soc. London A 348 (1976) 379–392.
[21] T.H. Kosel and T.S. Sriram, A study of erosive rebound parameters, Fossil Energy Materials Program Conference, US DOE, Oak Ridge National Laboratory, May 19–21 (1987).
[22] Y. Yokouchi, T.W. Chou and I.G. Greenfield, Metall. Trans. A 14 (1983) 2415–2421.
[23] S.P. Timoshenko and J.N. Goodier, Theory of Elasticity, 3rd Ed. (McGraw-Hill, New York, 1951).
[24] I.M. Hutchings and R.E. Winter, Wear 27 (1974) 121–128.
[25] R.E. Winter and I.M. Hutchings, Wear 34 (1975) 141–148.
[26] L.D. Dyer, ASM Trans. 58 (1965) 620.
[27] E.B. Iturbe, I.G. Greenfield and T.W. Chou, in: Proc. ELSI V, Cavendish Laboratory, Cambridge, England (1979) Article No. 30.
[28] I.G. Greenfield and R. Vignaud, in: Composites '86, eds K. Kawata, S. Umekawa and A. Kobayashi (Japan Soc. Comp. Materials, Tokyo, 1986).
[29] S.L. Rice, Wear 54 (1979) 291–301.
[30] K. Wellinger and H. Breckel, Wear 13 (1969) 257–281.
[31] I.G. Greenfield and E.B. Iturbe, J. Mater. Sci. 20 (1985) 4399–4406.
[32] I.G. Greenfield, K.R. Stull and R. Vignaud, in: Proc. 7th Int. Conf. on Erosion by Liquid and Solid Impact, Cambridge, England (1987) pp. 63/1–63/6.
[33] I.G. Greenfield, in: Proc. 10th Annual Discontinuously Reinforced MMC Working Group Meeting, Park City, UT, January 5–7 (1988) pp. 305–319.
[34] S.K. Hovis, J.E. Talia and R.O. Scattergood, Wear 107 (1986) 175–181.
[35] A.E. Miller and J.P. Coyle, Wear 47 (1978) 211–214.
[36] K.C. Goretta, W. Wu, J.L. Routbort and P.K. Rohatgi, in: Proc. Conf. on Tribology of Composite Materials, Oak Ridge, TN, May 1–3, 1990, eds P.K. Rohatgi, P.J. Blau and C.S. Yust (ASM, Materials Park, Ohio, 1990) pp. 147–155.
[37] W. Wu, K.C. Goretta and J.L. Routbort, J. Mater. Sci. & Eng. A 151 (1992) 85–95.
[38] S.K. Hovis, J.E. Talia and R.O. Scattergood, Wear 108 (1986) 139–155.
[39] D. Hull, An Introduction to Composite Materials (Cambridge University Press, Cambridge, 1981) p. 61.
[40] W.J. Head, L.D. Lineback and C.R. Manning, Wear 23 (1973) 291–298.

PART IV

*Tribology of Composite Materials
for Special Applications*

Advances in Composite Tribology
edited by K. Friedrich
© 1993 Elsevier Science Publishers B.V. All rights reserved

Chapter 13

Tribology of Composites for Magnetic Tape Recording

H. JACOBI

Institut für Werkstoffe, Ruhr-Universität Bochum, Bochum, Germany

U. NOWAK

BASF AG, Ludwigshafen, Germany

Contents

Abstract

Tribological tests using a modified videotape recorder and using a microscratch apparatus are carried out to clarify the friction and wear behaviour of particulate magnetic tapes, both in practice-oriented tests and in model systems under enforced conditions.

The traces of the friction coefficient recorded during long-time still-picture tests are combined with SEM micrographs of the wear tracks at different stages. This method enables simultaneous observation of both the friction and the wear behaviour of the material. Tracks with a width less than 1 μm result from short-time tests and are due to the sharp edges of the videoheads. Thus, making it possible to examine the influence of particular microstructural components. A top view and sectional view of the wear grooves assist in explaining the resistance of the magnetic layers to penetration of the heads.

Microscratch tests are performed with different weightings of the tribological mechanisms "adhesion" and "abrasion", using a steel ball and a diamond tip. The friction and wear behaviour of the layers are examined in view of analogies with the results of the still-picture tests. Trying to develop a quick and easy method to perform practice-oriented tests and to rank the tapes with respect to low friction and high wear resistance is the main object of this work.

Tapes with different amounts and sizes of Al_2O_3 filler particles and different calendering temperatures of the magnetic layers are investigated.

1. Introduction

1.1. General aspects

In addition to the development of semiconductors, the magnetic storage of information can be considered as one of the most important preconditions for the evolution of human culture and technology during the last decades. Never before has a method existed which enables the possibility of storing any kind of data or recording sound, pictures, and movies in such a quick and cheap manner. Moreover, magnetic mass memories are easy to handle and to transport Nowadays, even removable hard-disk drives are obtainable. As a result, magnetic media are well accepted and nearly all spheres of everyday life would be unthinkable without them.

Competing technologies, such as magneto–optical data storage and phase-change optical recording, do not represent formidable rivals at this moment, because the drives, media, and all other components operating according to these methods, are still too expensive relative to the storage capacity. Besides, not all the technical problems have been solved and the lack of standards, i.e. the incompatibility of the products designed by different factories, has very unpleasant consequences for the acceptance and opening of the market [1–7].

However, the potential for future developments in magnetic-media technology is far from being fully utilized yet. Metal and metal oxide thin-film media, thin-film heads, perpendicular recording, and digital audio and video techniques promise increasing read–write speeds, transfer rates, and track as well as linear recording densities [8–11].

1.2. Fundamentals of magnetic recording

The majority of the magnetic heads used in disk drives, computer tape drives, audio systems, and videotape recorders are of the inductive type, which allows the combination of both the read and the write function in one head (fig. 1). These read/write heads consist of a ring of high-permeability soft-magnetic material with an electrical winding and a gap in the magnetic material at or near the surface of the storage medium. Writing is accomplished by passing a current through the coil. The flux is confined to the magnetic core, except in the region of the small non-magnetic gap. The fringe field in the vicinity of the gap magnetizes the medium longitudinal to the plane of the coating moving over the write head. The magnetic medium consists of high-coercivity hard-magnetic material that retains its magnetization after it has passed through the field from the head gap.

The medium passes over the read head, which, like the write head, is a ring core with an air gap. Each particle in the medium is a miniature magnet and its flux lines will add up with those of other particles to provide an external medium flux, proportional in magnitude to the medium magnetization. The flux lines from the medium permeate the core and induce a voltage in the head's winding. This voltage, after suitable amplification, reproduces the original signal. A single head can be used for both the read and the write function. Because of the high relative motion and unit pressure between the counterparts in videotape recorders, high-temperature isostatic-pressed or single crystalline ferritic heads of the manganese–zinc and nickel-zinc type are mostly used [8,9,11–14].

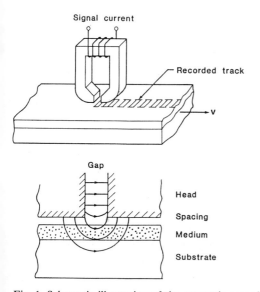

Fig. 1. Schematic illustration of the magnetic recording process, showing a three-dimensional view (a) and a cross section (b). (From Daniel and Mee [13].)

1.3. Tribology of magnetic storage systems

From the tribological point of view the most important parameter in designing magnetic storage systems is the distance between the magnetic heads and the coating. The signal amplitude decreases about two orders of magnitude when increasing the distance by about 1 μm. Furthermore, higher track and linear densities in the case of longitudinal recording, and new technologies like perpendicular recording, or the thin-film technique in head and media design require extremely small spacings. Thus, regarding only the magnetical restrictions, every fact points to a zero distance to optimize the properties of the systems. This requirement collides with the demand of avoiding direct contact to exclude friction and wear effects and, thereby, maximize the life expectancy of both heads and media [11].

These conflicting requirements led in hard-disk technology to the development of heads sliding on a hydrodynamic air bearing during running operation and only getting into contact with the magnetic layer under start/stop conditions [11,13–15].

Working with flexible media like tapes and floppy disks, local contacts between the heads and the asperities of the magnetic layer cannot be prevented. Videosystems, for example, operate with minimal spacings, with distances smaller than three times the standard deviation of the coating roughness [11,16–20].

Medium	(Head-to-medium distance)/ (Standard deviation of the surface roughness) [1]
Videotape	⩽ 3
Computer tape	> 4
Hard disk	⩾ 6

Physical and mechanical interactions of the participating surfaces cause dissipation of energy (friction) and dissipation of material (wear). Both effects influence each other, superimpose, and may lead to a reduction of the record and playback quality by changing the surface morphology of the media and the heads. To understand these interactions and improve the lifetime of both the magnetic layers and the heads, investigations on the tribological behaviour of the contacting surfaces are necessary.

2. Experimental

2.1. Material

Modern magnetic tapes are made of laminate composites, mainly consisting of a magnetic layer, the base film or substrate, which is usually a PET foil, and a back coating, composed of black carbon particles embedded in a polyurethane matrix to improve the running characteristics of the tape and to prevent the layer from electrostatic charging during operation. Liquid and solid lubricants, only as protec-

tive overcoats, are optional and depend on the field of application and the particular manufacturer. Two types of magnetic layers exist
– thin-film media and
– particulate media.

While metal or metal oxide thin-films have a thickness of 50 up to 500 nm, particulate layers are thicker and have thicknesses between 2 and 6 μm. These conventional coatings are particle reinforced composites composed of a polymeric matrix, like polyurethane, containing a fine dispersion of hard-magnetic particles such as Fe_2O_2, CrO_2, Fe, or $BaFe_{12}O_{19}$. The mixing of different types of magnetic particles and the addition of small amounts of non-magnetic ceramic fillers, like Al_2O_3 and ZrO_2, are common methods to improve the mechanical properties of the coating and to guarantee a significant head-cleaning effect [8,9,11,21–24].

Four series of magnetic tapes with Al_2O_3 as reinforcing filler component were investigated. The series differed in chemical composition. For type A, high-molecular binder with OH-groups based on bisphenol-a, with a high glass transition point, T_g, was used. The cobinder of the type-C, -D and -F tapes consisted of polyvinyl polyacetate with high T_g, partially modified with hydroxylic groups. For lubrication, fatty acids and esters of fatty acids were added to the dispersion. The coatings of all series were based on commercially available high-molecular polyurethane binder cross-linked with polyisocyanate.

The filler particles in tapes of type A consisted of Al_2O_3 with 0.7, 0.33, and 0.2 μm nominal particle size. The magnetic layers of the type-D tapes had different calendering temperatures, from 30 up to 90°C. The type-C and type-F tapes were filled with different amounts of alumina fillers with 0.2 and 0.7 μm size, ranging from 0 to 3.5 vol%.

Tape C/82 (3.5%, 0.2 μm, 80°C) is an example of the system of notation used in this chapter. This tape contains 3.5 vol% of Al_2O_3 filler particles with a nominal size of 0.2 μm. The magnetic layer was calendered at 80°C.

The magnetic particles of type-A and -F tapes consisted of cobalt-modified Fe_2O_3, while for types C and D 10% CrO_2 was added. For the substrate a standard video PET foil of 14.3 μm was used. As can be seen in table 1, the tapes of type A are the thickest. The unused surfaces of some of the investigated magnetic tapes are presented in fig. 2.

2.2. Practice-oriented still-picture tests

The still-picture tests were carried out with the same videoheads in laboratory air at both constant ambient temperature and relative humidity. The tests were performed with a modified VTR of the "Video 2000" format, using azimuth recording and a wrap angle of about 183° (figs. 3 and 4). The two multicrystalline manganese–zinc heads on the head wheel rotated with 1500 rpm under an angle of 2.7° with the edge of the tape. The Vickers hardness of the various counterparts and the tape components can be found in table 2. The diameter of the drum led to a relative velocity between the heads and the tape of 5.1 m/s. The width of the

TABLE 1

Composition of the tested particulate magnetic tapes.

Tape	Magnetic layer		Alumina filler particles		
	Thickness (μm)	Calendering temperature (°C)	Average diameter (μm)	Volume fraction (%)	Average distance (μm)
A/2	5.6	55	0.7	3.2	2.7
A/4	5.1	85	0.7	3.2	2.7
A/6	5.4	55	0.33	3.2	1.3
A/8	5.0	85	0.33	3.2	1.3
A/10	5.7	55	0.2	3.2	0.8
A/12	5.7	85	0.2	3.2	0.8
C/76	3.5	80	–	–	–
C/82	3.5	80	0.2	3.5	0.7
D/19	4.0	30	0.2	3.8	0.7
D/20	4.0	45	0.2	3.8	0.7
D/21	4.0	60	0.2	3.8	0.7
D/22	4.0	75	0.2	3.8	0.7
D/24	4.0	80	0.2	3.8	0.7
D/23	4.0	90	0.2	3.8	0.7
D/28	4.0	30	0.7	3.8	2.5
D/29	4.0	45	0.7	3.8	2.5
D/30	4.0	60	0.7	3.8	2.5
D/31	4.0	75	0.7	3.8	2.5
D/33	4.0	80	0.7	3.8	2.5
D/32	4.0	90	0.7	3.8	2.5
F/29	4.0	50	–	0	–
F/31	4.0	50	0.7	0.5	4.7
F/57	4.0	50	0.7	1.0	3.8
F/68	4.0	50	0.7	2.0	3.2
F/33	4.0	50	0.2	0.5	1.3
F/59	4.0	50	0.2	1.0	1.1
F/71	4.0	50	0.2	2.0	0.9
F/30	4.0	85	–	0	–
F/32	4.0	85	0.7	0.5	4.7
F/58	4.0	85	0.7	1.0	3.8
F/70	4.0	85	0.7	2.0	3.2
F/34	4.0	85	0.2	0.5	1.3
F/60	4.0	85	0.2	1.0	1.1
F/73	4.0	85	0.2	2.0	0.9

magnetic track recorded by one head was about 25 μm. The following modifications were made to the system [25–27]:
– permanent still-picture mode and
– constant tape tension during the tests.
    While performing a still-picture test, the voltage U, used to rotate the top of the drum, was measured and recorded versus time. Since the top of the drum rotated with a constant number of revolutions, the alterations in the trace of the electric

A/2   A/10   C/76   C/82

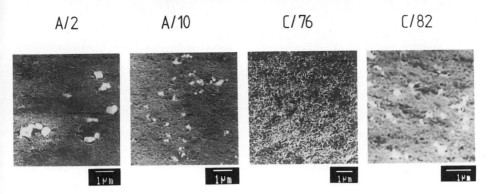

Fig. 2. Unused surfaces of some particulate magnetic layers. The longitudinal direction is from left to right.

energy are directly proportional to the changes in the friction coefficient, $\mu(t)$, between the heads and the tape and the rotating part of the drum and the tape, respectively. Constant current and resistance leads to the following relation:

$$\Delta U(t) \sim \Delta\mu(t). \tag{1}$$

This means, an increase (decrease) of the friction between the counterparts leads to an increase (decrease) of the voltage. The dimensionless coefficient $\mu'(t)$ was defined as the voltage $U(t)$ normalized by the no-load voltage U_0:

$$\mu'(t) := \frac{U(t) - U_0}{U_0} \sim \mu(t). \tag{2}$$

U was measured with no tape around the drum. By measuring the coefficient μ' it is possible to test and compare the dependence of the friction behaviour of different videotapes on time. The following parameters were maintained constant during all tests:
- tape tension: 0.4 N,
- number of revolutions: 1500 rpm,
- head protrusion: 30 μm.

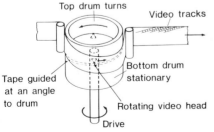

Fig. 3. Helical-scan rotating-head arrangement.

a)

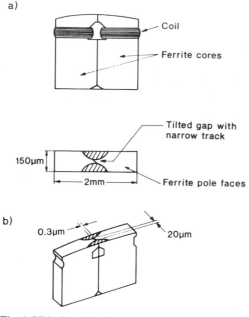

b)

Fig. 4. Videohead design for consumer tape recorder. (From Daniel and Mee [13].)

After recording a cross-hatch pattern, still-picture tests of 15, 30, 60, 90 and 120 min duration were performed. For each test a new piece of tape was used. The relative quality of the recorded test pattern was checked during the tests using a connected TV screen.

The influence of the rotating top part of the drum on the friction and wear behaviour was examined in long-time tests with the heads removed. For this purpose the heads were disassembled carefully after the termination of the still-picture test series.

TABLE 2

Chemical composition and hardness of the basic bodies and counterparts used in the different tribological systems.

Name	Material	Vickers hardness
I. Basic body		
Magnetic particle	CrO_2	800–1200
	$Co-\gamma-Fe_2O_3$	450– 550
Filler particle	Al_2O_3	1800
Magnetic layer	$Fe_2O_3 + PUR + Al_2O_3$	20– 25
II. Counterpart		
Videohead	MnZn–ferrite	600– 700
Steel ball	100Cr6	820
Diamond tip	C	10.000

Fig. 5. The diamond tip installed in the microscratch apparatus.

The worn parts of the tapes were examined in the scanning electron microscope without any surface preparation. In the case of tape C/76 (0%) an evaporation with gold was necessary, since this tape without alumina particles, exhibited deeper wear scars in the magnetic layer than all others.

2.3. Microscratch tests

The model tests were performed by using a specially constructed scratch apparatus (fig. 5). The specimens were attached to the surface of a carriage and moved with constant speed relative to the counterpart, which was fixed in the horizontal direction. After taring the lever arm, the penetrator was loaded with a discrete load. The tribological tests were carried out under the following conditions:

Counterpart	Shape	Load (N)	Velocity (μm/s)
Steel ball (100 Cr 6)	2 mm diameter	0.5	1.6
Diamond tip	115° nose angle	0.2	1.6

Both in the case of the spherical indenter and the diamond tip it was possible to make scratches in the magnetic layer without detaching it from its substrate or getting into contact with the polymeric foil (for the Vickers hardnesses of the counterparts see table 2).

The tangential force was measured and processed with a system consisting of a load cell, an analog to digital converter and a microcomputer with suitable software. This made it possible to take the inherent friction of the microscratch apparatus into account. Dividing the measured frictional force by the adjusted load yields to a trace of the friction coefficient versus the time or the course, respectively.

3. Main mechanisms of friction and wear

3.1. Still-picture tests

3.1.1. Influence of the counterparts
3.1.1.1. *Drum.* Since it is quite well known that γ-Fe_2O_3 particulate media, which do not contain any hard phase, exhibit a rather bad wear behaviour, the still-frame tests without magnetic heads were performed with the most sensitive tape, C/76 (0%, 80°C) [11,28].

Using only the rotating top part of the cylinder as counterpart, the average friction coefficient amounted to $\mu' = 0.003$, and after 90 min parallel microgrooves could be observed on the worn surface of the magnetic layer. The maximum value for both the width and the depth of the grooves was 3 μm (fig. 6).

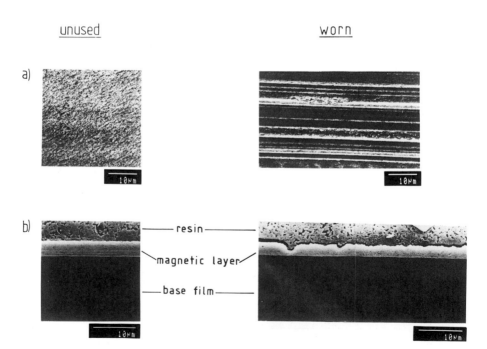

Fig. 6. Top view (a) and sectional view (b) of tape C/76 (0%, 80°C) in unused conditions and worn by the rotating top part of the drum.

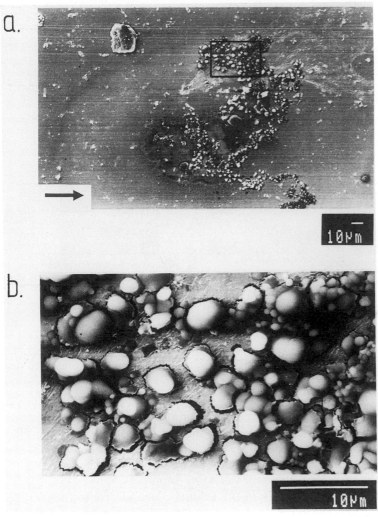

Fig. 7. Abrasive particles embedded in a layer of friction polymer on the surface of the rotating top part of the drum. Overall view and detail.

To find the reason for the primarily abrasive, and not adhesive, behaviour of the counterpart, the surface of the drum was examined by scanning electron microscopy and energy-dispersive X-ray analysis. As can be seen in fig. 7, accumulations of spherical Al_2O_3 particles with a planar extent of 10^3–10^4 μm^2 are located on the surface. The clusters are randomly arranged on the rotating part of the drum. The size of a single particle reaches up to a diameter of 5 μm.

The particles are embedded in a thin layer of a so-called "friction polymer", from the polyurethane of the tested magnetic layers. Since only a few spots of the substrate can be observed as the result of detached particles, the intertacial van

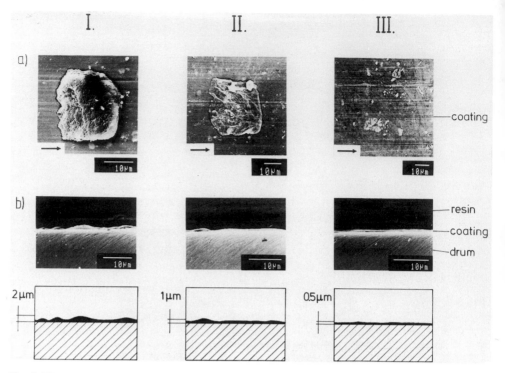

Fig. 8. The mechanism of tribologically induced layer formation on the surface of the drum. Top view (a) and sectional view (b) of the friction polymer.

der Waals bonds between the particles and the layer and the layer and its substrate seem to be very strong (see fig. 7). Therefore, the agglomerates of ceramic filler particles are able to act in a mainly abrasive way on the soft basic material. The thickness of the layer and the mechanism of its formation are shown in fig. 8, which represents a vertical embedding of the worn drum.

Particles of wear debris removed from the magnetic layers, stick to the metallic surface of the drum due to its high free surface energy. The high relative motion of about 5 m/s and repeated-pass sliding lead to a rise of the contact temperature in the interface and, thereby, to both chemical degradation of the polymer and plastic deformation of the debris. A transfer film with an average thickness of about 0.5 μm is formed on the countersurface. The layer is structurally and chemically different from the original polymer of the tape surface and is continuously regenerated as long as the tape remains in rubbing contact. It acts as a protective film, which reduces the free surface energy and, thereby, the adhesive part of the friction coefficient [11,29].

3.1.1.2. Videoheads.   At the beginning of a still-picture test there will be only direct contact between the edges of the heads and the asperities of the magnetic layer. The heads form a tent with the tape (fig. 9). Its volume, or "pump volume", depends on the head-tip penetration, the geometrical shape of the heads, and the

STILL PICTURE TEST

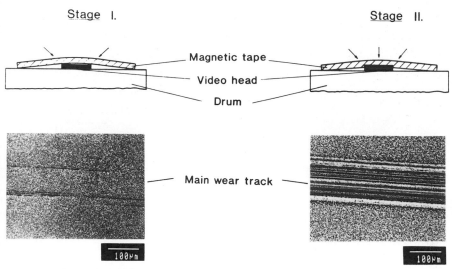

Fig. 9. The increase of the real area of contact between the rotating videohead and the magnetic tape with increasing still-picture test time.

stiffness of the tape, that means the elastic modulus of the PET foil which was used as the substrate [30].

With increasing test time the heads penetrate deeper into the layer and, thereby, the real area of contact extends, which leads in the case of a constant tape tension to a decreasing unit pressure. The sharp edges of unused magnetic heads benefit the formation of shavings on the surface of the tested particulate tapes (fig. 10). Abrasion is the dominating mechanism in the wear process at that moment. In this special case the size and the shape of the shavings can be simulated using the microscratch test apparatus and a diamond tip as the counterpart, although the relative motion in both systems differ by a factor of 3×10^6. The enormous breaking length of the shavings of some hundred micrometers is an indication for both the high tensile strength and the ductility of the layer material.

If the real area of contact between the heads and the tape is small, a high cyclic unit pressure and high transverse stresses are the result. In the read–write-gap area of the heads a decrease of the thickness of the heads from 154 down to 25 μm leads to a superposition of the transverse and normal forces and, primarily at that location, to the initiation of microcracks in the brittle manganese–zinc ferrite (fig. 11). Crack growth and the breaking up of head material follows. The amount of head wear decreases with increasing test time until the head shape fits the shape required by the tape stiffness and head protrusion [31].

During the running-in tests, a thin layer of polymer-based wear debris was built up on top of the head plateaus, according to the same principles as the transferred-layer formation on the surface of the rotating part of the drum (see fig. 8)

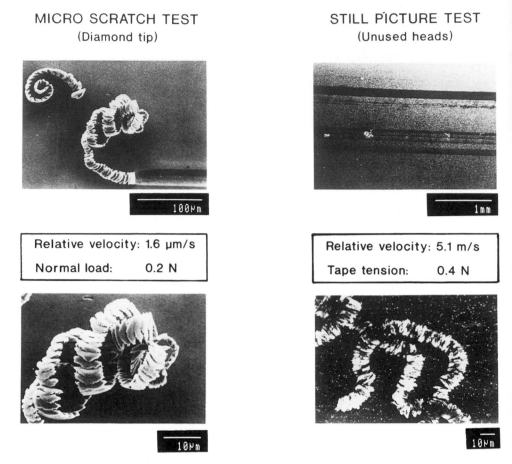

MICRO SCRATCH TEST
(Diamond tip)

STILL PICTURE TEST
(Unused heads)

Relative velocity: 1.6 µm/s
Normal load: 0.2 N

Relative velocity: 5.1 m/s
Tape tension: 0.4 N

Fig. 10. Simulation of the appearance of wear occuring in still-picture tests with unused ferritic heads by microscratch tests with a diamond tip as the counterpart.

[11,29]. In this case as well, the layer acts as a protective film and is continuously regenerated during operation. Because of the more intense wear of the protruding heads, the layer is thinner as in case of the drum. Therefore, the quality of the test pattern shown on the TV screen was not significantly influenced by the enlarged distance between the heads and the stored information (cf. section 1.3).

On the head plateau at the run-out zone of the tape a rather small "nose", consisting of tribopolymer, is formed by the rubbing contact (fig. 12). At the run-out zone of the entire head a larger "nose" is formed, containing other components, such as dust, filler particles, fragments of head material, and wear debris removed from the rollers and pinchers of the recorder.

Tribological tests with a reference tape assisted in determining the end of the running-in period of the heads and the drum. From that time onwards the friction

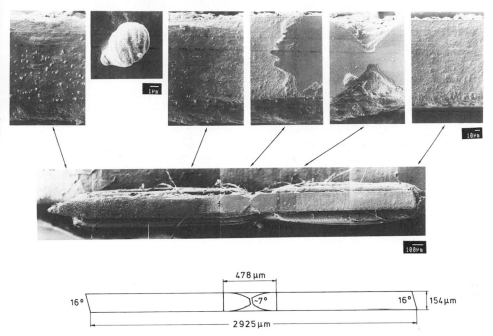

Fig. 11. Top view of a used videohead worn by particulate magnetic tapes after in total 250 h of still-picture test time.

and wear behaviour of the system could be considered as constant. Caused by the modifications of the counterparts, adhesion has become the main tribological mechanism [20,21,28,32].

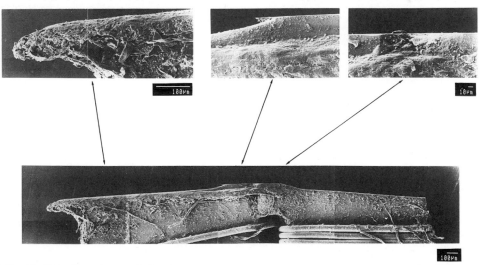

Fig. 12. Side view of a used videohead worn by particulate magnetic tapes after in total 250 h of still-picture test time.

3.1.2. Variation of microstructural parameters

3.1.2.1. Amount of Al_2O_3 *filler particles.*   To clear up the influence of the amount of Al_2O_3 particles on the friction and wear behaviour, still-picture tests with the tapes C/76 (0%, 80°C) and C/82 (3.5%, 0.2 μm, 80°C) were conducted. Scanning electron micrographs of the worn surfaces taken at 30 min intervals were combined with the trace of the friction coefficient. The sliding direction of the counterparts, "videoheads" and "drum", always was from left above to right below.

At the beginning of the still-picture tests local adhesive contacts occur between the heads and protruding asperities of the magnetic layer. The high relative motion of about 5 m/s causes a rise of the temperature in the interface above the softening point of the polyurethane, which is a polymer of the thermoplastically deformable rubber type. The cyclic loading and unloading of the asperities impedes the sinking of the heat and accelerates the softening of the material. Because of a small real area of contact the appearing transversal stresses exceed the yielding point of the heated polymer, leading to plastic deformation and, thereby, to an increase of the contact area. The magnetic material becomes smoother at the wear track. A steep increase of the friction coefficient is the result (fig. 13). The deeper penetration of the magnetic layer by the rotating heads

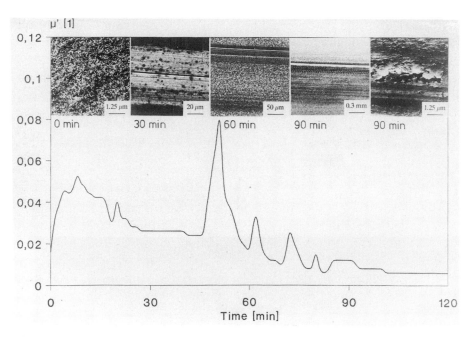

Fig. 13. Variation of the filler volume fraction: the trace of the friction coefficient of tape C/76 (0%, 80°C) combined with SEM micrographs of the wear tracks.

combined with an increase of the real area of contact leads to the following phenomena:

(1) The friction caused by adhesion increases until the real area of contact is equal to the apparent area of contact.

(2) The heat flux from the surface into the bulk material increases with decreasing roughness of the magnetic layer. As the temperature in the interface decreases, the plastical deformability of the polyurethane is reduced and the polymer exhibits more and more a rubber-like behaviour, i.e. elastic deformation as a reaction to stress.

(3) The tape fits closely to the drum and the unit pressure between the heads and the magnetic layer decreases since the tape tension maintains constant during the entire test (cf. section 2.2).

Hence it follows, that at reaching the maximum contact area the dissipation of energy caused by both adhesive contacts and microplastic deformations decreases. A reduction of the friction coefficient is the result (see fig. 13).

Since the relative soft composite, consisting only of polyurethane and γ-Fe$_2$O$_3$ magnetic particles, is not able to produce a significant head-cleaning effect, wear debris removed from the magnetic layer accumulates down at the heads during the tests. This so-called "head clogging" is supported by the lack of ceramic fillers and enlarges the spacing between the heads and the surface of the tape. The loss by increasing spacing increases until regular playback of the recorded test pattern is no longer possible.

On the other hand, the accumulation of wear debris slows down the rotation of the heads. The frictional force in the interface rises steeply until it overcomes the adhesion between the debris and the head. When the debris is removed, the coefficient μ' decreases quickly. The height of the frictional peaks decreases with the amount of magnetic material removed from the wear track (see fig. 13).

An increasing penetration of the heads in the magnetic layer causes a decrease of the space between the tape and the drum. If the space falls down below a critical value, the abrasive particles with maximum size on the surface of the drum come into contact with the tape and begin to groove the layer (fig. 14). As can be seen in section 3.1.1.1, its influence on the friction and wear is negligible if compared with the heads' effect.

The sectional view of a vertically embedded tape of the C/76 type shows a complete removal of the magnetic layer in the main wear track, right down to the substrate. After the dropping of the signal amplitude the automatic track guidance system failed and lateral motions of the heads could no longer be corrected. For this reason the main wear track is much wider than the videoheads (cf. fig. 11).

The use of 3.5 vol% of small Al$_2$O$_3$ particles with 0.2 μm nominal size leads to a slower increase of the real area of contact and, thus, to less adhesion between the rotating heads and the tape. A lower maximum of the friction coefficient, $\mu'(t)$, at the beginning of the still-picture test is the result (fig. 15).

For the whole duration of the test, the tape C/82 (3.5%, 0.2 μm, 80°C) shows a superior friction and wear behaviour to tape C/76 (0%, 80°C). This can be explained by the microstructural influence of the hard phase. Clusters and single

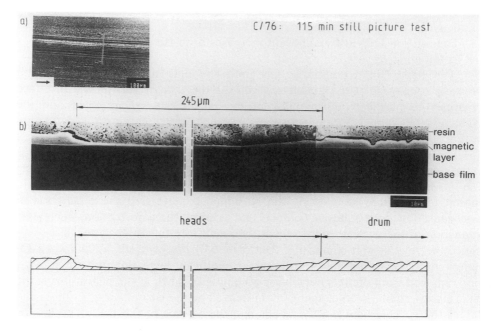

Fig. 14. Top view (a) and sectional view (b) of a wear track in the surface of tape C/76 (0%, 80°C) after 115 min of still-picture test time.

filler particles support the heads and reduce their interactions with the polymer of the matrix. The increase of the temperature in the contact zone is less steep than for the tape without any ceramic filler particles, and plastic deformations of the asperities are limited [33,34].

This prevents the magnetic heads from penetrating the magnetic layer. Consequently, after 300 min test time the magnetic material in the main wear track is removed by the heads only down to a depth of about 40 to 50% of the original coating thickness (fig. 16).

Since the heads are cleaned by the abrasive Al_2O_3 particles, the extremely detrimental head clogging does not appear. Without this accumulation of wear debris at the rotating heads the trace of the friction coefficient is smooth and frictional peaks do not exist. Using tape C/82 (3.5%, 0.2 μm, 80°C) the recorded test pattern could be observed until the end of the test with good quality and the signal amplitude had not decreased significantly.

The tribological behaviour of the hard phase changes with increasing test time. It can be subdivided into two different periods of time (cf. fig. 9): during the first stage of wear neither a demolition of clusters nor relative motions of single embedded filler particles could be observed. The distance between the particles and the agglomerates and their size are the important factors influencing friction and wear. The width of the wear tracks of C/76 (0%, 80°C) and C/82 (3.5%, 80°C) differ by a factor of 100 (fig. 17).

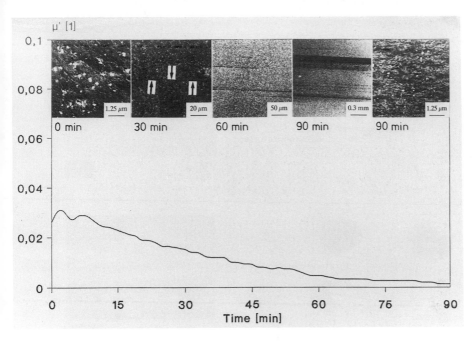

Fig. 15. Variation of the filler volume fraction: the trace of the friction coefficient of tape C/82 (3.5%, 2 μm, 80°C) combined with SEM micrographs of the wear tracks.

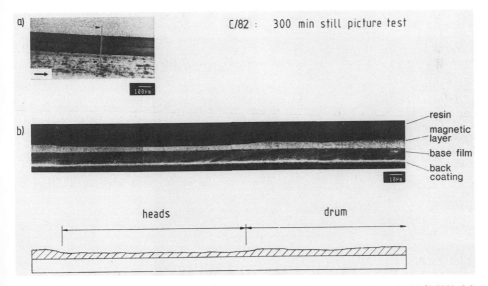

Fig. 16. Top view (a) and sectional view (b) of a wear track in the surface of tape C/82 (3.50%, 0.2 μm, 80°C) after 300 min of still-picture test time.

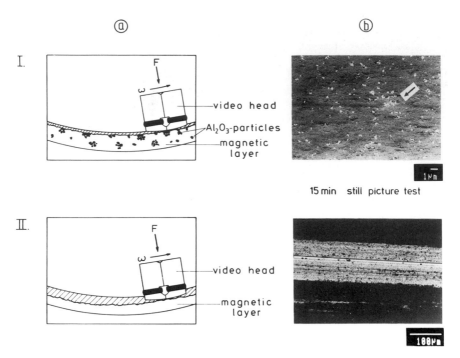

Fig. 17. Because of the supporting effect of clusters and single particles during the first stage of wear, tapes containing a higher amount of alumina show much smaller wear tracks than tapes without a reinforcing component.

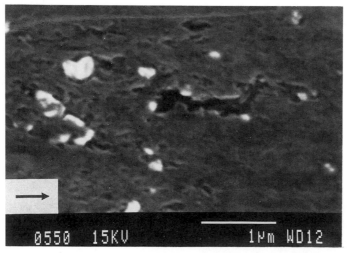

Fig. 18. Small filler particles, agglomerates and pores at the surface of the magnetic layer are covered with binder during the second stage of wear.

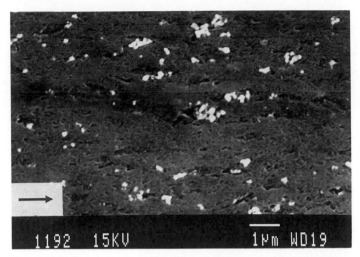

Fig. 19. After a certain incubation time, agglomerates of filler particles are demolished by the rotating heads during the second stage of wear.

Since the penetration velocity of the heads and the depth of the wear tracks in the magnetic layer is limited, the wear and friction are strongly reduced, but not completely stopped. Figure 18 shows that the supporting effect of the small Al_2O_3 particles with an average size of 0.2 μm is less efficient with increasing duration of the still-picture test (stage II). This is due to the filler particles being smeared up with binder. The tapes become smoother because the polyurethane flows over the particles [35]. As a consequence, the effective area fraction of the alumina particles is lowered in the surface of the layer. Since the real area of contact increases and the unit pressure decreases, this does not lead to a dramatic failure but to a constant reduction of the tape durability.

Another effect which increases the wear of the layer with increasing test time, is the demolition of clusters after a certain incubation time, beginning at the front areas of the agglomerates (fig. 19). The composite material locally depletes of alumina and the resistance at the rotating heads is reduced.

3.1.2.2. Size of Al_2O_3 filler particles. Two tapes of type A with alumina particles of 0.2 and 0.7 μm nominal size were used to examine the influence of the size on the lifetime of the particulate magnetic tapes. The total volume fraction of Al_2O_3 amounted in both cases to 3.2% (cf. table 1).

Tape A/2 (3.2%, 0.7 μm, 55°C), containing relatively large particles, shows a much steeper increase of the friction coefficient $\mu'(t)$ at the beginning of the still-picture test than tape A/10 (3.2%, 0.2 μm, 55°C) (figs. 20 and 21). During the whole test time the unsteady trace of the friction is on a higher level and superimposed by frictional peaks. The height of the peaks is much lower than for tape C/76, without alumina, described in section 3.1.2.1. The trace of $\mu'(t)$ for tape A/10 (3.2%, 0.2 μm, 55°C) is as smooth as it is for tape C/82. At the end of

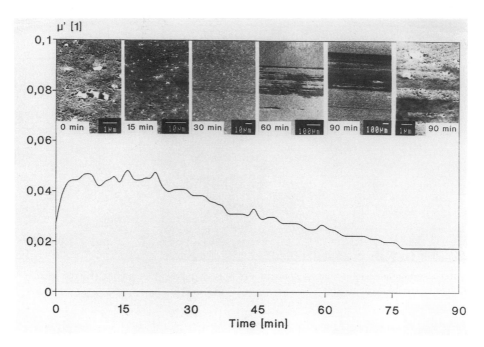

Fig. 20. Variation of the filler diameter: the trace of the friction coefficient of tape A/2 (3.2%, 0.7 μm, 55°C) combined with SEM micrographs of the wear tracks.

each of the documented test periods the wear of tape A/2 (3.2%, 0.7 μm, 55°C) is always more severe than the wear of tape A/10 (3.2%, 0.2 μm, 55°C).

On condition that the total amount of alumina remains constant, an increase of the size of the hard phase leads to a decrease of the tribological behaviour. The larger the particles, the stronger the similarity to the behaviour of magnetic layers without any alumina filler particles (cf. fig. 13).

Figure 22 will assist in explaining the role of the size of the alumina particles. During the first stage of wear, large particles ($D_p = 0.7$ μm) as well as small particles ($D_p = 0.2$ μm) and agglomerates of small particles support the video-heads and prevent them from penetrating the magnetic layer. Relative motion or a demolition of large particles does not appear.

The main parameter influencing the tribological behaviour is, therefore, the distance between the ceramic particles, i.e. the free path length of the rotating heads over the magnetic material. Thus, an identical width of the wear track, of 1 μ, is the result of a 2 min still-picture test in the case of tape A/2 (3.2%, 0.7 μm, 55°C), instead of 20 min in the case of tape A/10 (3.2%, 0.2 μm, 55°C). The distance between the alumina particles amounts to 2.07 and 0.77 μm, respectively. This is equivalent to particle densities of 0.13 and 1.01 μm^{-2}. The higher the density of the filler particles in the microstructure of the composite material, the higher the efficiency of minimizing the adhesive contacts between and plastic

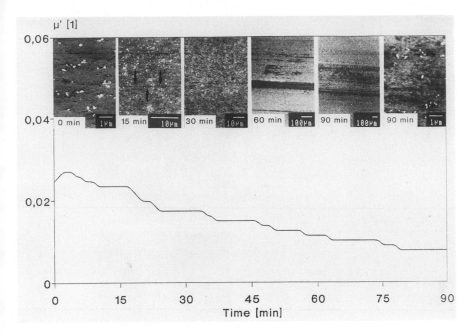

Fig. 21. Variation of the filler diameter: the trace of the friction coefficient of tape A/10 (3.2%, 0.2 μ, 55°C) combined with SEM micrographs of the wear tracks.

deformations of the basic body and the counterparts, and the higher the lifetime of the media.

During the second stage of wear, the resistance of the large particles ($D_p =$ 0.7 μm) to the wearing counterparts is reduced by three effects (figs. 23 and 24). Besides the covering up with binder, the interfaces between elements of the hard phase and the matrix material are damaged by repeat-pass sliding of the counterparts. The cyclic loading and unloading leads to relative motion of the large particles. After a certain incubation time the remaining bonds are too weak to prevent the particles from being broken out.

Well-embedded ceramic particles with an extremely high interfacial quality cannot compensate the peaks of the transverse stresses by relative motion. Microcracking of the brittle ceramic fillers is the consequence.

3.1.2.3. Calendering temperature of magnetic layers. When containing rather large alumina particles, magnetic layers calendered at low temperatures ($T_c = 30°C$) exhibit a time-dependent friction coefficient, $\mu'(t)$, which is significantly higher as for layers calendered at high temperatures ($T_c = 90°C$) (figs. 25 and 26). Moreover, the width of the main wear tracks produced by the rotating videoheads differ by a factor ten after a period of 30 min of still-picture test time. Another 30 min later the differences in friction and wear between the tapes D/28 (2.5%, 0.7 μm, 30°C) and D/32 (2.5%, 0.7 μm, 90°C) begin to become blurred. The width of the main wear track and the height of the friction coefficient $\mu'(t \geqslant 60$ min) are approxi-

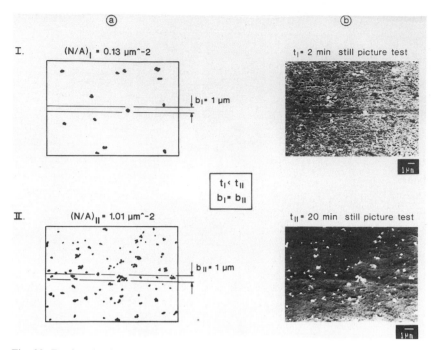

Fig. 22. During the first stage of wear the ability of supporting the heads and preventing them from penetrating the magnetic layer does primarily not depend on the size of the alumina particles but on the distance between the fillers.

Fig. 23. With increasing still-picture test time at the second stage of wear, the interfaces between badly embedded alumina particles and the matrix are damaged by the rotating videoheads. Relative motion and a subsequent breaking out of the fillers are the results.

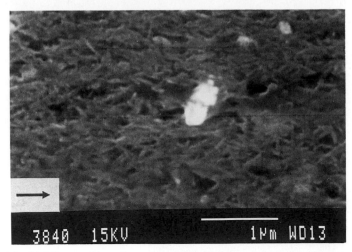

Fig. 24. Microcracking of a well-embedded filler particle during the second stage of a still-picture test.

mately alike. Summing up, it can be said that varying calendering temperatures of magnetic layers containing large alumina particles ($D_p = 0.7 \ \mu m$) have consequences on the tribological behaviour only during the first 60 min of a still-picture test.

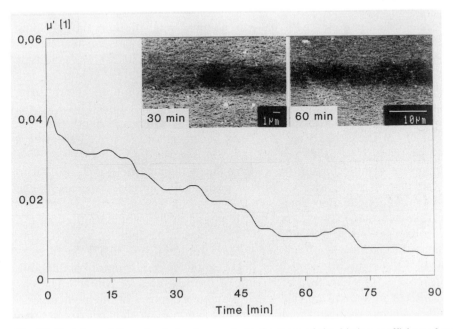

Fig. 25. Variation of the calendering temperature: the trace of the friction coefficient of tape D/28 (3.8%, 0.7 μm, 30°C) combined with SEM micrographs of the wear tracks.

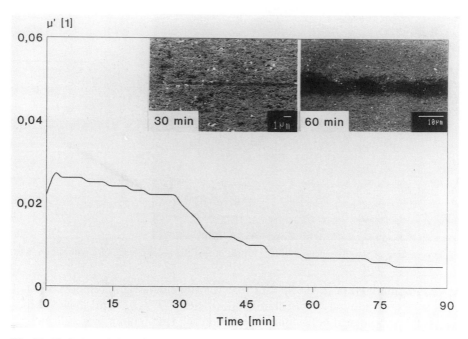

Fig. 26. Variation of the calendering temperature: the trace of the friction coefficient of tape D/32 (3.8%, 0.7 μm, 90°C) combined with SEM micrographs of the wear tracks.

Using small filler particles with 0.2 μm nominal size, there is only a slight difference in the trace of the friction coefficient of magnetic layers calendered at high ($T_c = 90°C$) and low ($T_c = 45°C$) temperatures (figs. 27 and 28). Still, the tape calendered at high temperature exhibits lower friction. After 30 min of test duration the widths of the wear tracks are similar and amount to about 1 μm. Differences occur in the following test period: after 60 min tape D/24 (2.5%, 0.2 μm, 80°C) shows a wear track nearly unchanged in width and depth, whereas the wear of D/20 (2.5%, 0.2 μm, 45°C) has become more severe, presenting a track of 12 μm width.

It is important to note that in the case of small filler particles ($D_p = 0.2$ μm) different calendering temperatures do not lead to significant consequences for the wear until a certain incubation time, about 30 min.

It is well known that magnetic layers calendered at high temperatures have a higher ultra-microhardness as well as a higher elastic modulus of compression, when compared to layers calendered at low temperatures [21,31,36,37]. An increasing hardness of the composite leads to a decrease of the plastic deformability and, at least if it comes to wear, to a reduction of the real area of contact. Thereby adhesive contacts are minimized. The greater amount of pores in the low temperature calendered layers represent weak points in relation to mechanical stresses and has unpleasant consequences on the tribological behaviour. Pores may act as nuclei for further deterioration of the microstructure.

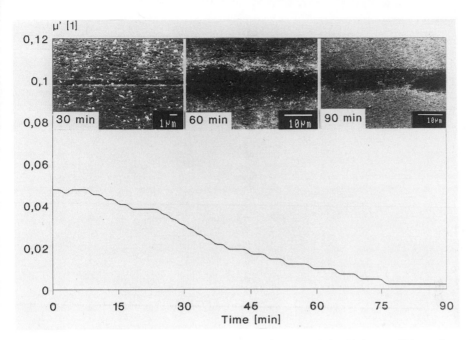

Fig. 27. Variation of the calendering temperature: the trace of the friction coefficient of tape D/20 (3.8%, 0.2 μm, 45°C) combined with SEM micrographs of the wear tracks.

As we have mentioned in section 3.1.2.2, an increasing distance between the Al$_2$O$_3$ particles leads to an increasing influence of the relative soft matrix material on the friction and wear behaviour during still-picture tests. The greater the distance between the single elements of the hard phase, the greater the tendency of showing a tribological behaviour like that of unreinforced magnetic layers (cf. figs. 20 and 21).

As a consequence, harder matrices containing large filler particles with 0.7 μm nominal size, exhibit lower friction and less wear than softer ones during the first 30 min of a still-picture test. With increasing test time the efficiency of the reinforcing component is influenced in different ways depending on the calendering temperature:

(1) The damage of interfaces between the hard phase and the matrix and the following breaking out of the alumina particles needs less energy if the matrix is soft.

(2) Magnetic layers calendered at high temperatures encourage microcracking of the Al$_2$O$_3$ particles when coming into contact with their counterparts. Peaks of transversal stresses can only be reduced to a limited degree, because of the small ductility of the matrix.

These microstructural effects form the basis of the similar resistance to wear after 60 min still-picture test time in the case of the tapes D/28 (2.5%, 0.7 μm, 30°C) and D/32 (2.5%, 0.7 μm, 90°C).

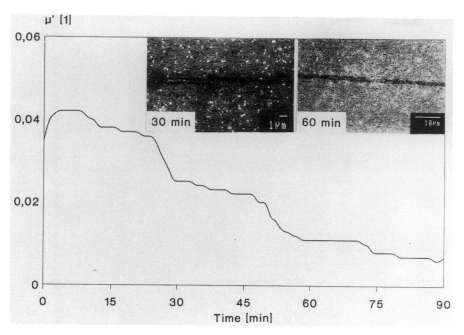

Fig. 28. Variation of the calendering temperature: the trace of the friction coefficient of tape D/24 (3.8%, 0.2 μm, 80°C) combined with SEM micrographs of the wear tracks.

If the free path length between the alumina particles is as small as it is in the case of the tapes D/20 (2.5%, 0.2 μm, 45°C) and D/24 (2.5%, 0.2 μm, 0°C), the influence of the relatively soft matrix is less important (cf. section 3.1.2.2). The higher area density of the hard phase supports the rotating heads and prevents them from penetrating the magnetic layer. With increasing test time the efficiency of the reinforcing component is reduced, depending on the calendering temperature of the magnetic layer:

(1) The covering up of single Al_2O_3 particles and agglomerates with binder needs less energy if the matrix material is soft, which leads to a higher likelyhood of this effect in the case of tapes calendered at low temperatures.

(2) The breaking up of clusters is mainly influenced by the strength of the van der Waals bonds between the single particles of an agglomerate and does not depend on the calendering temperature.

Therefore, the efficiency to prevent wear of small Al_2O_3 particles ($D_p = 0.2$ μm) inside magnetic layers calendered at low temperatures is reduced in a more severe way with increasing test time.

3.2. Microscratch tests

3.2.1. Spherical counterpart

On condition of both constant calendering temperature and size of the alumina filler particles, the coefficient of sliding friction can be decreased by increasing the

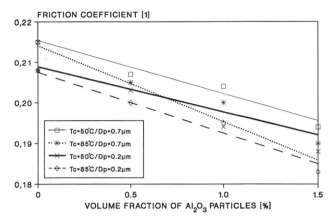

Fig. 29. The trace of the coefficient of sliding friction versus the volume concentration of the alumina filler particles for different calendering temperatures and varying Al_2O_3-particle sizes.

volume fraction of the hard phase (fig. 29). The rate of the slope is greater when small particles ($D_p = 0.2$ μm) are used, i.e. small particles exhibit a higher reinforcing efficiency

$$T_c = \text{const.:} \quad |\Delta\mu_{0.2\,\mu m}(\phi)| > |\Delta\mu_{0.7\,\mu m}(\phi)|.$$

In addition, rather large particles ($D_p = 0.7$ μm) always lead to a higher coefficient of fraction (fig. 30).

$$T_c = \text{const.:} \quad \mu_{0.7\,\mu m}(\phi) > \mu_{0.02\,\mu m}(\phi).$$

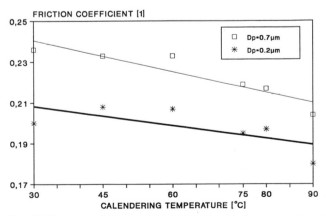

Fig. 30. The trace of the coefficient of sliding friction versus the calendering temperature for different sizes of the filler particles.

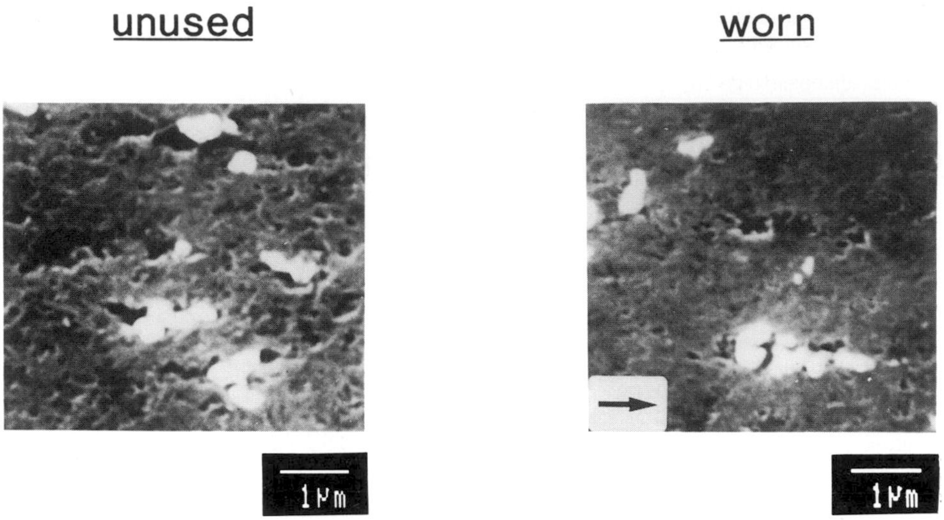

unused   ## worn

Fig. 31. The surface of a particulate magnetic layer in unused conditions and worn by a spherical indenter, using a relative velocity of 1.6 μm/s and a normal load of 0.5 N.

If one is observing the influence of the calendering temperature for low volume fractions of Al_2O_3 particles, it is obvious that a higher T_c in each case produces less friction, but the difference for tapes calendered at low temperatures decreases with the amount of alumina. A certain volume fraction exists, from which point onwards the influence of the calendering temperature is negligible and the friction and wear behaviour is dominated by the reinforcing component only. This point will be reached much earlier if small particles instead of larger ones are used.

During the single-scratch experiments, no interface phenomena appeared with respect to the alumina particles and the matrix material. The remaining deformation of the matrix is characterized by a compression of the pores similar to the calendering process. The rise of the heat at the asperities of the magnetic layer is below the softening point of the polyurethane and not dramatic therefore a covering up of Al_2O_3 particles with binder could not occur. Tribochemical effects do not play an important role in this tribological system [38,39] (fig. 31). Distinguishing between the locations of appearance, the following energy-dissipative processes can be determined as being the most important ones:

bulk material: • plastic deformation,
                • elastic–viscous deformation,
contact area: • adhesion.

The plastic deformation by compressing the layer is mainly influenced by the pore volume content, i.e. by the thermomechanical treatment during the production process of the tape. The higher the calendering temperature, the lower the pore concentration and the higher the microhardness of the layer. This leads to only a negligible dissipation of energy by plastic deformation of high temperature calendered layers [38].

As a reaction towards stress, the particle reinforced and thermoplastically deformable rubber matrix shows a time-dependent, reversible deformation behaviour controlled by the relaxation time τ. A difference in phases δ appears between the form change and the mechanical stress. This non-linear behaviour can be described by the complex modulus, E^* [34,40,41], where the real part represents the elastic term and the imaginary part the dissipative term:

$$E^* = E' + iE''. \tag{3}$$

The relative importance of the dissipative term with respect to the elastic term in the deformation process is indicated by the tangent of the loss angle, defined by

$$\tan \delta = E''/E'. \tag{4}$$

For the description of the damping properties of an elastic–viscous material, the complex modulus of compression, K^*, or the modulus of transverse elasticity, G^*, under oscillating compressive stress or torsional strain are determined in general [40,41].

The elasticoviscous behaviour of a particulate magnetic layer is influenced by the three parameters
– the calendering temperature,
– the volume fraction of the alumina filler particles, and
– the size of the alumina filler particles.
The dissipation of energy caused by the hysteresis effect during elastic–viscous deformation of particulate magnetic layers was suggested by Wierenga and Schaake [38] to be:

$$d\gamma_{\text{elasticoviscous}} \sim \frac{\tan \delta}{E'}, \tag{5}$$

and in general for particle reinforced polymers Briscoe [42] gave:

$$d\gamma_{\text{elasticoviscous}} \sim \frac{\tan \delta}{(E')^{1/3}}. \tag{6}$$

With increasing temperature of the calendering process the porosity of the magnetic layer decreases and the elastic term, E', decreases too. At the same time the loss tangent, $\tan \delta$, diminishes. As a consequence, the dissipation of energy produced by elastic–viscous deformation is minimized.

The pore concentration of the magnetic layer is the prevailing parameter, while the deformation properties of the bulk polymer are not significantly influenced by the calendering temperature, but only by the rate of mixture between the hard and soft segments of the utilized polyurethane type [38].

A higher amount of alumina filler particles gives rise to an increase of the modulus of compression, because on increasing the volume fraction and, thereby, decreasing the free path length, the flow and strain fields around particles interact. The composite becomes stiffer [42–44].

The elastic behaviour of a composite consisting of hard, spherical particles embedded in a soft, polymeric matrix, can be described using the empirically determined equations of Guth and Kerner [43]:

$$K' = K'_m \left[1 + 2.5\phi + 14.1\phi^2 \right],$$ (7)

$$K' = K'_m \left[1 + \frac{\phi}{1-\phi} \frac{15(1-\nu_m)}{(8-10\nu_m)} \right].$$ (8)

These correlations are only valid if the fillers' modulus of compression, K'_f, can be considered to be much higher than the modulus of the matrix material, K'_m. On condition of ideal interfacial quality an estimation of the expected increase of elasticity, $\Delta K'$, caused by the Al_2O_3 particles is possible. Volume fractions of the alumina fillers less than 10% lead to similar results in both equations, since for low concentrations the stiffening action of the filler is independent of its size. For example, $\Delta K'$ amounts to 15% if the layer contains 5 vol% of alumina.

The real area of contact between the magnetic layer and the spherical counterpart is inversely proportional to the elastic modulus of compression, K' [38]:

$$A_{real} \sim F_n/K'.$$ (9)

When increasing the volume fraction of the fillers, a lower real area of contact and, thereby, lower adhesion is the result. Summing up, the energy dissipative effects "plastic deformation", "elastic–viscous deformation", and "adhesion" superimpose and lead to a reduction of the friction coefficient.

No correlations exists to describe the influence of the particle size on the elastic–viscous behaviour of a particle reinforced composite. Referring to the well-known increasing tendency to the formation of clusters with decreasing size of the fillers, an estimation can be made using Mooney's formula [43]

$$K' = K'_m \exp\left(\frac{2.5\phi}{1-S\phi} \right).$$ (10)

The crowding factor, S, represents the ratio between the volume occupied by the filler particles and the true volume of the fillers. For close packed spheres $S = 1.35$. In the case of an ideal dispersion S amounts to 1.0.

As the particle size decreases, the surface area increases, providing a more efficient interfacial bond. This is also accompanied by a tendency for increased agglomeration of the particles. Since agglomerates are able to carry a larger proportion of the load then the primary particles, a higher modulus is the result [43].

Since the discussed microstructural mechanisms do not act contrary to the dissipation of energy, the law of mixtures in the calculation of the friction coefficient with respect to the tribological system "magnetic layer–steel ball" is given by [21,45–47]

$$\frac{1}{\mu(\phi, D_p, T_c)} = \frac{\phi}{\mu_f(D_p)} + \frac{(1-\phi)}{\mu_m(T_c)}.$$ (11)

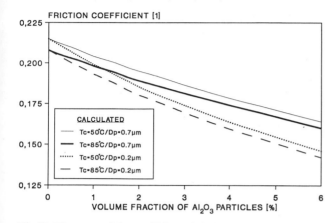

Fig. 32. The trace of the coefficient of sliding friction versus the volume concentration of filler particles calculated by the rule of mixtures.

The friction coefficient of the pure matrix material, dependent on the calendering temperature, and the friction coefficient of the reinforcing component, dependent on the size of the Al_2O_3 particles, have been calculated for tapes of type F by the use of linear regression analysis:

$$\mu_m(T_c = 50°C) = 0.215, \qquad \mu_f(D_p = 0.2 \ \mu m) = 0.024,$$
$$\mu_m(T_c = 85°C) = 0.208, \qquad \mu_f(D_p = 0.7 \ \mu m) = 0.035.$$

This method enables the possibility of extrapolating friction coefficients of tapes containing higher amounts of alumina than the tested ones but with similar chemical composition. An interpolation of the friction of magnetic layers calendered at temperatures between the peak values examined is possible (fig. 32).

3.2.2. Diamond tip

Tapes of type A with an average thickness of the magnetic layer of 5 to 6 μm were tested to guarantee that the bottom of the microgrooves does not contact the substrate. The penetration depth of the diamond tip amounted for each tape to about 4 μm (cf. fig. 10). The porous magnetic material inside the wear scar is compressed until the pore volume concentration in the near-surface area is reduced down to zero percent (figs. 33 and 34). If the maximum ability of plastic deformation is depleted, microcracks perpendicular to the sliding direction of the counterpart are the consequence in the case of magnetic layers containing large ($D_p = 0.7$ μm) filler particles. The polyurethane binder is only plastically deformable if it is heated, which in this tribological system plays no role. The rise of the temperature in the interface between the diamond and the magnetic layer is

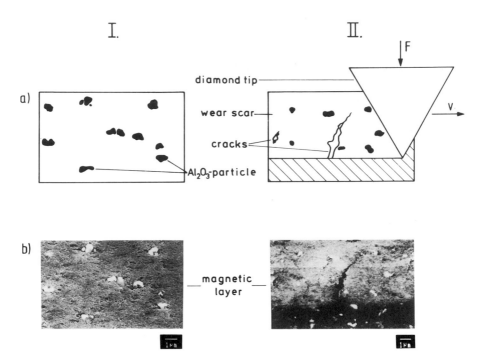

Fig. 33. Microcracking of the matrix material occurs during microscratch tests with a diamond tip if magnetic layers with large filler particles ($D_p = 0.7$ μm) are tested.

negligible. Both materials exhibit a minimal free surface energy and, therefore, minimal adhesion. The main mechanism of wear is abrasion. If large particles get into contact with the counterpart, the appearing transversal stresses lead to relative motion of the filler particles. Since at low temperatures the matrix material can only react with elastic–viscous deformations, crack initiation at the interfaces between the reinforcing component and the matrix material are the consequence. The crack propagation within the soft layer is encouraged by the great distance between the Al_2O_3 particles of about 2.7 μm (cf. table 1). Since the single particles are well embedded in general, removal of particles out of the composite could not be observed.

If small ($D_p = 0.2$ μm) particles are used, agglomerates of fillers in the flanks of the grooves are broken up if they get into contact with the diamond tip. The fragments on their part exhibit an abrasive behaviour and lead to third-body abrasion. With respect to the moving counterpart, this effect can be considered as similar to rolling friction, and leads to a decrease of the transversal stresses in the contact area. In combination with the small free path length between the single particles and clusters of about 0.8 μm, the likelihood of crack initiation is diminished (cf. table 1).

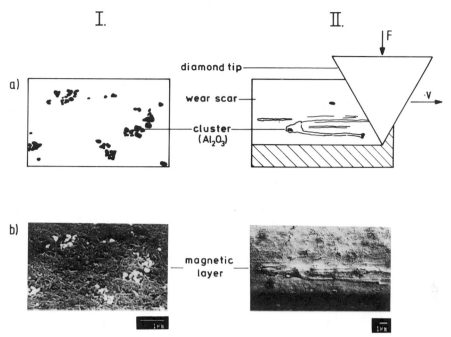

Fig. 34. Clusters of small particles ($D_p = 0.2$ μm) are broken up by the diamond tip. As a consequence, third-body abrasion occurs.

On condition of a constant amount of alumina, small filler particles exhibit a higher ability of mechanical support of the counterpart than large particles.

4. Concluding remarks

4.1. Microstructure of particulate magnetic layers

Magnetic layers consisting of cobalt doped γ-Fe_2O_3 particles embedded in a polyurethane matrix cannot be used as flexible media without a reinforcing component. The operativeness of a particular magnetic storage system is reduced by both the low lifetime of the layer and by the accumulation of wear debris at the heads.

If spherical Al_2O_3 particles are added to the dispersion, the negative effects of the fillers on the stored information and the positive effects on the tribological behaviour must be considered to find a compromise. Since the alumina particles do not show ferromagnetic behaviour, they act as voids with respect to the read/write process. The requirements and restrictions may be summed up as follows:

(1) *Magnetic restrictions.* The volume of the single particles and agglomerates used as the reinforcing phase is limited by the size of the basic bit-cell of the recorded data in order to minimize the likelyhood of signal losses and errors in the

read process. Considering the different fields of application, the length, width and unit area of one bit amounts in the case of particulate layers [8,9] to:

$$l_{\text{bit}}: \quad 0.2\text{--}2.0 \ \mu\text{m},$$
$$w_{\text{bit}}: \quad 10\text{--}100 \ \mu\text{m},$$
$$\rightarrow A_{\text{bit}}: \quad 2\text{--}200 \ \mu\text{m}^2.$$

In this connection, it is important to note that the pores act as failure spots with respect to the stored information too. From a pure magnetic point of view a high calendering temperature and, thereby, a high compression of the magnetic layer is recommended.

With decreasing size of the ceramic filler particles the volume fraction of agglomerates and the overall size of the clusters increase. Hence, an ideal dispersion of the reinforcing phase should be aspired to.

(2) *Tribological requirements.* The maximum size of the Al_2O_3 particles is restricted both by the free path length between the single particles and clusters supporting the wearing heads and by the amount of head wear needed to clean the countersurface. The minimum size of the alumina fillers depends on the quality of dispersion. Although agglomerates have a high short-time loading capacity, after a certain incubation time they will be demolished by the heads, beginning at the front areas. An ideal dispersion of the reinforcing phase is therefore desirable. Well-embedded single particles of rather small size are mainly covered with binder in the duration of a still-picture test. Microcracking of the ceramic particles appears above a critical size.

Magnetic layers containing Al_2O_3 particles with an average diameter of 0.2 μm show a good combination of high wear resistance, low friction,and sufficient head cleaning effect. Apart from the ratio of hardness, the difference in particle volume between the magnetic particles and the non-magnetic fillers is substantial (cf. table 2):

- γ-Fe_2O_3 magnetic particles: $1 \times 10^{-4} \ \mu\text{m}^3$,
- Al_2O_3 filler particles: $40 \times 10^{-4} \ \mu\text{m}^3$ $(D_{\text{p}} = 0.2 \ \mu\text{m})$,
                                        $200 \times 10^{-4} \ \mu\text{m}^3$ $(D_{\text{p}} = 0.33 \ \mu\text{m})$,
                                        $1800 \times 10^{-4} \ \mu\text{m}^3$ $(D_{\text{p}} = 0.7 \ \mu\text{m})$.

Considering the relevant magnetic and tribological parameters examined during practice-oriented still-picture tests and model tests, an ideal microstructure of a particulate layer for magnetic tape recording can be assumed to be as follows:
- volume fraction of the Al_2O_3 fillers: $\approx 2.0\%$,
- average diameter of the Al_2O_3 fillers: 0.2 μm,
- calendering temperature of the magnetic layer: 90°C.

Magnetic layers showing a composition similar to this one, exhibit an excellent compromise of an extended servicelife and good recording and playback quality.

4.2. Simulation of practice-oriented tribological tests

In the case of repeated-pass sliding during still-picture tests, the tribological behaviour of the tapes can be subdivided into two periods of time. The first stage is characterized by plastic deformation of the matrix asperities, high flash temperatures in the contact area, and strong adhesive forces between the videoheads and the coating. At that moment the influence of the rotating top part of the head drum on the friction and wear is insignificant.

The variation of the microstructural parameters "amount of alumina particles", "size of the alumina particles", and "calendering temperature" effects identical changes in the trace of the friction coefficient both during still-picture tests and scratch tests with a steel ball as the counterpart. The tribological effects induced by the spherical indenter are adhesion, elastic–viscous deformation, and plastic deformation by compressing the porous magnetic coating. The rise of the temperature in the contact zone is negligible.

Increasing either the amount of alumina or the calendering temperature or decreasing the size of the filler particles results in a decrease of the adhesion and the deformation term of the dissipated energy for both testing methods.

Since interfacial phenomena do not appear within the particle reinforced composite "magnetic layer" during the first stage of a still-picture test, the practice-oriented friction behaviour can be simulated correctly by microscratch tests and a spherical counterpart. At that time, the wear behaviour of different magnetic layers is proportional to their friction behaviour.

During the second stage of the still-picture tests the wear behaviour of the coatings is characterized by an increasing occurrence of interfacial phenomena, such as
- the covering of small particles and clusters of small particles with binder,
- the damage of interfaces between the reinforcing phase and the matrix and the subsequent breaking out of single Al_2O_3 particles exceeding a critical size,
- the demolition of clusters after a certain incubation time,
- the microcracking of well-embedded alumina particles which exhibit a rather high interfacial quality.

These effects and an increasing influence of the rotating top part of the drum on the trace of the friction coefficient lead to a different behaviour of the wear rate and the friction coefficient.

Microscratch tests with a diamond tip offer the possibility of simulating the demolition of agglomerates of small alumina particles and the wear behaviour of the pure matrix. Since neither adhesion nor a rise in temperature within the contacting surfaces appear during these model tests, further interfacial effects cannot be observed.

In the special case of new, unused videoheads with sharp edges, the wear mechanisms "abrasion" and "microshavings" as the appearance of wear can be simulated by using a diamond tip.

To sum up, it can be said that large-scale practice-oriented still-picture tests in the first stage can be simulated quickly by microscratch tests and a spherical counterpart. This statement is valid up to that moment when interfacial phenomena between the filler particles and the matrix begin to influence the friction and wear behaviour of the composite more intensely. During the second stage single energy-dissipating interfacial processes can be simulated with microscratch tests by using a suitable counterpart, such as a diamond tip.

Acknowledgements

The authors would like to thank their colleagues at BASF AG for providing the magnetic tapes. Furthermore, the authors thank Prof. Dr.-Ing. E. Hornbogen (Ruhr-Universität Bochum), Dr. H. Jakusch (BASF), and Dr. A. Koller (AGFA) for many helpful discussions and also Mr. S.W. Marsh (Marsh Dental Lab, Preston, UK) for carefully checking the manuscript.

List of symbols

A	area
A_{bit}	area of a bit-cell
A_{real}	real area of contact
b	width of a wear track
D_p	nominal size of the alumina filler particles
E^*	complex modulus of elasticity
E'	elastic term of the complex modulus
E''	dissipative term of the complex modulus
F_n	normal load
i	imaginary unit
K'	elastic term of the complex modulus of compression
K_m	modulus of the matrix material
K_f	modulus of the filler particles
l_{bit}	length of a bit cell
N	number of single filler particles and agglomerates
S	crowding factor
t	time
T_c	calendering temperature of the magnetic layer
T_g	glass transition point
U	voltage used to rotate the top part of the drum
U_0	no-load voltage
v	velocity
w_{bit}	width of a bit cell
$d\gamma_{elasticoviscous}$	dissipation of energy by elastic–viscous deformation
δ	difference in phases
$\tan \delta$	tangent of the loss angle
Δ	difference
ϕ	volume concentration of the alumina filler particles
τ	relaxation time
μ, μ'	friction coefficients
μ_f	friction coefficient of the filler particles
μ_m	friction coefficient of the matrix material
ν_m	coefficient of transversal contraction of the matrix material

References

[1] D.A. Harvey, State of the media, Byte 11 (1990) 275–281.
[2] C. Bonnebat, Information storage technologies – Some recent trends and current prospects analyzed from the media manufacturer standpoint, IEEE Trans. Magn. MAG-23 (1987) 9–15.
[3] D.J. Gravesteijn, C.J. van der Poel and P.M.L.O. Scholte, Phase change optical recording, Philips Tech. Rev. 44 (1989) 250–258.
[4] R. Wood, Magnetic megabits, IEEE Spectrum 5 (1990) 32–37.
[5] S. Apiki and H. Eglowstein, The optical option, Byte 10 (1989) 160–174.
[6] W. Lee, Thin films for optical data storage, J. Vac. Sci. Technol. A 3 (1985) 640–646.
[7] G. Bate, Materials challenges in metallic, reversible, optical recording media: A review, IEEE Trans. Magn. MAG-23 (1987) 156–161.
[8] F. Jorgensen, The Complete Handbook of Magnetic Recording, 3rd Ed. (TAB Books, Blue Ridge Summit, PA, 1988) pp. 1–437, 630–644.
[9] M. Camras, Magnetic Recording Handbook (Van Nostrand Reinhold, New York, 1988) pp. 15–261, 293–334, 504–537.
[10] T. Suzuki, Perpendicular magnetic recording, IEEE Trans. Magn. MAG-20 (1984) 675–680.
[11] B. Bhushan, Tribology and Mechanics of Magnetic Storage Devices (Springer, New York, 1990) pp. 1–991.
[12] J.C. Iwata, Magnetic recording, IBM Res. Mag. 12 (1985) 6–11.
[13] E.D. Daniel and C.D. Mee, Introduction, in: Magnetic Recording Volume I: Technology, eds C.D. Mee and E.D. Daniel (McGraw-Hill, New York, 1987) pp. 1–21.
[14] C.D. Mee, The Physics of Magnetic Recording (North-Holland, Amsterdam, 1986) pp. 85–235.
[15] W.D. Nix, Mechanical properties of thin films, Metall. Trans. A 20 (1989) 2217–2245.
[16] F.W. Hahn Jr, Wear of recording heads by magnetic tape, in: Tribology and Mechanics of Magnetic Storage Systems, ed. B. Bhushan, ASLE Special Publication SP-16 (ASLE, Park Ridge, IL, 1984) pp. 41–48.
[17] R. Kaneko, Micro-tribological approach to head–medium interfaces of magnetic recording, in: Proc. 5th Int. Congress on Tribology, EUROTRIB, eds K. Holmberg and J. Nieminen, Vol. 3 (Espoo, 1989) pp. 210–215.
[18] D.P. Smith, Reliability issues in data tape recording, in: Proc. 5th Int. Congress on Tribology, EUROTRIB, eds K. Holmberg and J. Nieminen, Vol. 3 (Espoo, 1989) pp. 216–221.
[19] B. Bhushan and K. Tonder, Roughness-induced shear and squeeze film effects in magnetic recording - Part I: Analysis, J. Tribol. 11 (1989) 220–227.
[20] E. Rabinowicz, The tribology of magnetic recording systems – An overview, in: Tribology and Mechanics of Magnetic Storage Systems, ed. B. Bhushan, ASLE Special Publication SP-21 (Park Ridge, IL, 1986) pp. 1–7.
[21] A. Broese van Groenou, On the interpretation of tape friction, IEEE Trans. Magn. MAG-26 (1990) 144–146.
[22] E.P. Wohlfahrt, Magnetic material for recording, Phys. Status Solidi 91 (1985) 339–350.
[23] S.B. Luitjens, Magnetic recording trends: media developments and future (video) recording systems, IEEE Trans. Magn. MAG-26 (1990) 6–11.
[24] P. Williams and S. Wales, Recent developments in particulate recording media, IEEE Trans. Magn. MAG-24 (1988) 1876–1879.
[25] J.F. Robinson, Videotape Recording, 3rd Ed. (Butterworths, London, 1975) pp. 19–54, 166–187.
[26] R.E. Jones Jr and C.D. Mee, Recording heads, in: Magnetic Recording, Volume I: Technology, eds C.D. Mee and E.D. Daniel (McGraw-Hill, New York, 1987) pp. 244–336.
[27] H. Jacobi and U. Nowak, The influence of Al_2O_3 particles on the friction and wear behavior of particulate video tapes, in: Tribology and Mechanics of Magnetic Storage Systems, ed. B. Bushan, ASLE Special Publication SP-29 (Park Ridge, IL, 1990) pp. 114–122.
[28] G. Steinberg, Tribology of magnetic media and its relation to media failure, IEEE Trans. Magn. MAG-23 (1987) 115–117.

[29] J.L. Lauer and W.R. Jones Jr, Friction Polymers, in: Tribology and Mechanics of Magnetic Storage Systems, ed. B. Bushan, ASLE Special Publication SP-21 (Park Ridge, IL, 1986) pp. 12–23.

[30] H.L. Zahn, Friction – Its influence in rotary magnetic tape recorders, SMPTE J. 7 (1989) 520–524.

[31] A. Broese van Groenou, Tape friction and head wear in video tape recording, Paper AD-89/008, Philips Research Laboratories, Research Group Magnetism (1988).

[32] B. Bhushan, B.S. Sharma and R.L. Bradshaw, Friction in magnetic tapes I: Assessment of relevant theory, ASLE Trans. 27 (1983) 33–44.

[33] A.A. Staals, M.C. van Houwelingen and H.F. Huisman, Localization and characterization of sub-surface particles in magnetic tape, IEEE Trans. Magn. MAG-23 (1987) 112–114.

[34] R.C.F. Schaake and H.F. Huisman, Headwear regulation of CrO_2 video tape, IEEE Trans. Magn. MAG-23 (1987) 103–105.

[35] K. Miyoshi and D.H. Buckley, Effect of wear on structure-sensitive magnetic properties of ceramic ferrite in contact with magnetic tape, in: Tribology and Mechanics of Magnetic Storage Systems, ed. B. Bhushan, ASLE Special Publication SP-19 (Park Ridge, IL, 1985) pp. 112–128.

[36] B. Bhushan, Analysis of the real area of contact between a polymeric magnetic medium and a rigid surface, J. Tribol. 106 (1984) 26–34.

[37] J.H.M. van der Linden, P.E. Wierenga and E.P. Honig, Viscoelastic behavior of polymer layers with inclusions, J. Appl. Phys. 62 (1987) 1613–1615.

[38] P.E. Wierenga and R.C.F. Schaake, The effect of mechanical and chemical surface properties on the friction of magnetic tapes, Wear 119 (1987) 29–50.

[39] J.J. Brondijk, P.E. Wierenga, E.E. Feekes and W.J.J.M. Sprangers, Roughness and deformation aspects in calendering of particulate magnetic tapes, IEEE Trans. Magn. MAG-23 (1987) 146–149.

[40] R.G. Stacer, C. Hübner and D.M. Husband, Binder/filler interaction and the nonlinear behavior of highly-filled elastomers, Rubber Chem. & Technol. 63 (1990) 488–502.

[41] C.M. Roland and G.F. Lee, Interaggregate interaction in filled rubber, Rubber Chem. & Technol. 63 (1990) 554–566.

[42] B.J. Briscoe, Interfacial friction of polymer composites – General fundamental principles, in: Friction and Wear of Polymer Composites, ed. K. Friedrich (Elsevier, Amsterdam, 1986) pp. 25–60.

[43] S. Ahmed and F.R. Jones, A review of particulate reinforcement theories for polymer composites, J. Mater. Sci. 25 (1990) 4933–4942.

[44] F.J. Guild and R.J. Young, A predictive model for particulate-filled composite materials – Part I: Hard particles, J. Mater. Sci. 24 (1989) 298–306.

[45] E. Hornbogen, Description of wear of materials with isotropic and anisotropic microstructures, in: Proc. Int. Conf. on Wear of Materials WEAR, Vancouver, ed. K.C. Ludema (1985) pp. 477–484.

[46] E. Hornbogen, Friction and wear of materials with heterogeneous microstructures, in: Friction and Wear of Polymer Composites, ed. K. Friedrich (Elsevier, Amsterdam, 1986) pp. 61–88.

[47] K. Friedrich, Wear of reinforced polymers by different abrasive counterparts, in: Friction and Wear of Polymer Composites, ed. K. Friedrich (Elsevier, Amsterdam, 1986) pp. 233–287.

Advances in Composite Tribology
edited by K. Friedrich

Chapter 14

Friction and Wear of Polymers, Ceramics and Composites in Biomedical Applications

G.W. STACHOWIAK

University of Western Australia, Department of Mechanical and Materials Engineering, Nedlands, WA 6009, Australia

Contents

Abstract

The tribological characteristics of non-metallic materials used in orthopaedic implants and restorative dentistry are described in this chapter. Polymers, ceramics and composites are used as articulating components in hip, knee, elbow, shoulder and finger prostheses, while in dentistry they perform as denture teeth, veneers and filling materials.

The wear and frictional characteristics of commonly used materials such as UHMWPE and alumina, tested on simple laboratory rigs and on joint simulators, are discussed and compared with the clinical performance of actual prosthetic parts. Despite the limitations with respect to the simulation of the real contact conditions in the body, the laboratory tests provide a relatively quick assessment of the suitability of particular materials for orthopaedic applications. Both the advantages and disadvantages of several articulating combinations, i.e. UHMWPE/metal, UHMWPE/alumina and alumina/alumina, are underlined by comparing the frictional levels, wear rates and topographical features on the worn surfaces of the tested pairs.

Certain weak points of the currently used articulating systems, such as the deterioration of UHMWPE properties, the increased wear after prolonged use under biological conditions and the brittleness of alumina, have prompted research efforts towards the development and optimization of substitutes. The tribological characteristics of short carbon fibre reinforced UHMWPE in laboratory tests and its clinical performance as tibial components are discussed. Although carbon fibre reinforced UHMWPE is characterized by better creep resistance, its wear behaviour is not improved in comparison to unmodified UHMWPE. The composite suffers from wear modes typical of matrix/filler interfaces, i.e. fibre pull-out, fibre breakage, crack initiation at the interface and pit formation. Although other composites, such as carbon fibre/polysulfone, carbon fibre/epoxy and carbon fibre/carbon composites, are being considered as replacements for metallic stems in a hip prosthesis, they are not considered for the use as articulating systems. The main barriers preventing the tribological application of modern composites are: manufacturing difficulties and the inadequate tribological characteristics of these materials, especially under lubricated conditions.

Partially stabilized zirconia (PSZ), which is a two-phase ceramic and displays a transformation toughening mechanism, has been proposed as an alumina substitute, and has already been applied in some selected clinical trials. Laboratory tests have shown that PSZ ceramics cause lower wear of UHMWPE in the articulating systems compared to traditional alumina.

Among the tribological properties of restorative dental materials and denture teeth their resistance to abrasion is of major importance. The abrasion resistance of acrylic resins, composite resins and ceramics tested on various test rigs, i.e. pin-on-disc, abrasion and scratch test rigs, is discussed and their clinical performance is evaluated. Acrylic resins and composite resins perform successfully as denture teeth, veneers and restorative materials for anterior teeth. The use of resins in posterior tooth fillings is still controversial due to limitations in their abrasive wear resistance. Traditional dental ceramics such as porcelain often cause excessive wear of the opposing natural enamel. Newly developed glass–ceramic materials are being evaluated as substitutes for porcelain.

1. Introduction

Non-metallic materials used in biomedical applications comprise polymers, ceramics and their composites. These biomaterials can be broadly divided into two groups: materials used directly in surgical reconstruction and materials used in medical devices. The scope of this chapter is directed to the tribological behaviour of the first group of biomaterials. There are two main areas of application in which the tribological characteristics of biomaterials are of primary importance: orthopaedic surgery (joint prostheses) and dentistry (restorations).

The artificial materials for use in a human body must satisfy the requirement of biocompatibility, must be non-toxic and must possess adequate physical and mechanical properties. Furthermore, the materials must be durable and stable enough to withstand the corrosive effects of body fluids, and their properties should remain unchanged over long periods of time, ideally for the life-time of a patient.

The material biocompatibility and toxicity are initially assessed in vitro by using cell culture methods and then tests are conducted in vivo on small animals. These tests provide information on the reaction of body tissue to foreign bulk material and small particles which might be generated by wear, corrosion or fracture of artificial implants. Since the implant components are repeatedly subjected to often large bending or torsional forces, high mechanical strength, ductility and good fatigue resistance of the material are required. For load-bearing articulating components extremely good tribological characteristics, i.e. low coefficient of friction and resistance to wear damage, are also required. The tribological properties of biomaterials are first assessed on simple laboratory test rigs and then on joint simulators. The final and most reliable evaluation is, however, provided by clinical analysis of real implants which have been functioning in the body over long periods of time.

In this chapter the knowledge available on the tribological characteristics of non-metallic materials used in orthopaedic surgery and dentistry is summarized. Tribological characteristics of polymers and ceramics and their composites used in orthopaedic and dental applications are briefly described. The performance of these materials in laboratory tests, on joint simulators and in real applications is compared and discussed. New advances in biomaterials, especially the introduction of composites, are also outlined.

2. Tribological characteristics of biomaterials in orthopaedic applications

A total joint replacement is regarded as one of the most significant and profound achievements in orthopaedic surgery. Many problems facing material scientists, technologists and surgeons had to be overcome before successful joint prostheses were developed and implanted. A wide range of research activities in these fields is being continued in an attempt to improve and update the material selection, designs and manufacturing technology of joint prostheses, as well as to improve the surgical procedures and techniques of their implantation. The main problem with artificial implants is, however, that in the long run they tend to fail. Aseptic loosening seems to be the main cause of the late failures and is related to many factors, such as material characteristics, prosthetic design and surgical techniques.

2.1. Natural joints

Natural load-bearing articulating joints are considered to be very efficient bearings which allow for the relative motion of bones under an applied load with minimal friction and wear of the contacting surfaces. The articulating surfaces are covered with a layer of cartilage, which is lubricated with synovial fluid to produce a coefficient of friction in the range of 0.005–0.025 [1]. Although several lubrication mechanisms operating in the synovial joints have been proposed in the past [2–10], none of them gives a complete explanation of all the phenomena observed.

The normal functioning of a joint can be impaired by numerous factors, such as degenerative disease (arthritis) or injury. When all the traditional treatments fail to improve the joint condition, a total joint replacement becomes the last chance to eliminate pain and restore the function of the diseased joints. The remarkable anatomical and physiological features of a natural articulating joint are impossible to reproduce artificially, and thus some compromise has to be reached in order to design and build an optimally functioning artificial implant. Worldwide, more than half a million joint prostheses are implanted in humans every year. Among them the hip and knee joints are the most important considering the high degree of disability which is caused by the impaired function of these joints.

2.2. Orthopaedic implants

The long-term performance of a total joint replacement prosthesis depends on:
– the mechanical properties of the materials used for the loaded components,
– the quality of implant fixation to the surrounding bony structure,
– and the rate of wear and creep of the articulating components.

Until recently, metallic alloys were the only materials used for femoral component applications. Now, the possibilities of replacing metals by non-metallic composite materials, such as fibre reinforced plastics (FRP) and carbon fibre reinforced carbon (CFRC), are studied intensely [11–15]. The metallic stems have a certain drawback since, due to their high elastic moduli, they are not biomechanically compatible with the host bone. The metallic stem geometry combined with its high stiffness results in stresses which are mainly carried by the implant, and this causes problems at the interface with the bone. The resulting stress distribution is responsible for stress shielding effects, bone resorption and aseptic loosening, which often lead to the premature failure of an implant [16]. Another problem associated with the use of metallic components is that the body tissues, especially those adjacent to the implant, are exposed to elevated concentrations of metallic ions leaching from the component surface. It has been shown that some of the ions can produce severe adverse tissue reactions [17–21].

As mentioned already, the loosening at the bone/implant interface is the major cause of long-term implant failure and there are a number of factors responsible. The stiffness mismatch between bone and implant causes detrimental bone remodelling and may result in prosthesis fracture. The loosening of the implant can also result from the disruption of the cement mantle due to inadequate properties of acrylic cement, or necrosis of surrounding tissues caused by elevated temperatures reached during its polymerization. Another very significant factor is associated with aseptic loosening due to the presence of wear particles. Finally, poor stabilization and relative motion at the interface between implant and tissue under load, especially during the early healing period, and the formation of interfacial soft connective tissue, also contribute to the loosening effects. Infection, although not as common, is also a very serious source of prosthesis failure, but it is usually manifested after short periods of time.

Good tribological characteristics of the articulating implant components are critical to proper implant functioning. The implant components which are susceptible to wear damage, are acetabular cups and femoral heads in total hip replacements and tibial, patellar or femoral components of knee prostheses. The materials and prosthesis designs in current use vary widely, but one of the four following material combinations is always used in highly stressed articulating contacts: metal/metal, metal/polymer, ceramic/polymer or ceramic/ceramic. A schematic diagram of bearing components in joint prostheses is shown in fig. 1. The contacting surfaces in a joint oscillate under cyclic load and mixed boundary and hydrodynamic lubrication [22,23]. The fluid lubrication is provided by the generation of a secondary synovial fluid at the implant site.

2.3. Tribological evaluation of biomaterials for orthopaedic applications

When evaluating the tribological characteristics of biomaterials, the whole tribosystem, which includes materials, environment, contact geometry, applied load, surface speed and temperature etc., must be considered. For example, the tribological characteristics of non-metallic biomaterials are particularly affected by

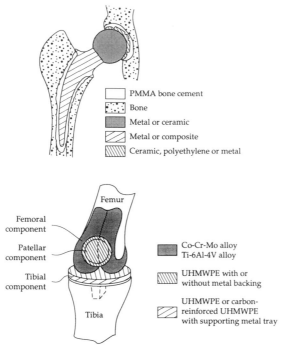

Fig. 1. Schematic representation of the total hip and knee joint replacements, showing bearing components.

the nature of the environment, whereas the tribological properties of metallic biomaterials are less susceptible.

A number of experimental methods have been developed for the evaluation of wear and frictional properties of biomaterials in vitro. Before final clinical trials, candidate materials are tested in two stages:
– on standard (simple) laboratory test rigs and
– on joint simulators.

The initial standard (simple) laboratory tests are conducted in order to: (i) preselect material combinations with tribological characteristics suitable in artificial implants, (ii) provide quantitative information on wear and friction behaviour and (iii) reveal the basic mechanism of wear involved. These tests are then followed by more sophisticated evaluation of selected materials on joint simulators. The information obtained from tests on joint simulators is used to predict the actual performance of artificial implants inside the human body and optimize the choice of materials, size of articulating components and their surface finish. The prediction of the long-term behaviour of implant components in in vivo conditions, based on the results obtained from the laboratory tests, can be very inaccurate, since material degradation in vivo and the body reaction to gradually accumulating wear debris are not well known.

2.3.1. Simple wear and friction tests

The most commonly used laboratory test rigs are listed in table 1. In the tests performed, the dependence of the wear and frictional coefficients on operating parameters such as load, surface speed, contact geometry, surface temperature, environment, etc., is determined for a particular combination of materials.

The results obtained from standard (simple) wear tests only provide information useful in the initial preselection of materials and material combinations. Unfortunately, none of the standard laboratory wear test rigs yield relevant information concerning wear and friction in the implant, because of the difficulties involved in reproducing the real contact conditions. Thus depending on the testing method used the results obtained can be different. Notwithstanding these facts, it has been demonstrated that in many cases these tests have provided accurate information about possible wear mechanisms involved in vivo, and the information gained has been used in the material selection for the actual application [32,33]. Despite difficulties in designing an adequate laboratory rig and test methodology for the evaluation of the tribological properties of materials used for artificial implants, an attempt has been made to standardize the procedure. A typical example of such a standardization is the ASTM F 732 standard, which covers the laboratory practice to evaluate the friction and wear properties of polymeric materials for use in total joint prostheses.

2.3.2. Joint simulators

Newly designed prosthetic components and whole prostheses are tested in joint simulators, in which the conditions of the load and motion cycles are faithfully reproduced to simulate the working cycle of a natural joint. Although these tests are more realistic, they also have some limitations. The typical life expectancy of an implant exceeds ten years, thus accelerated wear tests are conducted on joint simulators under more severe operating conditions than those encountered in natural joints. In the simulators the performance of the whole system is assessed. The information gained refers not only to the tribological characteristics of a material, but also to the design and manufacturing aspects of the whole prosthesis.

TABLE 1

Brief characteristics of commonly used laboratory wear tests.

Geometry	Motion	Load	Lubricant	Ref.
Disc-on-flat	unidirectional	constant	water	[24]
Washer-on-flat	oscillating	constant	blood plasma	[25]
Shaft on half journal	oscillating	oscillating	water	[26]
Journal	unidirectional	constant	dry, water	[27]
Pin-on-flat	oscillating	constant	serum, water, saline	[28]
			Ringer's solution	[29]
			water	[30]
Tri-pin-on-disk	unidirectional	constant	water	[31]
Cylinder-on-flat	oscillating	constant	Ringer's solution	[29]

The tribological evaluation of materials in joint simulators involves the measurement of wear rates and frictional torque and changes in the physical properties of the articulating components. After the test, the worn surfaces are examined and the morphology of the wear particles is studied. It has been demonstrated that simulator tests, if properly applied, can provide reliable information about the effects on friction and wear of any changes in the material manufacturing process, sterilization doses, surface finish of the counterface and modification of the prosthesis design.

Numerous types of joint simulators have been developed over the years. The most common are hip joint simulators, which are of a ball-in-socket design [e.g. 27,34–37], and knee joint simulators [e.g. 38–42].

2.3.3. Evaluation of wear on components in vivo and ex vivo

The operating conditions in simulated tests are only a close representation of the conditions operating in the body. In reality the loading pattern is more complex and the additional action of supporting muscles is very difficult to reproduce in test rigs. There is also no animal on which the prosthesis for humans could accurately be tested. Thus in the final stage the prostheses are evaluated in vivo in clinical trials on humans. The information on the tribological performance of the articulating components in real implants is also obtained from ex vivo examination. The wear of articulating components in vivo depends on many factors, such as material (its physical and chemical characteristics), mechanical (prosthesis design, its correct and stable positioning) and clinical (patient age and activity).

X-ray radiological techniques are used to evaluate the wear performance of the prosthetic articulating components in vivo. This information is supplemented by the examination of components removed from the body. The ASTM F 561 standard describes the procedure for the retrieval and analysis of orthopaedic implants. The removed components are examined by microscopic techniques and the wear is measured, for example, by surface profilometry techniques. In addition, histological examination is performed to determine the reaction of the surrounding body tissue to the implant material.

3. Tribological characteristics of polymers and polymer composites in orthopaedic applications

3.1. Polymers in orthopaedic implants

Currently, the most common design of a bearing in joint replacement prostheses consists of a polymer/metal or polymer/ceramic combination as the articulating pair (low friction arthroplasty). These combinations of materials exhibit coefficients of friction up to three times lower than the equivalent metal/metal pairs [27,35]. The polymeric component of the bearing is usually stationary and the metallic or ceramic components move against it. The lifetime of such an articulating joint is limited by the rate of penetration of the metallic or ceramic component into the polymer, which in turn depends on the combined effects of polymer creep and wear.

Among numerous polymeric materials currently available it appears that ultra-high molecular weight polyethylene (UHMWPE) offers the best combination of low friction, wear resistance and biocompatibility [43]. Thirty years of clinical experience have confirmed the suitability of UHMWPE for the bio-environment and UHMWPE is now used almost exclusively in total replacement joints. Apart from plain UHMWPE its composite containing short carbon fibres has also found applications in artificial implants, especially in knee prostheses. Among the other polymers, polyoxymethylene (Delrin) has been used with some success as an articulating hip component, but attempts to use polytetrafluoroethylene (PTFE) and polyester were unsuccessful and resulted in early prosthesis failures. Although the use of UHMWPE is widely accepted, the research efforts on further improvements of the material wear and creep resistance characteristics are continuing. The work is concentrated mainly in the areas related to experimentation with different technological processes and optimization of the irradiation sterilization procedure.

3.2. Wear and friction characteristics of polyethylene in laboratory tests

The tribological characteristics of UHMWPE are analyzed with respect to several factors, such as operating conditions, environment, type of counterface material, counterface roughness and the material properties of polyethylene.

3.2.1. Effects of operating conditions
The operating conditions which affect wear behaviour include sliding distance, contact load and speed. The wear rates of polyethylene depend on the sliding distance. Usually the rates are higher at the beginning of the test and then decrease as sliding continues [44,45] (fig. 2). The scatter between the results found on different test rigs for the sliding distance over which the wear rate of polyethylene initially increases is high. In some cases it is as long as 25 km [45] or it can be as short as 1.2–2.5 km [44]. After a very long sliding distance, the value of which depends on the applied load, the wear curve again changes its slope and this time the wear increase is attributed to the onset of the fatigue wear mechanism [32]. Thus, usually very long wear tests are required in order to obtain reliable data on the long-term wear rate of polyethylene.

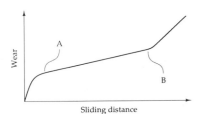

Fig. 2. Typical wear graph of UHMWPE sliding on a steel counterface.

The contact stresses affect the wear behaviour of polyethylene in a different way [44,46–49]. The variation in wear rate with contact pressure often follows an exponential relationship [46,48,49]:

$$W = W_0 \exp(AP),$$ (1)

where W is the wear rate, e.g. depth of the wear scar per unit sliding distance (mm/mm), P is the pressure (MPa) and W_0 and A are constants.

The wear rates of polyethylene are low at low loads and increase exponentially when the load is increased. The constants W_0 and A depend on the test conditions and the material properties of polyethylene [46,49]. The pressure–velocity (PV) limit, commonly used in engineering applications, is also used as a design criterion in artificial implants with polymeric parts. Since artificial joints are lubricated, the permissible PV-values are higher than for dry sliding. The problem, however, is that during walking short periods of lubricant starvation may occur, e.g. in the hip joint, and the critical PV-value may be exceeded [44]. Under high contact stresses there is a sharp increase in the wear of polyethylene, which is of great importance, especially when considering knee prostheses, where contact stresses above the yield stress can be encountered. This might be one of the reasons why the satisfactory wear performance of polyethylene in hip prostheses is often found to be inadequate in knee prostheses.

3.2.2. Effects of environment

The wear resistance of UHMWPE is usually lower under dry sliding than in the presence of aqueous lubricants [44]. The coefficient of wear for UHMWPE sliding in air on stainless steel is almost three times higher than the corresponding coefficients under water and blood plasma lubrication [44]. Although UHMWPE wear behaviour was shown in one study to be insensitive to the type of liquid environment [44], the contrary findings were reported elsewhere [28,50]. UHMWPE sliding against stainless steel exhibited higher wear and friction in water and saline solution than in blood serum. The presence of corrosion products was also recorded during these experiments [28,50]. The lowest amount of wear damage on both the steel and UHMWPE sliding surfaces was observed in bovine serum [28]. It appears that serum proteins form a boundary lubricating layer on the steel surface, reducing the adhesive forces and at the same time preventing polymer film transfer.

3.2.3. Type of counterface material

For many years the stainless steels AISI 316L and 316L VM (e.g. ASTM specification F 138) were the main metallic alloys used for femoral components. There were, however, persistent problems associated with stainless steel implants due to their susceptibility to corrosion in body fluids [51], and these materials have now been replaced by more suitable cobalt and titanium alloys in many prosthetic designs. Cast Co–Cr–Mo alloys (e.g ASTM specification F75) offer a superior combination of biocompatibility, corrosion resistance and mechanical strength compared to that of stainless steels. Titanium alloys such as Ti–6Al–4V (e.g.

ASTM specification F 136) exhibit even better biocompatibility and resistance to the highly corrosive body fluids, and furthermore have high mechanical strength and fatigue properties. Since the titanium alloys do not contain the known allergens such as nickel, cobalt, or chromium, any adverse tissue reaction to the leaching metallic ions should be reduced.

Generally the wear behaviour of UHMWPE against stainless steel, cobalt–chromium and titanium alloys is similar, provided the test conditions are comparable (fig. 3). A zero-wear model [52], defined as a wear magnitude which is not causing a significant change to the original surface roughness, was used to study the tribological behaviour of UHMWPE against all three metallic alloys mentioned above (i.e. stainless steel, Co–Cr–Mo and titanium alloys) [53]. All three pairs showed similar behaviour and gave the same maximum shear stress value for a zero-wear level. No significant difference in wear was noted for highly stressed UHMWPE contacts against stainless steel, cobalt–chromium and titanium alloys in blood serum [47]. There is, however, some uncertainty with reference to the wear characteristics of the UHMWPE/titanium alloy pair since both unsatisfactory [45,54,55] and satisfactory [56] wear performance of this system has been reported. The generation of a black deposit and its embedding in the polyethylene surface combined with damage to the titanium alloy surface was observed in laboratory tests in dry and water environments [45] and in saline solution [55]. Poor wear characteristics of UHMWPE against titanium alloys were also confirmed after wear tests where a counterformal knee prosthesis contact condition was applied [54]. The UHMWPE/Ti–6Al–4V combination tested exhibited high friction, and wear was detected on both components. The increased wear of titanium alloys is especially noticeable in the presence of abrasive media such as acrylic cement particles [47]. The problem seems to be associated with the disruption during sliding of a thin passivating titanium oxide film. Hence, there have been several attempts to generate a wear-resistant surface layer by various techniques, such as passivating [57], nitriding [47], TiN ion plating [54,58], nitrogen ion implantation [1,59,60] and oxygen diffusion hardening [58]. Although all these surface treatments usually resulted in reduced wear damage on titanium alloy samples, the durability of the wear-resistant layers generated was low.

Since the early 70s high-density alumina ceramic has been used as a replacement for metallic femoral articulating components in some of the prosthetic designs [61,62]. The tribological advantage of a polyethylene/alumina configuration over a polyethylene/metal articulating system has been demonstrated in many laboratory studies in dry and wet conditions [50,63–67]. In general UHMWPE showed lower friction and wear when slid against alumina than when slid against stainless steel [50,65], Co–Cr–Mo alloy [63] and Ti–6Al–4V alloy [67]. The observed reduction in wear varied from about 2 to 20 times. On the other hand, it has been demonstrated that when bovine serum is used as a lubricant, the wear of polyethylene against various polished ceramics is of similar magnitude to the wear against polished metallic alloys [47,68].

The recent development of new ceramics with better toughness properties than the usually brittle alumina paves the way for their use as alternative counterfaces

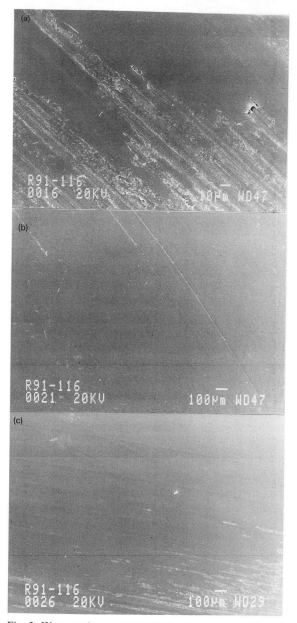

Fig. 3. Wear surface of UHMWPE after sliding against (a) nitrogen-strengthened high performance steel (Protasul-S30), (b) CoCrMo alloy (Protasul-2) and (c) alumina (Biolox) discs in blood serum (Courtesy of R.M. Streicher, Sulzer Medical Technology).

for UHMWPE in articulating joint prostheses [29,50]. The wear behaviour of UHMWPE against the variety of new ceramics such as zirconia-toughened alumina, aluminalon and polycrystalline tetragonal zirconia, was investigated and

TABLE 2

Selected friction and wear coefficients of polyethylene against smooth metallic and ceramic counterfaces.

Counterface	Wear coefficient $(mm^3\ N^{-1}\ m^{-1})$	Friction coefficient	Test conditions	Ref.
EN58J	$2.5-3.4\times10^{-7}$	–	4.0 MPa, 0.24 m s^{-1}, dry	[32]
EN58J	$0.9-2.0\times10^{-7}$	–	45 N, 0.02 m s^{-1}, dry	[32]
316L	1.7×10^{-7}	–	3.4 MPa, 0.04 m s^{-1}, water	[44]
316L	1.7×10^{-7}	–	3.4 MPa, 0.04 m s^{-1}, plasma	[44]
316L	5.0×10^{-7}	–	3.4 MPa, 0.04 m s^{-1}, dry	[44]
316L	9.1×10^{-9}	0.04–0.16	3.4 MPa, 0.05 m s^{-1}, serum	[28]
316L	8.9×10^{-9}	0.03–0.09	6.9 MPa, 0.05 m s^{-1}, serum	[28]
Co–Cr	7.5×10^{-9}	0.05–0.11	6.9 MPa, 0.05 m s^{-1}, serum	[28]
Alumina	6.2×10^{-9}	0.06–0.10	3.4 MPa, 0.05 m s^{-1}, serum	[47]
Zirconia	5.6×10^{-8}	0.04	3.4 MPa, 0.05 m s^{-1}, serum	[50]
Alumina	1.0×10^{-7}	0.05	3.4 MPa, 0.05 m s^{-1}, serum	[50]
316L	1.8×10^{-7}	0.07	3.4 MPa, 0.05 m s^{-1}, serum	[50]

compared with the wear of UHMWPE/alumina couples [29]. Zirconia-toughened alumina is a composite ceramic containing dispersed zirconia particles in an alumina matrix [69], aluminalon is a mixture of alumina and γ-aluminium oxinitride [70] and polycrystalline tetragonal zirconia is a fine-grained zirconia stabilized by yttria [69]. The studies conducted using pin-on-plate and cylinder-on-plate configurations and Ringer's solution as lubricant revealed that the wear of UHMWPE against zirconia and alumina-based ceramics was lower than that against pure alumina [29]. The advantage of yttria-stabilized zirconia over pure alumina in terms of the UHMWPE wear reduction was confirmed in another set of tests, where sliding was carried out in water, physiological saline and bovine serum [50].

The typical tribological characteristics of polyethylene rubbing against various metallic and ceramic counterfaces are compiled in table 2.

3.2.4. Effects of counterface surface roughness

The surface roughness of the hard counterface strongly affects the wear behaviour of polyethylene, but this effect is different in dry and wet conditions. Under dry sliding the coefficient of wear of polyethylene displays its minimum when the counterface surface roughness is about $R_a = 0.1$ μm [30,71,72]. The weak effect of surface roughness on wear in the dry environment has been attributed to the formation of a thin polymer transfer film on the metallic counterface. The film covers the underlaying surface marks and irregularities and gives a smooth surface. In liquid environments the film transfer formation is disrupted and the wear of polyethylene intensifies with increasing counterface surface roughness for both metallic [28,30,56,73] and ceramic [68,74] counterparts. The smoothing of the original grinding marks on metallic plates after contact with UHMWPE under bovine serum lubrication was observed and eventually the wear rates decreased again for long sliding distances [28]. The partial film formation, especially in water and Ringer's solution, also appears to be responsible for the similarity between the long-term wear rates of surfaces of different roughness [56]. For very smooth counterfaces there is practically no difference in the polyethylene wear under dry and wet conditions. The relationship between wear of UHMWPE in water and the counterface roughness has been described by the following expression [30]:

$$k = 4.0 \times 10^{-5} R_a^{1.2}, \tag{2}$$

where k is the wear coefficient (mm^3 N^{-1} m^{-1}) and R_a is the counterface surface roughness (μm).

The increased polyethylene wear in vivo has been attributed in some cases to the imperfections present on the metallic counterfaces. Some of these imperfections, e.g. scratches, may be introduced by contaminants during the manufacturing or even in the implantation process. The surface scratches and imperfections can critically affect the polymer wear. It has been shown in laboratory tests conducted on UHMWPE in water, that even a single transverse scratch on the polished stainless steel plate increases the polymer wear by an order of magnitude [75]. The pile-up of hard material along the scratch edges appeared to be responsible for the enhanced polymer wear.

3.2.5. Effects of the physical properties of polyethylene

The physical properities of polyethylene which affect its wear and friction coefficients include: crystallinity, molecular weight and material imperfections such as fusion voids. These properties can be affected strongly by the manufacturing parameters, doses of irradiation and storage conditions. It is well documented that the wear resistance of polyethylene improves with increase of its molecular weight [25,45,49]. Artificial implants are sterilized by gamma irradiation, whose effect on wear is still disputed. Some irradiation doses seem to have a beneficial effect and improve the wear resistance of UHMWPE, while others are detrimental. For example, after a low dose of irradiation of about 5 Mrad [48,76] or a high dose between 200 and 1000 Mrad [25] the wear resistance of a material, in comparison to its unsterilized state, is slightly improved. Contrary to this, a decrease in the wear resistance of UHMWPE was observed after irradiation with a dose of about 2.5 Mrad [47], 10 and 20 Mrad [77].

There have been some attempts to improve the wear resistance of polyethylene by carbon fibre reinforcement, but these have not been very successful. The data reported show either no clear advantage or else deterioration of the wear resistance of carbon fibre filled polyethylene [25,45,47]. The tribological characteristics of the polyethylene are also affected by manufacturing techniques. A wide variation in the molecular weight, its distribution and the presence of various internal defects can be obtained when applying different technological parameters. For example, the wear rate of moulded UHMWPE was found to intensify with increasing process temperature [45]. The presence of fusion defects was found to be responsible for the increased wear damage [78], whereas the completely fused UHMWPE exhibited enhanced tensile properties and improved fatigue, wear and creep performance [79]. The wear tests using completely fused polyethylene pins sliding against silicon carbide discs have shown a 50% increase in wear resistance compared to conventionally produced polyethylene. The strength, creep and frictional properties of UHMWPE can also be improved by new technological processes [80].

3.3. Wear of polyethylene in joint simulators

Joint simulators have been developed to reproduce as closely as possible the in vivo operating conditions, and commercially available and newly developed components for joint replacement are routinely evaluated on these machines. The wear resistance of polyethylene as a function of material properties such as density, crystallinity, molecular weight and its distribution, has been studied [81,82]. It was found that the original material properties, which are determined by the manufacturing process, doses of irradiation and the storage conditions, change during the simulation tests to various degrees [82]. The material-property changes, which comprise chain scission, oxygen degradation and the formation of low molecular weight parts, have a detrimental effect on the wear resistance of UHMWPE [81,82]. It was also found that there were differences in the polyethylene properties between the samples tested on a joint simulator [82] and those retrieved after operating in the human body [83]. Ideally the wear processes in vivo and in vitro would be the same, so that reliable data about the joint replacement can be obtained in the laboratory.

A wide range of data has been obtained from the UHMWPE wear tests performed on joint simulators [27,63,81,84,85]. For example, wear rates between 0.05 and 0.2 mm yr^{-1} were recorded for polyethylene/cobalt chromium alloy couples [27,63], between 0.035 and 0.1 mm yr^{-1} for polyethylene/stainless steel pairs [81] and 0.03 mm yr^{-1} for polyethylene/alumina [63]. These values are similar to the previously reported laboratory and clinical wear results [35,86–89], and the wide variation in obtained data indicates the effects of various factors on the wear rate of polyethylene. The dimensional changes of the polyethylene sockets due to wear were very small. When measured after the simulation tests it was clear that they were not large enough to impair the mechanical functioning of the prostheses for up to 10 yr in service. Since the wear on polyethylene cups is very low it seems that the issue of concern should be directed to the reaction of the body to accumulated wear debris.

The wear rates of UHMWPE hip and knee components tested on joint simulators, have shown a similar dependence on the counterpart material to that observed in simple laboratory wear tests [60,63,85,90]. The mean wear rates of polyethylene components were usually higher when run against titanium alloy as compared to wear against stainless steel and cobalt alloy components [60,85]. The damage to the titanium alloy parts in the form of scratching, especially in the presence of acrylic cement particles, was also more extensive than that found on other metallic surfaces [84]. However, a vast improvement in the wear resistance of the titanium alloy parts sliding against UHMWPE was observed when nitriding [85] or nitrogen ion implantation [60] was used to generate a wear resistant top layer on the titanium alloy. The simulation tests have also confirmed the advantages of ceramics as polyethylene counterfaces over the traditionally used metallic alloys [63,90]. In the tests conducted in a walk simulator the wear rate of a polyethylene socket run against an alumina head in water was 20 times lower than the wear against a Co–Cr–Mo alloy head [63]. The frictional torque and the amount of heat generated during the simulation tests in water was also higher for a cobalt alloy/UHMWPE pair than for an alumina/UHMWPE pair [90]. The modification of the lubricating medium by the addition of hyaluronic acid to water [90] or the replacement of water by synovial fluid [91] did not have a significant effect on the frictional torque. On the other hand, when acrylic cement particles were present in water the frictional torque increased sharply [90]. There was also a sharp increase in frictional torque in dry sliding. The analysis of the heat dissipation mechanisms demonstrated that the magnitude of the heat generated was sufficient to increase the temperature on the bearing surface and to accelerate wear, creep and oxidation degradation of the UHMWPE component [92].

The wear features observed on the surface of the polyethylene cups after simulation tests included polishing, peeling and pitting [81,82], with occasional folds, cracks and grooves present [27,35,81]. A thin polymer transfer film on the metallic femoral head in saline solution was observed [27], and shallow abrasive scratches and grooves were often visible on the polished metallic surfaces [27,81,85]. It seems that the metallic-surface scratching found in joint simulator tests is more extensive than the damage found on the equivalent retrieved implant components, and may be attributed to the presence of contaminants in vitro [27,85].

3.4. Wear and friction of alternate polymers

Apart from UHMWPE several alternate polymers have been tried in orthopaedic joint implant applications, but the final results are usually unsatisfactory.

Before the introduction of UHMWPE in the early 1960s, polytetrafluoroethylene (PTFE) articulating hip sockets were used in the first metal/polymer total hip replacements [93]. The PTFE cups were found to suffer from excessive wear, and these clinical findings were later substantiated by the results obtained from laboratory wear tests [33,47,55]. The wear of PTFE against stainless steel was more than three orders of magnitude higher than the wear of polyethylene under the same contact stress of 6.9 MPa and bovine serum lubrication [47]. The average wear rate for PTFE pins sliding on a stainless steel disc under various environments, such as water, Ringer's solution and bovine synovial fluid, was found to be about 3×10^{-5} mm^3 N^{-1} m^{-1} [33]. An increase in wear for samples tested in bovine synovial fluid and Ringer's solution compared to those tested in water was observed. The wear coefficients of PTFE obtained in simple laboratory tests were in good correlation with coefficients calculated on the basis of the observed penetration rate of the steel heads into PTFE cups in retrieved components [33].

Polyester has also been tried, but the performance was unsatisfactory [94]. The laboratory evaluation of the wear resistance of polyester tested against metallic counterfaces has confirmed that its wear resistance is much worse than that of polyethylene [25,47,95]. The published data show that the wear rate of polyester is about two to three orders of magnitude higher than that of polyethylene [25,47] and is strongly load dependent [47].

Polyoxymethylene, better known as Delrin (DuPont trademark), can be regarded as the only alternate polymer used in orthopaedic joint implants with some success. Delrin has higher hardness and lower creep than UHMWPE. It has been used in metal/polymer trunnion bearing type hip prostheses, mainly implanted in Scandinavia [96,97]. The use of Delrin for the acetabular cup was suggested in the mid-1960s following initial laboratory wear tests [98]. The biocompatibility tests conducted later confirmed the suitability of solid Delrin as an implant material [99]. The tribological evaluation of Delrin has, however, generated conflicting data. The wear behaviour of Delrin 550 against a cobalt–chromium alloy counterface in water, was found to be higher than the wear of UHMWPE tested under the same conditions [24]. The wear and friction properties of Delrin 150 against polished 316L stainless steel in dry, water and blood plasma environments were tested, and it was found that its wear resistance depended on the test environment [100]. The wear behaviour of Delrin 150 was better than that of UHMWPE in plasma, but in water under similar test conditions its wear behaviour was worse. In early hip simulation tests the wear of Delrin against cobalt–chromium–molybdenum alloy in plasma was found to be lower than the wear of UHMWPE in a similar configuration [34]. In contrast, the laboratory studies using Delrin/stainless steel pairs and bovine serum lubrication revealed a much lower wear resistance and higher friction of Delrin 150 compared to conventional UHMWPE [47].

The early examination of five retrieved trunnion hip prostheses, which have been in service for up to six years, revealed an acceptable wear performance of Delrin, with signs of wear found on only one component [101]. More recent clinical evaluation of hip prostheses with Delrin cups and trunnion sleeves, however, has shown a high incidence of aseptic loosening, especially of the sockets [102–105], with excessive wear damage found on some of the removed Delrin components [104]. The histological studies of the tissue reaction to wear debris from Delrin and UHMWPE acetabular cups also showed more extensive inflammation and necrosis associated with prostheses containing Delrin [105]. The examination of the tissue surrounding loosened hip prostheses demonstrated that large amounts of Delrin wear particles induced fibrinoid necrosis and macrophage activation [106]. It is still, however, unclear whether the loosening of the prostheses with Delrin components is predominantly due to the strong tissue reaction to Delrin wear particles [105,106] or can be attributed to design related factors [104,107,108].

The laboratory wear results obtained for conventional UHMWPE and alternate polymers are summarized in table 3.

It can be seen from table 3 that the wear resistance of UHMWPE is superior to other polymers under almost all test conditions applied. The superior wear resistance of the UHMWPE has also been confirmed clinically and this material is now used almost exclusively in articulating component applications. A slight advantage of Delrin over UHMWPE in plasma, which has been observed in

TABLE 3
Summary of the wear results for conventional UHMWPE and alternate polymers.

Polymer	Wear ($mm^3 N^{-1} m^{-1}$)	Test configuration	Lubricant	Ref.
UHMWPE	3.6×10^{-7}	disc on polymer flat, 5.5 MPa	water	[24]
Delrin 550	1.1×10^{-6}	disc on polymer flat, 5.2 MPa	water	[24]
UHMWPE *	1.5×10^{-7}	washer on polymer flat, 3.7 MPa	blood plasma	[25]
UHMWPE **	4.5×10^{-7}	washer on polymer flat, 3.7 MPa	blood plasma	[25]
Polyester AP-4	7.3×10^{-6}	washer on polymer flat, 3.7 MPa	blood plasma	[25]
Delrin 150	1.0×10^{-6}	washer on polymer flat, 3.7 MPa	water	[100]
Delrin 150	1.5×10^{-7}	washer on polymer flat, 3.7 MPa	blood plasma	[100]
UHMWPE	0.1×10^{-7}	polymer flat pin-on-flat, 3.45 MPa	bovine serum	[47]
Delrin 150	3.3×10^{-7}	polymer flat pin-on-flat, 3.45 MPa	bovine serum	[47]
Polyester AP-4	9.7×10^{-7}	polymer flat pin-on-flat, 3.45 MPa	bovine serum	[47]
PTFE	1.4×10^{-5}	polymer flat pin-on-flat, 6.9 MPa	bovine serum	[47]
PTFE	2.0×10^{-5}	tri polymer pin-on-disc, 40 N	distiled water	[33]
PTFE	3.4×10^{-5}	tri polymer pin-on-disc, 40 N	synovial fluid	[33]

* Molecular weight 3×10^6.
** Hercules 1900.

laboratory tests, has not been confirmed in practical applications. The other two polymers, PTFE and polyester AP-4, although used in artificial implants, demonstrated clearly in the laboratory tests an inferior wear behaviour, and their use in implant applications has been terminated. Although the simple laboratory tests do not fully simulate the real conditions in the body, these tests can successfully be used as a screening procedure for materials suitable for orthopaedic applications.

3.5. Polymer composites in orthopaedic implants

In the context of the limitations of the biomaterials currently used in joint replacements there has been a considerable interest in the development of polymer-based composites with a more adequate combination of strength, stiffness, toughness and environmental-resistance properties [109]. The addition of fillers in the form of fibres, whiskers or dispersed particles to traditional polymers improves their strength and stiffness properties. The mechanical properties of the polymer-based composites can be modelled up to the level adequate even for their use in load-bearing orthopaedic applications. Interest in polymer composites is additionally stimulated by the possibility of controlling Young's modulus in these materials in such a way that biomechanical compatibility with the host bone is achieved. Young's modulus of bone is low compared to the metallic alloy implants, and from a mechanical viewpoint the combination of materials with low and high Young's moduli has severe disadvantages.

Recently, a number of fibre reinforced polymer composites, such as carbon fibre/epoxy [110], carbon fibre/PEEK [111], graphite/polysulfone, glass/epoxy and aromatic polyamide fibre/polypropylene [112], have been evaluated as potential candidates for applications in surgical orthopaedics. The inertness of polymeric biomaterials such as polysulfone (PS) and polyetheretherketone (PEEK) and their composites has been proven in in vitro and in vivo biocompatibility tests [111,113,114]. Excellent in vitro cellular biocompatibility of PS and PEEK composites was demonstrated in cell culture studies using mouse fibroblast cells [114]. It was found that the strength properties of carbon fibre/PEEK composites were not affected by aging in vitro and in vivo and the material did not elicit an adverse muscular tissue reaction, although released carbon particles were observed in the capsule [111]. The mechanical properties of graphite/polysulfone, glass/epoxy and polyamide/polypropylene composites were stable after immersion in the simulated physiological environment and short-term autoclave sterilization [115].

Since poor mechanical properties of plain polyethylene were prohibitive for its application as a metallic-component replacement in joint implants, a short graphite fibre reinforced polyethylene composite has been evaluated for these applications [116,117]. Although the strength of the graphite fibre reinforced polyethylene increased considerably compared to plain polyethylene, it was still three times lower than that of bone, and therefore was not sufficient for the material to be used for bone fixation devices or stem components in joint implants [116]. However, since its wear resistance was comparable to that of plain polyethylene and creep was more than an order of magnitude less, the material was proposed for use as an articulating surface, especially in knee prostheses [116]. The comparison of the mechanical behaviour of plain UHMWPE and carbon fibre/UHMWPE tibial components of total knee prostheses showed that under static load the increased stiffness of carbon fibre/UHMWPE resulted in smaller contact areas and higher stresses measured on the tibial components [118]. The presence of high contact stresses on the polyethylene composite surface was regarded as a detrimental factor, since the probability of pitting-type wear damage could increase [118]. On the other hand, the higher stiffness of carbon fibre/UHMWPE was desirable, since the time-dependent deformation of reinforced polyethylene parts was significantly reduced. This gave better resistance of these components to indentation from the matching metallic femoral surfaces [118,119].

Recently, carbon fibre reinforced polymers such as carbon fibre/polysulfone [13,15,120] and carbon fibre/epoxy [121], have been seriously considered as replacements for the metallic stems in joint implants. The strength and stiffness of the advanced composites can be tailored to satisfy specific requirements of the hip stem components. The mechanical properties can be altered by varying the fibre content for short-fibre reinforcement and by changing the fibre orientation and stacking sequence of continuous fibre reinforced prepregs. The lower stiffness of the composites complies better with the bone properties and may reduce the occurrence of detrimental effects such as stress shielding, leading to bone resorption at the implant/bone interface, etc. The main concern, however, is whether these composite materials maintain their mechanical properties over a period of time sufficiently long for practical orthopaedic applications. The mechanical and, in particular, the fatigue behaviour of fibre reinforced materials is governed by the fibre/matrix interfacial bond, which for some composites can be adversely affected in the physiological environment [122]. The interfacial bond strength preservation depends on the composite constituents, and it appears that polymers reinforced by carbon fibres display better mechanical stability in a physiological environment than polymers reinforced with polyaramide fibres [122]. A prototype hip prosthesis consisting of a carbon fibre/polysulfone core and braid and polysulfone exterior coating was developed and successfully implanted in a canine model [15]. A remodelling of the surrounding bone after implantation was found to be constructive and no adverse tissue response to the composite material was observed [15]. The tribological characteristics of this composite were not evaluated since the bearing system consisted of a titanium alloy head articulating against a natural hip socket and the composite material was used for the stem.

3.6. Wear and friction of polymer composites in orthopaedic implants

While improvements of the mechanical properties of carbon fibre reinforced polyethylene have been achieved and are well documented, its tribological advantages over plain polyethylene remain controversial. Comparative wear studies between plain polyethylene and polyethylene reinforced with 20 and 40 wt% high-modulus carbon fibres were conducted on a polishing machine with the polymer and composite samples loaded against a rotating, smooth, stainless steel plate, and a solution of a polishing powder in water was added [116]. The tests conducted did not reveal any remarkable difference in their resistance to abrasive wear. A small decrease in the wear coefficient, compared to plain polyethylene, was observed for polyethylene composites moulded with 10 vol% of carbon fibres in wear tests against stainless steel lubricated with blood plasma [25]. The carbon fibre/UHMWPE composite Poly Two (Poly Two is a registered trademark of Zimmer, USA) showed the same or slightly lower wear than plain polyethylene in tests conducted by the manufacturer [117,123]. On the other hand, lower wear resistance of Poly Two was recorded in sliding contacts against metallic counterfaces lubricated by bovine serum [47]. The wear rate of UHMWPE reinforced by graphite powder showed a strong time dependence, and the wear of filled polyethylene was superior to that of plain polyethylene for relatively short sliding distances up to 1×10^5 m [24]. For longer sliding distances this was reversed, and the wear rate of the reinforced polyethylene was almost an order of magnitude higher than that of plain polyethylene [45].

Microscopic examination of the wear damage on carbon fibre reinforced UHMWPE samples after simple laboratory [47] and knee simulator [124] tests showed that most surface damage was related to the presence of carbon fibres. The originally smooth and shiny surface of composites became dull and roughened. There was also evidence of fibres being worn out [47,124]. In some places the removal of the UHMWPE matrix from the fibres and the formation of pits was observed, and this process was associated with poor fibre–UHMWPE adhesion and/or fatigue crack initiation at the carbon fibres [124]. The removed and/or broken fragments of carbon fibres were also found in the lubricating medium [47]. The presence of scratches on the components was attributed to the abrasive action of broken carbon fibres [124]. Examination of the wear particles from polyethylene and carbon reinforced polyethylene acetabular cups run on a hip simulator, revealed that large particles from UHMWPE cups had a plate-like morphology whereas carbon fibres of various lengths were dominant in the debris generated from carbon reinforced components [125].

Despite the disputable character of the tribological performance of carbon fibre reinforced UHMWPE over unmodified UHMWPE this composite material has found commercial application mostly as the tibial component in articulating joint implants. Clinical performance data obtained from the examination of retrieved parts of tibial components made of plain and carbon filled UHMWPE, revealed that both materials experienced a similar amount of wear damage, which was directly related to the length of time of implantation [126].

Another group of polymer composites evaluated as possible candidates for orthopaedic applications includes polymers containing polytetrafluoroethylene (PTFE) fillers. Plain PTFE is a well-known solid lubricant, and the addition of this material to polymers was expected to improve the tribological characteristics of the composite materials. A number of commercially available polymer composites, such as SP-211 (Du Pont trademark), which is a polyimide containing 10% PTFE and 15% graphite, Torlon (Amoco trademark), which is a (poly(amide-imide)) resin with small amounts of PTFE, and other additives and Delrin AF (Du Pont trademark), which is a polyoxymethylene containing 22% PTFE, were tested in a thrust-washer-type wear rig [127,128]. The wear resistance of Torlon and SP-211 against a polished 316L stainless steel counterface was found to be better than that of plain UHMWPE in dry conditions [127]. Both composites, however, suffered 100 times higher wear when tested in water. The wear of SP-211 was also increased, between 5 and 10 times, in blood plasma [127]. The wear and friction behaviour of Delrin AF sliding against 316L stainless steel under dry and lubricated by blood plasma conditions was of a similar magnitude to the wear of plain UHMWPE tested under similar conditions [128]. When tested under lubrication with water, the wear of Delrin AF increased ten times [128]. After a series of intense tests, Torlon and SP-211 were finally rejected as potential biomaterials for joint implant applications, while Delrin AF appears to be worth further consideration [127,128].

The wear coefficients of plain polyethylene and selected polymer composites, obtained under similar test conditions using a thrust-washer wear tester, are compared in table 4.

The main problem with the tribological behaviour of polymer composites seems to be associated with the presence of the fibre/matrix interface, which is a weak point in the system. The wear damage usually starts at the interface, especially when alternating stresses are involved. Additional detrimental effects are observed in the presence of liquid environments. Polymer composites, contrary to plain polymers, often exhibit better wear resistance under dry than under wet conditions [127,129]. Water and aqueous lubricating systems appear to cause serious degradation in cases when strain, incompatibility and decohesion develop between the base material matrix and the filler material. Apart from improving the tribological characteristics of composites, a substantial amount of work is still needed in the area of manufacturing technology, so that composite materials which would give

TABLE 4
Summary of the wear results of polymer composites.

Composite	Lubricant	Wear coefficient	Ref.
Plain UHMWPE	dry	5.0×10^{-7}	[44]
Torlon	dry	6.7×10^{-8}	[127]
SP-211	dry	5.4×10^{-8}	[127]
Delrin AF	dry	1.7×10^{-7}	[128]
Plain UHMWPE	water	1.7×10^{-7}	[44]
Torlon	water	6.7×10^{-6}	[127]
SP-211	water	5.7×10^{-6}	[127]
Delrin AF	water	5.8×10^{-6}	[128]
Plain UHMWPE	blood plasma	1.6×10^{-7}	[44]
SP-211	blood plasma	4.2×10^{-7}	[127]
Delrin AF	blood plasma	1.9×10^{-7}	[128]
UHMWPE + 10% graphite	blood plasma	2.9×10^{-7}	[25]

reliable performance over a specific length of time, can be produced. Until the problems outlined above are resolved, the application of polymer composites in articulating contacts in joint prostheses is unlikely to move beyond its present experimental stage.

4. Tribological characteristics of ceramics and ceramic composites in orthopaedic applications

There are several different types of ceramic materials currently used in orthopaedic applications. They include:
 (i) oxide ceramics,
 (ii) synthetic carbons,
(iii) calcium phosphate ceramics and
(iv) various types of glasses and glass–ceramics.

Among these the oxide ceramics, with high-density alumina (Al_2O_3) as their main representative, are best known for their applications in joint replacement prostheses as articulating components. Polymeric carbons, pyrolitic carbons and carbon–carbon composites are still at the stage of evaluation for possible orthopaedic applications. The various calcium phosphate ceramics have proved to be satisfactory biomaterials for bone substitutes. Phosphate ceramics, glasses and glass–ceramics are also used as coatings on metallic femoral stems. The application of these coatings to the metallic stems seems to combine the strength and stiffness of a metallic alloy with the ability of a ceramic to form a biological attachment to the bone. Since good tribological properties are not essential for these specific applications the calcium phosphate and glass–ceramics will not be discussed further.

4.1. Oxide ceramics

The beginning of the use of oxide ceramics in orthopaedic implants can be dated to the early 1970s [130,131], when the first alumina hip prostheses were implanted in France. Oxide ceramics in general have high strength in compression and good wear and corrosion resistance properties. They are also inert, hence some of them offer an exceptionally good biocompatibility [132–134]. Among many potential ceramic materials for artificial implants, high-purity dense aluminium oxide (Al_2O_3), produced in accordance with e.g. the ASTM F 603-78 standard, has been selected for applications in prostheses of the heavy load-bearing joints, such as knee and hip joints. The material selected is hard, wear and corrosion resistant and has very good biocompatibility, which has been extensively tested on small animals [134–137]. Because of its unique properties the material seems to be ideal for artificial implants. Some thought was given to manufacturing the whole implant from this material, but the inherent brittleness of alumina ceramic and the associated low level of fracture toughness prohibited the application of this material as a complete replacement of the metallic components in orthopaedic implants. A compromise was found which uses ceramic heads mounted on metallic stems and articulating against polymer or ceramic sockets. The majority of joint prostheses with ceramic/ceramic and ceramic/UHMWPE combinations are being used in Japan and Europe, where the implantations are carried out in major medical centres in France, Switzerland and Germany.

To ensure the successful performance of the ceramic components of the implant, the material is manufactured according to very stringent specifications referring to its chemical purity, density, grain size and mechanical properties [138]. The in vivo mechanical and tribological behaviour of the ceramic components largely depend on the chemical composition and impurity contents of the material, and the manufacturing procedure, which affects the number and size of flaws, the grain size and its distribution [63,139]. Other aspects considered for the successful functioning of the implant are related to the design, surface finish and mechanical tolerances of the articulating components [140]. The main advantage of the use of alumina/alumina or alumina/UHMWPE configurations is that these systems exhibit much lower coefficients of friction than their metallic equivalents [27,63, 90,141]. The wear rates of polyethylene against a ceramic counterface are also significantly lower than in metal/UHMWPE systems [50,63–65,142]. Prostheses with both ceramic articulating components [130] were introduced in order to further reduce the generation of wear debris, the amount of which has been positively correlated with late prosthesis failure. The apparent advantage of ceramics in orthopaedic prosthetic applications is attributed to their unique properties such as inertness, which results in better corrosion resistance, mirror surface finish, good wettability and good resistance to abrasion, which is usually much higher than that of metallic alloys. The superior tribological performance of ceramics over the metallic alloys has also been confirmed in partial joint replacements (hemiarthroplasty) where implant materials in the shape of cups articulate against natural acetabular sockets [143].

The brittleness of high-density alumina is one of its drawbacks, since it is responsible for occasional in vivo fractures of femoral heads coupled with tapered cones of metallic stems [144]. For this reason the minimum diameter of most alumina heads is limited to 32 mm. In an attempt to overcome this limitation much

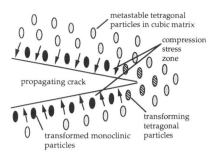

Fig. 4. Schematic diagram of the stress-induced transformation toughening process in PSZ ceramics.

tougher zirconium oxide based ceramics were considered. Toughened zirconia ceramics include magnesia partially stabilized zirconia (Mg-PSZ) [145,146] and tetragonal zirconia polycrystals [147]. The microstructure of the Mg-PSZ ceramic consists of small tetragonal precipitates in a cubic matrix, while the TZP ceramic is composed wholly of small tetragonal grains. Due to the transformation strengthening mechanism taking place under stress conditions, toughened zirconia ceramics exhibit superior bending strength and toughness properties, compared to the alumina ceramic [145,146,148]. The metastable tetragonal phase in zirconia ceramics transforms into the monoclinic phase under stress. This transformation is accompanied by a positive volume change and the generation of a compressive stress field which strengthens the usually brittle ceramics. The transformation is usually triggered by tensile stresses in the vicinity of crack tips and the resultant new stress field acts as a barrier to the propagating crack by decreasing the tensile stresses around the tip. The development of the crack is significantly slowed down. The transformation toughening mechanism contributes to the increase of the strength and fracture toughness of partially stabilized zirconia ceramics. The strengthening process is shown schematically in fig. 4.

It seems that the application of tougher ceramics for a femoral head could result in a reduction in the incidence of component fracture and allow for smaller femoral heads to be implanted [149]. Thus the PSZ and TZP ceramics were considered as a potential alumina substitute for the femoral heads. The biocompatibility of Mg-PSZ ceramics has been assessed in in vitro and in vivo experiments [150,151]. No degradation of the mechanical and structural properties was found in these ceramics after contact with living tissue and no adverse tissue response was recorded [151]. Only a small decrease in bending strength ($\sim 6\%$) due to stress corrosion was observed after immersing the PSZ samples in a boiling saline solution. The affinity of the implanted tetragonal zirconia polycrystal (TZP) ceramics and cell culture cytotoxicity tests have also shown a good biocompatibility. No adverse tissue reaction was observed [152]. Quantitative histomorphometry was applied to assess the short-term biocompatibility of TZP ceramics implanted in rats and to compare it with the tissue response to the control alumina ceramic [148]. No significant differences were found between TZP and alumina ceramics with respect to membrane thickness and cell distribution surrounding the implants, although the tissue reaction induced was slightly higher for alumina samples. The experiments conducted on the stability of the mechanical properties of TZP in liquid environments have also yielded encouraging results [148,153]. No deterioration in the bending strength and only slight microstructural changes were observed after in vivo and in vitro (37°C) testing over a 12 month period [153]. The toughness of TZP ceramics was also not affected by gamma-ray sterilization and aging in Ringer's solution at 37°C for up to 100 days [148].

4.2. Wear and friction of oxide ceramics

The wear and frictional characteristics of ceramic/ceramic systems have been evaluated on various laboratory test rigs and joint simulators. Load, speed and environmental conditions were carefully selected to closely represent those encountered in the body. The results obtained were then compared with the tribological characteristics of traditionally used metal/metal or metal/polymer systems. The superior wear resistance of an alumina/alumina pair over metal/metal and metal/polymer combinations was recorded in laboratory tests using a rolling cylinder against a stationary conformal surface [27]. The pair also exhibited the lowest coefficient of friction of 0.1, when lubricated with water. With diminishing lubrication this coefficient rose rapidly, reaching a value of 0.7 under conditions of dry sliding. It was thus anticipated that some problems might be experienced with alumina/alumina systems in cases of starved lubrication.

The subsequently conducted tests on a walk simulator confirmed the excellent wear and frictional performance of the alumina/alumina system [63,90]. It was found that after 3000 h of continuous water lubricated sliding, which corresponds roughly to about 10 yr of operation, the wear rates of the alumina ball on alumina socket were about 50% lower then with alumina/polymer systems tested under the same conditions. By comparison, the same wear rate was reached by the metal/metal pair only after 150 h of sliding. The frictional torque and heat generated were also lower for this system than for any other material combination tested [90]. The amount of heat generated in a joint, especially during more strenuous exercises, is important, as the increased temperature can result in accelerated wear, creep and oxidation, especially of polymeric articulating components such as UHMWPE sockets. In general, the results obtained seem to indicate that ceramic systems offer some clear advantages over metallic systems in total joint replacement prostheses.

Although the zirconia-based ceramics are very much tougher than alumina and also appear to be more resistant to abrasive wear [154,155], their tribological behaviour is still disputable. It was found that the dry abrasive wear resistance of TZP ceramics is superior to that of Al_2O_3 when sliding on coarse-grid SiC abrasive paper, but the situation was reversed when a fine-grid SiC paper was used [156]. In the comparative sliding wear studies using a pin-on-ring tribometer alumina out-performed zirconia in the self-mated configuration [157] (fig. 5). The wear loss of self-mated Mg-PSZ was about three orders of magnitude larger than that of self-mated Al_2O_3 in tests in water [157]. In the experiments performed on a

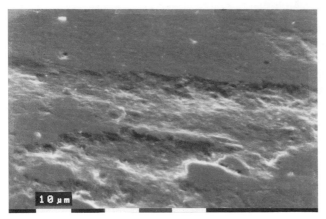

Fig. 5. Wear damage on zirconia ceramics in self-mated configuration after sliding in water.

ring-on-disk test rig under Ringer's solution lubrication [158] the wear resistance of an alumina/alumina system was found to be superior to that of a zirconia/zirconia system. The wear resistance of the latter pair was so low that the zirconia ceramics were pronounced unsuitable for orthopaedic applications in the zirconia/zirconia configuration. Instead, the use of polymer sockets as counterparts for zirconia balls has been suggested. The experiments conducted showed that, indeed, the wear of UHMWPE against zirconia-based ceramics was lower than that against alumina [29,50]. The frictional torque levels of metal, alumina and PSZ against UHMWPE sockets measured in a hip joint simulator were also found to be lowest for a zirconia/polymer system [149]. It appears that in a polymer/zirconia configuration the mechanical advantages of toughened zirconias are properly utilized and at the same time excessive wear of the ceramic is avoided.

4.3. Carbon materials and their composites

In recent years synthetic carbons which include pyrolitic carbon, glassy carbon, carbon fibres and carbon fibre reinforced carbon (CFRC) composites have been gaining much attention as potential candidates for biomedical applications. Carbon materials can be produced either by carbonization of organic polymeric precursors (glassy carbon and carbon fibres) [159,160] or by thermal cracking of gaseous hydrocarbons and deposition of a thin carbon layer on a core of a soft elec-tographite (pyrolitic carbon) [161,162]. The carbon materials have excellent bio-compatibility [163,164] and mechanical properties (especially a low stiffness similar to that of bone), and so are suitable for applications such as dental and or-thopaedic implants and heart valves. They have been used in the form of thin fibres in the repair of tendons and ligaments [165], or as loosely woven carbon fibre patches in the regeneration of the articular surfaces of osteoarthritic joints [166].

So far, clinical experience with pyrolitic carbon in joint prostheses has been limited to thumb and big toe total or partial arthroplasty [167]. In orthopaedic

surgery of highly stressed joints carbon fibre reinforced carbon (CFRC) composites have been proposed for use as bone pins [168] and hip prosthesis stems [169,170]. It has also been found that among CFRC composites, only materials with a pyrolitic carbon matrix are suitable for the manufacture of hip stems [170]. The mechanical evaluation of CFRC hip stems showed that they possess a mechanical strength similar to 316L stainless steel [169] and that the CFRC stems perform better than metallic alloy stems [170]. The lower Young's modulus of carbon materials, which allows better stress transfer characteristics to the bone, is particularly beneficial [171]. In addition, lower rigidity together with open porosity of CFRC materials encourage bone ingrowth and attachment to the implant [170,171]. In the novel designs of hip prosthesis utilizing this material, however, the femoral stem is usually made of carbon composite, while the articulating components are made of more traditional metallic or ceramic material. There are two reasons limiting the application of carbon materials for articulating surfaces:
- difficulty in the manufacture of bulky artefacts and
- poor tribological properties.

A limited clinical trial with CFRC stems with alumina heads on UHMWPE sockets was conducted in France [170]. Some attempts were also made to manufacture glassy carbon femoral heads to be fitted on titanium or CFRC shafts [172], but these were unsuccessful due to material fissuring. One of the problems associated with carbon materials is that when implanted in the body all of them easily release particles. These particles do not seem to cause any immediate adverse effects and the studies conducted have confirmed the absence of any inflammatory tissue response to carbon particles [165,168,171], which are often transported to the nearest lymph node without traces of inflammation or foreign-body reaction, although their presence is still of great concern.

The mechanical and tribological properties of glassy carbon are inferior to pyrolitic carbon, but bulky components can be manufactured from the former material whereas pyrolitic carbon is only suitable for thin coatings. A general comparison between the properties of glassy and pyrolitic carbons is given in table 5 [173]. It has to be emphasized, however, that the mechanical properties are directly related to the quality and heat treatment and can significantly vary [159,160].

The literature on the tribological behaviour of carbon materials refers mainly to the engineering applications. Excellent wear resistance in the dry sliding of glassy carbon on itself has been recorded in the ball-on-disk configuration under low load [174]. The monolithic glassy carbon materials, sliding against themselves, exhibited

TABLE 5
A general comparison of the properties of glassy and pyrolitic carbon [173].

	Density (g cm^{-3})	Young's modulus (GPa)	Modulus of rupture (MPa)	Hardness (GPa)
glassy carbon	1.5	24	200	2.0
pyrolitic carbon	2.0	35	400	2.5

wear coefficients in the vicinity of 10^{-8} mm^3 N^{-1} m^{-1} and coefficients of friction in the range of 0.02–0.06 for low loads, but there was an increase in the wear coefficient with load. When glassy carbon balls were slid against austenitic stainless steel discs an increase of about two orders of magnitude in the wear coefficient and friction coefficient up to the level of 0.2 was observed [174]. Slightly lower wear was recorded when carbon matrix composite was used in the same configuration [175]. A nitrogen atmosphere seems to have some beneficial effects since glassy carbon displayed lower friction and better wear resistance than in air. Higher friction coefficients of about of 0.2 and wear coefficients in the range 10^{-6} to 10^{-8} mm^3 N^{-1} m^{-1} have been reported in earlier studies [176,177]. It has been concluded that, in general, glassy carbons are suitable for wear resistance applications where contact stresses are low.

A number of attempts have been made to improve the wear resistance of glassy carbons through various surface treatments, such as ion implantation with helium, argon and nitrogen [178] or impregnation with ceramics [179]. Increased hardness and resistance to 1 μm diamond polishing of glassy carbon after helium-ion implantation has been achieved [173,178]. These apparent property improvements were associated with structural changes due to electronic energy loss and collision damage processes. Further studies have shown that nitrogen-ion implanted glassy carbons are 100 times more wear resistant than unimplanted glassy carbons and about 15 times more wear resistant than pyrolitic carbons to 1 μm diamond polishing [180]. The nitrogen-ion implanted glassy carbons performed equally well in sliding tests with distilled water lubrication in the ball-on-disc configuration with a ruby ball and a glassy carbon disc [180].

The wear resistance of a pyrolitic carbon and its very smooth surface finish have been found to be sufficient for low loaded total or hemiprosthetic arthroplasty in hand and foot [167]. In these applications the wear of pyrolitic carbon was found to be negligible and the frictional levels no higher than those encountered in ceramics, polymers or metal alloys. The manufacturing difficulties combined with excessive wear at higher loads are the main reasons why carbon materials cannot be used for high-loaded joint prostheses. In laboratory trials carbon materials in bulky samples tend to crack whilst in the form of a thin coating they quickly wear off [181]. The wear resistance and surface finish of CFRC composites are also not satisfactory for articulating component applications [181].

An impregnation of carbon materials with ceramics has also been suggested as a possible way of improving the wear resistance. Among many carbon/ceramic composites carbon/silicon carbide (C/SiC) has been selected and evaluated as a potential biomaterial for orthopaedic applications. The biocompatibility studies reveal similar though slightly stronger early tissue reaction to C/SiC implants as compared to pure carbon materials and an acceptable response to the C/SiC wear debris [179]. The problem is that, although the carbon/SiC composites are stronger than carbon/carbon equivalents, their lower porosity prohibits bone ingrowth [171], and there also might be some difficulties in finding material suitable for the mating surface. This seems to be a serious limitation of these materials in biomedical applications.

5. Evaluation of polymer, ceramic and composite biomaterials performance in the body

It has been almost 30 years since major breakthroughs in the applications of new materials, improved designs and better surgical techniques were achieved and cleared the way for the successful joint replacement. Continual improvements in materials, design and surgery over this period of time significantly improved the reliability and lifetime expectancy of joint prostheses. The success rate achieved with modern prostheses varies between different orthopaedic centres, with medium-term (10 yr) follow-up (clinical) evaluation reported from 30% [182] and 60% [183] to more than 90% [184,185] as success rate in hip prostheses. Despite high success rates achieved on the basis of short and medium terms, there is still the unresolved problem of long-term follow-up failure. The weak point of the system appears to be material related, especially its tribological performance.

Wearing through of the polymeric socket and tibial components is seldom observed and only single cases, usually in metal-backed designs, have been reported. Gradual wear at the articulating surfaces of the artificial prosthesis, however, can lead to restricted motion of the articulating hip components and promote subluxation or loosening of the implant. On the other hand, excessive wear on the bearing surfaces of the knee prosthesis increases the contact conformity and exerts higher implant/bone interface forces. The higher forces acting at the interface may destabilize the implant fixation. The most important problem is associated, however, with the generation of wear debris and its release to the surrounding tissue. The living body has some capacity to deal with a certain amount of foreign particles, but when this amount is exceeded adverse tissue reaction is observed, often leading to prosthesis failure.

5.1. In vivo wear of UHMWPE acetabular cups

In vivo wear damage to the hip joint prostheses employing metal or ceramic femoral heads and polymer acetabular cups, is almost always associated with the cups. It is measured as the rate of penetration of the head into the cup and is usually expressed in mm yr^{-1}. Despite earlier claims regarding their inaccuracy [186], radiographic methods or techniques are commonly used to assess the average rate of penetration of the head in the cup. The penetration rate determined by these methods includes the contribution of both wear and creep components.

The first long-term clinical evaluation of wear in vivo, based on the roentgenographic measurements of thickness changes in polyethylene acetabular cups sliding against stainless steel femoral heads gave an average rate of penetration of 0.12 mm yr^{-1} [86]. The later revision of these results produced a slightly higher average penetration rate of 0.15 mm yr^{-1}, with a variation from 0 to 0.45 mm yr^{-1} [87]. A much lower average penetration rate of 0.07 mm yr^{-1}, but with values ranging from 0.054 to 0.38 mm yr^{-1} was reported in further studies [88]. The results subsequently obtained from radiographic measurements were more consistent and the average rate of penetration measured was in the range of 0.1 to 0.2 mm yr^{-1} [187–191]. A small percentage of components, however, experienced wear two to four times higher than the average. The overall accuracy of the radiographic technique was confirmed later, when the results obtained were compared with direct measurements on the retrieved components [88,89,192,193]. Although wear results obtained in in vivo conditions show a reasonable agreement with the wear rates recorded in tests on joint simulators, they are much higher than the wear rates obtained from laboratory tests.

5.2. Wear in retrieved articulating prosthetic components

The first low-friction arthroplasty of the hip based on a stainless steel femoral head and PTFE acetabular cup was not successful due to the excessive wear of the polymer. PTFE acetabular cups were quicky penetrated by steel heads and the measured average rate of penetration was as high as 2.26 mm yr^{-1}, with the values measured ranging from 0.91 to 6.04 mm yr^{-1} [194]. The penetration included both wear and creep, but it was concluded that the contribution of creep to the overall damage of PTFE was minimal.

In subsequent designs the PTFE acetabular cups were replaced initially by polyethylene cups and later by UHMWPE cups. Despite a high success rate with the latter material, the clinical evaluation shows that UHMWPE often wears more quickly in vivo than indicated otherwise by laboratory tests. There are two main factors responsible for the accelerated wear of the polymer cups: in vivo degradation of the polyethylene properties and abrasion caused by fragments of acrylic cement. The analysis of the physical properties of the retrieved UHMWPE hip and knee components showed that the polymer undergoes a time-dependent biodegradation in the body environment, due to combined chemical and mechanical factors [83,195,196]. The structural changes, which include molecular weight decrease, crystallinity, density and stiffness increase, seem to enhance the probability of crack development in service and also increase wear [1,83,195–198]. The additional contribution to the increased wear of UHMWPE is related to the abrasion caused by the acrylic cement particles [1,199,200]. The hard abrasive particles embedded in the soft UHMWPE surface can produce grooves and scratches on the hard counterface. They can also leave pits when released from the polyethylene [1,200]. The scratches formed on the metallic counterfaces can in turn accelerate the wear of UHMWPE sockets [89,200].

The spread in the clinical wear rate data obtained is reflected in the morphological variation of the worn UHMWPE acetabular cup surfaces, where damage ranging from mild polishing to severe surface disruption has been observed. Almost all retrieved acetabular cups display the presence of two distinct regions on the articulating surface (fig. 6): a smooth area of high penetration rate and a rough low-loaded region still showing some of the original machining marks [1,193,199–202]. These two areas are separated by a low ridge and both display evidence of abrasive damage. Adhesive wear dominates in the high-wear area [32,199,200], while cracks, mainly due to plastic deformation, are occasionally recorded in the

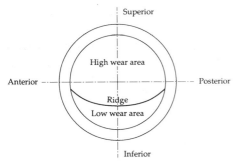

Fig. 6. Distinct wear areas on retrieved acetabular sockets [200]. ˙

vicinity of the ridge [200]. Distinctive wear-surface features include scratches, pits, craters, folds and embedded acrylic cement particles [32,199–201,203]. Some cases of severe wear are attributed to the abrasive action of acrylic cement [1,199,200,204] and others to the fusion or other defects of the polyethylene [200,203]. In the absense of bone, cement wear is often mild, and surface features such as polishing, shallow scratches and folds prevail.

Although similar wear features are observed on the retrieved articulating tibial and patellar components from total knee prostheses, the overall wear damage appears to be higher on the knee parts as compared to hip sockets [205–207]. This could be attributed to the different degree of conformity between hip and knee articulating systems, much higher contact stresses encountered on the tibial polyethylene parts, and wear debris, which is often trapped in the knee joint. The amount of wear damage on knee components has been positively correlated with patient weight and service time of an implant [205–207]. More damage has been observed on thinner and flatter parts of the prosthesis [207]. The characteristic wear features found on UHMWPE tibial and patellar components include burnishing, scratching, pitting, delamination and cracks [26,205,208]. Three-body abrasion associated with the. presence of broken acrylic bone cement particles, manifested by the presence of scoring marks, dents, craters and shredding, is also observed [26,199,203]. The accelerated wear on the metal-backed tibial and patellar components may even result in component fracture [209,210]. The contribution of creep and/or plastic deformation to the overall wear damage of knee polyethylene parts also seems to be significant [205–208].

The improved mechanical properties of carbon fibre reinforced UHMWPE as compared to plain UHMWPE have prompted the application of this composite in hip, knee and ankle articulating components in some of the designs [126]. The articulating components made of these materials showed similar wear resistance [126]. This confirmed the earlier laboratory results, which demonstrated that carbon fibre reinforced UHMWPE does not offer a better wear performance over plain UHMWPE. The amount of surface damage on both materials increased with service time of an implant and the mechanisms of wear occurring were similar, with prevailing features such as burnishing, scratching and pitting [126]. The only

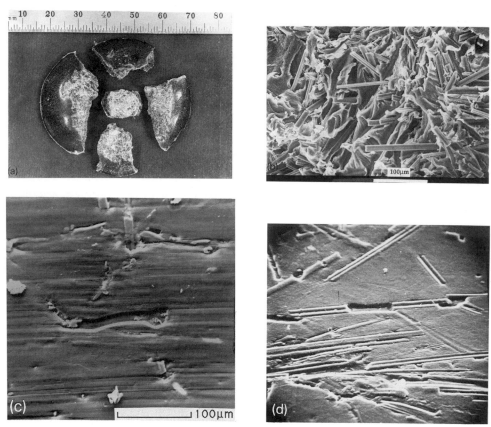

Fig. 7. Wear damage on carbon fibre reinforced UHMWPE tibial components (Courtesy of C.M. Rimnac, Dept. Biomechanics, Hospital for Special Surgery, New York).

difference was the occurrence of delamination wear in carbon fibre reinforced components, and this mode of damage was associated with the presence of carbon fibres [211] (fig. 7).

The amount of wear of the polyethylene articulating components seems to be partly influenced by the choice of the counterface material [1], a confirmation of the laboratory findings. An average rate of penetration of 0.2 mm yr^{-1} has been quoted for polyethylene cups matched with Co–Cr–Mo alloy femoral heads and lower rates of 0.1 mm yr^{-1} in cases of pairing with alumina heads [1]. Laboratory findings describing poor tribological performance of a polyethylene/Ti–6Al–4V alloy pair have been confirmed by reported failures of total hip replacement (THR) incorporating this articulating system [21,212]. The excessive wear of titanium alloy femoral heads, causing a blackening of the surrounding tissues, resulted in severe pain and necessitated revision surgery [21,212].

In order to reduce the overall amount of wear debris released to the surrounding tissue and at the same time maintain the low coefficient of friction, alumina ceramics were used for both articulating components in the total hip replacement. The survival rate of alumina/alumina hip prostheses appears to be similar to that of metal/polymer equivalents [139,213]. Microscopic examination of retrieved alumina components revealed a wide spread of wear damage, ranging from negligible to catastrophic [139]. After the implantation, alumina heads and sockets undergo a normal running-in wear process, which is characterized by relief polishing and very little wear, e.g. about 0.2 μm after 8 yr of implantation [139,214]. Higher wear rates, ranging from 6 to 9 μm yr^{-1}, measured on retrieved alumina heads and sockets have also been reported [141], but they are still much lower than the usually cited wear rate of 100–200 μm yr^{-1} for a UHMWPE/metal combination. Despite the generally good wear resistance of alumina ceramic, incidents of increased wear in vivo have also been reported. The incidents of severe wear were related either to material properties and the finishing quality of the alumina components [140,144,215,216] or mechanical problems associated with misalignment, e.g. incorrect relative position of the articulating head and socket surfaces [139,214,217,218]. Grain excavation [144] and corrosion at alumina grain boundaries contaminated by alkali and silica [216] were found to initiate wear. The wear process, once initiated, was accelerated by three body abrasion.

Among 1330 cases of retrieved alumina components only six cases of severe wear were attributed to a large deviation in the sphericity and incorrect tolerances between articulating components [140]. Severe wear marks were also found on the majority of the retrieved alumina head and socket pairs which were removed due to socket loosening [139,214]. It was concluded that deep wear tracks, of about 4–80 μm, observed on the articulating surfaces occurred as a consequence of the socket loosening and tilting from its correct position. A number of other incidents of increased wear on alumina components were also related to exceptional cases of incorrect positioning of the head and socket [217] during or after the implantation. Incorrect positioning of the socket at implantation or after aseptic loosening alters the stress distribution at the head/socket interface. This usually results in a local increase of the contact stresses on the ceramic surfaces, which initiates abnormal wear of the material by the development of cracks, grain pull out and then three body abrasion.

The spread of penetration rates and clinical wear coefficients reported in the literature is very wide. The wear rates range from practically zero damage to severe wear disruption of the articular surfaces. There is a number of contributing factors to the observed variation in the reported results. The main contributing factors are: patient characteristics such as age, weight and activity, prosthesis design, time of implantation, structural variation of the biomaterial properties, the presence of acrylic debris, the surface roughness of a hard counterface, methodological errors and, finally, the chaotic character of the wear process itself. It thus appears very difficult if not impossible, to give one representative value of the wear rate which would adequately characterize the tribological behaviour of a particular biomaterial.

5.3. Body reaction to wear debris

Despite the low average wear rates of articulating surfaces, a certain amount of wear particles is still generated in vivo, and the amount of debris released to the body accumulates with time. As the average life of modern prostheses becomes longer, the problems associated with the presence of wear debris in the body are of major concern. Apart from the wear particles generated by the articulating surfaces, there are also broken fragments of acrylic cement, particles detached from metallic and ceramic femoral stem coatings, and particles released due to abrasion between a cement mantle and a metallic stem, all of these contributing to the overall amount of particulate debris. The gradual accumulation of the debris can result in an adverse reaction in the surrounding tissues [219–223].

This adverse tissue reaction to wear debris has been suggested as an important mechanism responsible for pain, bone lysis, loosening at the prosthesis/bone interface, and ultimate implant failure [221,224–226]. Hence, the rate of debris generation and their morphology and chemical composition are very important from the clinical viewpoint.

It has been widely documented that the foreign-body or inflammatory reactions are evoked by the presence of metallic [21,212,226,227], polyethylene [220,221,228–232] and acrylic cement particles [220,233–235]. Although there is no general agreement to what wear debris material causes the most severe body response [221,226,229], it appears that milder inflammation is often noted in the presence of ceramic wear debris whereas the tissue reaction to the particles generated in metal/UHMWPE prostheses is higher [139].

The wear particles released to the body by articulating surfaces and fragments of acrylic cement accumulate first in the newly formed joint capsule. It has been observed that the foreign-body reaction to the presence of wear particles results in the formation of a granulation tissue in the joint capsule [220]. The granulation tissue displays a tendency towards fibrosis and the subsequently formed scar tissue can impair joint mobility [220]. Large volumes of metallic and polymer wear particles present in joint capsules can cause severe pain, even when the whole prosthesis is firmly fixed [21]. The cell response is affected by the size of the particle, with macrophages activated by smaller debris and giant cells found around bigger particles [220,221]. The severity of the reaction increases with increases in the amount of wear debris [134,139,220,236]. It is often found to be lower for press-fit prostheses and higher for cemented and/or porous prostheses [237–240].

The wear particles can be distributed even to the distant places around the joint stem and often metallic and acrylic particles are found together with wear debris at the femoral bone/implant interfaces [241,242]. The wear debris generated by the articulating surfaces can be transported by a fibrous membrane separating the implant from the bone and in cases of cemented prostheses can reach the femoral bone through disruptions in the cement. The presence of wear particles at the interface induces bone resorption (osteolysis), which leads to implant loosening, and presents a severe problem for the subsequent revision surgery [223–

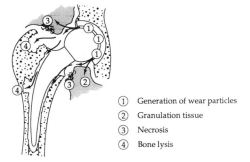

① Generation of wear particles
② Granulation tissue
③ Necrosis
④ Bone lysis

Fig. 8. Body reaction to wear debris (after ref. [220]).

225,227,243,244]. Large osteolytic lesions can even cause fracture of a femoral shaft [245]. Bone resorption has been found around both cemented [237,246] and cementless joint prostheses [227,247], caused by the presence of wear particles. Elevated levels of the specific agents, such as prostaglandin E2 (PGE2) and collagenase, released by the fibrous membrane surrounding the implant as a responce to wear debris, are claimed to stimulate the bone lysis [248–254]. The increased production of PGE2 and other substances is activated by the presence of particulate metallic and polymeric debris [251,252,255] and is found around loosened prostheses in particular [256,257] (fig. 8).

The clinical evidence reported seems to indicate that the presence of particulate debris in the body is the main reason for late prosthesis failure. The solution to this problem seems to be in finding a means to reduce the amount of particles released to the body, both from the articulating surfaces as wear products, and from cement and stems. At present, the complete elimination of the particles in the implant systems appears to be remote, but the the problem could be minimized through:

– wear debris reduction and
– sealing off the wear debris access to the bone/implant interface.

Reduction of wear debris in the system can be achieved through the application of materials with better tribological characteristics which must also fulfill the requirements of biocompatibility. A number of materials have been developed over the years for applications in artificial joint prostheses. Among currently used materials for low-friction articulating surfaces the UHMWPE/ceramic combination seems to give the least amount of debris. The possibility of further improvement in wear resistance of the UHMWPE is not particularly high and a better substitute for this material is yet to be found. The existing ceramic materials are still being improved and new ceramics developed. Metal/metal systems, despite their drawbacks, are characterized by very low wear, and in some specific cases their application could be considered. By applying smooth femoral stems metallic-debris generation can further be reduced [258].

In cases of cemented prostheses the integrity of the cement mantle and the absence of cracks and openings is essential, as this can prevent the wear debris from entering the bone/cement interface. The use of collarless prostheses which are self-wedged against a cement mantle, helps to close the gap between the stem and cement, thus inhibiting the transportation of wear debris around the implant [258,259].

6. Tribological characteristics of polymeric and ceramic biomaterials used in restorative dentistry

The area of restorative dentistry includes conservation of decayed teeth, restoration of defective, missing or traumatically injured teeth and partial or full dentures. Dental materials used for restorative applications must fulfill the specific requirements of biocompatibility, mechanical properties and aesthetics. The common non-metallic dental materials used for restorative dental work are listed in table 6 [260].

The data on the wear and frictional behaviour of dental materials are used to predict the amount of wear of restorations during mastication or normal tooth contacts and the level of frictional forces which are transmitted to the supporting oral tissues. The major wear mechanisms operating in the mouth are two- and three-body abrasion. The former process takes place during physiological rubbing of teeth in saliva and the latter involves the presence of abrasive particles from food and/or toothpaste [261,262].

Among the tribological properties of restorative materials and denture teeth their resistance to abrasion is of major importance. Typical laboratory tests for the evaluation of the abrasion resistance of a candidate dental material are: abrasion, pin-on-disc (plate) and scratch tests [263].

In the abrasion tests the materials of interest are rubbed against abrasive papers under water or slurry lubrication and the amount of lost volume is measured [264]. Less severe tests, with abrasives such as calcium carbonate [262,265] or an alumina/glass mixture [266], are sometimes used since they seem to provide more realistic simulation. A number of special tests have also been devised to evaluate the resistance of restorations to abrasiveness of dentifrices [267–269].

TABLE 6
Types of non-metallic restorative dental materials [260].

Material type	Applications	Advantages over metallic equivalents
Resin composites (macrofilled, micro-filled, hybrid)	Anterior and posterior restorations	Aesthetics, absence of mercury, corrosion resistance
Acrylic resin	Dentures and artificial teeth	Aesthetics
Ceramics (feldspathic, alumina/glass)	Partial and full dentures	Aesthetics, high hardness

The pin-on-disc type test rigs are adapted to evaluate the wear between two sliding bodies, one of which is usually a loaded pin made of human enamel and the other is a disc or plate made of the dental material tested [270–272]. The depth of the wear track on the discs and/or volume lost by the pin are measured, and the wear rates or wear coefficients determined.

The scratch test method involves the determination of the material resistance to penetration. A diamond hemisphere [273,274] or a cone [275] is slid across the surface of the specimen under specified load and in a selected environment and the scratch width is measured. The scratch width depicts the extent of surface failure and can then be correlated with the wear resistance. The mode of failure, i.e. elastic, ductile or brittle, and the transition load between ductile deformation and brittle fracture are determined by subsequent microscopic examination of the wear scars.

The major problem with the laboratory tests, however, is associated with the difficulty in extrapolating the obtained results to the material behaviour in vivo. Thus, a clinical evaluation of the material performance in real dental applications is conducted after the initial laboratory tests. This involves visual observation, exploration, photographic assessment of restorations and scanning electron microscopy (SEM) examination of their replicas. The first guidelines for the clinical evaluation and criteria to measure the degree of degradation of restorations were introduced in the early 1970s by the US Public Health Service [276]. Subsequently, special methods have been devised to facilitate the wear measurements of restorations in vivo [277–282]. The character of some of the methods is qualitative, in which the wear of replicas of new and worn tooth fillings which have worked in the mouth for a specific period of time, is compared. The replicas are examined by SEM and the wear mechanism and surface changes are studied [283]. The other methods have more quantitative character, in which the lost volume on the replicas is determined [280,281,284].

6.1. Wear of dental resins and their composites

Acrylic resins and, later, composite resins with macro-, micro-, colloidal and hybrid fillers have gained wide acceptance as denture teeth, veneers and restorative materials, mostly for anterior teeth. Their advantages over traditionally used amalgams in restoration include improved aesthetics and elimination of potentially dangerous mercury present in amalgams. A serious limitation of dental composites, however, is their wear resistance, which is often not adequate for their use in posterior tooth fillings, where higher mastication stresses are encountered.

6.1.1. Wear in vitro

The in vitro tribological characteristics of early unfilled and filled resins have been compared in abrasion [285–287] and scratch tests [287–289]. Abrasion tests on silicon carbide paper showed that the composite resins with a silanated filler had the highest wear resistance and that the acrylic resin was more resistant to wear than the diacrylate resin [285]. The wear resistance of dimethacrylate resins was a function of the type of abrasive paper used, and was lower on SiC paper and higher on alumina and quartz paper [287]. The resistance to penetration, frictional torque and wear modes in water of various filled and unfilled restorative resins was also compared, using a scratch test [288,289]. The tests confirmed that the resistance to penetration of a resin could be improved by the application of appropriate silanated fillers. The mode of deformation depended on the matrix material and the wear damage was more severe for cross-linked diacrylate resin than for linear acrylic resin [288,289]. The deformation mode also varied between various dimethacrylate resins, and ranged from plastic deformation to brittle fracture [287].

Over the years the composite resins have undergone various modifications in an attempt to improve their wear resistance [290]. The hybrid composite resins, which are reinforced by a mixture of submicron and conventional-sized macrofiller particles, have shown better in vitro wear resistance than conventional resins [291,292]. The wear of small-particle, microfilled and hybrid resins was compared in abrasion tests with silicon carbide and calcium carbonate as abrasive [292]. The hybrid resin displayed the best wear resistance, which was not affected by storage in water, contrary to results obtained for the other two resins [292]. During the wear tests all composite resins absorb water, which seems to negatively affect the matrix/filler bonds [292–294].

6.1.2. Wear in vivo

The in vivo wear performance of dental composites in posterior restorations has been compared with that of amalgams [295–298]. It was found that the conventional composite resins were less wear resistant than amalgams and the wear degradation started shortly after the restoration took place [295–297, 299]. Although the recent development of improved composites has led to their more confident and frequent use for posterior tooth restorations, the evaluation tests still show higher wear of these materials compared to that of amalgams [298,300–303]. The in vivo wear of composite resins proceeds in two stages [292,297,304]. In the first stage, the polymeric matrix is abraded, exposing filler particles. In the second stage, the forces acting on the exposed particles cause their rotation, chipping and dislodgement. The comparison between in vivo and in vitro wear patterns has confirmed the difficulty in realistic simulation of wear behaviour of dental materials in laboratory tests [292].

6.2. Wear of dental ceramics

The main applications of porcelains in dentistry include denture teeth and aesthetic coatings used in crown and bridge prosthetics, where the close resemblance to natural teeth porcelain is utilized. The tribological properties of porcelains, which are of major importance in these applications, have been evaluated in vitro. The frictional properties of porcelains were found to be strongly affected by the type of environment [305,306]. For example, the coefficient of friction of porcelain sliding against porcelain was lower in air than in saliva [305]. Lower friction in air than in water was also found in scratch tests where a diamond hemisphere was run across as-glazed porcelain samples [306]. Greater surface damage was also recorded in water than in air for as-glazed porcelain. Surface coatings applied to the ceramics affect their tribological behaviour to various degree. Thin gold coating on porcelains seems to eliminate the dependence of their tribological behaviour on the environment and gave slightly reduced frictional forces [306]. On the other hand, ion plated chromium coating provided some improvements in porcelain wear resistance in water [306]. The character of the wear damage observed on porcelain samples was usually brittle, even under loads which produced ductile deformation on acrylic samples [306–308].

There is also another major problem, related to the excessive wear of natural enamel in the contact with ceramic teeth. Traditional feldspathic dental porcelains usually cause higher wear of the opposing natural enamel than the other restorative materials [272,309–311]. Newly developed glass–ceramic materials have been evaluated with respect to wear of the opposing natural enamel using a pin-on-disc wear rig [272]. The experiments showed that the wear of enamel pins depended on the type of ceramic disc and was about two times lower for glass–ceramic discs than for feldspathic ceramic discs [272]. The results obtained from pin-on-disc tests were in agreement with previously reported results from simulation tests using an "artificial mouth" [312].

7. Conclusions

During the past thirty years impressive progress has been made in the development and implantation of biomaterials into the human body with the aim of reconstructing or restoring the normal function of diseased or damaged body parts. In the field of total joint replacement the introduction of low-friction arthroplasty greatly increased the useful life of joint prostheses [93]. The much lower friction of polymer/metal combinations compared to previously used metal/metal systems reduced the possibility of prosthesis loosening due to high mechanical forces in the tribosystem. The introduction of ceramic articulating components further improved the performance of joint prostheses, due to better compatibility of ceramics with living tissue and better tribological characteristics of ceramic/ceramic or ceramic/polymer systems (lower friction, lower heat generation and lower polymer wear). However, the much higher wear of a polyethylene cup compared to the wear of its metallic equivalent seems to be a major drawback, and reconsideration of a metal/metal combination has been forecast [1]. The biomaterials currently used as articulating components seem to be adequate for short-term use. Failure to achieve long-term life of artificial prostheses is partly a consequence of the excessive wear of the articulating surfaces and the subsequent adverse reaction of body tissue to the presence of foreign particles. This reaction leads to the failure of implant fixation and often causes severe pain. The properties of some materials, like polyethylene, can also deteriorate in a body environment over long periods of time, which further contributes to the problems encountered.

The improved reliability and life of joint prostheses could make the current goal of feasible operations on younger and more active patients achievable. Prostheses with much longer life expectancy, of about 30–40 yr, are required, as opposed to currently available life expectancy of about 10–15 yr. This requires the development of both biomaterials with better mechanical and tribological properties and the improvement of the implant/bone interface. Among the tested new biomaterials, ceramic and polymer composites, such as carbon/carbon and carbon/polysulfone, seem to be promising candidates for application as femoral components in joint prostheses. The apparent advantage of composite material is that it can be developed with mechanical properties (e.g. elastic modulus) closely matching those of a bone, and therefore minimizing the stress concentration at the interface with bone. However, due to their poor tribological characteristics these materials seem to be at their present stage unsuitable for articulating surfaces. The attempts to use carbon fibre reinforced UHMWPE for articulating surfaces did also not show any significant reduction in the amount of generated wear debris.

Much wider use of composite biomaterials is observed in dentistry, where composite resins have found applications as restorative materials for anterior and posterior teeth. At present, resins reinforced by particulate fillers are most popular. Their clinical performance in very demanding posterior restorations is ranked as acceptable, and they now replace traditionally used amalgams more frequently. Currently, development work is carried out in an attempt to develop continuous fibre reinforced plastics for potential dental applications such as dental teeth, since fibre reinforced acrilic resins have much better mechanical properties than traditional unfilled acrylic.

Acknowledgements

I wish to thank firstly my wife Grazyna for her help in very meticulous selection and search of the relevant literature for this chapter. I would also like to thank Dr. R.M. Streicher from Sulzer Medica, Switzerland, for very valuable discussion and micrographs, Professor P. Christel from Laboratoire de Recherches Orthopédiques for his comments, Dr. C.M. Rimnac from the Hospital for Special Surgery, New York, and Dr. H.A. McKellop from the University of Southern California, Department of Orthopaedics, Los Angeles, for the micrographs, Professor K. Friedrich from the Institute for Composite Materials, University of Kaiserslautern, for constructive discussion. Finally, I would like to thank the Department of Materials, Ecole Polytechnique Fédérale de Lausanne and the Department of Mechanical and Materials Engineering, University of Western Australia, for their help during the preparation of the manuscript.

List of symbols

A constant
k wear coefficient ($mm^3 N^{-1}m^{-1}$)
P pressure (MPa)

R_a surface roughness (μm)
W wear rate (mm/mm)
W_0 constant

References

[1] R.M. Streicher, Sulzer Tech. Rev. 1 (1991) 13–17.
[2] D. Dowson, V. Wright and M.D. Longfield, Biomed. Eng. 4 (1964) 160–165.
[3] F.C. Linn and E.L. Radin, Arthritis Rheum. 11 (1968) 674–682.
[4] R.S. Fein, Proc. Inst. Mech. Eng. London 181 (1967) 125–128.
[5] M.D. Longfield, D. Dowson, P.S. Walker and V. Wright, Biomed. Eng. 9 (1969) 517–522.
[6] D. Dowson, A. Unsworth and V. Wright, J. Mech. Eng. Sci. 12 (1970) 364–369.
[7] A. Unsworth, D. Dowson and V. Wright, J. Lubr. Technol. 97 (1975) 369–376.
[8] P.A. Torzilli, in: Handbook of Engineering in Medicine, eds D.G. Fleming and B.N. Feinberg (Chemical Rubber Co., Cleveland, OH, 1976) pp. 225–251.
[9] W.H. Davis, S.L. Lee and L. Sokoloff, Arthritis Rheum. 21 (1978) 754–756.
[10] D. Dowson and Z.M. Jin, Eng. Med. 15 (1986) 62–65.
[11] J.L. Katz, in: Contemporary Biomaterials, eds J.W. Boretos and M. Eden (Noyes Publications, Park Ridge, NJ, 1984) pp. 453–475.
[12] P. Christel, P.-F. Bernard and A. Meunier, in: Proc. 2nd World Congress on Biomaterials, Washington, DC, April 27–May 1 (1984) p. 178.
[13] M. Roffman, D.G. Mendes, Y. Charit and M.S. Hunt, in: Proc. 2nd World Congress on Biomaterials, Washington, DC, April 27–May 1 (1984) p. 177.
[14] R. Huiskes, in: Frontiers in Biomechanics, eds G.W. Schmid-Schonbein, S.L.-Y. Woo and B.W. Zweifach (Springer, Berlin, 1986) pp. 245–262.
[15] A.M. Weinstein, F.P. Magee, J.B. Koeneman, J.A. Longo and R.A. Yapp, in: Proc. 3rd World Biomaterials Congress, Kyoto, Japan, April 21–25 (1988) p. 225.
[16] H. Kusswetter, E. Gabriel, T. Stuhler and L. Topfer, in: The Cementless Fixation of Hip Endoprostheses, ed. E. Morscher (Springer, New York, 1984) pp. 17–20.
[17] A.M. Pappas and J. Cohen, J. Bone & Jt. Surg. 50-A (1968) 535–547.
[18] M. Mital and J. Cohen, J. Bone & Jt. Surg. 50-A (1968) 547–556.
[19] L.J. Bearden and F.W. Cooke, J. Biomed. Mater Res. 14 (1980) 289–309.
[20] R. Michel, J. Hofmann, F. Loer and J. Zilkens, Arch. Orthop. Trauma Surg. 103 (1984) 85–90.
[21] J. Black, H. Sherk, W.R. Rostoker and J.O. Galante, in: Proc. 3rd World Biomaterials Congress, Kyoto, Japan, April 21–25 (1988) p. 343.
[22] B. Weightman, S. Simon, I. Paul, R. Rose and E. Radin, J. Lubr. Technol. 94 (1972) 131–135.
[23] A. Unsworth, D. Dowson, V. Wright and D. Koshal, J. Lubr. Technol. 97 (1975) 377–381.
[24] J.O. Galante and W. Rostoker, Acta Orthopaed. Scand. 145 (1973) 1–46.
[25] J.H. Dumbleton, C. Shen and E.H. Miller, Wear 29 (1974) 163–171.
[26] P.S. Trent and P.S. Walker, Wear 36 (1976) 175–187.
[27] H. Beutler, M. Lehmann and G. Stahli, Wear 33 (1975) 337–350.
[28] H. McKellop, I.C. Clarke, K.L. Markolf and H.C. Amstutz, J. Biomed. Mater. Res. 12 (1978) 895–927.
[29] A. Ben Abdallah and D. Treheux, Wear 142 (1991) 43–56.
[30] D. Dowson, M.M.E. Diab, B.J. Gillis and J.R. Atkinson, in: Polymer Wear and Its Control, ed. L. Lee, Am. Chem. Soc. Symp. Ser. 287 (1985) pp. 171–187.
[31] J.R. Atkinson, K.J. Brown and D. Dowson, J. Lubr. Technol. 100 (1978) 208–218.
[32] K.J. Brown, J.R. Atkinson, D. Dowson and V. Wright, Wear 40 (1976) 255–264.
[33] D. Dowson and N.C. Wallbridge, Wear 104 (1985) 203–215.
[34] I. Duff-Barclay and D.T. Spillman, Proc. Inst. Mech. Eng. London 181, 3J (1966–1967) 90–103.
[35] B.O. Weightman, I.L. Paul, R.M. Rose, S.R. Simon and E.L. Radin, J. Biomech. 6 (1973) 299–311.
[36] D. Dowson, P.S. Walker, M.D. Longfield and V. Wright, Med. Biol. Eng. 8 (1970) 37–43.
[37] J.H. Dumbleton, D.A. Miller and E.H. Miller, Wear 20 (1972) 165–174.
[38] J.A. Shaw and D.G. Murray, Clin. Orthop. 94 (1973) 15–23.

[39] N.J. Zachman, ASME Paper No. 78-DET-59 (1978).
[40] K.W. Greer, Trans. 11th Int. Biomaterials Symposium, Vol. 3 (1979) p. 80.
[41] M.J. Pappas and F.F. Buechel, Trans. 11th Int. Biomaterials Symposium, Vol. 3 (1979) p. 101.
[42] B.M. Hillberry, J.A. Schaaf, C.D. Cullom and D.B. Kettelkamp, in: Biomaterials '84, Trans. 2nd World Congress, Vol. VII (1984) p. 149.
[43] J.R. Atkinson, J.M. Dowling and R. Cicek, Biomaterials 1 (1980) 89–96.
[44] J.H. Dumbleton and C. Shen, Wear 37 (1976) 279–289.
[45] W. Rostoker and J.O. Galante, J. Biomed. Mater. Res. 10 (1976) 303–310.
[46] W. Rostoker and J.O. Galante, J. Biomed. Mater. Res. 13 (1979) 957–964.
[47] H. McKellop, I. Clarke, K. Markolf and H. Amstutz, J. Biomed. Mater Res. 15 (1981) 619–653.
[48] R.M. Rose, W.R. Cimino, E. Ellis and A.M. Crugnola, Wear 77 (1982) 89–104.
[49] R.M. Rose, H.V. Goldfarb, E. Ellis and A.M. Crugnola, Wear 92 (1983) 99–111.
[50] P. Kumar, M. Oka, K. Ikeuchi, K. Shimizu, T. Yamamuro, H. Okumura and Y. Kotoura, J. Biomed. Mater. Res. 25 (1991) 813–828.
[51] A. Weinstein, H. Amstutz, G. Pavon and V. Franceschini, J. Biomed. Mater. Res. Symp. 4 (1973) 297–325.
[52] R.G. Bayer, W.C. Clinton, C.W. Nelson and R.A. Schumacher, Wear 5 (1962) 378–391.
[53] S.H. Rhee and J.H. Dumbleton, Wear 36 (1976) 207–224.
[54] T. Murakami and N. Ohtsuki, in: Proc. 3rd World Biomaterials Congress, Kyoto, Japan, April 21–25 (1988) p. 340.
[55] P. Kumar, M. Oka, Y. Kotoura, K. Ikeuchi, Y. Murai, T. Yamamuro, H. Okumura and Y. Nakayama, in: Proc. 3rd World Biomaterials Congress, Kyoto, Japan, April 21–25 (1988) p. 339.
[56] D.A. Miller, R.D. Ainsworth, J.H. Dumbleton, D. Page, E.H. Miller and C. Shen, Wear 28 (1974) 207–216.
[57] W. Rostoker and J.O. Galante, Biomaterials 2 (1981) 221–224.
[58] R.M. Streicher, H. Weber, R. Schon and M. Semlitsch, Biomaterials 12 (1991) 125–129.
[59] R.A. Buchanan, E.D. Rigney Jr and J.M. Williams, J. Biomed. Mater. Res. 21 (1987) 355–366.
[60] H.A. McKellop and T.V. Rostlund, J. Biomed. Mater. Res. 24 (1990) 1413–1425.
[61] M. Salzer, K. Zweymuller, H. Locke, A. Zeibig, N. Stark, H. Plenk and G. Punzet, J. Biomed. Mater. Res. 10 (1976) 847–856.
[62] H.P. Sieber and B.G. Weber, in: Ceramics in Surgery, ed. P. Vincenzini (Elsevier, Amsterdam, 1983) pp. 287–294.
[63] M. Semlitsch, M. Lehmann, H. Weber, E. Doerre and H.G. Willert, J. Biomed. Mater. Res. 11 (1977) 537–552.
[64] D. Dowson and I.W. Linnett, in: Mechanical Properties of Biomaterials, eds G.W. Hastings and D.F. Williams (Wiley, Chichester, 1980) ch. 1, pp. 3–26.
[65] D. Dowson and P.T. Harding, Wear 75 (1982) 313–331.
[66] P. Kumar, M. Oka, Y. Kotoura, K. Ikeuchi, T. Yamamuro, H. Okumura and Y. Nakayama, in: Proc. 3rd World Biomaterials Congress, Kyoto, Japan, April 21–25 (1988) p. 587.
[67] T. Tateishi, Y. Shirasaki, A. Terui and H. Yunoki, in: Proc. 3rd World Biomaterials Congress, Kyoto, Japan, April 21–25 (1988) p. 341.
[68] H. McKellop, I. Clarke, K. Markolf and H. Amstutz, in: Trans. 25th Annual Meeting Orthop. Res. Soc., San Francisco, California, February 20–22 (1979) p. 212.
[69] G. Orange, K.M. Liang and G. Fantozzi, in: Proc. 14th Int. Conf. on Science of Ceramics, Canterbury, Kent, September 1987, ed. D. Taylor (Fairey Tecramics, UK, 1987) p. 709.
[70] D. Goeuriot-Launay, P. Goeuriot, G. Orange, G. Fantozzi and D. Treheux, Silic. Ind. 1–2 (1989) 3–9.
[71] M.A. Swikert and R.L. Johnson, NASA Technical Note D-8174 (1976).
[72] D. Dowson, J.M. Challen, K. Holmes and J.R. Atkinson, in: The Wear of Non-Metallic Materials, Proc. 3rd Leeds-Lyon Symp. on Tribology, Leeds, 1976, eds D. Dowson, M. Godet and C.M. Taylor (Mechanical Engineering Publications, London, 1978) pp. 99–102.
[73] B. Seedhom, D. Dowson and V. Wright, Wear 24 (1973) 35–51.
[74] H. Oonishi, H. Igaki and Y. Takayama, in: Proc. 3rd World Biomaterials Congress, Kyoto, Japan, April 21–25 (1988) p. 337.
[75] D. Dowson, S. Taheri and N.C. Wallbridge, Wear 119 (1987) 277–293.

[76] H. Oonishi, H. Igaki and Y. Takayama, in: Proc. 3rd World Biomaterials Congress, Kyoto, Japan, April 21–25 (1988) p. 588.
[77] R.M. Rose, E.V. Goldfarb, E. Ellis and A.M. Crugnola, J. Orthop. Res. 2 (1984) 393–400.
[78] H.J. Nusbaum, R.M. Rose, I.L. Paul, A.M. Crugnola and E.L. Radin, J. Appl. Polym. Sci. 23 (1979) 777–789.
[79] A.E. Zachariades, in: Proc. 3rd World Biomaterials Congress, Kyoto, Japan, April 21–25 (1988) p. 222.
[80] S. Li, A new enhanced ultra high molecular weight polyethylene for orthopaedic applications: A technical brief, DePuy Technical Brief (1989).
[81] R.M. Rose, H.J. Nusbaum, H. Schneider, M. Ries, I. Paul, A. Crugnola, S.R. Simon and E.L. Radin, J. Bone & Jt. Surg. 62-A (1980) 537–549.
[82] P. Eyerer, M. Kurth, H.A. McKellop and T. Mittlmeier, J. Biomed. Mater. Res. 21 (1987) 275–291.
[83] P. Eyerer and Y.C. Ke, J. Biomed. Mater. Res. 18 (1984) 1137–1151.
[84] H. McKellop, J. Kirkpatrick, K. Markolf and H. Amstutz, in: Trans. 26th Annual Meeting Orthop. Res. Soc., Vol. 5 (1980) p. 96.
[85] H. McKellop, A. Hosseinian, K. Burgoyne and I. Clarke, in: Proc. 2nd World Congress on Biomaterials, Washington, DC, April 27–May 1 (1984) p. 313.
[86] J. Charnley and Z. Cupic, Clin. Orthop. 95 (1973) 9–25.
[87] J. Charnley and D.K. Halley, Clin. Orthop. 112 (1975) 170–179.
[88] M.J. Griffith, M. Seidenstein, D. Williams and J. Charnley, Clin. Orthop. 137 (1978) 37–47.
[89] J.R. Atkinson, D. Dowson, G.H. Isaac and B.M. Wroblewski, Wear 104 (1985) 225–244.
[90] J.A. Davidson, G. Schwartz, G. Lynch and S. Gir, J. Biomed. Mater. Res. (Appl. Biomater.) 22 (1988) 69–91.
[91] J.A. Davidson and G.E. Lynch, in: Proc. 12th Annual Meeting, American Society of Biomechanics (University of Illinois, Urbana, 1988) pp. 4–5.
[92] J.A. Davidson, S. Gir and J.P. Paul, J. Biomed. Mater. Res. (Appl. Biomater.) 22(A3) (1988) 281–309.
[93] J. Charnley, Low Friction Arthroplasty of the Hip (Springer, Berlin, 1979).
[94] M. Semlitsch, Eng. Med. 3 (1974) 1–10.
[95] A. Cappozzo, L. Cini, A. Pizzoferrato, C. Trentani and S.S. Cortesi, J. Biomed. Mater. Res. 11 (1977) 657–669.
[96] T. Christiansen, Acta Chir. Scand. 135 (1969) 43–46.
[97] T. Christiansen, Acta Chir. Scand. 140 (1974) 185–188.
[98] A. Poli, Proc. Inst. Mech. Eng. London 181 (1966–67) 140–141.
[99] J.S. Fister, V.A. Memoli, J.O. Galante, W. Rostoker and R.M. Urban, J. Biomed. Mater. Res. 19 (1985) 519–533.
[100] C. Shen and J.H. Dumbleton, Wear 38 (1976) 291–303.
[101] J.H. Dumbleton, in: Corrosion and Degradation of Implant Materials, eds B.C. Syrett and A. Acharya (ASTM, Philadelphia, PA, 1979) pp. 41–60.
[102] G. Josefsson, B. Eriksson, M. Fagerlund, G. Gudmundersson, H. Helgasson, O. Karlsson, B. Nordstrom, B.-O. Olnas and S. Wijkstrom, Acta Orthop. Scand. 52 (1981) 696. Abstract.
[103] A. Alho, O. Soreide and A.J. Bjersand, Acta Orthop. Scand. 55 (1984) 261–266.
[104] E. Sudmann, L.I. Havelin, O.D. Lunde and M. Rait, Acta Orthop. Scand. 54 (1983) 545–552.
[105] E.B. Mathiesen, J.U. Lindgren, F.P. Reinholt and E. Sudmann, J. Biomed. Mater. Res. 21 (1987) 459–466.
[106] A. Ohlin and L.-G. Kindblom, Acta Orthop. Scand. 59 (1988) 629–634.
[107] A. Ohlin and T. Erikson, Acta Orthop. Scand. 55 (1984) 101. Abstract.
[108] E.B. Mathiesen, J.U. Lindgren, F.P. Reinholt and E. Sudmann, Acta Orthop. Scand. 57 (1986) 193–196.
[109] J.A. Davidson, J. Comp. Tech. Res. 9 (1987) 151–161.
[110] K. Tayton, C. Johnson-Nurse, B. McKibbin, J.S. Bradley and G. Hastings, J. Bone & Jt. Surg. 64-B (1982) 105–111.
[111] D.F. Williams, A. McNamara and R.M. Turner, J. Mater. Sci. Lett. 6 (1987) 188–190.
[112] G.B. McKenna, W.O. Statton, H.K. Dunn, K.D. Johnson, G.W. Bradley and A.U. Daniels, in: Proc. SAMPE 21st National Symposium, April (1976) pp. 232–240.

[113] C.A. Behling and M. Spector, J. Biomed. Mater. Res. 20 (1986) 653–666.

[114] L.M. Wenz, K. Merritt, S.A. Brown, A. Moet and A.D. Steffee, J. Biomed. Mater. Res. 24 (1990) 207–215.

[115] G.B. McKenna, G.W. Bradley, H.K. Dunn and W.O. Statton, J. Biomed. Mater. Res. 13 (1979) 783–798.

[116] E. Sclippa and K. Piekarski, J. Biomed. Mater. Res. 7 (1973) 59–70.

[117] R. Ainsworth, G. Farling and D. Bardos, in: Trans. 23rd Orthop. Res. Soc. Symp., Vol. 2 (1977) p. 120.

[118] T.M. Wright, T. Fukubayashi and A.H. Burstein, J. Biomed. Mater. Res. 15 (1981) 719–730.

[119] W. Rostoker and J.O. Galante, J. Biomed. Mater. Res. 13 (1979) 825–828.

[120] F.P. Magee, A.M. Weinstein, J.A. Longo, J.B. Koeneman and R.A. Yapp, Clin. Orthop. 235 (1988) 237–252.

[121] F-K. Chang, J.L. Perez and J.A. Davidson, J. Biomed. Mater. Res. 24 (1990) 873–899.

[122] R.A. Latour Jr, J. Black and B. Miller, in: Proc. 3rd World Biomaterials Congress, Kyoto, Japan, April 21–25 (1988) p. 470.

[123] Zimmer USA, Poly Two Carbon Polyethylene Composite, Technical Report (Zimmer, Warsaw, IN, 1977).

[124] C.D. Peterson, B.M. Hillberry and D.A. Heck, J. Biomed. Mater. Res. 22 (1988) 887–903.

[125] L.S. Stern, M.T. Manley and J. Parr, in: Proc. 2nd World Congress on Biomaterials, Washington, DC, April 27–May 1 (1984) p. 66.

[126] T.M. Wright, C.M. Rimnac, P.M. Faris and M. Bansal, J. Bone & Jt. Surg. 70-A (1988) 1312–1319.

[127] C. Shen and J.H. Dumbleton, Wear 40 (1976) 351–360.

[128] C. Shen and J.H. Dumbleton, Wear 40 (1976) 371–382.

[129] J.K. Lancaster, Wear 20 (1972) 315–333.

[130] P. Boutin, Presse Méd. 79 (1971) 639–640.

[131] P. Boutin, Rev. Chir. Orthop. 58 (1972) 229–246.

[132] S.F. Hulbert, F.A. Young, R.S. Mathews, J.J. Klawitter, C.D. Talbert and F.H. Stelling, J. Biomed. Mater. Res. 4 (1970) 433–456.

[133] H. Kawahara, Clin. Mater. 2 (1987) 181–206.

[134] J. Harms and E. Mausle, J. Biomed. Mater. Res. 13 (1979) 67–87.

[135] W.C. Richardson, J.J. Klawitter, B.W. Sauer, J.R. Pruitt and S.F. Hulbert, J. Biomed. Mater. Res. 9 (1975) 73–80.

[136] P. Griss, H. von Andrian-Werburg, B. Krempien and G. Heimke, J. Biomed. Mater. Res. Symp. 4 (1973) 453–462.

[137] P. Griss and G. Heimke, in: Biocompatibility of Clinical Implant Materials, Vol. I, ed. D.F. Williams (CRC Press, Boca Raton, FL, 1981) pp. 155–198.

[138] A. Zeibig and H. Luber, in: Ceramics in Surgery, ed. P. Vincenzini (Elsevier, Amsterdam, 1983) pp. 267–275.

[139] P. Boutin, P. Christel, J.-M. Dorlot, A. Meunier, A. de Roquancourt, D. Blanquaert, S. Herman, L. Sedel and J. Witvoet, J. Biomed. Mater. Res. 22 (1988) 1203–1232.

[140] P. Boutin, in: Orthopaedic Ceramic Implants, eds H. Oonishi and Y. Ooi, Proc. Jpn. Soc. Orthop. Cer. Implants 1 (1981) 11–18.

[141] P. Boutin and D. Blanquaert, Rev. Chir. Orthop. 67 (1981) 279–287.

[142] K. Kawauchi, Y. Kuroki, S. Saito, H. Ohgiya, S. Sato, S. Kondo, I. Hirose, S. Obara, H. Aso, K. Yamano, S. Sasada and S. Norose, in: Orthopaedic Ceramic Implants, eds H. Oonishi and Y. Ooi, Proc. Jpn. Soc. Orthop. Cer. Implants 4 (1984) 253–257.

[143] S. Nakagawa, M. Oga, K. Hayashi, N. Matsuguti and Y. Sugioka, in: Proc. 3rd World Biomaterials Congress, Kyoto, Japan, April 21–25 (1988) p. 346.

[144] W. Plitz and P. Griss, in: Implant Retrieval: Material and Biological Analysis, eds A. Weinstein, D. Gibbons, S. Brown and W. Ruff NBS SP-601 (US Dept. of Commerce, Washington, DC, 1981) pp. 131–156.

[145] R.H.J. Hannink and R.C. Garvie, J. Mater. Sci. 17 (1982) 2637–2643.

[146] M.V. Swain, R.C. Garvie and R.H.J. Hannink, J. Am. Ceram. Soc. 66 (1983) 358–362.

[147] T.K. Gupta, F.F. Lange and J.H. Bechtold, J. Mater. Sci. 13 (1978) 1464–1470.

[148] P. Christel, A. Meunier, M. Heller, J.P. Torre and C.N. Peille, J. Biomed. Mater. Res. 23 (1989) 45–61.

[149] T. Tateishi and H. Yunoki, Bull. Mech. Eng. Lab. Jpn. 45 (1987) 1–9.
[150] S.A. Bortz and E.J. Onesto, Composites 5 (1974) 151–156.
[151] R.C. Garvie, C. Urbani, D.R. Kennedy and J.C. McNeuer, J. Mater. Sci. 19 (1984) 3224–3228.
[152] T. Ohashi, S. Inoue, K. Kajikawa, K. Ibaragi, M. Tada, M. Oguchi, N. Kanematsu, A. Kamegai, Y. Muramatsu and K. Kondo, in: Proc. 3rd World Biomaterials Congress, Kyoto, Japan, April 21–25 (1988) p. 405.
[153] K. Shimizu, P. Kumar, M. Oka, Y. Kotoura, Y. Nakayama, T. Yamamuro, T. Yanagida and K. Makinouchi, in: Proc. 3rd World Biomaterials Congress, Kyoto, Japan, April 21–25 (1988) p. 406.
[154] W. Dawihl and G. Altmeyer, Z. Werkstofftech. 7 (1976) 208.
[155] M.V. Swain and M.F. Goss, CSIRO, Advanced Materials Laboratory Report No. ALM-87–10 (1981).
[156] D. Holz, R. Janssen, K. Friedrich and N. Claussen, J. Eur. Ceram. Soc. 5 (1989) 229–232.
[157] K.-H. Zum Gahr, Wear 133 (1989) 1–22.
[158] A. Toni, A. Sudanese, G.L. Cattaneo, D. Ciaroni, T. Greggi, D. Dallari and A. Giunti, in: Proc. 3rd World Biomaterials Congress, Kyoto, Japan, April 21–25 (1988) p. 338.
[159] F.C. Cowlard and J.C. Lewis, J. Mater. Sci. 2 (1967) 507–512.
[160] G.M. Jenkins and K. Kawamura, Polymeric Carbons – Carbon Fibre, Glass and Char (Cambridge University Press, Cambridge, 1976).
[161] J.C. Bokros, Carbon 15 (1977) 355–371.
[162] A.D. Haubold, W.S. Shim and J.C. Bokros, in: Biocompatibility of Clinical Implant Materials, Vol. II, ed. D.F. Williams (CRC Press, Boca Raton, FL, 1981) pp. 3–42.
[163] P. Christel, B. Buttazzoni, J.L. Leray and C. Morin, in: Biomaterials 1980, eds G.D. Winter, D. Gibbons and H. Plenk Jr (Wiley, New York, 1982) pp. 87–96.
[164] P. Christel, CRC Crit. Rev. Biocompatibility 2 (1986) 189–218.
[165] A.A. Marino, S. Fronczak, C. Boudreaux, D.N. Liles, E.M. Keating and J.A. Albright, IEEE Eng. Med. Biol. 5 (1986) 31–34.
[166] R.J. Minns, in: Proc. 2nd World Congress on Biomaterials, Washington, DC, April 27–May 1 (1984) p. 50.
[167] S.L. Kampner and A.M. Weinstein, in: Proc. 2nd World Congress on Biomaterials, Washington, DC, April 27–May 1 (1984) p. 228.
[168] G.M. Jenkins and F.K. de Carvalho, Carbon 15 (1977) 33–37.
[169] P. Christel, P.-F. Bernard and A. Meunier, in: Proc. 2nd World Congress on Biomaterials, Washington, DC, April 27–May 1 (1984) p. 178.
[170] P. Christel, A. Meunier, S. Leclercq, Ph. Bouquet and B. Buttazzoni, J. Biomed. Mater. Res. Appl. 21(A2) (1987) 191–218.
[171] P. Christel, in: Proc. 3rd World Biomaterials Congress, Kyoto, Japan, April 21–25 (1988) p. 38.
[172] G.M. Jenkins and C.J. Grigson, J. Biomed. Mater. Res. 13 (1979) 371–394.
[173] M. Farrelly and J.T.A. Pollock, Mater. Forum 10 (1987) 198–201.
[174] R.A. Burton and R.G. Burton, J. Tribol. 112 (1990) 68–72.
[175] R.A. Burton and R.G. Burton, IEEE Trans. 12 (1989) 224–228.
[176] J.K. Lancaster, ASLE Trans. 18 (1975) 187–201.
[177] J.K. Lancaster, ASLE Trans. 20 (1977) 43–54.
[178] J.T.A. Pollock, R.A. Clissold and M. Farrelly, J. Mater. Sci. Lett. 6 (1987) 1023–1024.
[179] P. Christel, M. Homerin and A. Dryll, in: Proc. 2nd World Congress on Biomaterials, Washington, DC, April 27–May 1 (1984) p. 72.
[180] J.T.A. Pollock and L.S. Wielunski, Mater. Res. Soc. Symp. Proc. 110 (1989).
[181] P. Christel, private communication (1991).
[182] C.J. Sutherland, A.H. Wilde, L.S. Borden and K.E. Marks, J. Bone & Jt. Surg. 64-A (1982) 970–982.
[183] R.N. Stauffer, J. Bone & Jt. Surg. 64-A (1982) 983–990.
[184] J. Older, Clin. Orthop. 211 (1986) 36–42.
[185] L.P. Brady and J.W. McCutchen, Clin. Orthop. 211 (1986) 51–54.
[186] I.C. Clarke, K. Black, C. Rennie and H.C. Amstutz, Clin. Orthop. 121 (1976) 126–142.
[187] Z. Cupic, Clin. Orthop. 141 (1979) 28–43.
[188] E.A. Salvati, P.D. Wilson Jr, M.N. Jolley, F. Vakili, P. Aglietti and G.C. Brown, J. Bone & Jt. Surg. 63-A (1981) 753–767.

[189] J.P. Clarac, P. Pries, L. Launay, P. Martin, H. Freychet and P. Nonet, Rev. Chir. Orthop. 72 (1986) 97–100.

[190] B.M. Wroblewski, Clin. Orthop. 211 (1986) 30–35.

[191] T.H. McCoy, E.A. Salvati, C.S. Ranawat and P.D. Wilson Jr, Orthop. Clin. North Am. 19 (1988) 467–476.

[192] B.M. Wroblewski, J. Bone & Jt. Surg. 67-B (1985) 757–761.

[193] C.M. Rimnac, P.D. Wilson Jr, M.D. Fuchs and T.M. Wright, Orthop. Clin. North Am. 19 (1988) 631–636.

[194] J. Charnley, A. Kamangar and M.D. Longfield, Med. Biol. Eng. 7 (1969) 31–39.

[195] E.S. Grood, R. Shastri and C.N. Hopson, J. Biomed. Mater. Res. 16 (1982) 399–405.

[196] P. Eyerer, in: Proc. 2nd World Congress on Biomaterials, Washington, DC, April 27–May 1 (1984) p. 68.

[197] M.L. Jacobs and J. Black, J. Biomed. Mater. Res. Symp. 6 (1975) 221–225.

[198] B. Weightman, D.P. Isherwood and S.A.V. Swanson, J. Biomed. Mater. Res. 13 (1979) 669–672.

[199] W. Rostoker, E.Y.S. Chao and J.O. Galante, J. Biomed. Mater. Res. 12 (1978) 317–335.

[200] J.R. Atkinson, D. Dowson, G.H. Isaac and B.M. Wroblewski, Wear 104 (1985) 217–224.

[201] P.S. Walker and B. Gold, J. Lubr. Technol. 95 (1973) 333–341.

[202] J.M. Dowling, in: Biocompatible Polymers, Metals, and Composites, ed. M. Szycher (Technomic Publishing, Lancaster, PA, 1983) ch. 19, pp. 407–425.

[203] R.M. Rose, A. Crugnola, M. Ries, W.R. Cimino, I. Paul and E.L. Radin, Clin. Orthop. 145 (1979) 277–286.

[204] R.M. Rose, A.M. Crugnola, W.R. Cimino and M.D. Ries, in: Implant Retrieval: Material and Biological Analysis, eds A. Weinstein, D. Gibbons, S. Brown and W. Ruff, NBS SP-601 (US Dept. of Commerce, Washington, DC, 1981) pp. 3–28.

[205] R.W. Hood, T.M. Wright and A.H. Burstein, J. Biomed. Mater. Res. 17 (1983) 829–842.

[206] M. Landy and P.S. Walker, Trans. Orthop. Res. Soc. 10 (1985) 96.

[207] T.M. Wright and D.L. Bartel, Clin. Orthop. 205 (1986) 67–74.

[208] G.W. Hastings, Wear 55 (1979) 1–9.

[209] J.C. Bayley, R.D. Scott, F.C. Ewald and G.B. Holmes, J. Bone & Jt. Surg. 70-A (1988) 668–674.

[210] A.V. Lombardi, G.A. Engh, R.G. Volz, J.L. Albrigo and B.J. Brainard, J. Bone & Jt. Surg. 70-A (1988) 675–679.

[211] T.M. Wright, D.J. Astion, M. Bansal, C.M. Rimnac, T. Green, J.N. Insall and R.P. Robinson, J. Bone & Jt. Surg. 70-A (1988) 926–932.

[212] H.J. Agins, N.W. Alcock, M. Bansal, E.A. Salvati, P.D. Wilson, P.M. Pellicci and P. Bullough, J. Bone & Jt. Surg. 70-A (1988) 347–356.

[213] H.S. Dobbs, J. Bone & Jt. Surg. 62-B (1980) 168–173.

[214] J.-M. Dorlot, P. Christel and A. Meunier, J. Biomed. Mat. Res. 23(A3) (1989) 299–310.

[215] W. Plitz and H.U. Hoss, in: Biomaterials 1980, eds G.D. Winter, D.F. Gibbons and H. Plenk Jr (Wiley, New York, 1982) pp. 187–196.

[216] A. Walter and W. Plitz, in: Ceramics in Surgery, ed. P. Vincenzini (Elsevier, Amsterdam, 1983) pp. 253–259.

[217] P. Griss and G. Heimke, Arch. Orthop. Traumat. Surg. 98 (1981) 157–164.

[218] J.-M. Dorlot, P. Christel, L. Sedel, J. Witvoet and P. Boutin, in: Biological and Biomechanical Performances of Biomaterials, eds P. Christel, A. Meunier and A.J.C. Lee (Elsevier, Amsterdam, 1986) pp. 495–500.

[219] H.G. Willert and M. Semlitsch, Sulzer Tech. Rev. 2 (1975) 119–133.

[220] H.G. Willert and M. Semlitsch, J. Biomed. Mater. Res. 11 (1977) 157–164.

[221] J.M. Mirra, H.C. Amstutz, M. Matos and R. Gold, Clin. Orthop. 117 (1976) 221–240.

[222] A. Pizzoferrato, Biomater. Med. Devices Artif. Organs 7 (1979) 257–262.

[223] J.M. Mirra, R.A. Marder and H.C. Amstutz, Clin. Orthop. 170 (1982) 175–183.

[224] D. Hierton, G. Blomgren and J.U. Lindgren, Acta Orthop. Scand. 54 (1983) 584–588.

[225] N.A. Johanson, J.J. Callaghan, E.A. Salvati and R.L. Merkow, Clin. Orthop. 213 (1986) 189–196.

[226] U.E. Pazzaglia, L. Ceciliani, M.J. Wilkinson and C. Cell'Orbo, Arch. Orthop. Traumat. Surg. 104 (1985) 164–174.

[227] A.V. Lombardi, T.H. Mallory, B.K. Vaughan and P. Drouillard, J. Bone & Jt. Surg. 71-A (1989) 1337–1342.
[228] F.C. Ewald, C.B. Sledge, J.M. Corson, R.M. Rose and E.L. Radin, Clin. Orthop. 15 (1976) 213–220.
[229] P.A. Revell, B. Weightman, M.A.R. Freeman and B.V. Roberts, Arch. Orthop. Traum. Surg. 91 (1978) 167–181.
[230] J.K. Maguire, M.F. Coscia and M.H. Lynch, Clin. Orthop. 216 (1987) 213–223.
[231] D.W. Howie, B. Vernon-Roberts, R. Oakeshott and B. Manthey, J. Bone & Jt. Surg. 70-A (1988) 257–263.
[232] S.B. Goodman, V.L. Fornasier and J. Kei, Contemp. Orthop. 17 (1988) 59–63.
[233] S.B. Goodman, V.L. Fornasier and J. Kei, Clin. Orthop. 232 (1988) 255–262.
[234] S.M. Horowitz, C.G. Frondoza and D.W. Lennox, J. Orthop. Res. 6 (1988) 827–832.
[235] E. Barth, T. Sullivan and E.W. Berg, in: Proc. 37th Annual Meeting of the ORS, California (1991) p. 187.
[236] H.G. Willert, G. Buchhorn and U. Buchhorn, in: Implant Retrieval: Material and Biological Analysis, eds A. Weinstein, D. Gibbons, S. Brown and W. Ruff, NBS SP-601 (US Dept. of Commerce, Washington, DC, 1981) pp. 239–267.
[237] W.H. Harris, A.L. Schiller, J.M. Scholler, R.A. Freiberg and R. Scott, J. Bone & Jt. Surg. 58-A (1976) 612–618.
[238] I.W. Brown and P.A. Ring, J. Bone & Jt. Surg. 67-B (1985) 218–221.
[239] L.C. Jones and D.S. Hungerford, Clin. Orthop. 225 (1987) 191–206.
[240] E. Letournel, in: Non-Cemented Total Hip Arthroplasty, ed. R. Fitzgerald Jr (Raven Press, New York, 1988) pp. 318–350.
[241] V.L. Fornasier and H.U. Cameron, Clin. Orthop. 116 (1976) 248–252.
[242] P.P. Anthony, G.A. Gie, C.R. Howie and R.S.M. Ling, J. Bone & Jt. Surg. 72-B (1990) 971–979.
[243] B.M. Wroblewski, J. Bone & Jt. Surg. 61-B (1979) 498–500.
[244] T.A. Gruen, L.D. Dorr, R. Bloebaum, J. Emmanual and M.T. Saberi, in: Proc. 37th Annual Meeting of the ORS, California (1991) p. 528.
[245] U. Pazzaglia and P.D. Byers, J. Bone & Jt. Surg. 66-B (1984) 337–339.
[246] K. Tallroth, A. Eskola, S. Santavirta, T.S. Lindholm and Y.T. Konttinen, J. Bone & Jt. Surg. 71-B (1989) 571–575.
[247] S. Santavirta, V. Hoikka, A. Eskola, Y.T. Konttinen, T. Paavilainen and K. Tallroth, J. Bone & Jt. Surg. 72-B (1990) 980–984.
[248] S.R. Goldring, A.L. Schiller, M. Roelke, C.M. Rourke, D.A. O'Neill and W.H. Harris, J. Bone & Jt. Surg. 65-A (1983) 575–584.
[249] S.R. Goldring, M. Jasty, M. Roelke, C.M. Rourke, F.R. Bringhurst and W.H. Harris, Arthritis Rheum. 29 (1986) 836–842.
[250] S.E. Mather, J. Emmanual, F.P. Magee, T.A. Gruen and A.K. Hedley, Trans. Orthop. Res. Soc. 14 (1989) 498.
[251] S.B. Goodman, R.C. Chin and S.S. Chiou, in: Proc. 15th Annual Meeting Soc. Biomaterials, Florida (1989) p. 92.
[252] D.W. Murray and N. Rushton, J. Bone & Jt. Surg. 72-B (1990) 988–992.
[253] T.S. Thornhill, R.M. Ozuna, S. Shortkroff, K. Keller, C.B. Sledge and M. Spector, Biomaterials 11 (1990) 69–72.
[254] K.J. Kim, P. Greis, S.C. Wilson, J.A. D'Antonio, E.J. McClain and H.E. Rubash, in: Proc. 37th Annual Meeting of the ORS, California (1991) p. 191.
[255] N.E. Bennett, J.T. Wang, C.A. Manning and S.R. Goldring, in: Proc. 37th Annual Meeting of the ORS, California (1991) p. 188.
[256] M. Jasty, S.R. Goldring and W.H. Harris, Trans. Orthop. Res. Soc. 9 (1984) p.125.
[257] S.B. Goodman, R.C. Chin, S.S. Chiou, D.J. Schurman, S.T. Woolson and M.P. Masada, Clin. Orthop. 244 (1989) 182–187.
[258] J.L. Fowler, G.A. Gie, A.J.C. Lee and R.S.M. Ling, Orthop. Clin. North Am. 19 (1988) 477–489.
[259] G. Bannister, Orthop. Clin. North Am. 19 (1988) 567–573.
[260] R.G. Craig, in: Handbook of Biomaterials Evaluation, ed. A.F. von Recum (Macmillan, New York, 1986) pp. 55–72.

[261] A. Harrison, Dent. Advertiser 39 (1984) 8–11.
[262] K.D. Jorgensen, Scand. J. Dent. Res. 88 (1980) 557–568.
[263] A. Harrison, Dent. Advertiser 40 (1984) 10–14.
[264] A. Harrison and T.T. Lewis, J. Biomed. Mater. Res. 9 (1975) 341–353.
[265] K.D. Jorgensen, Aust. Dent. J. 27 (1982) 153–158.
[266] L. Ehrnford, T. Derand, L.-A. Larsson and A. Svensson, J. Dent. Res. 59 (1980) 716–720.
[267] G.K. Stookey and J.C. Muhler, J. Dent. Res. 47 (1968) 524–532.
[268] J.R. Heath and H.J. Wilson, J. Oral Rehabil. 3 (1976) 121–138.
[269] J.F. Roulet and T.K. Roulet-Mehrens, J. Periodontol. 53 (1982) 257–266.
[270] W.F. Bailey, S.L. Rice, R.J. Albert and S.C. Temin, J. Dent. Res. 60 (1981) 914–918.
[271] J.E. McKinney, Wear 76 (1982) 337–347.
[272] R.R. Seghi, S.F. Rosenstiel and P. Bauer, J. Dent. Res. 70 (1991) 221–225.
[273] J.M. Powers and R.G. Craig, J. Dent. Res. 51 (1972) 168–176.
[274] J.M. Powers and R.G. Craig, J. Dent. Res. 51 (1972) 605–610.
[275] A. Harrison and R. Huggett, Dent. Techn. 29 (1976) 4.
[276] J.F. Cvar and G. Ryge, US Public Health Service publication No. 790–244 (US Government Printing Office, San Francisco, 1971).
[277] J.W. Osborne, M.L. Swartz, C.J. Goodacre, R.W. Phillips and E.N. Gale, J. Prosthet. Dent. 40 (1978) 413–417.
[278] A.K. Abell, K.F. Leinfelder and D.T. Turner, J. Dent. Res. 60 (1981) 323. Abstract.
[279] G. van Groeningen and J. Arends, J. Dent. Res. 61 (1982) 571. Abstract.
[280] J.T. Atkinson, D. Groves, M.J. Lalor, J. Cunningham and D.F. Williams, Wear 76 (1982) 91–104.
[281] P. Lambrechts, M. Vuylsteke, G. Vanherle and C.L. Davidson, J. Dent. 12 (1984) 252–267.
[282] K.F. Leinfelder, D.F. Taylor, W.W. Barkmeier and A.J. Goldberg, Dent. Mater. 2 (1986) 198–201.
[283] T.H. Hirt, F. Lutz and J.F. Roulet, J. Oral Rehabil. 11 (1984) 511–520.
[284] P. Glentworth, A. Harrison and G.E. Moores, Wear 93 (1984) 53–62.
[285] J.M. Powers, L.J. Allen and R.G. Craig, J. Am. Dent. Assoc. 89 (1974) 1118–1122.
[286] A. Harrison, R. Huggett and R.W. Handley, J. Biomed. Mater. Res. 13 (1979) 23–34.
[287] J.M. Powers, W.H. Douglas and R.G. Craig, Wear 54 (1979) 79–86.
[288] J.M. Powers, J.C. Roberts and R.G. Craig, J. Dent. Res. 55 (1976) 432–436.
[289] J.M. Powers, J.C. Roberts and R.G. Craig, Wear 39 (1976) 117–122.
[290] K.F. Leinfelder, Composite resins, Dent. Clin. North Am. 29 (1985) 359–371.
[291] A.J. De Gee, C.L. Davidson and M.J. Frijn, J. Dent. Res. 64 (1985) 370. Abstract.
[292] D.C. Sarrett, K.-J.M. Soderholm and C.D. Batich, J. Dent. Res. 70 (1991) 1074–1081.
[293] K.-J. Soderholm, M. Zigan, M. Ragan, W. Fischlschweiger and M. Bergman, J. Dent. Res. 63 (1984) 1248–1254.
[294] G.M. Montes-G and R.A. Draughn, Dent. Mater. 2 (1986) 193–197.
[295] R.W. Phillips, D.R. Avery, R. Mehra, M.L. Swartz and R.J. McCune, J. Prosthet. Dent. 30 (1973) 891–897.
[296] K.F. Leinfelder, T.B. Sluder, C.L. Sockwell, W.D. Strickland and J.T. Wall, J. Prosthet. Dent. 33 (1975) 407–416.
[297] R.P. Kusy and K.F. Leinfelder, J. Dent. Res. 56 (1977) 544.
[298] D.F. Williams and J. Cunningham, in: Proc. 2nd World Congress on Biomaterials, Washington, DC, April 27–May 1 (1984) p. 53.
[299] W.B. Eames, J.D. Strain, R.T. Weitman and A.K. Williams, J. Am. Dent. Assoc. 89 (1974) 1111–1117.
[300] J.R. Sturdevant, T.F. Lundeen, T.B. Sluder, A.D. Wilder and T.F. Taylor, Dent. Mater. 4 (1988) 105–110.
[301] H. Shintani, N. Sator and J. Sator, J. Prosthet. Dent. 62 (1989) 627–632.
[302] R.D. Norman, J.S. Wright, R.J. Rydberg and L.L. Felkner, J. Prosthet. Dent. 64 (1990) 523–529.
[303] G.H. Johnson, D.J. Bales and G.E. Gordon, J. Dent. Res. 70 (1991) 344. Abstract.
[304] W.J. O'Brien and J. Yee, Oper. Dent. 5 (1980) 90–94.
[305] A. Koran, R.G. Craig and E.W. Tillitson, J. Prosthet. Dent. 27 (1972) 269–274.
[306] G.R. Miller, J.M. Powers and K.C. Ludema, Wear 31 (1975) 307–316.

[307] C.N. Raptis, J.M. Powers and P.L. Fan, Wear 67 (1981) 177–185.
[308] R. Delong, W.H. Douglas, R.L. Sakaguchi and M.R. Pintado, Dent. Mater. 2 (1986) 214–219.
[309] J.A. Mahalick, F.J. Knap and E.J. Weiter, J. Am. Dent. Assoc. 82 (1971) 154–159.
[310] A. Ekfeldt and G. Oilo, Acta Odontol. Scand. 46 (1988) 159–169.
[311] M.G. Wiley, J. Prosthet. Dent. 61 (1989) 133–137.
[312] R. Delong, C. Sasik, M.R. Pintado and W.H. Douglas, Dent. Mater. 5 (1989) 266–271.

Advances in Composite Tribology
edited by K. Friedrich

Chapter 15

High-Speed Tribology of Polymer Composites

IKUO NARISAWA

Department of Materials Science and Engineering, Yamagata University, 4-3-16 Jonan, Yonezawa-city 992, Japan

Contents

Abstract

Phenolic-based composites for automobile friction materials contain a variety of inorganic fillers and modifiers in addition to the base resin and fibres (mostly asbestos). The friction behavior at low speeds does not provide a sufficient foundation for the prediction of the friction and wear behavior at high speed, since the temperature rise at high speeds causes chemical changes in some of the friction constituents of the composite system.

The purpose of this chapter is to provide a clear interpretation of the frictional performance of phenolic-based composites at high speeds. The friction behavior of commercial brake material at high speeds involves changes in the friction coefficient with sliding distance: an initial decrease, then an increase and finally a constant value with frequent fluctuations, with increasing distance. Finally, the friction tends to have a rather constant value although sporadic fluctuations still remain. The final constant friction behavior is closely related to decomposition of the resin due to the temperature rise at the contacting surface. The critical temperature to reach constant friction depends on the sliding speed and normal load.

At high-speed friction of composites simply consisting of a phenolic base resin and carbon, glass, or aramid fibres, the amount of fibres affects the friction behavior. Generally, incorporation of fibres in the resin reduces the friction coefficient. Glass fibre is most effective in decreasing the friction coefficient, and relatively higher amounts of fibres are needed for carbon and aramid fibres to minimize the friction coefficient. The fluctuations in the friction disappear when putting fibres into the resin. This result suggests that reduction of the deformation at high temperatures and products of fibre wear during sliding are responsible for the low friction coefficient of the composites at high speeds.

1. Introduction

There is a desire to use polymer composites in many mechanical components which are required to operate in high-speed conditions. A typical example is automotive friction materials, which usually consist of phenolic resins with fibres. During the performance of high-speed friction materials heat is produced at the contact surface. This temperature rise affects the friction materials and plays an important role in the friction behavior for high-speed friction. The reason for the use of phenolic-based composites under high-speed conditions is that they exhibit more stable thermal and mechanical properties than other polymers, such as thermoplastics.

Generally, phenolic-based composites for friction materials for high speed contain a variety of inorganic fillers and modifiers in addition to the combination of the base resin and fibres (mostly asbestos). Hypotheses concerning the friction under low speed conditions cannot provide a sufficient foundation for the prediction of the friction and wear behavior at high speeds, since the temperature rise may cause chemical changes in some of the friction constituents of the system, even if the resin and fibres are thermally stable. There have been several studies aimed at understanding these chemical changes [1–6], to enable a clear interpretation of the friction performance of phenolic composites as a disk brake pad. Most of the friction tests in these studies have been carried out at high temperatures instead of under high-speed conditions for simplicity of the testing techniques. The comprehensive mechanism explaining the large decrease in the frictional coefficient at elevated temperatures, which is believed to correspond to friction at high speeds, is the decomposition of organic materials in the composite system.

However, the friction and wear behavior of the composites under high-speed conditions is still not understood, because friction is a complex process that is characterized by the mechanical–molecular structure of the composite and is affected by various factors. The need for an understanding of the friction at high speeds requires actually carrying out high-speed friction tests and to record quantitatively the temperatures as well as the friction and wear during the contact period.

This chapter is concerned with changes in the friction and wear behavior of phenol composites at high-speed friction. The performance of a simple composite system consisting of resin and carbon, glass, or aramid fibres was studied at various speeds. The attention was focused on the role of the fibres in the friction and wear under high-speed conditions compared with that at a low speed.

2. Experimental procedure and materials

The high-speed friction testing machine shown in fig. 1 was used. A small specimen with a cylindrical friction surface of 12 mm diameter and 8 mm long was rubbed against a gray cast iron disk (Japanese Industrial Standard FC20). The friction surface of the specimen was finished with 1500 grade abrasive paper. The specimen was conditioned for more than 48 h at room temperature before testing. The surface of the iron disk was polished with 100 grade abrasive paper and, in addition, it was cleaned by rubbing with an acetone moistened cloth immediately prior to testing.

The specimen was mounted in the specimen holder, which is connected to an arm having a linear motion bearing. The other end of the arm is fixed to a plate to which a strain gauge is attached. The specimen can be pressed to the disk and the contact load can be determined by using a weight. The frictional force was

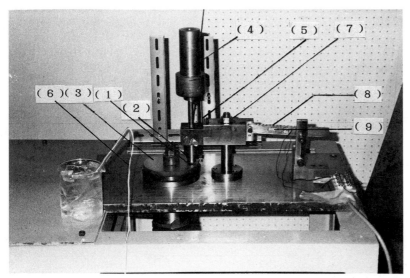

Fig. 1. High-speed fiction testing machine. (1) Specimen, (2) specimen holder, (3) disk (cast iron), (4) weight, (5) linear motion bearing,(6) thermocouple, (7) bearing, (8) plate for torque measurement, (9) strain gauge.

TABLE 1
Experimental conditions. *

| Material | Testing condition | |
(pin)	Load (N)	Sliding speed (m/s)
Commercial	38	3.9
brake pad	69	
	88	
	38	7.9
	69	
	88	
	38	15.2
	69	
	88	
Carbon, glass,	38	2.0
or aramid fibre		5.2
composites		15.2

* Disk: gray cast iron (FC20 Japanese Industrial Standard). Initial sliding: 38 N and 5.2 m/s.

recorded using an electromagnetic oscillograph recorder. The rotational speed of the disk was controlled using an inverter controller.

The specimen was initially rubbed against the disk at a speed of 5.2 m/s under a load of 38 N throughout this work. This pre-rubbing was carried out to obtain equal initial conditions for the tests. After pre-rubbing, the friction tests were carried out under the test conditions shown in table 1. The temperature of the specimen surface was continuously recorded by using a copper–constantan thermocouple, which was placed in the specimen 2 mm above the contacting surface. The length of the friction contact was 9360 m for all tests. The wear was measured by weighing the specimen after the test. To study the worn surface, the specimen was examined in a scanning electron microscope. A thermogravimetric analysis was also done, to obtain the thermal properties of the specimen.

The materials used were phenolic resin based composites containing various types of short fibres, shown in table 2. The phenolic resin was a commercial grade of phenolic–aralkyl resin, Milex XL-325M (Mitsui Toatsu Chem. Inc.) containing less phenolic hydroxyl groups than a typical phenolic novalak, which improve the heat resistance of the resin. Curing of the mixture was done by adding hexamethy-

TABLE 2
Fibres contained in composites.

Fibre	Manufacturer	Shape and dimensions
Pitch-based carbon fibre	Osaka Gas Co. Ltd.	Milled fibre (diameter: 13 μm, length: 0.5 mm)
Glass fibre	Nittobo Co. Ltd.	Powder (diameter: 10 μm, 50–100 mesh)
Aramid fibre	Teijin Co. Ltd.	Pulp (diameter: 10 μm, length: 3 mm)

lene tetramine to the base resin. The mixture of the resin, curing agent and fibres was cured at 100°C for 30 h and post-cured at 150°C for 3 h and 180°C for 5 h. For comparison purposes, a commercial brake pad, being molded from phenolic resin and asbestos, with the addition of mineral fillers and modifiers, was also used.

3. General aspects of the tribology of brake materials at high-speed friction

3.1. Friction behavior

The formulation of commercial brake material is complicated, usually being composed of a phenolic resin, asbestos, and a variety of inorganic fillers and modifiers. The characteristics of brake friction are such that the sliding speed varies over a wide range from a very high value to a low value. Figure 2 shows some typical friction force–time curves at various sliding speeds. At the relatively low speed of 3.9 m/s the friction force remains constant throughout sliding. On the other hand, at the high speed of 15.2 m/s the friction force varies remarkably with time.The friction behavior at this high sliding speed can be characterized by the following three stages:

(1) a first stage, in which the friction force decreases with time, intermittently fluctuating to the high side,

(2) a second stage, in which the friction force increases with time, covering up the intermittent fluctuation, and

(3) a third stage, in which the force reaches a stationary constant value.

At the medium speed of 7.9 m/s the friction behavior shows only the first stage of high-speed friction. Thus, it is clear that the sliding speed markedly influences the friction behavior of the brake materials, even under the same normal load. Figure 3 shows the relative coefficient of friction at the end of the test versus the sliding speed. Here, the relative coefficient of friction is defined as

$$\hat{\mu}_f = \mu_t / \mu_i,$$

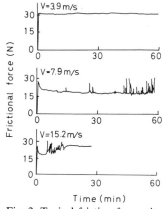

Time (min)

Fig. 2. Typical friction force–time curves of the commercial brake material at various sliding speeds. The normal load is 69 N.

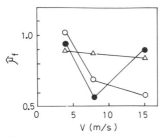

Fig. 3. Relation between the relative friction coefficient at the end of sliding and the sliding speed. Load: ○, 38 N; ●, 69 N; and △, 88 N.

where μ_t and μ_i are the friction coefficients at time t and in the pre-rubbing period, respectively. The value of $\hat{\mu}_f$ is affected by the normal load. Figure 4 shows the relation between the friction coefficient and the load at various sliding speeds. At the low speed of 3.9 m/s the frictional coefficient decreases slightly with increasing load. At the intermediate speed of 7.9 m/s the coefficient of friction shows lower values than at low-speed sliding. First it slightly decreases with increasing load and then it increases. At the high speed of 15.2 m/s the friction coefficient initially increases with increasing load and then it slightly decreases. There is little effect of the sliding speed on the value of $\hat{\mu}_f$ at the large normal load of 90 N. Figure 5 shows the variation in the relative friction coefficient with temperature, T_p, of the specimen surface. At the relatively low speed of 3.9 m/s there is small temperature rise with time during sliding. The relative coefficient of friction remains constant with temperature. On the other hand, at the high speed of 15.2 m/s the temperature rise is large, up to above 300°C under a normal load of 88 N. The value of the friction coefficient falls with temperature in the initial stage, increases after showing a minimum value, and finally reaches a stationary value. At the intermediate speed of 7.9 m/s the friction shows low- or high-speed-type behavior depending on the normal load. Under normal loads of 69 and 88 N the friction behavior is similar to that under high-speed conditions. Conversely, under a load of 38 N the friction shows low-speed-type behavior. The experimental results indicate that there are two mechanisms of friction, depending on both the sliding speed and the normal load. When the normal load is low the friction shows

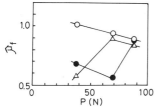

Fig. 4. Relation between the relative friction coefficient and the normal load, P. Sliding speed: ○, 3.9 m/s; ●, 7.9 m/s; and △, 15.2 m/s.

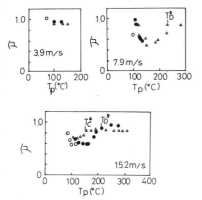

Fig. 5. Variation of the relative friction coefficient with the pin temperature, T_p. Normal load: ○, 38 N; ●, 69 N; and △, 88 N.

low-speed-type behavior even if the sliding speed is high. There is a critical temperature, T_p^*, at which the value of the friction coefficient reaches a stationary value under high speed and large normal load conditions. However, the value of T_p^* is not constant, it is affected by both the sliding speed and the normal load. Figure 6 shows the relation between the critical temperature T_p^* and the sliding speed under a normal load of 88 N. The value of T_p^* decreases linearly with the sliding speed and the curve can be expressed as

$$T_p^* = T_d - CV,$$

where V is the sliding speed and C is a constant which depends on the normal load. Extrapolation of the curve to $V = 0$ gives the thermal decomposition temperature (T_d), as demonstrated by the thermogravimetric curve shown in fig. 7. Apparently, the sliding speed and normal load are accelerating factors for thermal decomposition. It is well known that the frictional behavior at high-speed friction is closely related with the thermal and mechanical properties of materials. The

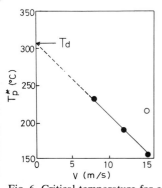

Fig. 6. Critical temperature for constant friction, T_p^*, as a function of the sliding speed.

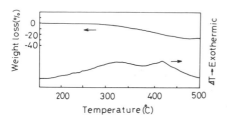

Fig. 7. Thermogravimetric curves for the commercial brake material used in this experiment.

effect of decomposition of the phenolic resin at friction has been well discussed by several investigators. According to their discussion, decomposition of the resin by heat during sliding produces liquid products which create conditions of boundary or semifluid friction. The deformation of the material at elevated temperatures also plays an important role in increasing the coefficient of friction, because the resin softens with increasing temperature. At high-speed friction, the result that the coefficient of friction increases with increasing temperature prior to the stationary friction above the decomposition temperature can be attributed to an increase in the elastic deformation at high temperature.

3.2. Wear and topography of worn surfaces

The rate of wear, W, is plotted as a function of the normal load at various sliding speeds in fig. 8. At the low speed of 3.9 m/s the value of W is independent of the normal load. At the high speed of 15.2 m/s, however, the value of W increases with increasing normal load. At the intermediate speed of 7.9 m/s the wear behavior changes with the normal load. Under normal loads of 38 and 69 N the wear behavior is almost the same as that for low-speed friction, but under the large load of 88 N the behavior is similar to that for high-speed friction.

The optical-microscope observations on worn specimen surfaces after a sliding distance of 9360 m under various conditions are summarized in fig. 9. In figs. 9f to i, in which the morphology under high-speed conditions are shown, very similar features can be seen. A black spot can be observed on these surfaces. On the other hand, there is no black spot on the worn surfaces for low-speed friction (figs. 9a to d), but many cracks, both in the matrix material and the fillers, normal to the

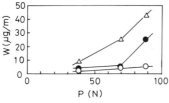

Fig. 8. Relation between the wear rate, W, and the normal load, P. Sliding speed: ○, 3.9 m/s; ●, 7.9 m/s; and △, 15.2 m/s.

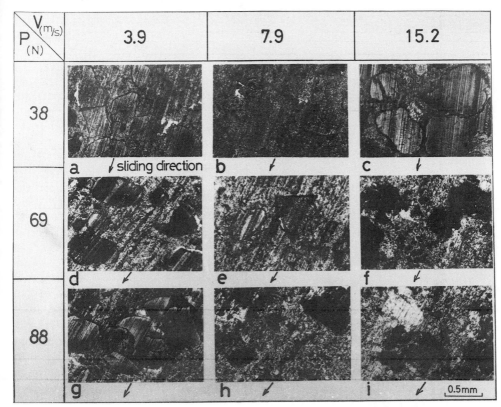

Fig. 9. Optical microphotographs of worn surfaces. The arrow indicates the sliding direction.

sliding direction can be observed. Thus, there is a clear distinction between the worn surfaces for high-speed and low-speed friction.

4. Tribology of carbon-fibre composites at high-speed friction

4.1. Friction behavior

The friction behavior of phenolic composites which contain different amounts of short pitch-based carbon fibres, will be described. For the purpose of comparison the friction of the neat phenolic resin will be also shown. Figure 10 shows the variations of the coefficient of friction and specimen temperature of the neat resin and composites with sliding distance at the low speed of 2.0 m/s. As shown in fig. 10a, the friction coefficient of the neat resin exhibits a relatively high value in the initial stage and then slightly decreases to a constant value with increasing sliding distance, although there is a sudden change in a later stage of sliding. The change in the specimen temperature corresponds well with the frictional changes. The coefficient of friction for the carbon-fibre composites is approximately constant, as

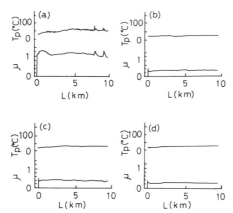

Fig. 10. Typical friction force–time and temperature–time curves for carbon-fibre composites at the low speed of 2.0 m/s. fibre content: (a) basic resin, (b) 35.2 wt%, (c) 55.9 wt%, and (d) 74.8 wt%.

shown in figs. 10b, c, and d. At the extremes they show a low friction coefficient of one-third to a quarter of that of the neat resin. The temperature rise of the specimens is low and also nearly constant. There is little effect of the fibre contents on the friction behavior.

Figure 11 shows the variations of the coefficient of friction and specimen temperature with sliding distance at the medium speed of 5.2 m/s. As shown in fig. 11a, the friction coefficient of the neat resin decreases rapidly following high friction in the initial stage and then again increases with increasing sliding distance showing a large number of fluctuations in the curve. The temperature of the specimen rises rapidly in the initial stage and then increases continously with

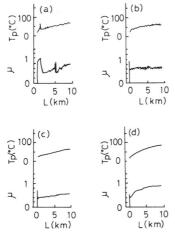

Fig. 11. Typical friction force–time and temperature–time curves for carbon-fibre composites at the medium speed of 5.2 m/s. fibre content: (a) basic resin, (b) 35.2 wt%, (c) 55.9 wt%, and (d) 74.8 wt%.

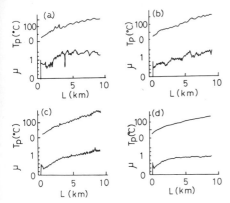

Fig. 12. Typical friction force–time and temperature–time curves for carbon-fibre composites at the high speed of 15.2 m/s. fibre content: (a) basic resin, (b) 35.2 wt%, (c) 55.9 wt%, and (d) 74.8 wt%.

increasing sliding distance, corresponding with the change in the friction coefficient. The fluctuations of the friction coefficient of the composite specimen which contains 35.2 wt% of fibres, are much smaller than those of the neat resin, as shown in fig. 11b. The coefficient of friction increases slightly with increasing sliding distance and reaches a constant value near the final stage. The temperature of the specimen rises gradually with increasing sliding distance and finally attains a constant value. The friction coefficients of the specimens which contain higher amounts of fibres, increase smoothly, without significant fluctuations with increasing sliding speed, as shown in figs. 11c and d.

In fig. 12 are shown the variations of the friction coefficient and specimen temperature with sliding distance at the high speed of 15.2 m/s. The neat resin shows more unstable friction than the friction observed at low speeds. The friction coefficient increases showing considerable fluctuations, following an initial decrease. It seems to reach a constant value, although the fluctuations still remain large. The specimen temperature also increases with sliding distance showing fluctuations. There are also similar friction increases for the composite specimens. Although fluctuations of the friction coefficient are still observed for the specimen which contains the highest amount of fibres (74.8 wt%), the variation is more stable than for the other two composite specimens, which contain lower amounts of fibres.

Figure 13 shows the relation between the average minimum coefficient friction and the fibre content at a sliding distance of 9000 m. The incorporation of fibres in the basic resin significantly reduces the friction coefficient at the low sliding speed of 2.0 m/s. At the high sliding speed of 15.2 m/s the composite specimens show higher friction than those obtained in the low-speed region. The reduction in the friction coefficient by the incorporation of fibres is small. At the medium speed of 5.2 m/s the behavior is similar to that at low speed below a fibre content of about 60 wt%, but at the high content of 74.8 wt% the coefficient of friction approaches that for high-speed friction. Thus, the friction behavior at medium sliding speed

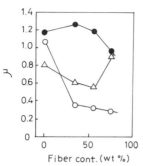

Fig. 13. Average friction coefficient of carbon fibre composites at 9000 m sliding distance as a function of the fibre content. Sliding speed: ○, 2.0 m/s; △, 5.2 m/s; and ●, 15.2 m/s.

involves two types, those for low and high speeds, depending on the fibre content. Figure 14 shows the fluctuation of the friction coefficient which is defined as

$$\overline{\Delta\mu} = \sum_{i=1}^{10} (\mu_{max} - \mu_{min})/10,$$

where $\overline{\Delta\mu}$ is the average fluctuation and μ_{max} and μ_{min} are the maximum and minimum friction in the ith stage. The curves were devided into ten regions ($i = 1$ to 10). The average fluctuation, which increases with the sliding speed, decreases with increasing fibre content. It is clear that the fibres play an important role in the reduction of the friction fluctuations.

4.2. Wear and topography of worn surfaces

Figure 15 shows the wear rate as a function of the fibre content under various sliding-speed conditions. The wear rate, W, decreases when incorporating fibres in the basic resin, at the low speed of 2.0 m/s and there is little effect of the fibre content. By contrast, the value of W increases with increasing fibre content at the high speed of 15.2 m/s. At the medium speed of 5.2 m/s, the wear rate first decreases and then increases in a similar way as for high speed friction. Thus, the

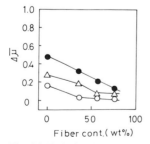

Fig. 14. Relation between frictional fluctuations of the carbon-fibre composites and the fibre content. Sliding speed: ○, 2.0 m/s; △, 5.2 m/s; and ●, 15.2 m/s.

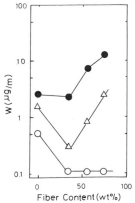

Fig. 15. Wear rate of carbon-fibre composites as a function of the fibre content. Sliding speed: ○, 2.0 m/s; △, 5.2 m/s; and ●, 15.2 m/s.

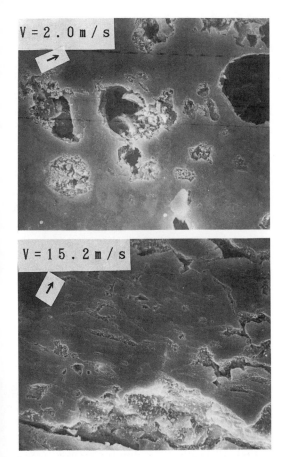

Fig. 16. SEM microphotographs of the worn surface of neat resin. The arrows indicate the sliding direction.

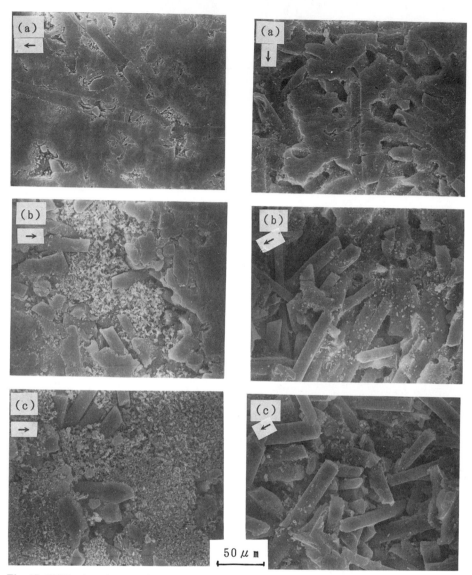

Fig. 17. SEM microphotographs of the worn surfaces of carbon fibre composites. Sliding speed: left: 2.0 m/s, right: 15.2 m/s. fibre content: (a) 35.2 wt%, (b) 55.9 wt%, and (c) 74.8 wt%. The arrows indicate the sliding direction.

two types of wear behavior can be seen at the medium speed of friction, depending on the fibre content.

Figures 16 and 17 show scanning electron microphotographs of the worn surfaces of the neat resin and the composite containing 74.8 wt% fibres, respectively. At a low speed of friction for the neat resin there are many cavities, where cohesive failure leading to flaking occurred. With increasing sliding speed, numer-

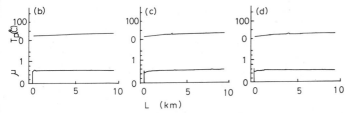

Fig. 18. Typical friction force–time and temperature–time curves for glass-fibre composites at the low speed of 2.0 m/s. fibre content: (b) 45.6 wt%, (c) 66.2 wt%, and (d) 82.0 wt%.

ous discontinuous cracks are formed and the matrix detaches from the surface. There was a considerable change in color of the surface material after high-speed sliding. The process is considered to be delamination wear, in which detachment of the polymer occurs, followed by cracking of the deteriorated surface layers due to the temperature rise. On the other hand, the worn surfaces of the composites show after low-speed friction features with a considerable amount of wear particles, which are produced by cracking and flaking of the carbon fibres. The number of wear particles decreases with increasing sliding speed. Pull-out of the carbon fibres occurs at high-speed friction. The presence of wear particles on the surface explains the result of a low friction coefficient for low-speed friction.

5. Tribology of glass-fibre composites at high-speed friction

5.1. Friction behavior

The friction behavior of glass-fibre phenolic composites will be described in this section. Figure 18 shows the variations of the friction coefficient and specimen temperature with sliding distance at the low speed of 2.0 m/s. The friction coefficients are low and constant, following an initial slight increase. The temperature rise is also stable throughout sliding. There are no significant effects of the fibre content on the friction behavior. The behavior is quite similar to that of the carbon-fibre composites, except for the slightly higher friction coefficients. Figure 19 shows the friction behavior under high-speed conditions, i.e. 15.2 m/s. In contrast to the carbon-fibre composites, the behavior becomes stable after an

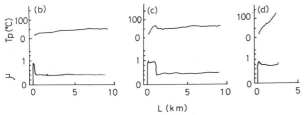

Fig. 19. Typical friction force–time and temperature–time curves for glass-fibre composites at the high speed of 15.2 m/s. fibre content: (b) 45.6 wt%, (c) 66.2 wt%, and (d) 82.0 wt%.

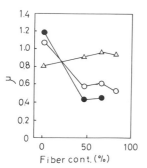

Fig. 20. Average friction coefficient of glass-fibre composites at 9000 m sliding distance as a function of the fibre content. Sliding speed: ○, 2.0 m/s; △, 5.2 m/s; and ●, 15.2 m/s.

initial rise during a short distance. The friction fluctuations disappear during sliding. The temperature rise is not very large, except for the specimen containing the highest content of glass fibres.

Figure 20 shows the relation between the average minimum coefficient of friction and the fibre content at a sliding distance of 9000 m. The dependence of the friction coefficient on the glass-fibre content is different from that for carbon-fibre composites at higher-speed friction. At the high speed of 15.2 m/s, the friction coefficient decreases with increasing fibre content, similar as that seen under the low-speed conditions. However, under medium-speed conditions, the coefficient increases with increasing fibre content. This is closely related to the surface morphology of the specimen during friction. The fluctuations of the friction coefficient during sliding are much smaller for glass-fibre composites than for carbon-fibre composites, as shown in fig. 21. There is no significant effect of the sliding speed on the friction fluctuations.

5.2. Wear and topography of worn surfaces

Figure 22 shows the relation between the wear rate and the sliding speed. The rate of wear, W, increases with increasing sliding speed. Filling of the resin with a

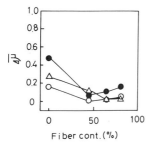

Fig. 21. Relation between the frictional fluctuations of glass-fibre composites and the fibre content. Sliding speed: ○, 2.0 m/s; △, 5.2 m/s; and ●, 15.2 m/s.

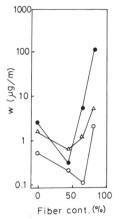

Fig. 22. Wear rate of glass-fibre composites as a function of the fibre content. Sliding speed: ○, 2.0 m/s; △, 5.2 m/s; and ●, 15.2 m/s.

high amount of fibres accelerates the wear rate at all sliding conditions. The composites containing a medium amount of glass fibres, have a reduced wear rate at low and medium sliding speeds.

Figure 23 shows scanning electron microphotographs, in which worn surfaces of the specimen containing 82.0 wt% glass fibres are compared with surfaces before sliding. It is possible to recognize fine fragments of the glass fibres over a wide area of the worn surface after sliding at a low speed. They are attached to, but not embedded in, the surface. Similar features can be observed in the surface worn at a high speed, although fractured and pulled-out glass fibres can also be observed. The features of the surface morphology of the specimen worn at a medium speed are quite different from these two specimens, which show a lower friction coefficient. The lower friction at low and high speeds is attributed to lubricating action of the fine fragments formed at the contacting surfaces.

6. Tribology of aramid-fibre composites at high-speed friction

6.1. Friction behavior

The friction behavior of aramid-fibre phenolic composites will be described in this section. Figure 24 shows the variations of the friction coefficient and specimen temperature with sliding distance at the low speed of 2.0 m/s. The friction coefficients for these aramid-fibre composites are constant throughout sliding, following an initial slight increase. The temperature rise is also stable throughout sliding. There are no significant effects of the fibre content on the friction behavior. The behavior is quite similar to that of carbon- and glass-fibre composites. Figure 25 shows the frictional behavior at the high speed of 15.2 m/s. When the fibre content is low the friction coefficient increases with increasing sliding distance, showing small fluctuations, as the neat resin does. With increasing fibre

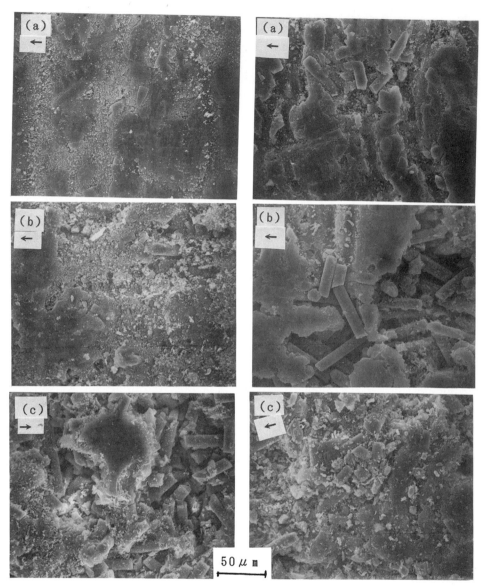

Fig. 23. SEM microphotographs of the worn surfaces of glass-fibre composites. Sliding speed: left: 2.0 m/s, right: 15.2 m/s. fibre content: (a) 45.6 wt%, (b) 66.2 wt%, and (c) 82.0 wt%. The arrows indicate the sliding direction.

content, the friction behavior becomes stable after an initial rise during a short distance and the frictional fluctuation disappears during sliding. The temperature rise is not very large, except for the specimen containing the lower content of fibres.

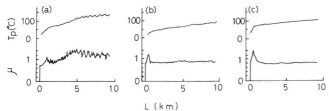

Fig. 24. Typical friction force–time and temperature–time curves for aramid-fibre composites at the low speed of 2.0 m/s. fibre content: (a) 31.4 wt%, (b) 51.7 wt%, and (c) 71.4 wt%.

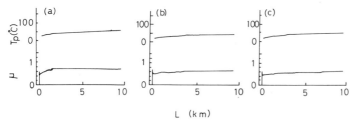

Fig. 25. Typical friction force–time and temperature–time curves for aramid-fibre composites at the high speed of 15.2 m/s. fibre content: (a) 31.4 wt%, (b) 51.7 wt%, and (c) 71.4 wt%.

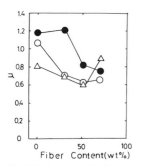

Fig. 26. Average friction coefficient of aramid-fibre composites at 9000 m sliding distance as a function of the fibre content. Sliding speed: ○, 2.0 m/s; △, 5.2 m/s; and ●, 15.2 m/s.

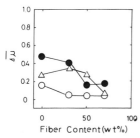

Fig. 27. Relation between the frictional fluctuations of aramid-fibre composites and the fibre content. Sliding speed: ○, 2.0 m/s; △, 5.2 m/s; and ●, 15.2 m/s.

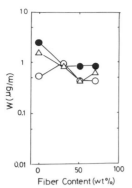

Fig. 28. Wear rate of aramid-fibre composites as a function of the fibre content. Sliding speed: ○, 2.0 m/s; △, 5.2 m/s; and ●, 15.2 m/s.

Figure 26 shows the relation between the average minimum coefficient of friction and the fibre content at a sliding distance of 9000 m. At the high speed of 15.2 m/s the friction coefficient decreases above a certain fibre content. However, at low- and medium-speed conditions the coefficient decreases with increasing fibre content. The frictional fluctuations during sliding are plotted as a function of the fibre content in fig. 27. The fluctuations decrease with increasing fibre content, and they are intermediate between those for glass-fibre and carbon-fibre composites.

6.2. Wear and topography of worn surfaces

Figure 28 shows the relation between the wear rate and the sliding speed. In contrast to the carbon- and glass-fibre composites, the rate of wear decreases with increasing fibre content and the sliding speed does not significantly affect the wear rate.

Figure 29 shows scanning electron microphotographs of the worn surfaces of the specimens containing various amounts of aramid fibres. The fibres cannot be seen on the worn surface of the specimen containing a small amount of aramid fibres (31.4 wt%), but many cracks are formed normal to the sliding direction. On the other hand, fibres attached to the resin form the surface layer of the composite containing the high fibre content of 71.4 wt%. Many fibres and their fibrils can be observed on the worn surface. The presence of a mixed layer of fibres and resin at the contacting surface is responsible for the lower friction coefficient of composites which have a high fibre content.

7. Discussion of a physical model of the tribology in composites

As shown in the preceding sections, the frictional phenomena for a composite at high-speed friction are complex. In the studied range of variation in the type and amount of fibres in the composites, the most important factor is the sliding speed,

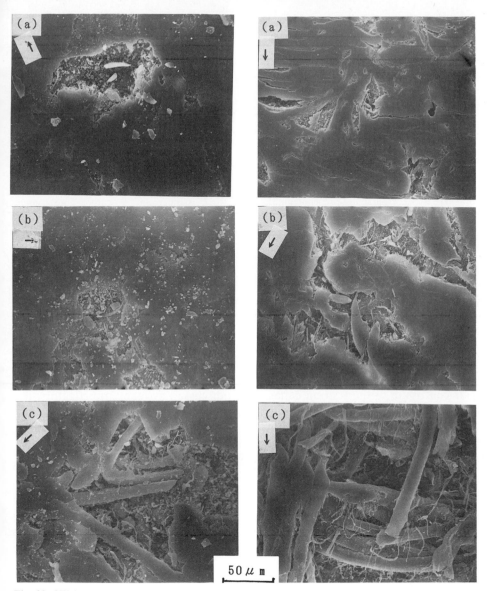

Fig. 29. SEM microphotographs of the worn surfaces of aramid-fibre composites. Sliding speed: left: 2.0 m/s, right: 15.2 m/s. Fibre content: (a) 31.4 wt%, (b) 51.7 wt%, (c) 71.4 wt%. The arrows indicate the sliding direction.

which affects the variation of the friction behavior with sliding distance. As shown in fig. 30, the sliding distance and amount of fibres have no significant effect on the average frictional coefficient when the sliding speed is low. Incorporation of fibres in the resin can significantly reduce the coefficient of friction. The incorpo-ration of carbon fibres is the most efficient way to decrease the frictional coeffi-

I. Narisawa

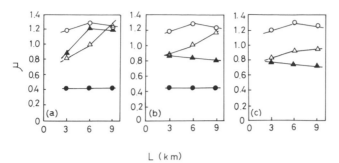

L (km)

Fig. 30. Friction coefficients of the carbon, glass, and aramid composites as a function of the sliding distance at the low speed of 2.0 m/s. (a) ○: basic resin, △: carbon-fibre content 35.2 wt%, ●: glass-fibre content 45.6 wt%, ▲: aramid-fibre content 31.4 wt%, (b) ○: basic resin, △: carbon-fibre content 55.9 wt%, ●: glass-fibre content 66.2 wt%, ▲: aramid-fibre content 51.7 wt%, (c) ○: basic resin, △: carbon-fibre content 74.8 wt%, ●: glass-fibre content 82.0 wt%, ▲: aramid-fibre content 71.4 wt%.

cient. Glass fibres and aramid fibres have the same effect on the coefficient of friction when the content of fibres is high. On the other hand, the incorporation of glass fibres, is the most effective way to reduce the friction coefficient at high-speed friction, as shown in fig. 31. The sliding distance and amount of glass fibres do not affect the coefficient of friction. In corporation of a small amount of carbon or aramid fibres is not enough to reduce the friction coefficient. Filling the resin with high amounts of these fibres is needed to decrease the friction coefficient.

It is well known that the adhesion, the surface roughness, and deformation effects contribute to the friction behavior. As discussed by many investigators, the liquid and gaseous products produced by decomposition of a basic phenolic resin due to a temperature rise at the contacting surface, contribute considerably to the friction coefficient at high-speed friction. As shown by the friction of a commercial brake material, the critical temperature at which the decomposition may make the

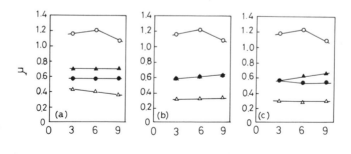

L (km)

Fig. 31. Friction coefficients of the carbon, glass, and aramid composites as a function of the sliding distance at the high speed of 15.2 m/s. (a) ○: basic resin; △: carbon-fibre content 35.2 wt%, ●: glass-fibre content 45.6 wt%, ▲: aramid-fibre content 31.4 wt%, (b) ○: basic resin, △: carbon-fibre content 55.9 wt%, ●: glass-fibre content 66.2 wt%, ▲: aramid-fibre content 51.7 wt%, (c) ○: basic resin, △: carbon-fibre content 74.8 wt%, ●: glass-fibre content 82.0 wt%, ▲: aramid-fibre content 71.4 wt%.

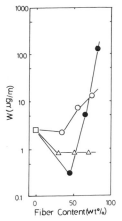

Fig. 32. Wear rate of the basic resin, carbon-, glass-, and aramid-fibre composites after 9000 m sliding at the high speed of 15.2 m/s. ○: basic resin, △: carbon-fibre composites, ●: glass-fibre composite, ▲: aramid-fibre composite.

coefficient of friction stable and constant is affected by mechanical factors such as the normal load and the sliding speed. This suggests that deformation effects, especially at high temperatures, greatly contribute to the friction behavior. At high-speed friction of the basic matrix resin, the friction coefficient increases with increasing temperature. It is believed that the resin softens and the contact area increases with increasing temperature. However, as demonstrated by the results of friction of composites, the friction coefficient after a certain sliding distance is not sensitive to a temperature rise, since incorporation of fibres in the base resin improves the mechanical properties of the material. The products of wear can also affect the friction behavior at high-speed friction. Small particles on the contact surface produced by fracture of fibres, may reduce the coefficient of friction. This mechanism is related to the strength of the fibres used. The result that glass fibre is most effective in reducing the friction coefficient can be associated with the property that it is most the susceptible one to fracture of the three types of fibres used here. As shown in fig. 32, the results of the comparison of the wear rate after 9000 m sliding support this suggested mechanism.

8. Concluding remarks

The high-speed friction of a commercial brake material and composites composed of a phenolic resin and carbon, glass, or aramid fibres varies considerably with sliding distance. The commercial brake material, which is composed of resin, asbestos, and a variety of fillers and modifiers, shows high friction in the initial stage of friction. The friction tends to decrease first and then increase again with frequent fluctuations with increasing sliding distance. Finally, the friction tends to have a rather constant value, although sporadic fluctuations still remain. The final constant friction is closely related to decomposition of the resin due to the temperature rise at the contacting surface. The critical temperature to reach constant friction depends on the sliding speed and the normal load. Therefore, deformation effects at elevated temperatures play an important role in increasing the coefficient of friction before constant friction.

At high-speed friction of composites simply consisting of a phenolic base resin and carbon, glass, or aramid fibres, the amount of the fibres affects the friction behavior. Generally, incorporation of fibres in the resin reduces the friction. Glass fibre is most effective in decreasing the friction coefficient, and relatively higher amounts of fibres are needed for carbon and aramid fibres to realize low friction. The fluctuations of the friction disappear when putting fibres into the resin. This result suggests that reduction of the deformation at high temperatures and products of fibre wear during sliding are responsible for the low friction of the composites at high-speed friction.

List of symbols

C constant
P normal load (N)
T temperature (°C)
V sliding speed (m/s)
W wear rate (g/m)
μ coefficient of friction
$\hat{\mu}$ variation of the frictional coefficient

Subscripts
f   Relative value
i   pre-rubbing period
t   time
d   decomposition
p   sliding surface
*   constant value

References

[1] K. Tanaka, S. Ueda and N. Noguchi, Wear 23 (1973) 349–365.
[2] J. Bros, Wear 41 (1977) 271–286.
[3] L.S. Bark, D. Morgan and S.J. Percival, Wear 34 (1975) 131–139.
[4] L.S. Bark, D. Morgan and S.J. Percival, Wear 41 (1977) 309–314.
[5] Y. Mizutani, K. Kato and Y. Shimura, Proc. JSLE Int. Tribology Conf. (1985) pp. 489–494.
[6] M. Inoue, Trans. JSME-C 56 (1985) 1433–1439.

Advances in Composite Tribology
edited by K. Friedrich

Chapter 16

Tribology of Polymer Composites Used as Frictional Materials

G. CROSA

FRANCISCO STEDILE S.A., Caxias do Sul, RS, Brazil

I.J.R. BAUMVOL

Instituto de Física, UFRGS, Porto Alegre, RS, Brazil

Contents

Abstract

We present here the tribological phenomena that occur in the rubbing of asbestos-free frictional material polymer composites against cast iron metallic counterparts, under conditions of pressure, velocity, and temperature similar to those in real use in the automotive industry. The results of some current research projects in this area are also presented.

1. Introduction

The formulation and production of frictional materials for the braking systems of automobiles, trucks, airplanes, racing cars, and other vehicles have undergone major changes in the least two decades. There has been a shift in the direction of a better heat resistance, a more stable coefficient of friction, and extended durability.

Before this, the most significant evolution in frictional materials occurred some seventy or eighty years ago with the use of asbestos fibres, a natural occurring mineral fibre that acts as an ideal ingredient due to its outstanding properties, like mechanical strength, heat resistance, and low Mohs hardness, together with low cost and easy processability. Thermosetting phenol-formaldehyde resins reinforced with asbestos of different fibre lengths, formed the basis for the development of frictional materials in the automotive industry.

The thermal stability of asbestos is limited to about 500°C. Above this temperature it gives off water and is converted to forsterite or olivine, depending on the amount of iron oxide present, which gives as a consequence changes in the friction properties of the composite. Of the different kinds of asbestos, chrysotile is normally used in the industry. Moreover, asbestos is considered to be a health hazard, since occupational exposure can lead to some forms of the lung disease called asbestosis.

Although asbestos is a well-suited fibre for composites for frictional materials, because of its mechanical and heat resistance properties and low cost, the industry has been actively attempting to replace asbestos because of its hazard problems [1]. Nevertheless, replacing asbestos in frictional materials has proven to be a difficult task, because asbestos has both long and short fibre segments. So far, no other single synthetic or natural fibre can by itself replace asbestos in frictional materials, since no other single material has a morphology like asbestos or the same combination of frictional performance, thermal stability, and low cost. Manufacturers of frictional materials are trying combinations of different fibres to replace the asbestos, and one common approach is to use blends of long and short fibres. The short fibre is needed to reinforce the areas between and around the long fibres and to contribute to other important mechanical, thermal, and frictional properties to the composite. Steel, aramid, glass, ceramic, mineral, and carbon fibres are typical examples [2].

Apart of fibres, another distinct class of materials are necessarily used in frictional material formulations. These are the friction modifiers and lubricants

Fig. 1. Different types of molded, rolled, and sintered frictional materials commonly used for automotive, railways, and industrial applications.

(including elastomers), added to improve the mechanical and frictional characteristics. Abrasive powders, like aluminium oxide and chromium oxide, are commonly used to make the friction coefficient higher. Brass, copper, or other types of metal chips are added to improve the dissipation of the heat generated during braking. Lubricants, such as graphite, are added to moderate the friction and wear.

In order to complete the description of the ingredients that make up the composites used for frictional materials, it is necessary to mention here two other classes, namely the binder component and the fillers. The binder is the glue that holds the other ingredients together. The binder mostly used is a phenol-formaldehyde thermosetting resin, like the novolac type [3]. The fillers are generally minerals such as barytes, clay, or calcium carbonate, that are added with the main purpose of controlling the costs.

Thus, there are four primary categories of ingredients, with occasional overlap of functions. The selection of materials in these four categories is the challenging task facing anyone formulating a frictional material composite [4].

The basic manufacturing process consists of the mixing of the ingredients in the formulation, this mixing being either dry or wet, followed preferably by cold mechanical performing, and curing in molds heated at temperatures in the range 130 to 160°C and at pressures high enough to assure the desired level of packing. Hereafter the frictional material composite is submitted to thermal treatment, cooling, and finishing operations (cutting, drilling, etc.). In fig. 1 are shown different examples of molded and rolled frictional materials commonly used for automotive applications.

It is the aim of this chapter to discuss the tribological phenomena occurring in the rubbing of asbestos-free frictional material composites against metallic counterparts, usually grey cast iron, under conditions of pressure, velocity, and temperature similar to those in real use in the automotive industry. Furthermore, we will describe the results of some current research projects, being put forward by the authors.

In section 2 the working parameters and testing conditions of frictional materials for automotive applications are presented, with some examples of testing equipment and typical results. In section 3 the basic tribological concepts occurring in testing and use of frictional materials are presented, emphasizing the peculiarities brought about by the unusually severe requirements imposed by the load-carrying capacity and temperature ranges involved. In section 4 the results of research work in progress in the areas of abrasives and fibres for frictional material composites are discussed.

2. The working parameters and the testing of frictional materials

The stopping of any automotive vehicle is the result of the joint action of the friction forces generated by the brakes, the resistance to rolling, the resistance to movement due to the air, and the internal resistance of the engine. The magnitude of the friction force depends on the design of the brakes, on the force used to actuate the brakes, and on the nature of the system used to transmit this force. In any case, the resulting force cannot overcome the value allowed by the adherence of the tyres to the road. In summary, the stopping of a vehicle is a complex phenomenon that depends on the vehicle, the road, and the driver.

During braking, the velocity as well as the acceleration vary, and it is usual to plot the retarding acceleration as a function of time, as illustrated in fig. 2 [3]. The analysis of fig. 2 allows us to divide the braking procedure in four main phases: (i) the *initial response phase* (A to B), from the beginning of the actuation force until the onset of the braking force, (ii) the *pressure build-up phase* (B to C), from the onset of the braking force to the moment when it reaches its stabilized value, (iii)

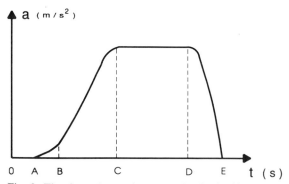

Fig. 2. The four phases that occur in the braking process: (i) initial response phase (A to B), (ii) pressure build-up phase (B to C), (iii) active braking phase (C to D), and (iv) final response phase (D to E).

the *active braking phase* (C to D), from the beginning of the stabilized braking force until its cessation, and (iv) the *final response phase* (D to E), from the disactivation until the disappearing of the braking force.

2.1. Brake assemblies

The brakes used are either hydraulic or pneumatic. They must be capable of stopping the vehicle, quickly and reliably under varying conditions. For instance, in a car, when the pads touch the disc, the cohesion due to the friction tends to shear the frictional material attached to the back-plates. The brake is a combination of components designed to obtain the maximum yield for the braking action, being the mechanism that transforms the applied pressure to mechanical forces that retard the wheel movement until final stopping [3,5–10].

In drum brakes the friction forces are generated on the inner surface of a drum. Figure 3 illustrates the forces developing in a drum-brake assembly [5]. One illustrative example is the S-cam drum brake for commercial vehicles, of which a schematic drawing is shown in fig. 4a. Two shoes, referred to as the leading and the trailing one, are linked by anchor pins, which prevents them to move relative to each other. At the upper side both shoes can be pressed apart by a cam device, having the form of an S. When the brake is applied this cam rotates, pressing the shoes, fitted with a lining, against the drum. Figure 4b shows the hydraulic equivalent of the S-cam drum brake or simplex brake [7]. When the vehicle is moving forward, the leading shoe is more demanded than the trailing one. When the vehicle is moving to the rear the situation is reversed. The forces developing in the leading and trailing shoes are such that when the leading shoe is energized, the trailing one is de-energized, resulting in a braking force that keeps a ratio of about 2/1 between the two shoes.

The disc brake has been an important progress in the development of brake assemblies, as a viable alternative to drum brakes. The main components which make up the disc brake assembly are the caliper, the brake disc, and the pads. The

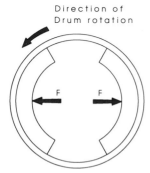

Fig. 3. Brake action in a drum brake. This is a kind of friction brake in which the friction forces are generated on the inner surface of a drum. When the brake is applied, the shoes, fitted with the linings, are pressed against the inner surface of a rotating drum.

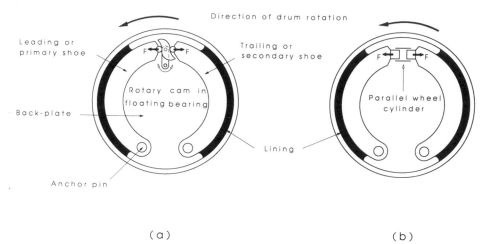

(a)   (b)

Fig. 4. Schematic drawing of a S-cam drum brake: (a) the mechanical system with the metallic drum, the frictional material linings, the leading or primary shoe, the anchor pin, and the trailing or secondary shoe; (b) the hydraulic equivalent.

caliper is built up with one or more cylinders, with one piston for each cylinder. A piston seal is placed between the piston and the cylinder.

Between the piston and the brake disc the pads are placed. The pads are attached to the caliper by retaining pins, which permit movement of the pads during brake application. Figure 5 is a schematic illustration of this [9,10]. The

BRAKING PRESSURE

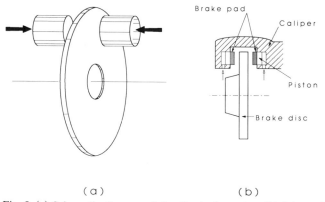

(a)   (b)

Fig. 5. (a) Schematic diagram of the disc brake action. (b) Schematic diagram of a disc brake system. The pads are placed between the pistons and the brake disc. The pads are attached to the caliper by retaining pins, which allow to move the pads against the disc during brake application.

functioning of a disc-brake assembly is relatively simple. In operation, movement of the brake pedal forces a piston to move in the master cylinder. This causes a pressure on the liquid ahead of the piston, forcing the liquid under pressure to the caliper. The pistons in the caliper are activated and the pads are forced against the brake disc.

After releasing the hydraulic pressure, the caliper pistons return to their resting position, holding the pads close to the disc, ready for a new brake application.

2.2. Brake drums and discs

In this chapter we do not make a very strict distinction between either drums and discs or between disc brake pad and drum brake lining composites, since the basic tribological concepts remain the same, regardless of the application. However, it is useful to make some specific remarks concerning the metallic drum on the one hand, which is the tribological counterpart of the lining polymer composite, and the metallic disc on the other, which is the tribological counterpart of the pad.

The drum, more by its shape than by the material used to fabricate it, has to resist the radial forces resulting from the contact with the friction material. In this way, inhomogeneous stresses (including those of thermal origin) will develop, as shown in fig. 3, and in order to resist these the drum must be rigid enough. On the other hand, it cannot be so rigid to the point of not being able to adapt to the deformations caused by its tribological counterpart. The material used to make the drums plays an important role regarding the friction and wear properties. Fatigue strength, low thermal capacity, and high thermal conductivity are also essential characteristics of the drums. The material most commonly used, fulfilling these requirements, is automotive pearlitic cast iron with a lamellar pearlite matrix with a Brinell hardness in the range HB(170–230), for normal service conditions.

Concerning disc brakes, the progress in the developments in the automotive industry led to higher speeds, resulting in a great quantity of heat to be absorbed and dissipated during braking. This, of course, caused higher temperatures to be reached in the disc brake. It has been found that pearlitic grey cast iron resists scouring and heat spotting while maintaining mechanical strength over a wide range of temperatures and has a good thermal conductivity. This material is recommended as mating surface for disc brakes. Nowadays, high-grade, grey iron alloys are being used for making discs [5,12].

2.3. The requirements for frictional materials

Undoubtedly a suitable friction coefficient is the main requirement for a frictional material. It is desirable to have the friction coefficient as constant as possible in the appreciably wide interval of temperatures that ranges from below 0°C in the winter to well above 600°C. The friction coefficient has to be stable not only in these working temperatures, but also regarding different velocities, pressures and environments. Typical values for the friction coefficient are between 0.3

and 0.5. The complete set of intervals for all these working parameters will be given in section 3.

Following the friction coefficient very closely, the requirements on the wear of the frictional materials and of the metallic counterface are also very important, since they determine the useful lifetime of the component. The wear of the frictional material is also related to the friction coefficient, not only through the basic tribological correlations between friction and wear, but also due to the influence of the wear debris on the friction coefficient. The working temperatures and surface finish are also very important factors in determining the wear rate of a frictional material. However, a moderate amount of wear is absolutely necessary in order to renew the rubbing surface, otherwise the surface becomes carbonized, with a complete change of the tribological characteristics. Typical wear rates at moderate temperatures (up to 300°C) are between 2 and 8×10^{-5} mm^3 N^{-1} m^{-1}.

Further requirements are dimensional stability with increasing temperature, mechanical strength, especially compressive and tensile strength to be able to deal with the forces normal to the surfaces, and resistance to shear stress caused by the tangential forces due to rotation. Typical values for the normal tensile strength are between 3.5 and 13.5 MPa, and internal shear strengths of at least 10 MPa are common values.

It is a common feature of the majority of the frictional materials that the friction coefficient decreases when the temperature increases, which is well known by the name *fade*. By definition, *fade* is a temporary reduction of the braking effectiveness due to a loss of friction between the braking surfaces, resulting from heat. This loss in friction coefficient must be kept between acceptable limits, such that the brake system as a whole still can present a good effectiveness. By cooling down the frictional material the friction coefficient should increase again with the decreasing temperature, up to a value similar to that before *fade*, and this phenomenon is usually called *recovery*.

2.4. Friction and wear tests

The testing machines used to evaluate the frictional material composites are designed to resemble closely the working conditions in the field. They can be essentially classified in equipment to accomplish quick testing of samples (usually of reduced size) of the frictional materials and equipment to do long-lasting tests on the pads and brake linings with regular sizes and mounting conditions. Examples of equipment commonly used for the quick tests are the *FAST* and the *SAE J661b* machine [11]. For real-scale tests there are several different models of *inertial brake dynamometers*. We here describe briefly one machine of each kind and we discuss the typical outputs of tests done with them.

The *FAST* (friction assessment and screening test) machine was developed by the research staff of Ford Motor Company for evaluation of frictional materials [13,14]. The loading conditions as well as the temperature variations should be correlated with conditions occurring in automotive use.

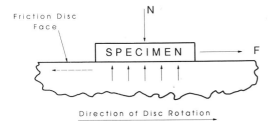

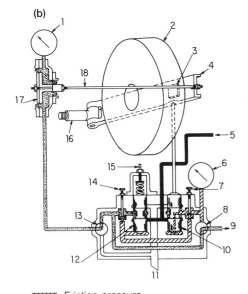

Friction pressure

Clamping pressure

Fig. 6. (a) The basic principle of elemental friction in a FAST testing machine: N is the normal force, F is the friction force acting on a specimen, and μ is the dynamic coefficient of friction referred to in the test as coefficient of friction. (b) Schematic diagram of a constant friction mode test in a FAST machine. Since this is a constant output mode of operation, any variation of the friction force from equilibrium results in a displacement of the clamping spool valve. This causes the clamping pressure to increase so as to provide a friction force that re-establishes the original equilibrium. In this way the control system maintains the friction force constant during the duration of the test: (1) clamping pressure gage, (2) motor-driven friction disk, (3) test specimen, (4) load arm, (5) pump supply connection, (6) friction pressure gage, (7) needle valve (for damping), (8) selector valve, (9) clamping pressure to transducer, (10) friction valve spool, (11) internal drain to reservoir, (12) clamping valve spool, (13) load dump valve, (14) needle valve (closed), (15) load control screw, (16) universal pivot, (17) clamping assembly, (18) tension rod.

The best known test procedure for the FAST machine is the friction versus temperature test. In this mode of operation both the friction force and the sliding velocity are maintained at constant values. Figure 6 shows a schematic diagram of the FAST machine. The discs used for testing are usually 180 mm pearlitic cast

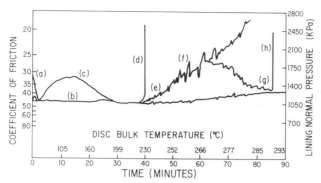

Fig. 7. Typical output traces of the FAST machine. The significance of the characteristic regions of the plot (from a to h) are explained in the text.

iron discs, with a hardness in the the range 170–230 HB. The frictional material samples to be tested have the dimensions $12.7 \times 12.7 \times 3.18$ mm.

In the test described here, the specimen is pressed against a rotating disc running at 870 rpm at constant drag, using a clamping load necessary to maintain a constant friction force of 77.5 N, during 90 min, with a constant sliding speed of 7 m s^{-1}. Since this is a constant-output mode of operation, any variation of the friction force from equilibrium results in a displacement of the clamping valve spool. This causes the clamping pressure to increase so as to provide a friction force that re-establishes the original equilibrium. In this way the control system maintains the friction force constant during the total duration of the test. The wear is evaluated by measurements of the thickness and by weighing the frictional material sample and disc after testing [13].

A typical output trace of the FAST machine is given in fig. 7. We notice in this plot several important aspects, indicated in fig. 7:

(a) is the bedding, the break-in period, which typically lasts about 2 or 3 min;

(b) is the trace of a typical brake lining with a coefficient of friction which varies slightly with temperature;

(c) is the trace for a brake lining which shows an early decrease in the coefficient of friction;

(d) this is the trace of a material with very high *fade*;

(e) this trace indicates a pronounced friction reduction with increased temperature (*fade*);

(f) spikes of this magnitude generally indicate unstable friction;

(g) such an increase in friction often is associated with a tendency to self-destruction at high-temperature operation;

(h) is a trace obtained when a specimen wears out and the load arm contacts the wear stop [13,15].

The *inertial brake dynamometer* provides laboratory facilities for the development and performance control of brake linings, pads, and brake systems, simulating the actual conditions of service. The friction and wear performances can only

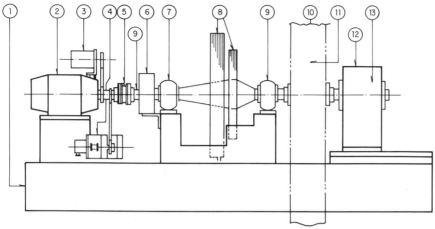

Fig. 8. Schematic drawing of an inertial brake dynamometer: (1) base, (2) motor, (3) motor fan, (4) tachometer generator, (5) flexible coupling, (6) auxiliary flywheel, (7) bearing, (8) flywheels (discs of inertia), (9) bearing, (10) cooling air duct, (11) brake tested, (12) tailstock, (13) torquemeter unit.

be adequately assessed if the test conditions reproduce the real conditions of use, i.e. the rubbing speed, specific pressure, temperature, and sequence of loading must be maintained between one test and the next or between the vehicle test and the dynamometer test.

A schematic view of the inertial brake dynamometer is given in fig. 8. A machine in current use at FRAS-LE is shown in fig. 9. The braking force is expressed here by the occurring torque, the load to be braked by the moment of inertia, and the speed by the number of revolutions per time of the main axle. The machine allows to test disc as well as drum brakes.

As an example of a brake test procedure a very simplified version is presented here. The purpose of this test is to assess and compare the performance of brake linings before, during, and after *fade* and *recovery* tests. Figure 10 shows the output of the computer-driven system that controls the test.

3. The basic tribology of frictional materials

3.1. Load-carrying capacity and temperature

The testing and use of frictional materials involve always the tribological couple formed by the polymer composite, usually frictional material attached to a metallic back-plate, and the metallic counterpart, which for most applications is pearlitic cast iron. In automotive applications the frictional material is the fixed or fluctuating component and the iron counterpart is the mobile (rotating) component. Aeronautical, railways, or industrial applications may involve different arrangements, which should be considered with their peculiarities.

According to the explanations given in the previous section, there are many different and rather sofisticated mechanisms of bringing the friction material in

Fig. 9. An inertial brake dynamometer in use at the testing laboratory. This dynamometer system provides laboratory facilities for the development and testing the performance of brake lining materials and braking systems. The latest digital computer technology is used to simplify the test setup and system operation, and to allow accurate and highly repeatable reproduction of the braking conditions.

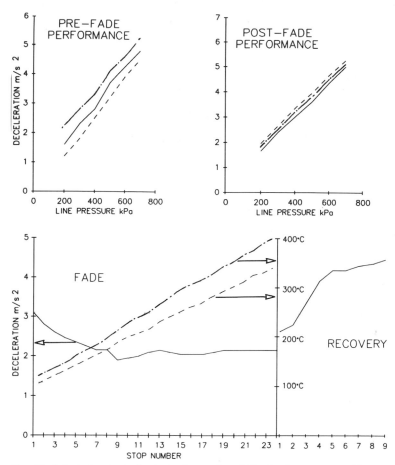

Fig. 10. A typical output of a test using an inertial brake dynamometer. The purpose of this test is to assess and compare the performance of the frictional material before and after a *fade* and *recovery* test. In the pre-*fade* performance there is a spread of the deceleration values of 40 (solid line), 60 (dash–dotted), and 80 km h^{-1} (dashed); in the post-*fade* performance this spread is reduced. The graph describing the *fade* step, includes also the lining temperature (dashed line) and the drum temperature (dot–dashed line), for both see the right-hand temperature scale. In the *recovery* test was found a small over-recovery in the deceleration value as compared with the deceleration at the first stop in the *fade* test.

physical contact with the surface of the cast iron rotor, as well as applying the adequate load to the system in order to obtain the necessary friction force and braking effect. The efficiency and the useful lifetime of the frictional material depends very much on the conditions during use, since the behaviour of the tribological couple concerning friction, wear, fracture, and even mechanical break-down, can vary dramatically with changes in the temperature, pressure, velocity, and external or environmental factors.

The load-carrying capacity or *pv* factor is the basic parameter that has to be considered here to characterize the working conditions. The *pv* factor in automotive applications for drum brakes varies between 0.3 and 20 MPa m s^{-1}, with a tangential rubbing velocity between 1 and 13 m s^{-1} at the effective radius of the brake and pressures between 0.3 and 1.5 MPa. For disc brake applications the pv factor varies between 0.3 and 100 MPa m s^{-1}, with a tangential rubbing velocity between 1 and 16 m s^{-1} at the effective radius of the brake and pressures between 0.3 and 6 MPa.

At the low load-carrying capacity side very moderate temperatures (60 to 150°C) will be generated and, consequently, moderate wear rates are expected to occur (typically 150×10^{-3} mm kW^{-1} h^{-1} or about 40×10^{-9} mm N^{-1} m^{-1}). However, at the high pv side higher temperatures will be generated in the rubbing surfaces, sometimes above 600°C, giving as a consequence wear rates that may be two or more orders of magnitude larger than at the low pv side.

The demand for a stable friction coefficient for all load-carrying capacities and temperature ranges imposes further conditions on the structure of the frictional material composite. Reinforcing additives will be necessary to withstand the high pressures at the contact asperities, preventing severe scuffing by the metal counterpart asperities, which would lead to an undesired, anomalously high ploughing-component contribution to the friction coefficient. Moreover, additives will be necessary to extend the elastic properties and agglutination capabilities of the matrix to a range of temperatures far beyond the characteristics of the phenolic resin alone.

One good example of trying to reinforce the polymer composite is the addition of very hard, finely divided powders, like alumina, chromium oxide, or quartz. A stable friction coefficient is shown in fig. 11 as a function of the concentration of two different kinds of alumina powders. However, fig. 11 also shows that one of the two kinds of aluminas cannot be used as a safe reinforcement to maintain friction stability, because its use above a certain concentration leads to very severe wear of both the frictional material pad and the metallic disc. In order to control this accelerated wear, it is necessary to add other ingredients that can control the wear without leading to the composite losing its load-carrying capacity and frictional characteristics.

Many other factors complicate the understanding of the tribological situation at hand. One worth to mention here is the action of the wear debris as a third body that can generate further abrasion on the one hand, or act as a lubricant that decreases the friction coefficient and wear rate on the other hand. Another important factor interfering and modifying the rubbing surfaces is the water that so many times enters the pad–disc system, producing several effects, like decreasing of the friction coefficient and corrosion. The high stresses and temperatures involved in the system may lead to cracks and fractures, especially near the edges of the friction linings. Figure 12 is an illustration of macroscopic fracture occurring after an inertial brake dynamometer test of a disc pad. The high temperatures that can be generated in the rubbing surfaces may produce hot-spots in the metallic disc, which produce modifications in the structure of the metallic material that can lead to cracks.

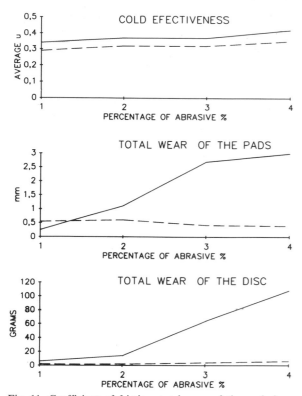

Fig. 11. Coefficient of friction, total wear of the pads in mm, and total wear of the discs in gr, as measured in a test using an inertial brake dynamometer. The discs were the usual grey cast iron discs and the pads were made of basic semi-metallic formulations reinforced with different proportions of two kinds of aluminas (one giving rise to the solid lines in the graphs and the other one to the dashed lines).

A comparison can be made between the typical load-carrying capacities and temperatures occurring in automotive applications of friction materials like those mentioned above and several other different applications of polymer composites, like vanes and gears in pumps, roll-neck bearings, chute liners, lubricated and dry bearings in general, and many others [16]. Figure 13 shows typical pv bands for several different reinforced polymers [17]. From this comparison it is easily verified that friction materials in automotive applications require load-carrying capacities much higher (up to a factor of one hundred) than all other tribologic situations were other polymer composites are used. As a direct consequence the temperatures involved in these frictional material applications are also systematically higher than in all other applications of polymer composites. The temperatures in the surfaces of the pads and discs may reach 700°C or more in some situations, whereas in all other applications the temperatures are normally well below 400°C.

The comparison just made is very useful to clarify the basic tribology of frictional materials, revealing that this class of polymer composites is unique in the

Fig. 12. Macroscopic fracture occurring in polymer composite frictional material pads. This is a pad with a crack transversal to the friction grooves that appeared after testing in an inertial brake dynamometer.

sense that it must have stable friction coefficients and wear rates comparable to many other different polymer composites, working at far higher load-carrying capacities, temperatures, and sliding distances. Furthermore, these extremely severe working conditions have also be made compatible with the metallic counter-part, which is a relatively soft steel or gray casting iron disc or drum.

To find solutions to fulfill the requirement of working under stable conditions of load-carrying capacity, temperature, and stress in situations that would certainly lead to failure or rupture of all other polymer composites is a matter of very skilled semi-empirical work. But it would be also extremely desirable to establish some basic tribological knowledge on this very complex situation.

3.2. The tribological couple friction material–metallic counterpart

Friction and wear characteristics are primarily determined by the nature of interactions between the two friction surfaces in contact. The nature and degree of interactions between the two surfaces are dependent on the properties of the surfaces as well as those of the bulk materials comprising the couple. The relevant surface properties include surface geometry, surface energy, chemical reactivity, and physical and mechanical properties of the surface under given conditions of temperature, velocity, pressure, and environment. For example, the wear of polymeric composite linings tends to decrease with increasing thermal conductivity and decreasing surface roughness of the rotor. Also, the wear of the linings can be

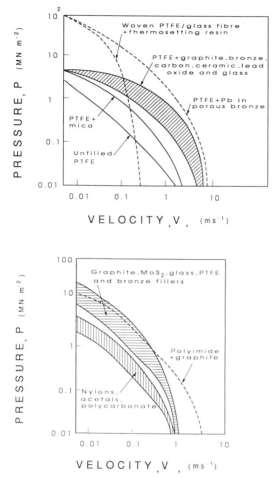

Fig. 13. Load-carrying capacity ($p-v$) relationships for several different reinforced thermoplastic and thermosetting polymer composites (from ref. [17]).

significantly reduced when the composition of the rotor friction surface is slightly modified [18,19].

In the case of a tribological couple used as frictional materials one can consider the friction force as a result of the combined effects of the deformation of the asperities, the adhesion force in the adhesive contacts, and the ploughing of the hard asperities, hard particles, and wear debris. The relative contribution of each one of these effects depends on the topography of the two surfaces, the history of the sliding movement, the specific materials used, and the environment.

The nature of the contact interface between frictional material composites and disc surfaces is such that it cannot be classified either as a metal–metal or a metal–elastomer combination. The wear phenomena associated with the frictional material–metal system reflect mechanisms which are characteristic of both types of interfaces. Five basic wear mechanisms may be distinguished: thermal, abrasive, adhesion-tearing, fatigue, and macro-shear wear.

Thermal wear deserves special attention because it encompasses a group of physical and chemical reactions in the course of which inter-atomic bonds are continuously broken. These reactions include pyrolysis (or thermal decomposition), oxidation, particulation, explosion, melting, evaporation, and sublimation. The rates at which they occur increase exponentially with temperature. Pyrolysis occurs predominantly at the centres of linings and pads, and to a lesser extent at corners and edges. Oxidation, on the other hand, predominates at corners and edges, being less severe near the centre. Explosive reactions can occur under highly severe braking conditions, where the rate of heat input is so high that solids are converted into gases well beneath the surface: because these gases are greater in volume than the solids which they replace, they create instantaneous pressures which break the linings and pads in an explosive manner. Flash temperatures created by the sudden rupture of welded asperity tips may be as high as 760°C, and this causes rapid decomposition of the organic compounds used in frictional materials, as well as transforms the cast-iron mating surfaces from pearlite to martensite (hot-spots).

Of the remaining types of wear in frictional materials, abrasive wear includes ploughing and grinding by wear debris and foreign particles. Fatigue wear occurs in two forms, namely thermal and mechanical. Thermal fatigue is caused by repeated heating and cooling, which induces cyclic stresses in the surface material and steep thermal gradients. A special case of thermal fatigue is called thermal-shock cracking, occurring as a result of a single, abusive loading. Mechanical fatigue is caused by repeated mechanical stressing of the frictional material. Macro-shear is a relatively sudden failure of a frictional material which has been previously weakened by heat, and it is most likely to occur at elevated temperatures and under severe braking conditions.

All these forms of wear can be controlled or minimized by the rapid removal of frictional heat, so that the sliding interface is kept as cool as possible. Thermal wear in particular can also be controlled by the use of polymers, fibres, and friction modifiers that have high thermal stabilities. A brake design should eliminate, if possible, high local pressures, especially at the corners of linings and pads; it should also block the passage of road dirt into the interface and yet permit the removal of wear debris.

Much has been written already about the mechanisms of friction and wear, and the reader may refer to many of these well-established basic texts. However, in the tribology of frictional materials the different additives used to make the composites include many fibres and powders that are much harder than the polymer resin and the metal surface. These hard particles are responsible for most of the abrasion occurring, either when firmly attached to the composite surface or when released in the form of wear debris or even when transferred from the polymer composite to the surface of the metallic counterpart. This justifies the inclusion here of some basic facts about the contribution of abrasion to friction and wear.

The contribution of abrasion to friction is manifested in the ploughing component of friction. The ploughing of a surface is the situation in which the harder

FRICTION   AND   WEAR

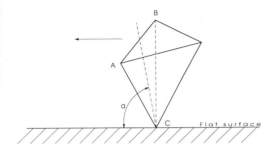

ABRASIVE   WEAR   MECHANISMS

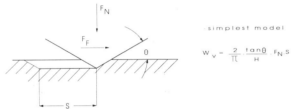

Fig. 14. (Top) Schematic illustrations of the ploughing component of friction. (Bottom) Schematic illustration of the abrasive wear mechanism, including an expression for the amount of wear in the simplest model (from refs. [6,7,10]).

protrusions initially indent the softer surface under the applied load and after that, due to the applied tangential force, produce a groove in this surface. Figure 14 illustrates the situation, and it has been shown that the friction coefficient depends on the fracture strength, the elastic modulus, the hardness, and the relative angle between the normal to the surface and the hard tip.

The harder surface asperities pressing into the softer surface produce plastic flow of the softer surface around the asperities of the harder surface. When a tangential motion is imposed, the harder surface removes the softer material by the combined effects of "micro-ploughing", "micro-cutting" and "micro-cracking". In the simplest model of abrasive wear processes the wear volume is related to the asperity slope of the penetrating abrasive particle and the hardness of the abraded material [20]. The wear produced by hard abrasive particles is seen to depend strongly on the shape of the abrasive, the relative hardness with respect to the abraded surface, the grain size, and the grain density.

In the typical tribological situation of a polymer composite with a significant amount of abrasives in its formulation rubbing against a metallic counterpart, one has to consider initially the action of the abrasives using the two-body abrasion mechanism. So, the abrasives usually increase the load-carrying capacity of the composite [16], as well as constitute a set of protrusions harder than the metal. Depending upon the amount of abrasive in the composite one expects to have a reduction of the wear rate of the composite, an increase of the friction coefficient, and an increase of the wear of the metal. These are the expected consequences in the most simple case, where the dissipative forces do not generate high temperatures in the rubbing surfaces. If the temperature increases above a certain value (typically above 300°C for phenolic resin composites) this simple picture may be changed completely. The softening of the phenolic resin matrix due to heating results in changes in the mechanical properties of the surface and also in the loss of cohesion of the abrasive powders to the matrix. The abrasive particles that abandon the composite due to the loss of adhesion, will act as hard third-body abrasives. The obvious consequences are an increase of the wear of the composite and, if the third body is continuously supplied, an increase of the friction coefficient. The wear of the metallic counterpart will also increase due to the action of a third body which is harder than the metallic surface.

Due to the complexity of the formulation of a typical polymer composite used as frictional material, where not only hard abrasives take part but also soft metallic fillers and lubricants, the third body action is not always the same. This is nicely illustrated by the rubbing of a certain kind of composite pad against a cast iron disc under pv conditions that cause temperatures in the surfaces in the range between 450 and 600°C. It was observed that after a short time of working in these conditions the friction coefficient was reduced to a value lower than the required one. The wear of disc and pad were also significantly reduced. So, everything

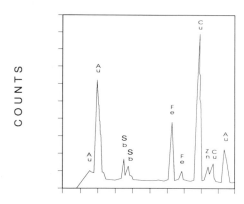

Fig. 15. The spectra of characteristic X-rays observed in electron microprobe analyses of the polymer composite pad surface before testing. We notice that the polymer composite pad contained appreciable amounts of Fe, Cu, and Zn, apart of also significant amounts of Al, Sb, and S. The Au peaks that appear are due to the usual coating of the specimen used in scanning electron microscopy.

happened as if a lubricant was being introduced in this tribological situation. Indeed, a visual inspection of the cast iron surface showed that it was covered by a very adherent black layer.

Electron microprobe analyses of the pad before testing are shown in fig. 15. We can see that the polymer composite pad formulation contained appreciable amounts of Fe, Cu, and Zn, apart of also significant amounts of Al, Sb, and S. The relative proportion of these elements in the surface of the pad are altered after testing. The disc composition before testing was the expected one, containing mostly Fe and a minor concentration of Si. After testing, the surface of the disc showed also significant amounts of Cu and Zn as well as minor concentrations of Sb and Al.

Scanning electron micrographs of the disc before and after testing are shown in fig. 16. The surface of the disc before testing shows only the usual finishing tracks. However, after testing the micrographs show that the disc surface is covered by a homogenous layer of a very finely divided powder, with grain sizes around 0.3 μm or less.

This finely divided powder layer that was covering the disc surface after testing, had the above-mentioned composition, which means that it was rich in Fe, Cu, and Zn. So, this layer was transferred from the pad to the disc during the testing, and, since the surface temperatures were above 450°C, it is reasonable to assume that these elements were fully oxidized. A layer of very finely divided grains of metallic oxides can be a kind of ideal solid lubricant, a fact that is well known for a long time.

In summary, the explanation for the loss of friction and the reduction of wear in this case is the formation of solid lubricant layers, very adherent and constantly renewed, on the surface of the disc due to material transferred from the composite pad at high temperatures.

4. The tribological effects of different fibres and abrasives in basic non-asbestos formulations

In this section we report on some recent results of friction and wear measurements made on basic non-asbestos polymer composites in common use in the automotive industry to which several different fibres and abrasives were added in controlled quantities. Surface characterization and surface profilometry of the frictional materials and of the metallic counterparts are also presented and the correlations with the friction and wear behaviour are discussed.

The fibres and abrasives that were used in this work are also described here, together with the available information on their physical properties [21,22].

4.1. Fibres

We describe here some of the fibres that are being evaluated as possible alternatives to replace asbestos in frictional materials. All of them are organic or inorganic man-made fibres with several different lengths and aspect ratios. Table 1 contains information on the physical properties of the fibres.

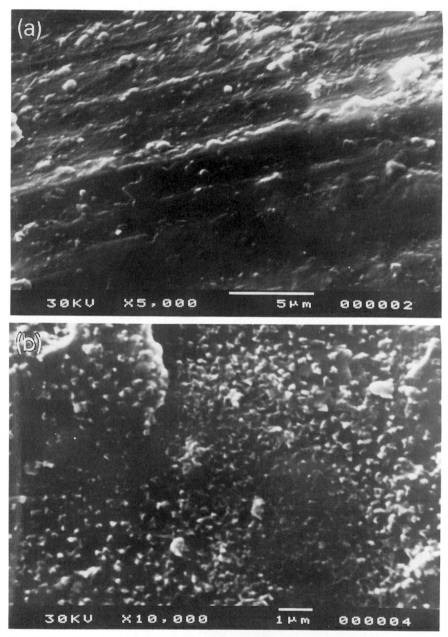

Fig. 16. Scanning electron micrographs of the disc before and after testing. The surface of the disc
before testing (a) shows only the usual finishing tracks. However, after testing (b) the micrographs show
the disc surface covered by a homogenous layer of a very finely divided powder, with grain sizes around
0.3 μm or less.

TABLE 1
Selected physical properties of the different fibres used in the present work.

Physical property	Fibre							
	Chrysotile asbestos	Aramid fibre	Acrylic fibre	Carbon fibre	Glass fibre	Steel fibre	Mineral fibre	Ceramic fibre
Tensile strength (MPa)	2100	2750	880	1300	3400	950	1500	1100
Modulus of elasticity (GPa)	11.7	62.0	17.7	30.0	72.0	11.0	70.0	152.0
Mohs hardness	2.5–4.0	–	–	6.0	6.5	5.0	6.0	6.0
Specific gravity (g/cm^3)	2.4–2.6	1.44	1.18	2.6	2.5	7.5	2.7	1.7
Diameter (μm)	16–30	12	10	10	10	120	5	10
Elongation to break (%)	–	3.3	15	2.0	4.8	7.0	–	1.0

As mentioned before, asbestos is a natural mineral fibre. Chrysotile asbestos fibre is a hydrous magnesium silicate having the chemical composition $Mg_3(Si_2O_5)(OH)_4$. An approximate chemical analysis is SiO_2 (36–42%), MgO (38–42%), H_2O (12–15%), FeO (traces to 6%), Fe_2O_3 (traces to 5%), Al_2O_3 (traces to 3%), and CaO (traces to 1%). The fibre lengths can vary from 100 to 6000 μm and the aspect ratios from 10 to 500. Figure 17 is a photograph of natural occurring asbestos rock.

Asbestos shows significant weight loss around 540°C. Above this temperature the water of hydration is driven out, accompanied by a loss in strength and embrittlement of the fibre. At about 800°C, the chrysotile is converted to a refractory silicate.

4.1.1. Aramid fibre

Aramid fibres are industrial fibres made of organic material. The chemical designation is poly-para-phenylene terephtalamide. The monomers that they are made of are para-phenylene, diamine, and terephtaloyl dichloride. The melting point is far above the decomposition point (around 500°C). This fibre has an almost 100% paracrystalline structure and a very high degree of orientation of the molecules (fibrils) along the fibre axis. Figure 18a is a SEM image of aramid fibre, showing that it is made up of fibrils which resemble asbestos.

4.1.2. Acrylic fibre

Acrylic fibres are based on polymers and copolymers of acrylonitrile. The properties of the final polymer are governed to a considerable degree by the molecular weight (degree of polymerization), as well as by the amount and type of comonomers and the production process to make the fibres. It is possible to control several physical properties of acrilic fibres by modifying the production process. The SEM image in fig. 18b shows that fibrillated acrylic fibre is made up of fibrils and also resembles asbestos, like the aramid fibre does. Due to the high

Fig. 17. Natural occurring asbestos rock. The various fibre lengths, characteristic of asbestos, are clearly seen in this photograph.

polyacrylonitrile content, the fibre is not thermoplastic and does not melt or smear on heating. Instead, it undergoes cyclization and carbonization when exposed to temperatures above 300°C.

4.1.3. Carbon fibre

Carbon fibres are normally made from polyacrylonitrile fibres which are heat treated at 2000°C. The SEM image in fig. 18c shows that carbon fibres are made up of smooth cylindrical filaments.

4.1.4. Glass fibre

Glass fibre continuous filaments are made from glass (type E) which is melted in a platinum bushing, in the bottom of which there are a large number of orifices. The molten glass flows through the orifices and, depending on the speed, temperature and other variables, it is possible to vary the fibre diameter. The continuous filaments can be cut to obtain shorter fibre lengths. The SEM image in fig. 18d shows that glass fibre is made up of bundles of single cylindrical and smooth filaments.

4.1.5. Steel fibre

Steel fibres are made by scraping a steel wire (normally 1010 or 1020). After this the filaments are milled or cut to obtain different fibre lengths. The SEM image in fig. 18e shows that milled steel fibre is made up of filaments with approximately the same diameter. The fibres are smooth and many are twisted.

4.1.6. Mineral fibre

Mineral fibres are manufacturated from rocks. The fibre usually looks like a comet, where the tail is the fibre and the head is the shot, which is a hard particle. In order to be able to use this kind of fibre in frictional materials it is necessary to control the amount and size of these shots. The SEM image in fig. 18f shows that mineral fibres are made up of smooth and more or less cylindrical filaments.

4.1.7. Ceramic fiber

Ceramic fibres are made in a process similar to that to make mineral fibres but using different raw material. Ceramic fibre is mainly made of SiO_2 and Al_2O_3. The SEM image in fig. 18g shows that ceramic fibre is made up of smooth and more or less cylindrical filaments. A shot is also shown in detail. The same kind of shot may also be present in mineral fibres.

4.2. Tribological effects

The evaluation of the tribological effects of adding the different kinds of fibres mentioned above was done using a basic non-asbestos polymer composite formulation to which the different fibres were added. We used two characteristic groups of fibers: (i) those that are mostly used for reinforcement of the polymer matrix (glass, steel, mineral, and ceramic) were added to the basic formulation in concentrations of 10 and 20 wt% and (ii) those that are used mostly as a processing aid (aramid, carbon, and acrylic), which were added in concentrations of 4 and 8 wt%.

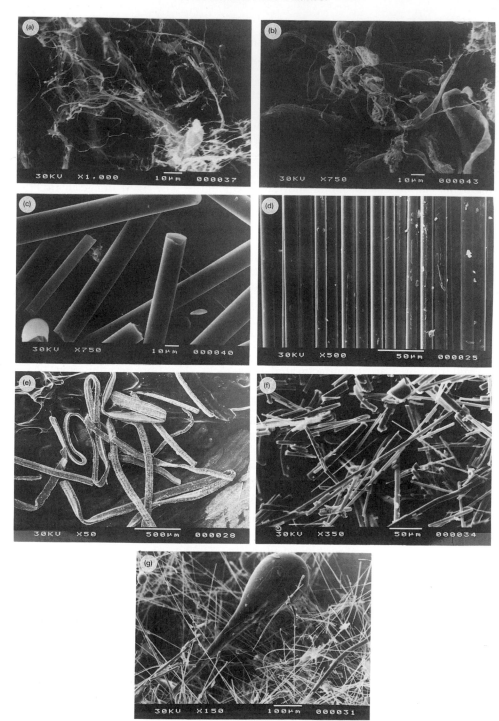

Fig. 18. Scanning electron microscope images of the different kinds of fibres commonly used in frictional material composites: (a) aramid fibre, (b) acrylic fibre, (c) carbon fibre, (d) glass fibre, (e) steel fibre, (f) mineral fibre, (g) ceramic fibre.

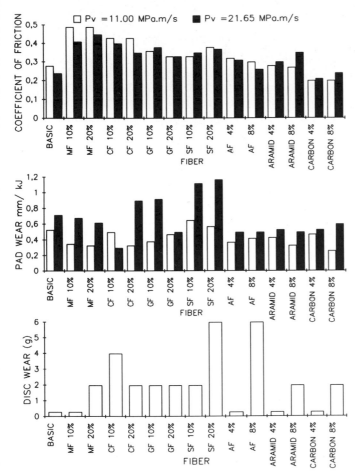

Fig. 19. Coefficient of friction, wear of the pads, and wear of the discs at two different load-carrying capacities for the various fibre-added frictional material polymer composites. MF – mineral fibre; CF – ceramic fibre; GF – glass fibre; SF – steel fibre; AF – acrylic fibre. The wear of the discs is the total wear for the two load-carrying capacities.

The coefficient of friction and the wear of the fiber-filled composite pads were compared with those of the unfilled basic composite pads, as well as the wear of the corresponding metallic discs. All tests were carried out in an inertial brake dynamometer fitted with a passenger-car disc brake, and two different tests were done, both at the same braking temperature of 230°C (measured in the disc): (i) with a pv factor equal to 11 MPa m s^{-1} and a rubbing velocity of 6.10 m s^{-1} and (ii) with a pv factor of 21.65 MPa m s^{-1} and a rubbing velocity of 7.5 m s^{-1}.

Figure 19 shows the results of the tests of the different composites, expressed as the friction coefficient and the wear of the pads for the two load-carrying capacities mentioned above, as well as the wear of the discs.

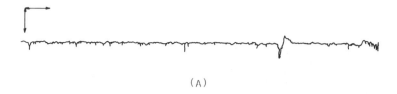

(A)

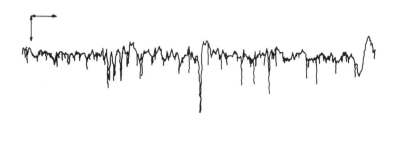

(B)

Fig. 20. Profile of the disc surfaces after rubbing against frictional material polymer composite pads with the addition of mineral (a) and ceramic (b) fibres ($\downarrow$ 10 μm; $\rightarrow$ 2500 μm).

The friction coefficient of all the fiber-added composites increased with respect to the basic formulation. The only exception was the carbon-fiber composite, which can be explained by the fact that carbon fibres have lubricating properties. Indeed, a large amount of material is observed to be transferred from the pad to the disc, as shown in fig. 22c, and it is basically the solid lubricant introduced in the system that reduces the friction coefficient.

The increasing of the friction coefficient with the addition of the different fibres can be understood if one remembers that the friction coefficient in these systems is mostly governed by the ploughing component of friction, and the addition of these hard fibres enhances the ploughing action of the polymer composite surface on the metallic disc surface.

An interesting comparison can be made between the friction and wear effects of adding mineral and ceramic fibres. As discussed in the previous section these two fibres are similar to comets, with a tail and a head or "shot". In particular, it is known that the shots are very hard particles. An inspection of fig. 19 reveals that a mineral-fiber composite has a higher friction coefficient than a ceramic-fiber one. The mineral fibre has a higher concentration of shots, but these are smaller and harder than those of the ceramic fibre. Indeed, the profilometry of the disc surfaces, shown in fig. 20, shows that the grooves produced by the mineral fibres (fig. 20a) in the disc surface are much shallower than those produced by the ceramic fibres (fig. 20b).

Fig. 21. Scanning electron microscope images of the fibre-added composite pads after testing against grey cast iron discs: (a) glass-fibre composite pad; (b) steel-fibre composite pad.

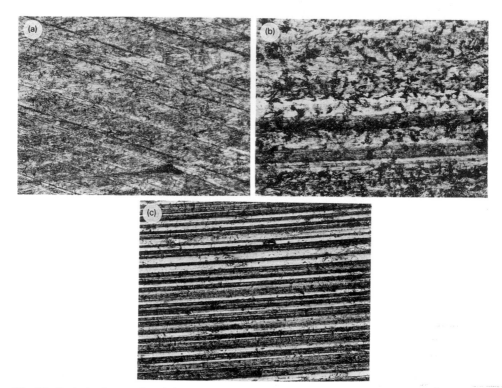

Fig. 22. Optical micrographs of the surface of the metallic discs after being tested against polymer composite pads made by the addition of aramid (a), acrylic (b), and carbon fibres (c). The carbon films transferred to the surfaces of discs tested against pads reinforced with carbon and acrylic fibres, can be seen in the micrographs.

Figure 21 shows that glass fibre is broken during the testing, generating wear debris, which will increase the ploughing action and so increase the friction coefficient, increasing also the amount of wear. Figure 21 also shows that steel fibres do not break during the testing but tend to lift off, increasing the angle between these particles and the surface of the disc, which will also increase the ploughing action. The effects on the friction coefficient and wear can be verified in fig. 19.

The three organic fibres (aramid, carbon, and acrylic) were the ones that presented less pad wear. For carbon and acrylic fibres this can be attributed to the lubricating action of the films of carbon transferred to the metallic disc surfaces, as shown in fig. 22. In the case of aramid fibre no significant material transfer to the metallic counterpart surface is observed. In this case, the wear resistance of the pads is probably associated with the mechanical resistance of this fibre, and the consequent increase that it produces in the load-carrying capacity of the pads.

4.3. Abrasives

The abrasives used in frictional materials are basically finely divided powders of refractory materials. The grains are usually very hard (6 to 10 on the Mohs scale). Natural abrasives include diamond dust, silica, corundum, and others. The most important synthetic ones are silicon carbide and calcined and fused alumina. Abrasive powders are used in frictional materials to increase the friction coefficient and to reduce the wear. The abrasive powders have various shapes, grain–sizes, hardnesses, and so on. We describe here some of the most used abrasives in frictional materials for automotive applications. Table 2 gives some physical properties.

4.3.1. Calcined alumina

This type of alumina comes from bauxite as a raw material, which contains approximately 50% Al_2O_3. After processing, the material is submitted to calcination at temperatures between 1300 and 1400°C, and as a result a material very pure (99% Al_2O_3) and very rich in α-alumina (90%) is obtained. Other peculiar characteristics of this material are the high hardness and very small particle sizes, as shown in fig. 23a.

4.3.2. Fused alumina

Fused alumina is known as dark brown Alundum abrasive. It is electrically fused from a mixture of bauxite, coke, and iron fillings. The processing temperature is around 2200°C and the blocks are transformed into fine powder by hammer milling. The shape and particle size distributions can be found in fig. 23b.

4.3.3. Green chrome oxide

Green chrome oxide is obtained by burning chromic acid in a closed furnace at 1000°C, having air as oxidizing agent. The product is ground in a hammer mill to obtain the desired particle sizes. The purity is around 97% Cr_2O_3 and the particle shape and size distribution is shown in fig. 23c.

TABLE 2
Selected physical properties of the abrasives used in the present work.

Physical property	Abrasive				
	Calcined alumina	Fused alumina	Green chrome oxide	Zirconite	Quartz
Mohs hardness	9.0	9.0	7.0	7.5	7.0
Melting point (°C)	2000	2000	1830	1785	1713
Specific gravity	3.8	3.95	4.8–5.1	4.25	2.5–2.8

4.3.4. Zirconite

Zirconite (zirconium silicate) is obtained by gravimetric separation from cassiterite and purification is accomplished by applying magnetic and electrostatic separation. Small particle sizes are obtained by grinding in hammer mills. The

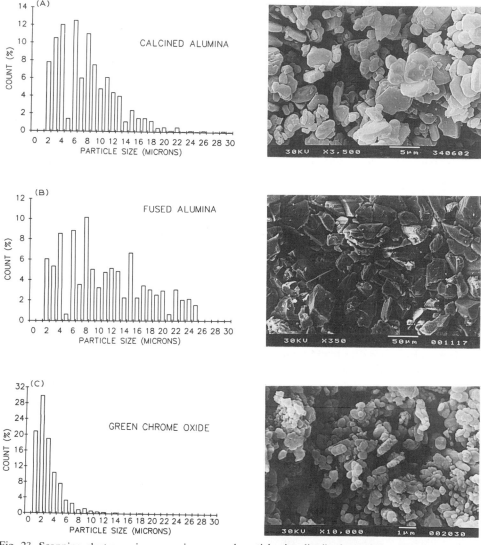

Fig. 23. Scanning electron microscope images and particle size distributions of the different abrasive powders commonly used in frictional material polymer composites: (a) calcined alumina, (b) fused alumina, (c) green chrome oxide, (d) zirconite, (e) quartz.

Fig. 23 (continued).

approximate chemical composition is ZrO_2 (54 to 58%). SiO_2 (31 to 35%), and Fe_2O_3 (1%). The particle shape and size distribution are shown in fig. 23d.

4.3.5. Quartz

The form of the quartz normally used as an abrasive, is crystallized silicon dioxide, with colour white to reddish vitreous luster. It is a natural occurring mineral, milled in a hammer mill. It is usually very pure and the typical particle shape and grain sizes are shown in fig. 23e.

4.4. Tribological effects

Abrasives are an important class of fillers that are added to the basic phenolic resin–fiber formulations, capable of providing suitable friction and wear characteristics. They are usually finely divided powders, much harder than the metallic counterface, capable of scouring it and, by consequence, increasing the friction coefficient of the metal–composite couple.

It is hoped that by controlling the amount, shape, and grain size of the different types of abrasives one can optimize the friction requirements and the life wear of both the pad and the disc. We present here some results obtained in the study of the effect of abrasive fine powders on the tribological properties of a non-asbestos semi-metallic disc pad basic formulation with 30 wt% of steel fibre, commonly used for automotive applications. The main reasons for choosing this basic formulation were:

(i) The basic composite should be as simple as possible in order to highlight the effect of the abrasive itself, avoiding possible combined effects that would occur if other fibres and fillers were present.

(ii) The basic composite should have a medium to low and stable friction coefficient against the disc, which is made of grey casting ion,

(iii) It is desired to use a composite with a formulation as close as possible to a real one.

In this study we have given emphasis upon three aspects of the effect of abrasives on friction materials, namely:

(i) the effect of different grain sizes of abrasives of a given chemical composition and shape,

(ii) the effect of abrasives of different chemical nature and of approximately the same size, and

(iii) the effect of varying the concentration of abrasives in the composite.

The samples used for the friction and wear measurements were composite pads rubbing against grey cast iron discs. The basic composite pads were made with 10 wt% of phenolic resin and 30 wt% of steel fibre and the rest are fillers. Abrasive powders of fused alumina and green chrome oxide were added as specified in the next section.

The tests were made in a fully automatic inertial brake dynamometer fitted with an Alfred Teves M48 disc brake with effective radius of 0.095 m, momentum of inertia of 39 kg m^2 and rolling radius of 0.285 m. The area of the pads was 25 cm^2.

The total amount of wear during a certain interval of time was determined externally by weighing or measuring the variation in thickness.

The tests were done in three steps, which are: (i) *cold effectiveness*: rubbing velocity 7.5 m s^{-1}, disc temperature 60°C, and variable pressure 2–10 MPa, (ii) *fade test*: rubbing velocity 7.5 m s^{-1} and pressure 6 MPa, leaving the temperature to vary freely in the interval 100–500°C, and (iii) *recovery test*: rubbing velocity 7.5 m s^{-1} and pressure 6 MPa, and cooling down the pads in steps from 300 to 100°C. The coefficients of friction were recorded continuously while the wear was measured by interrupting the tests at every 200 stops at the temperatures of 150, 250, and 350°C.

In fig. 24 we show the SEM images and the particle size distributions of the two fused alumina powders A and B as obtained from the image analyzer. The shape of the particles in both powders is almost pyramidal, which is a result of the milling process. The grain size distributions are characteristically different, with the grains of fused alumina A much larger than those of B. Electron microprobe EDAX analyses revealed a minimum of 98% purity of the alumina powders.

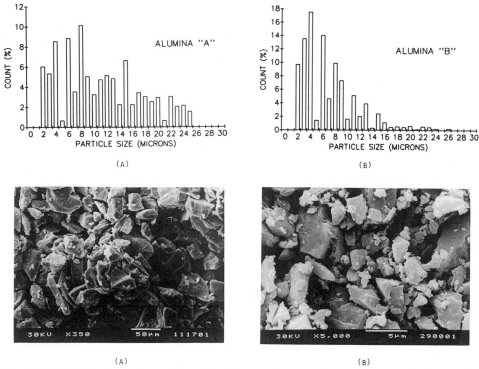

Fig. 24. Particle size distributions and scanning electron microscope images of the two alumina powders A and B.

The friction coefficients as measured by the inertial brake dynamometer, the total wear of the pads at three different temperatures, and the overall total wear of the pads are shown in fig. 25. We included the results of the tests done with pads consisting of the basic composite formulation, without abrasives, for comparison.

The total wear of the grey cast iron discs, optical microscope images of the wear tracks in the disc surface, and profilometry of the disc surfaces after testing are shown in fig. 26. Again, here we give the results of the discs tested against the basic composite pads for comparison.

The magnitude of the friction coefficient of the basic composite pads can be considered as medium to low in terms of the usual requirements for automotive applications. The wear of the basic composite pads was extremely high, far above the tolerable values, especially at the highest temperature of 350°C. The wear of the disc was very low, with the disc surface presenting very shallow grooves.

The addition of 2 wt% of fused alumina A to the basic composite increases the coefficient of friction in the *cold effectiveness* test and in the *fade* test significantly. The total pad wear is reduced by a factor of three, mainly because of the great reduction at higher temperatures. However, the disc wear is greatly enhanced, and large grooves can be seen in the microscope picture as well as in the profilometer trace.

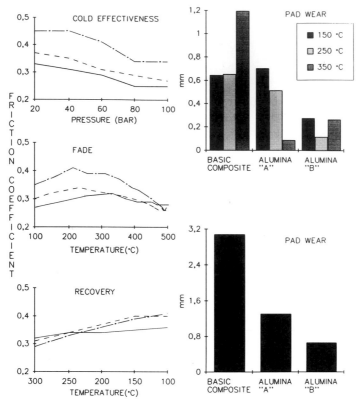

Fig. 25. Coefficient of friction under different testing conditions, wear of pads at different temperatures, and overall wear of pads of the basic composite unfilled and filled with aluminas A and B. In the plots of the coefficient of friction the full lines refer to the basic composite, the dashed lines to alumina A and the dash–dotted lines to alumina B.

The addition of 2 wt% of the smaller-sized fused alumina B also increases the friction coefficient significantly in the *cold effectiveness* and *fade* tests, but not as much as the fused alumina A. On the other hand, the pad wear is even smaller than that of the composite with the fused alumina A, and the disc wear is almost nil. Moreover, the optical microscope picture and the profilometer trace in fig. 26 reveal that the disc surface is polished by this abrasive powder.

The effects of varying the concentrations of fused alumina A and B are shown in fig. 27. The changes in the friction coefficient are moderate in all tests, but the wear of the pads and of the discs are completely different. The wear of the pads and discs is very little influenced by the increase of the concentration of fused alumina B, but the increase in the concentration of fused alumina A brings about a dramatic increase in both the wear of the pads and the discs.

In fig. 28 we show a comparison between the friction coefficients and the wear of the pads and the discs for composites made by the addition to the basic formulation of 2 wt% of fused alumina B and another abrasive, namely green

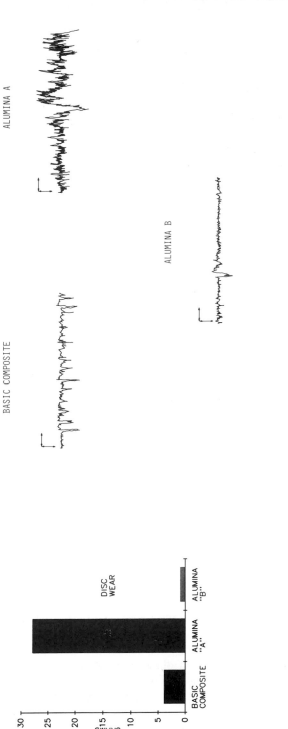

Fig. 26. Total wear of the discs, profilometer trace, and optical microscope images of discs worn against the basic composite pads unfilled and filled with aluminas A and B. In the profilometer traces the scales are: ↓ 10 μm; → 2500 μm.

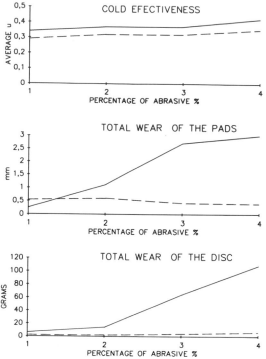

Fig. 27. Coefficient of friction, disc wear, and pad wear as a function of the concentrations of alumina A and B in the polymer composite. The full lines represents the results for alumina A and the dashed lines for alumina B.

chrome oxide, whose SEM image and grain size distribution is shown in fig. 23c. As one can see, the friction coefficient of the pads with green chrome oxide addition is much lower in all tests, being comparable to the values for the basic composite in all cases. On the other hand, the wear of the pads is reduced, even when compared to fused alumina B, while the wear of the discs is larger.

The friction behaviour of the composite pads rubbing against grey cast iron discs is governed by the ploughing component of the friction. So, the friction coefficient increases when one of the pyramidal shaped aluminas A or B is added to the basic composite because the ploughing effectiveness is increased. The ploughing component of the friction coefficient is expected to increase with increasing abrasives grain size, which is indeed observed.

At higher temperatures the friction behaviour is also strongly influenced by thermal degradation of the phenolic resin and by softening of the composite matrix. The result is a larger amount of penetration of the abrasive grains in the composite matrix under the applied pressures and the consequent loss of ploughing effectiveness.

By changing from Al_2O_3 to Cr_2O_3 not only the chemical composition (and physical properties) are changed but also the grain shape (pyramidal to round) and

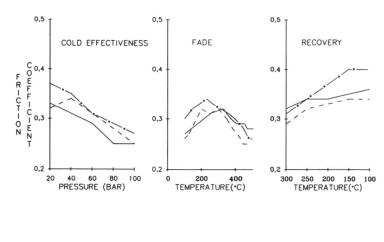

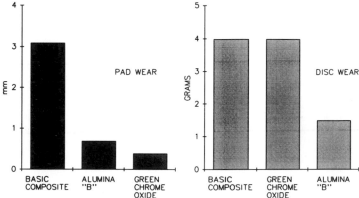

Fig. 28. Coefficient of friction, wear of the pads, and wear of the discs for polymer composite pads unfilled and filled with alumina B and with green chrome oxide. The full lines represent the results for the basic composite, the dashed lines the results for green chrome oxide and the dash–dotted lines the results for alumina B.

size (slight decrease). The result is a reduction in the ploughing effectiveness, causing the observed reduction in the friction coefficient.

The dependence of the friction coefficient on the concentration of abrasive aluminas A and B at low temperatures is consistent with the fact that enhancement of the surface density of sharp-edged abrasives increases the ploughing effectiveness and so the friction coefficient.

The wear of the pads is mainly a consequence of the abrasive component of the wear, although fatigue wear must also be taken into account. Since the basic composite does not contain particles appreciably harder than the disc counterface, the pads suffer abrasion from the hard asperities of the disc. The reinforcing action of the alumina powders added to the basic composite, increases the load-carrying capacity of the composite and, consequently, its wear resistance.

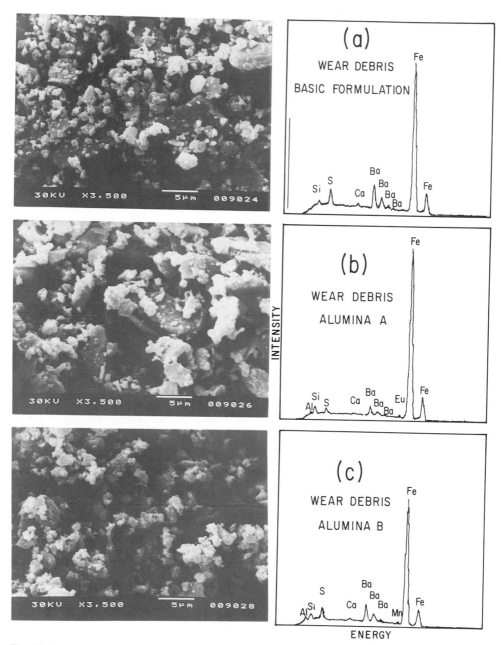

Fig. 29. Scanning electron microscope images and X-ray electron microprobe analyses of the wear debris resulting from the tests (at 150°C) of the basic-formulation pads and of the pads with alumina A and B.

TABLE 3
Concentrations in weight percent of the different elements in the near debris of the pads made of the basic formulation and of the basic formulation reinforced with alumina A or B.

Element	Concentration (wt%)		
	Basic composite	Alumina A	Alumina B
Si	0.41	1.32	0.50
Al	–	0.53	1.05
S	2.68	0.88	2.25
Ca	0.74	0.36	0.51
Ba	15.40	5.64	14.15
Fe	80.77	91.28	81.54

The wear of the discs is also dominated by the abrasive wear caused by the hard components of the pads, which is strongly dependent on grain size. The effect of increasing the surface concentration of the larger-sized alumina A has a larger influence on the wear than on the friction coefficient, while the influence of increasing the surface density of alumina B is not so pronounced both in friction and wear.

Analyses of the wear debris resulting from the tests of the different composite pads in the present work, reveal new aspects of the friction and wear mechanisms acting in the tribological system, especially the third-body abrasion. Figure 29 shows SEM images of the wear debris collected after testing at 150°C the basic composite pads and the pads with an addition of 2 wt% of alumina A and B, as well as X-ray electron microprobe spectra of the wear debris. The corresponding compositions are given in table 3.

We notice that the grain sizes of the wear debris coming from alumina B are smaller than those coming from the basic composite pad, whereas the wear debris resulting from alumina A are significantly larger. Furthermore, the compositions given in table 3 show that the case of the wear debris of the basic composite pads there is a large proportion of material coming from the pad itself (as indicated by the proportion of Ba in the wear debris, due to the loss of baryte from the pad). The same is seen for the wear debris of the pads reinforced with alumina B. Differing from them, the wear debris resulting from the alumina A pads has a much lower concentration of material coming from the pads, being mainly composed of Fe and Si, which are the elements that result from disc wear.

So, the third-body abrasion of the larger-sized wear debris resulting from the alumina A pads, contributes to the enhancement of the wear of both the grey cast iron disc and the composite pad itself, consistent with the results of the tests. The smaller-sized wear debris resulting from the alumina B pads, acts as a third-body abrader as well, but due to the small size it is responsible for the observed polishing of the disc.

The larger concentration of Fe and Si in the wear debris of the alumina A pads is a result of the larger wear of the discs tested against these pads, as compared to

(a)
Wear Debris
Cr_2O_3 - 150°C

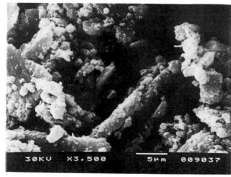

(b)
Wear Debris
Cr_2O_3 - 350°C

Fig. 30. Scanning electron microscope images of the wear debris resulting from the tests (at 150 and 350°C) of the pads with the addition of green chrome oxide.

the wear of the discs tested against the basic composite pads and the alumina B composite pads.

The effect of the temperature on the wear debris (and, consequently, on the friction and wear mechanisms) is nicely illustrated in fig. 30, where SEM images of the wear debris resulting from the tests done with pads reinforced with green chrome oxide at 150 and 300°C are shown. One can easily see that the grain sizes of the wear debris from the tests at 300°C are much larger than from the tests at 150°C. Also, the composition of the wear debris, as determined by X-ray photo-electron microprobe, reveals that for the tests done at 150°C the wear debris has a composition that resembles that of the debris coming from the basic composite pads, whereas the wear debris resulting from the tests done at 300°C has a much larger contribution of the Fe and Si coming from the disc. These features justify the friction and wear results of fig. 28, where we notice a poor *recovery* perfor-mance of the pads made by addition of green chrome oxide, as well as a larger pad wear. The much larger overall disc wear observed even if the grain sizes of the green chrome oxide raw material are smaller than that of alumina B, is due to the third-body abrasion of the large wear debris particles released at higher temperatures.

5. Concluding remarks

This chapter presented the fundamentals of the manufacturing and use of the polymer composites for frictional materials in current use in automobile, railway, and industrial applications.

Also discussed was the basic tribology involved in the characteristic situation of a polymer composite frictional material rubbing against a metallic counterpart, as well as were introduced the relevant aspects of some research efforts being put forward by the authors concerning the tribology of fibre and abrasive reinforcement of these polymer composites. It was made clear that a very complex situation is involved in all these cases, making it more or less impossible to design frictional materials based only on scientific knowledge. A lot of empiric expertise is always necessary to obtain adequated formulations.

Nevertheless, scientific work is growing up in this area, making possible a better control of some relevant parameters, like the fibres and abrasives adressed here, the role of the wear debris, material transfer, and others. Moreover, new materials are being introduced, either as reinforcements in the polymer composites or as alternatives for the metallic counterpart, and the tribology of these is still open to investigation.

Acknowledgements

The authors want to express their gratitude for the technical and scientific support received from Engs. Nelio Enderle, Hamilton Leal and Antonio Oliveira all along the development of the present work.

Our special and warm gratitude also to Dr. Fernanda C. Stedile for the many technical discussions during the work as well as the critical and careful review of the manuscript.

List of symbols

A	factor
B	factor
C	factor
D	factor
E	factor
S-cam	type of brakes
pv	load capacity
μ	dynamic coefficient of friction
a	acceleration
t	time

F	force
HB	brinell hardness
rpm	revolutions per minute

Subscripts

MF	mineral fibre
CF	ceramic fibre
GF	glass fibre
SF	steel fibre
AF	acrylic fibre

References

[1] F.J. Washabaugh, SAE Technical Paper 860630 (1986).
[2] H.Y. Loken, SAE Technical Paper 800667 (1980).
[3] M. Duchene and M. Charloteaux, Le Freinage (Editions Plantyn, Antwerp, 1975).
[4] Y. Yamashita, H. Asano, M. Kawase and K. Iwata, SAE Technical Paper 890861 (1989).
[5] A.K. Baker, Vehicle Braking (Pentech Press, London, 1987).
[6] T.P. Newcomb and R.T. Spurr, Commercial Vehicle Braking (Newnes–Butterworths, London, 1981).
[7] D. Goodsell, Dictionary of Automotive Engineering (Butterworths, London, 1989).
[8] W.H. Crouse, Automotive Mechanics (McGraw-Hill, New York, 1970).
[9] W. Alley and W.E. Billiet, Disc and Drum Brake Service (American Technical Society, Chicago, 1976).
[10] Anonymous, Bosch Automotive Handbook (Robert Bosch GmbH, Stuttgart, 1976).
[11] Anonymous, SAE Handbook, Vol. 2 (SAE, Warrendale, PA, 1990).
[12] V. Chiaverini, Aços e Ferros Fundidos (ABM, Sao Paulo, 1982).
[13] Link Engineering Company, Catalogues (Detroit, 1989).
[14] Greening Associates, Catalogues (Detroit, 1989).
[15] A.E. Anderson, S. Gratch and H.P. Hayes, SAE Technical Paper 670079 (1967).
[16] K. Friedrich, in: Friction and Wear of Polymer Composites, ed. K. Friedrich (Elsevier, Amsterdam, 1986) ch. 8.
[17] A.D. Sarkar, Friction and Wear (Academic Press, London, 1980).
[18] S.K. Rhee, R.T. DuCharme and W.M. Spurgeon, SAE Technical Paper 720056 (1972).
[19] S.K. Rhee, SAE Trans. 80 (1971) 362.
[20] H. Czichos, A System Approach to Friction and Wear (Elsevier, Amsterdam, 1980).
[21] A. Betejtin, Mineralogy (MIR Editorial, Moscow, 1977).
[22] Anonymous, Encyclopedia of Polymer Science and Engineering, Vol. 6 (Wiley, New York, 1990).

PART V

Tribological Response of Composites in Relation to Other Requirements

Advances in Composite Tribology
edited by K. Friedrich
© *1993 Elsevier Science Publishers B.V. All rights reserved*

Chapter 17

Rolling Contact Fatigue of Polymers and Polymer Composites

T.A. Stolarski

Department of Mechanical Engineering, Brunel University, Uxbridge, Middlesex UB8 3PH, UK

Contents

Abstract

The volume fatigue of polymeric materials has been studied with increasing interest in the last 40 years. It is a general observation, in case of volume fatigue, that failure at repeated loading occurs as a result of thermal softening, excessive creep or flow and/or the initiation and propagation of fatigue cracks. On the other hand, the problem of surface fatigue of polymers induced by a rolling contact has received very little, if at all, attention. However, as polymers and their composites are being increasingly used for load-bearing elements subjected to fatigue conditions, because they offer a number of obvious benefits, the need for more information and better understanding of polymer rolling contact fatigue is obvious.

This chapter discusses some fundamental problems related to polymer rolling contact fatigue and presents limited experimental results, illustrating the performance of engineering polymers in rolling contact. These results clearly show that typical surface fatigue damage can occur in certain polymers. Lubrication, in general, improves the performance of all polymers tested. The overal conclusion is that some polymers and their composites have a potential for use in rolling contact applications, provided that the load and speed are carefully selected.

1. Introduction

1.1. General considerations

The question of fatigue in engineering materials is not new; it has been recognized as an important problem for at least the last 200 years. It came into prominence when the steam engine revolutionized transportation and fatigue in rolling-stock axles became a serious problem. Thus the stage was set for the scientific and applied study of fatigue in ferrous materials, which has been continued since 1870. A particularly strong incentive to research on fatigue in ferrous materials has been given by the advent of steel hulls, crude oil pipelines and jet aircrafts.

Fatigue can be defined as the loss of strength or other important properties by a material as a result of stressing fluctuating in magnitude and direction over a period of time. Another plausible definition describes fatigue as the ultimate failure of a material or component at the application of a varying load whose maximum amplitude, if continuously applied, is insufficient to cause failure. It is a very common phenomenon in most engineering materials, including polymers and their composites. Various atomic or molecular processes, often poorly understood, may take place during the fatigue phenomenon. Some of them may be beneficial and some deleterious. The deterioration usually prevails over the strengthening, and failure, usually as a result of stresses that are small in comparison to those required for failure resulting from a static stress, occurs. A material subjected to a cyclic load will ultimately fail. Usually, the higher the applied load, the shorter the time to failure (fig. 1). In case of certain materials there may also be a clearly defined so called "endurance limit". This is a limiting low stress below which failure does not occur within any practical time. Apart from fracture, failure may also be defined in terms of the loss of functionality as defined by failure to meet certain design criteria, such as required strength, stiffness or integrity of shape.

Fatigue failure in polymers is controlled by a number of competing factors, such as loading conditions, material structure and morphology, composition and time, which is characteristic of polymers as a class of materials. However, since polymers and their composites are being increasingly used for load-bearing elements subjected to fatigue conditions, the engineer responsible for component design and materials selection must also know the fatigue characteristics and have some understanding of the effects of loading and material variables. This is dictated mainly by the nature of the fatigue process and the much more complex structure

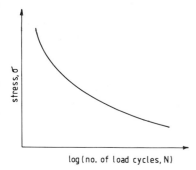

Fig. 1. Typical relationship between the applied stress and the number of load cycles to failure.

of polymers compared to ferrous materials. The problem is further aggravated by the fact that some engineering plastics are sensitive to fatigue conditions and that the existence of cyclic loading in service is not always recognized. For example, teeth in plastic spur gear are subjected to loading–unloading every revolution, although the torque applied is constant. The same happens in the case of a rolling contact bearing in which the inner and outer races are made of plastic. Pressure fluctuations in plastic pipe may cause premature brittle failure in a normally ductile polymer at very low loads. It could be said that some applications of modern engineering polymers are inherently of fatigue nature.

In materials selection and component design for rolling contact applications, the empirical characterization of fatigue in polymers is useful but not sufficient. Growing sophistication in these functions requires more fundamental understanding of the mechanisms underlying the overall fatigue process. Although stress versus the number of load cycles curves are useful when selecting a material for a particular application, they usually tell nothing about the initiation and propagation of crazes or cracks, yielding and drawing, or the storage, release and dissipation of energy within the rolling contact. The state of the art in the fatigue of polymers, however, is not nearly as well advanced as in the case with ferrous materials, in spite of the obvious need for better understanding demanded by the increase in use of engineering polymers in applications where fatigue is a predominant concern. With some exceptions, the phenomenon of fatigue in polymers and their composites has received relatively little attention. The situation is even worse in the case of fatigue resulting from a rolling contact. A great deal of empirical work has been done, but until recently studies of fundamental and applied nature have been almost exclusively restricted to rubbers and fibres. For example, it was said twenty years ago that fatigue in polymers constitutes one of the unsolved problems in polymer science [1]. Obviously, our understanding has advanced significantly since then, although there are still many unsolved problems.

1.2. Approaches to polymer fatigue

A number of major problems might be identified as being responsible for hindering the understanding of fatigue process in polymers. The traditional continuum-mechanical approach to fracture developed by physicists and mechanical engineers excludes atomic and molecular processes taking place during polymer fatigue. Therefore, there are certain difficulties in communication between polymer scientists and engineers. The ultimate fracture is usually accompanied by large-scale, irreversible, and non-linear deformations. On the other hand, polymer scientists are accustomed to measure small-scale, reversible, and linear deformations in order to characterize the effects of molecular properties and composition of a polymer. Besides, the exact nature of the competitive processes that are active during fatigue and are responsible for the damaging effect of intermittent loading, is not sufficiently known. The only way forward is to accept the duality of continuum and molecular descriptions as representing two different aspects of the same reality and to investigate the limits of using a linear approximation to model non-linear systems such as polymers. Nevertheless, the linear elastic fracture mechanics approach traditionally used for ferrous materials has proved to be quite promising when applied to polymers, although non-linear models are also being developed [2,3].

1.3. Loading conditions in common rolling contact applications

One of the promising new applications of polymers and their composites is as a material for rolling contact bearings. As a matter of fact, the contact between the rolling elements and the races or rings consists more of sliding than of actual rolling. The condition of no interfacial slip is seldom maintained, because of material elasticity and geometric factors. Moreover, inherent to the state of loading on rolling elements and inner and outer rings is cyclic loading, although the external load applied to a bearing has a static nature. The usual assumption is, therefore, that rolling contact bearing failures are fatigue failures.

The fatigue life of a rolling contact bearing is a function of a number of factors which are interwoven in a highly complex manner. The effect of increasing speed on the fatigue life, for instance, is chiefly an increase in the operating load and the clearance due to the centrifugal force, with a corresponding reduction of the loading zone. The contact angles also change, that at the inner raceway increasing and that at the outer raceway decreasing with rising speed.

Another common example of a nominally rolling contact are gears. The fatigue failure of toothed gearing may occur in any one or a combination of three basic modes:

(i)   Tooth bending, which, if continued for a sufficient number of cycles results in fracture at the root.

(ii) Surface distress as a result of mesh stresses exceeding the compressive fatigue limit in some localized areas of the tooth face. This condition may be caused by some sort of misalignment of tooth faces in the mesh. If caused by an initial defect, the defect may have been on the surface, or subsurface, but within the volume of highly stressed material.

(iii) Pitting resulting from rolling/sliding action in the contact region of two meshing teeth.

2. General aspects of fatigue of polymers and their composites

Polymeric materials subjected to strong mechanical and environmental excitation show – as many other materials – gradual deterioration of their performance including eventual failure. If the changes in properties are mostly due to chemical reactions we can speak of corrosion or of radiative degradation. If the deterioration in material properties is caused by the repeated cyclic or random application of mechanical stresses we have the case of fatigue. The volume fatigue of polymeric materials has received increased attention in the last 40 or so years. Reviews and monographs on this complex subject have appeared from time to time [4–6].

Polymers and their composites exhibit a much more complex behaviour when subjected to fatigue loading than ferrous materials. The net effect of the several competing processes depends on a number of factors, which include the temperature, time, environment, basic molecular properties of the polymer and the composition. The most important factors determining the fatigue behaviour of a polymer are:
- thermal effects during the loading–unloading cycle,
- morphological changes within the polymer,
- transition phenomena,
- molecular characteristics (molecular weight, thermodynamic state),
- chemical changes (degradation of bonds),
- homogeneous deformations,
- inhomogeneous deformations.

It can be said that the deformation and flow in polymeric material depend on several characteristics listed above. However, the molecular structure, molecular weight, composition and morphology are especially important in this respect. Polymers under stress are almost invariably in a non-equilibrium state, an equilibrium state or steady-state response being attained only under special test conditions. In fact, the mechanical properties, fatigue included, desired of polymeric materials in most engineering applications result, in large measure, from their non-equilibrium response to applied stress.

At elevated temperatures resulting, for instance, from frictional heating within the rolling contact area, polymeric chains continuously undergo random changes in configuration. The rate of such diffusional processes, which depends strongly on temperature, is the dominant factor that affects the response to stress.

Essentially all available information on polymer fatigue is empirical in nature. General data on fatigue of engineering polymers are arbitrarily summarized in table 1 on the basis of whether the $S–N$ curve passes above or below the point 28 MPa and 10^4 load cycles.

2.1. Physical states of polymers under stress

At an elevated temperature, a polymer is either an elastic solid or an elastic liquid, depending on whether or not it is cross-linked. When a constant stress is

TABLE 1
Fatigue strength for common engineering polymers.

Above	At 10^4 cycles (MPa)	Below	At 10^4 cycles (MPa)
Polyimides	38	Polystyrenes	14
Phenolics, molded	34	Vinyls, rigid	21
Polysulfones	28	Acetals	24.5
Acrylics	49	Polyethers	21
Polycarbonates	31.5	Polypropylene	21
Nylons	49	Polyethylene	11

applied to an elastic solid, the specimen rapidly attains a constant deformation; when the stress is removed, the specimen returns rapidly to its initial shape. This is not precisely true since many elastomers, even at elevated temperature first deform rapidly under a constant stress, but then creep very slowly, even when the network is chemically stable [7]. Under a constant stress, a non-cross-linked polymer undergoes flow much like a low molecular weight liquid. However, some energy is stored elastically, because the molecular chains have been deformed, on the average, from their most probably configurations. Hence, when the stress is removed, some elastic recovery occurs and, therefore, the material is called an elastic liquid because energy is both stored and dissipated during flow.

The volume of an amorphous polymer decreases in a near-linear manner with decreasing temperature (fig. 2). Thus, the free space available to the molecular chains is reduced, and consequently it is more difficult for configurational changes to occur or, in other words, the mobility is less.

With progressive decrease in temperature, the mechanical properties become highly time dependent (viscoelastic) and change from elastic-like to leathery. In some narrow temperature range, long-range configurational changes take place only very slowly and below a temperature called the glass temperature, T_g, long-range motions become frozen and only short segments and side groups continue to execute thermally activated motions. This is demonstrated by the thermal expansion coefficient, which is roughly three times greater above T_g than below. Below T_g, the polymer is called a glass, although glassy polymers usually show some form of ductility.

Semicrystalline polymers have both a glass temperature and a crystalline melting point (fig. 2). When such a polymer is slowly heated, the volume–temperature relation indicates that melting occurs over a temperature range. The temperature, T_m (fig. 2), is that at which the largest and most perfect crystallites melt.

Above T_m, a crystalline polymer has properties similar to an amorphous polymer and at temperatures considerably above T_m it is either an elastic solid or an elastic liquid, depending on whether it is cross-linked or not. In the range $T_g < T < T_m$, a semicrystalline polymer consists essentially of crystalline domains and amorphous material intermixed. Below T_g, both crystallites and amorphous glassy material exist.

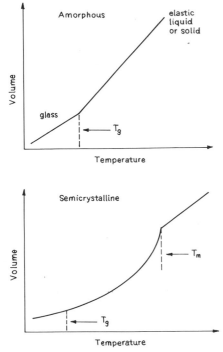

Fig. 2. Illustration of the volume–temperature relationship for amorphous and semicrystalline polymers.

2.2. *Response to applied stress of polymeric materials*

Under applied surface traction, an elastic solid undergoes, in general, a change in both shape and volume. According to the classical theory of elasticity, a measure of the resistance to a change in shape at constant volume, for a homogeneous isotropic material, is the shear modulus, G. Similarly, the measure of resistance to a volume change at constant shape is the bulk modulus, K. The resistance to a change in length is the tensile or Young's modulus, E. Because such a deformation usually involves both shape and volume changes, the tensile modulus is a function of G and K. The three moduli are comprehensively defined in fig. 3. The reciprocal of each modulus is a compliance.

Linear elastic behaviour in either shear, tension or bulk can be represented schematically by a spring. The spring symbolizes that stress is proportional to strain, the response to stress is instantaneous, no permanent deformation occurs, and the energy to deform the spring is stored completely, i.e. there is no dissipation.

A Newtonian fluid is usually represented by a dashpot for which: the stress is proportional to the shear or strain rate, no elastic recovery occurs when the stress is removed, and the energy-affecting flow is entirely dissipated.

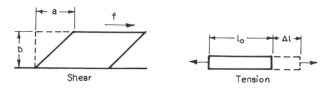

Fig. 3. Elastic moduli and compliances for isotropic elastic material under small deformations.

	Shear	Tension	Bulk
Stress	$\tau = f/A$	$\sigma_t = f/A_0$	Pressure
Strain	$\gamma = \tan^{-1}(a/b)$	$\epsilon = \Delta l/l_0$	$-\Delta V/V_0$
Modulus	$G = \tau/\gamma$	$E = \sigma_t/\epsilon$	$K = -\dfrac{p}{(\Delta V/V_0)}$
Compliance	$J = 1/G$	$D = 1/E$	$B = 1/K$

Polymers are classified as viscoelastic materials. The energy required to deform a viscoelastic material is partially stored and partially dissipated, therefore it exhibits characteristics of both an elastic solid and a liquid. Under sufficiently small deformations, the behaviour is linear, although only special tests can reveal whether or not the viscoelastic response is linear.

There are several mathematical ways to represent the linear viscoelasticity [8–11]. One of them is a mechanical model consisting of arrays of linear (Hookean) springs and linear (Newtonian) dashpots. The generalized Voight model consists of a large (often an infinite) number of Voight elements connected in series (fig. 4). A Voight element is constructed from a spring and a dashpot in parallel. The other model often used to represent the response of a polymeric material to a prescribed

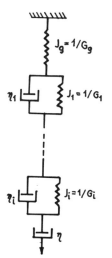

Fig. 4. Generalized Voight model to represent linear viscoelastic behaviour. For illustration, the shear behaviour and response to constant shear stress are considered.

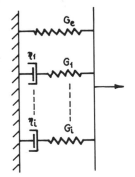

Fig. 5. Generalized Maxwell model to represent linear viscoelastic behaviour. For illustration, the shear behaviour and response to constant shear strain are considered.

strain–time history is the generalized Maxwell model (fig. 5). This model consists of a large (or infinite) number of Maxwell elements connected in parallel. Each Maxwell element represents a spring and a dashpot connected in series.

2.3. Phenomenological description of the fatigue process

In fundamental terms fatigue is due to the irreversible processes which take place when a cyclic load is applied to a polymeric material. The extent of fatigue damage and its importance is dependent primarily on the stress level at which irreversible damage occurs relative to the stress for complete failure. Consider two materials with stress–strain curves as shown in fig. 6. One material is almost perfectly elastic to fracture, whereas the other undergoes viscoelastic flow at about

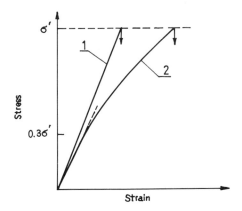

1–elastic; 2–viscoelastic

Fig. 6. Stress–strain curve. A is almost perfectly elastic to fracture; B undergoes plastic of viscoelastic flow.

$0.3\sigma^*$. The elastic material will be insensitive to alternating stresses below σ^*, and the stress–number of load cycles curve will be almost flat, pointing to a very good fatigue resistance. The other material will start to deform at relatively low stresses and fatigue damage will develop continuously. The amount of damage will increase with increasing load and the material will have a poor resistance to fatigue.

When this simple principle is applied to composite materials, it is clear that the response depends on the fibre arrangement and volume fraction as well as on the matrix and the fibre properties. All these factors determine the way that the load is distributed between the fibre, matrix and fibre–matrix interface. For example, in an aligned carbon fibre material the load is carried almost entirely by the fibres and little irreversible damage occurs until fibre fracture is initiated. Subsequent unloading and loading cycles result in a small redistribution of the load in the region of the broken fibres, and some fatigue damage may develop. This will occur only at stresses close to ultimate fracture, so that the fatigue resistance is considered as very good.

In a random glass fibre material microcracking by debonding occurs in transversely oriented fibre bundles at, usually, relatively low stresses, and this marks the onset of clearly defined irreversible damage. Cracking may be preceded by resin flow. During fatigue cycling the transverse cracks propagate and this eventually leads to resin cracking, which can occur at stresses well below those observed at monotonic loading.

Although the nature of the irreversible processes depends on the composite material and the loading conditions, the principle of progressive fatigue damage can be understood by reference to the simple example illustrated in fig. 7. The sketch represents part of a fibre bundle or lamina oriented with the fibre axis normal to the applied load. The bundle or lamina is constrained by adjacent bundles, so that the growth of a crack under monotonic loading occurs only under increasing strain conditions. Suppose that the bundle is loaded by a cyclic load as illustrated in fig. 8. In the first half cycle the deformation and fracture processes which occur depend on the stress amplitude and presumably the initial response is linear. As the stress increases, non-linear effects arise due to the viscoelastic properties of the resin and, at a later stage, to debonding and resin cracking. Thus, at very low stress amplitudes, when the response is fully elastic, fatigue damage will not develop.

When the stress amplitude is increased, viscoelastic flow occurs preferentially between closely spaced fibres, because of the high strain magnification in these regions. This flow will not be fully reversible when the stress is reduced, and during cycle loading additional stress and strain concentrations develop, which lead to the initiation of debonding at applied stresses below those observed in monotonic loading (see fig. 7). The debonding cracks grow during cyclic loading because some flow occurs at the crack tip during the loading half of the cycle, which is not fully reversed on unloading. As in uniaxial tensile tests, the cracks nucleate and propagate in regions of closely spaced fibres by the growth and coalescence of individual fibre debonds. When the fibres are widely spaced, the growth of the

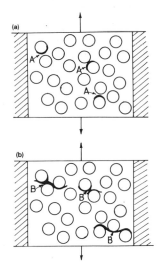

A - viscoelastic flow; B - debonding crack

Fig. 7. Schematic representation of the nucleation and growth of fatigue cracks in a transversely oriented bundle or lamina surrounded by differently oriented bundles or laminae.

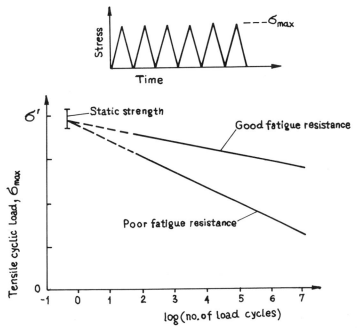

Fig. 8. Schematic representation of $S-N$ curves. Note that the alternating stresses may be tensile, compressive or shear.

crack from one fibre to the next depends on the resistance to fatigue crack growth of the matrix.

It is obvious from this simplified presentation of the mechanism that the fatigue properties will depend on the temperature and the cyclic loading frequency, since both these factors affect the amount of matrix flow. An important additional effect is viscous dissipative heating during cyclic loading, leading to a rise in temperature which can be in the region of 25 to 50°C, depending on loading frequency. The magnitude of the rise depends on the specimen geometry and the efficiency of heat dissipation to the surroundings. Thus, carbon fibre composite materials show lower temperature rises because of the relatively high thermal conductivity of the fibre.

A similar mechanism can be used to explain the fatigue behaviour of more complicated fibre arrangements. Progressive damage may involve intralaminar and interlaminar processes and the rate of damage build-up depends on the effective stresses causing different forms of microdamage.

A common feature of the fatigue failure of composite materials, especially when the element is uniformly stressed, is the occurrence of damage over a large volume of the material. The large-scale damage usually produces three important effects:
– the modulus of elasticity decreases progressively during the fatigue life,
– the hysteresis or damage loop during cyclic loading becomes progressively more pronounced,
– the residual strength of the material decreases progressively with the number of cycles.

In the case of a rolling contact configuration the failure almost exclusively develops from local regions of high stress or strain associated with the nature of the contact. The micromechanism of failure in these regions will be the same as that described above.

3. Behaviour of polymers and their composites in rolling contact

At first sight, engineering polymers may seem to be unlikely materials for rolling contact applications, such as bearings, gears, etc., particularly in view of the generally accepted premise that polymers are weak and soft. However, modern engineering polymers and their composites have physical and mechanical properties that can be considered as very attractive for rolling contact applications.

The main benefits resulting from using polymers in rolling contact applications can be summarized as follows:
(i)   corrosion resistance – many engineering polymers are considered to be chemically inert and therefore capable of operating in environments hostile to ferrous materials,
(ii) ability to operate without lubrication or to be lubricated with process fluids,
(iii) many potentially useful polymers are available at a much lower price than traditional engineering materials,
(iv) ease of processing and cost of manufacture make polymers the most competitive material for rolling contacts in underwater, marine, chemical and processing industries,

(v) injection moulding and extrusion eliminate the need, and therefore lower the cost, of finishing and other surface treatments, since elements and components manufactured with the help of these techniques come finished to dimensions and ready for assembly. For example, moulded toothed gears or components for rolling contact bearings can be produced at a very low unit cost.

These obvious benefits can be readily obtained, provided the application is characterized by light loads and low to moderate speeds. This is not only because of the fatigue strength of polymeric materials, but also the thermal softening and general decrease in mechanical strength at elevated temperatures.

As it was mentioned earlier, three principal mechanisms may contribute to the volume fatigue of polymeric materials. Although this view is now commonly accepted, the case of surface fatigue of polymers probably requires a modified approach. Despite the lack of sufficient understanding of surface fatigue of polymers, they have been successfully used as a material for gears, and recently for rolling contact bearings [12]. The present practice is to use polymers for the inner and outer races while retaining steel balls and rollers.

3.1. Characteristic of a rolling contact

Pure or free rolling is the most elemental form of rolling motion and is probably most nearly approached in the case of a cylinder or ball rolling without constraint in a straight line along a plane. Since all bodies are made from deformable materials, some deformation must ensue when they come into contact under load. The shape and size of the area of contact will depend upon such factors as the individual geometry, the load and the deformation characteristics of the materials.

The mechanism of rolling friction has been attracting little work of a fundamental nature, apart from the pioneer studies of Reynolds [13], Stribeck [14], Goodman [15], Palmgren [16], Tomlinson [17], Tabor [18] and Johnson [19]. Most of the work has been concerned with perfectly elastic materials, and results for rolling contact of polymers and their composites are relatively scarce.

Basically, there are two scales of size in the problem related to the contact of real engineering materials:
  (i) the bulk (nominal) contact dimensions and compressive deformations, which could be calculated by the Hertz theory and
  (ii) the height and spatial distribution of surface asperities.

For the situation to be amenable to quantitative analysis these two scales of size should be very different. In other words, there should be many asperities lying within the nominal contact area. When the two bodies are pressed together, true contact occurs only at the tips of the asperities, which are compressed as elastic or viscoelastic solids.

In the case of rolling contact of elastic materials, the process consists of two main effects. The first, which is usually predominant during the early stages of rolling, is primarily concerned with the plastic displacement of material from the path of the rolling element, and the resistance to rolling is essentially due to the work of plastic deformation of the material. The second effect predominates after repeated traversals of the same track; plastic displacement gradually comes to an end and the deformation becomes primarily elastic. The primary source of the rolling resistance is elastic hysteresis losses within the contacting materials.

The situation is quite different in the case of rolling contact between a hard smooth cylinder and a viscoelastic material. Tomlinson [17] explained his experimental observations in terms of molecular adhesion between lightly loaded rolling surfaces. According to him, the surface atoms are pulled away from their equilibrium positions until the displacement exceeds a certain distance; they then flick back to their old equilibrium positions and, in the process, energy is dissipated. Although this theory was not accepted it is quite clear that the idea of contribution of interfacial effects to the friction in rolling contacts between viscoelastic materials should be seriously considered in quantitative terms.

3.2. Thermodynamic equilibrium of a system created by rolling contact

The system consisting of two bodies in contact over an area A is considered. It can represent, for instance, a micro-contact created by a pair of asperities located on the surfaces of bodies in rolling contact. Adhesive junctions formed by the contacting asperities are ruptured due to tension exerted on them by the rolling motion (see fig. 9). The system can exchange work and heat with the surroundings, but no matter. A force P, representing tensile load on the contact and resulting from rolling motion, can be applied either by a fixed load, as in fig. 10a, or by a more complex loading system as shown in fig. 10b. It must be emphasized, however, that:

(i) The area of contact, A, is allowed to vary, so that the geometry of the system can change and linear elasticity must be excluded.

(ii) Variation of A may be accomplished independent of the load P, or of the elastic displacement δ, so that the state of the system depends in general on two independent variables, i.e. P and A or δ. The area of contact can decrease at fixed P or fixed δ, until separation of the two bodies occurs, corresponding to the rupture of a joint under fixed load or fixed grips conditions. The reduction in the contact area will be considered here as a propagation of crack in mode I, i.e. the opening mode. The energy of the system shown in fig. 10,

$$U = U(S, \delta, A),$$

is a function, besides the entropy S, of δ and A, and can be divided into elastic energy U_e and interfacial energy U_s. The interfacial energy is solely a function of A and can be written as:

$$U_s = -(\gamma_1 + \gamma_2 - \gamma_{12})A = -wA,$$

where γ_1 and γ_2 are the surface energies of the two bodies in rolling contact, γ_{12} is their interfacial energy and w is Dupre's energy of adhesion or the thermodynamic work of adhesion. The first differential of the energy is

$$dU = \left(\frac{\partial U}{\partial S}\right)_{\delta,A} dS + \left(\frac{\partial U}{\partial \delta}\right)_{S,A} d\delta + \left(\frac{\partial U}{\partial A}\right)_{S,\delta} dA$$

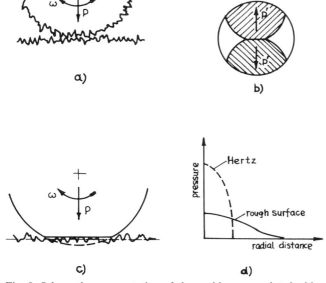

Fig. 9. Schematic representation of the problems associated with rolling contact. (a) Rolling contact between two rough surfaces under normal load. (b) Due to rolling motion the adhesive junction formed between the contacting asperities is subjected to tensile load. (c) Contact of a smooth sphere with a nominally flat rough surface. (d) Contact pressure as a function of radial distance from the centre of contact.

or

$$dU = T\,dS + P\,d\delta + (G - w)\,dA,$$

where

$$\left(\frac{\partial U}{\partial \delta}\right)_{S,A} = \left(\frac{\partial U_e}{\partial \delta}\right)_{S,A} = P,$$

$$\left(\frac{\partial U}{\partial A}\right)_{S,\delta} = \left(\frac{\partial U_e}{\partial A}\right)_{S,\delta} + \left(\frac{\partial U_s}{\partial A}\right)_{S,\delta} = G - w.$$

G denotes the variation of elastic energy with A at constant δ, and is called the strain energy release rate.

The three equations of state,

$$S = S(\delta, A),$$

$$P = P(\delta, A),$$

$$G = G(\delta, A),$$

contain the same information as the fundamental equation $U = U(S, \delta, A)$. For

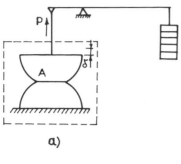

a)

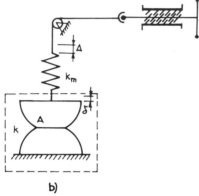

b)

Fig. 10. Study of a system modelling a pair of surface asperities in contact under a tensile load resulting from rolling motion. (a) State of equilibrium at fixed load condition. (b) State of equilibrium at fixed grips conditions.

equilibrium under various constraints, the following thermodynamic potentials can be used:

- At constant temperature, the Helmholtz free energy,

$$F = U - TS,$$
$$dF = -S\ dT + P\ d\delta + (G - w)\ dA.$$

- At fixed load, the enthalpy,

$$H = U - P\delta,$$
$$dH = T\ dS - \delta\ dP + (G - w)\ dA.$$

- At fixed load and temperature, the Gibbs free energy,

$$B = U - TS - P\delta,$$
$$dB = -S\ dT - \delta\ dP + (G - w)\ dA.$$

The problem can be simplified by assuming that thermal effects are negligible $(T = 0;\ S = 0)$. Thus,

$$dF = dU = d(U_e + U_s) = P\ d\delta + (G - w)\ dA,$$
$$dB = dH = d(U_e + U_s + U_P) = -\delta\ dP + (G - w)\ dA$$

noting that $U_P = -P\delta$ is the potential energy of the load P.

Appropriate relations can be written for $G - w$, leading to

$$G = \left(\frac{\partial U_e}{\partial A} \right)_\delta = \left(\frac{\partial U_e}{\partial A} \right)_P + \left(\frac{\partial U_P}{\partial A} \right)_P.$$

Furthermore,

$$\left(\frac{\partial G}{\partial \delta} \right)_A = \left(\frac{\partial P}{\partial A} \right)_\delta$$

and

$$\left(\frac{\partial G}{\partial P} \right)_A = - \left(\frac{\partial \delta}{\partial A} \right)_P.$$

Equilibrium at fixed load conditions $(\mathrm{d}P = 0)$ corresponds to an extremum of F or U and equilibrium at fixed grips condition $(\mathrm{d}\delta = 0)$ to an extremum of B or H. In either case, equilibrium is given by

$$G = w.$$

This equilibrium relation links two of the three variables δ, A and P of the equation of state, so that the equilibrium curves $\delta(A)$, $A(P)$ and $P(\delta)$ are a function of w. If $G \neq w$, the area of contact will spontaneously change so as to decrease the thermodynamic potentials. If $G < w$, it is clear that A must increase, and the crack recedes. Conversely, if $G > w$ the area of contact must decrease to have $\mathrm{d}F < 0$ or $\mathrm{d}B < 0$, the crack extends. $G\,\mathrm{d}A$ is the mechanical energy released when the crack extends by $\mathrm{d}A$. The breaking of interfacial bonds requires an amount of energy $w\,\mathrm{d}A$, and the excess $(G - w)\,\mathrm{d}A$ is converted into kinetic energy if there is no dissipative factor. G is often called crack driving force, but strictly speaking the crack driving force is $G - w$, which is zero when $G = w$ at equilibrium.

The equilibrium can be stable, unstable or indifferent. A thermodynamic system under a given constraint is stable if the corresponding thermodynamic potential is a minimum, i.e. if its second derivative is positive. Stability at fixed grips conditions $(\mathrm{d}\delta = 0)$ corresponds to

$$\left(\frac{\partial G}{\partial A} \right)_\delta > 0$$

and stability at fixed load conditions $(\mathrm{d}P = 0)$ corresponds to

$$\left(\frac{\partial G}{\partial A} \right)_P > 0.$$

The stability problem can be studied with an experimental setup having a finite stiffness, k_m. Figure 10 is a schematic representation of such a machine, and k_m is the stiffness characteristic of the coil spring. A displacement can be obtained by turning the screw (which is thermodynamically equivalent to placing a load P onto the spring and then clamping it without external work) and is divided into elastic displacement, δ_m, of the spring and elastic displacement, δ, of the two solids in contact. The spring exerts a force $P = k_m \delta_m$ on the two bodies in contact, and one has

$$\Delta = \delta + \delta_m = \delta + P/k_m.$$

The stability of the system involving the two elastic bodies in contact and the spring at fixed crosshead displacement, Δ, can now be studied. The energy of the system includes the elastic energy, U_m, of the spring:

$$U = U_e(A, \delta) + U_m(\delta_m) + U_s(A),$$

and its first differential is:

$$dU = \left(\frac{\partial U_e}{\partial A}\right)_\delta dA + \left(\frac{\partial U_e}{\partial \delta}\right)_A d\delta + \frac{dU_m}{d\delta_m} d\delta_m + \frac{dU_s}{d\alpha} dA$$

$$= G \, dA + P \, d\delta + P \, d\delta_m - w \, dA = P \, d\Delta + (G - w) \, dA.$$

Equilibrium at fixed Δ is still given by $G = w$, and the stability by

$$\left(\frac{\partial G}{\partial A}\right)_\Delta > 0,$$

but this stability depends on the stiffness k_m of the apparatus. Intuitively, one can see the spring as a reservoir that provides energy for crack propagation at constant Δ. It is interesting to compute

$$\left(\frac{\partial G}{\partial A}\right)_\Delta$$

as a function of

$$\left(\frac{\partial G}{\partial A}\right)_\delta$$

by considering $G[A, \Delta(\delta, A)]$ as a function of $G[A, \delta(\Delta, A)]$:

$$\left(\frac{\partial G}{\partial A}\right)_\Delta = \left(\frac{\partial G}{\partial A}\right)_\delta + \left(\frac{\partial G}{\partial \delta}\right)_A \left(\frac{\partial \delta}{\partial A}\right)_\Delta.$$

By differentiating the expression for Δ and rearranging the last equation one can obtain

$$\left(\frac{\partial G}{\partial A}\right)_\Delta = \left(\frac{\partial G}{\partial A}\right)_\delta - \left(\frac{\partial P}{\partial A}\right)_\delta^2 \left[k_m + \left(\frac{\partial P}{\partial \delta}\right)_A\right]^{-1}$$

The quantity

$$\left(\frac{\partial P}{\partial \delta}\right)_A = k$$

is the stiffness of the two elastic solids in contact, and is positive. Therefore,

$$\left(\frac{\partial G}{\partial A}\right)_\Delta$$

can be zero, whilst

$$\left(\frac{\partial G}{\partial A}\right)_{\delta}$$

is still positive. It can be, therefore, concluded that the stability range monotonically increases with the stiffness from the fixed-load case ($k_m = 0$) to the fixed grips case ($k_m = \infty$). When $k_m \to 0$ the fixed-load conditions are approached.

Estimation of the adherence or the normal load which the contact formed by a pair of surface asperities can support, is important for the assessment of the rolling friction. Knowing the statistical properties of a rough surface and the mechanism of and the force required to cause junction failure, an estimate of the desired forces may be made. A detailed analysis of this important problem is presented elsewhere [20].

3.3. Rolling contact mechanics of polymeric materials

The stress in polymeric materials in rolling contact is influenced by the rate of strain and, therefore, the contact stresses and deformation will depend upon the speed of rolling. The simplest way to account for time-dependent characteristics of a polymeric material is to model it as a linear viscoelastic material. This has been discussed in section 2.2 in relation to the general response of polymers to applied stresses. However, application of the linear theory of viscoelasticity to rolling contact is not simple, since the situation is not one in which the viscoelastic solution can be obtained directly from the elastic solution. It is not difficult to appreciate the reason for that. During rolling the material lying in the front half of the contact is being compressed, whilst that at the rear is being relaxed. With perfectly elastic material the deformation is reversible, so that both the contact area and the stresses are symmetrical about the centre line. Viscoelastic material, such as a polymer, however, relaxes more slowly than it is compressed, so that the two bodies in contact separate at a point closer to the centre line than the point where they first make contact. This is illustrated in fig. 11, where $z_1 < z_2$ and recovery of the surface continues after contact has ceased. The geometry of rolling contact of a polymeric material is different from that of the perfectly elastic case and, therefore, the viscoelastic solution cannot be obtained directly from the

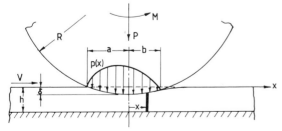

Fig. 11. Rolling of a rigid cylinder on a viscoelastic material.

elastic solution. Moreover, the point at which separation occurs ($x = z_2$) cannot be defined in advance. Usually it has to be located there where the contact pressure drops to zero.

In what follows, a one-dimensional model of the contact between polymeric material and a rigid cylindrical body of radius R shall be presented (see fig. 11). The polymer will be modelled by a simple viscoelastic foundation of parallel compressive elements which do not interact with each other. The rolling velocity is V and the cylinder makes first contact with the polymer substrate at $x = -z_1$. Since there is no interaction between the elements of the foundation, the surface does not depress ahead of the roller. Assuming that $z_1 \ll R$, the compressive strain in an element of the foundation at x is given by:

$$\varepsilon = -\left(\frac{\delta - x^2}{2R}\right)\frac{1}{h}, \quad -z_1 \leqslant x \leqslant z_2,$$

where δ is the maximum depth of penetration of the roller.

In the case of a perfectly elastic foundation characterized by a modulus K, the contact would be symmetrical ($z_1 = z_2$) and the stress σ in each element would be K_ε. The pressure distribution under the roller and the total load would then be given by equations proper for elastic contact. For a viscoelastic material the elastic modulus, K, is replaced by a relaxation function $\Psi(t)$. Thus, the stress in the viscoelastic element at x is given by:

$$p(x, t) = -\sigma = -\int_0^t \Psi(t - t')\frac{\partial \varepsilon(t')}{\partial t'}\, dt'.$$

At steady rolling $\partial/\partial t = V\,\partial/\partial x$, and, therefore, from the equation for the strain,
$$\partial \varepsilon/\partial t = Vx/Rh.$$

Substituting this in the equation for stress in a viscoelastic element and changing the variable from t to x gives

$$p(x) = -\frac{1}{Rh}\int_{-x}^x x'\Psi(x - x')\, dx'.$$

Further analysis requires specification of the relaxation function for the polymer. We use a simple delayed-elasticity model, for which the relaxation function is given by

$$\Psi(t) = K(1 + \beta\, e^{-t/T}).$$

The material described by the above relaxation function has a dynamic modulus $K(1 + \beta)$. Under static conditions the modulus is K, and T denotes the relaxation time. Combining the relaxation function with the equation for stress in a viscoelastic element gives

$$p(x) = \frac{Ka^2}{Rh}\left\{\tfrac{1}{2}(1 - x^2/z_1^2) - \beta\delta(1 + x/z_1^2) + \beta\delta(1 + \delta)[1 - e^{-(1 + x/z_1)/\delta}]\right\},$$

where $\delta = VT/z_1$ represents the ratio of the relaxation time of the material to the time taken for an element to travel through the distance equal to half of the contact width. The contact pressure is equal to zero at $x = -z_1$ and decreases to zero again at $x = z_2$.

The normal load is obtained from the expression

$$P = \int_{-z_1}^{z_2} p(x) \, dx = \frac{Ka^3}{Rh} F_p(\beta, \delta).$$

Because the pressure is distributed asymmetrically, the rolling motion is resisted by a moment given by

$$M = -\int_{-z_1}^{z_2} xp(x) \, dx = \frac{Ka^4}{Rh} F_m(\beta, \delta).$$

Finally, the coefficient of rolling friction can be expressed as

$$\mu_r = \frac{M}{PR} = \frac{a}{R} F_r(\beta, \delta).$$

Examination of the above equations reveals that at slow rolling speeds, when the contact time is long compared with the relaxation time of the material, the pressure distribution and load are quite close to that for a perfectly elastic material with modulus K. At the same time, the moment M is near zero.

At high speeds, however, the pressure distribution and load are close to the elastic-contact results, but with a dynamic foundation modulus $K(1+\beta)$. It is obvious that the relaxation effects play an important role in the behaviour of a contact only when the contact time is approximately equal to the relaxation time of the material. Only then, the contact becomes appreciably asymmetric and a maximum friction moment arises.

3.4. Fatigue considerations in rolling contact of polymers and their composites

The characteristic feature of polymer composite materials which defines, to a large extent, their fatigue behaviour is that they, unlike metals, are inhomogeneous on a gross scale and anisotropic. They tend to accumulate damage in a general rather than a localized way. Moreover, failure does not usually occur by propagation of a single macroscopic crack. The mechanisms of damage accumulation, including fibre and matrix cracking, debonding, transverse ply cracking and delamination, occur sometimes independently and sometimes interactively, and the predominance of one or an other of them may be strongly influenced by both the material variables and the testing procedure and conditions.

At the early stage of the life of a composite material subjected to a cyclic loading, one can expect damage of a very general type only. It will usually be distributed throughout the stressed region and will not have any immediate effect on the strength, although it could reduce the stiffness. The slight reduction, if any, in strength in the early stage of life (also called "wear-out") might be compensated by the so called "wear-in" process leading to a slight increase in strength. This increase in strength of a composite material may arise from an improved fibre

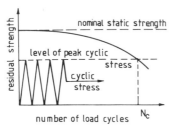

Fig. 12. Effect of wear-out on the residual strength.

alignment allowed for by small, stress-induced viscoelastic deformations in the matrix. With the passage of time, the amount of accumulated damage in a certain region may be sufficiently large as to reduce the load bearing capacity of the composite in that region to the level of peak applied load in the fatigue cycle (see fig. 12). This inevitably leads to a gross failure. When the gross failure is a result of a gradual process, it is referred to as degradation. On the other hand, catastrophic failure is usually termed "sudden-death". Changes of this type are not always related to the propagation of a single crack, and this must be remembered when an attempt is made to interpret fatigue data obtained during testing of composite materials by methods specifically developed for metallic materials.

The presence of a crack in a composite material does not necessarily mean that it will propagate under cyclic loading conditions, since crack propagation depends, to a significant extent, on the nature of the composite material. In glass fibre polymer laminates containing woven roving reinforcement, crack tip damage may remain localized by the complex geometry of the fibre array, and the crack may proceed through this damaged zone in a fashion analogous to the propagation of a crack in plastically deforming metals.

One noticeable feature of fatigue test results is their variability. This variability stems not only from the statistical nature of the progressive damage which leads to fracture of a composite, but more specifically from the highly variable quality that is usually found in many commercial composite materials. Another important problem is associated with the definition of the criterion of failure of a composite material. A traditional definition of failure in fatigue testing of metals states that failure occurs when a complete separation of the broken halves of a sample occurs. This can be quite meaningless in the case of a composite material when the sample has lost its shape integrity and its ability to sustain an applied stress as a result of extensive resin cracking.

The fatigue strength of many composite materials is strongly affected by the loading frequency since it has direct influence on temperature increases caused by hysteresis losses. Heat dissipation by conduction is virtually impossible in many reinforced polymers. In polymeric materials, even quite small temperature increases may lead to significant changes in the mechanical properties, while excessive heating will certainly cause thermal degradation. The response of composite materials to cyclic loading depends also on the size of the tested sample. This is especially evident when the inhomogeneity resulting from the distribution of the reinforcing agent is so great that it influences the fatigue response of a sample which size is comparable with the scale of the inhomogeneity. Therefore, in laminates with a woven roving structure, the width of a test piece should be sufficient in order to include several repeats of the weaving pattern. In a hybrid laminate it is important to make sure that a test sample is an order of magnitude wider than the width of individual fibre tows in the composite.

The role played by holes and notches in fatigue of a composite material is still a matter of discussion. It was found [21], for example, that sharp notches were more harmfull than drilled holes, although notches in general had little effect on the fracture strength due to the large number of debonding sites present in the material. A study by Owen and Bishop [22], on the other hand points to the fact that holes are fully effective in initiating fatigue damage, however, they do not necessarily affect the final failure.

The ability of composite materials to arrest a crack which stems directly from their inhomogeneity on a fine scale (interface between fibre and matrix) as well as on a gross scale (laminated structure), makes it very often difficult to apply fracture mechanics principles to fatigue life prediction and design. However, in composites containing woven cloth or in mouldings containing randomly distributed chopped fibres, notch-like defects tend to propagate in a more "normal" fashion. In such a case it is possible to use fracture mechanics principles in order to predict the crack growth rate.

The damage progressively accumulated in the sample of a composite material during cyclic loading will inevitably affect the macroscopic mechanical properties of the material to an extent depending on the composite geometry and mode of loading.

Cumulative damage of polymer composites resulting from cyclic loading has been subject of a number of studies [23]. All these studies aimed at obtaining a useful working design relationship similar to that proposed by Miner and Palmgren. For resin cracking during cyclic loading, the following non-linear damage law, independent of the stress level, has been proposed [23],

$$\Delta = \sum \left[A \left(\frac{n}{N} \right) + B \left(\frac{n}{N} \right)^2 \right],$$

where n is the number of cycles sustained by the composite at a stress level which would normally produce failure after N cycles. A and B are constants. B is negative and A is equal to unity at failure.

The above model was not unreservedly accepted, and an alternative one has been put forward [24]. This is, basically, a single-parameter, stress-independent damage model, where the damage factor, Δ, is given by

$$\Delta = \frac{e^{kx} - 1}{e^k - 1},$$

and $k = 1$ and $0 < x < 1$.

The model implies that the rate of damage accumulation, $d\Delta/dt$, depends, to a first approximation, linearly on time, t, and t in turn is the non-logarithmic cycle ratio n/N. For different values of k this function produces a family of exponential curves.

It is reasonable to expect an improved fatigue performance from composites reinforced with high-strength, but brittle, fibres, like glass and carbon. This is because the fibres should carry the major part of the load. As cyclic loading continues, however, even small movements in the matrix, resulting from its viscoelastic nature, can lead to local redistributions of stress which allow for some random fibre damage to occur by a normal brittle fracture process. Similar damage will also occur in the surroundings of any stress concentration. The damage accumulation, being rather a function of time than a function of the number of stress cycles, will build up to some critical level when the overall strength of a composite is below the applied peak stress and will lead, eventually, to failure.

When the applied stress is not uniquely uniaxial and aligned with the reinforcing fibres, then more severe loading conditions are imposed on the matrix and the fatigue strength of a composite material is reduced. A stress system of compression combined with shear, which is characteristic for rolling contact, is especially harmful to composites containing aramid fibres [25]. In this loading mode, the matrix and interface would normally be expected to control the fatigue performance of a laminated composite, but not the fibre. On the other hand, the overall fatigue behaviour of hybrid composites ough to be governed by the extent to which the particular fibre mix controls the strain levels in the composite.

It has been demonstrated by a number of researchers that the matrix and interface are the weak links within the polymer composite as far as fatigue strength is concerned. The results suggest that lower-reactivity resins are performing better when exposed to low-stress cycling load [26]. Any treatment or processing leading to an improved resistance of the matrix material to crack propagation will undoubtedly improve the fatigue strength. On the other hand, exposure of reinforced polymers to water and other surfactants often results in some degree of plasticization of the matrix [27] and weakening of the interfacial bonds. Therefore, one can expect that the fatigue strength of some reinforced polymers will be environmental sensitive when the behaviour is not governed exclusively by the fibres.

The fatigue strength of short-fibre composites is considerably lower than that of polymers filled with high volume fractions of continuous, aligned and rigid fibres. This is mainly because the weaker matrix has to support a much greater proportion of the cyclic loading. Localized matrix failures which are easily initiated, can destroy the integrity of the composite even though the fibres remain intact. Since the interfacial shear stresses probably reverse their direction at each load cycle, the interface region is especially vulnerable to fatigue crack initiation and eventual damage. An important cause of damage in reinforced polymers is thermal degradation resulting from large hysteresis losses and subsequent heat dissipated within the material characterized by low thermal conductivity. A fatigue strength of reinforced nylon as low as half the fatigue strength of unfilled nylon was reported [28].

4. Behaviour of polymers and their composites in rolling contact configuration

4.1. Experimental apparatus

The experimental configuration used to test rolling contact fatigue of some polymers consists of three steel balls in contact with a polymer test piece in the form of a hemispherically ended pin. The configuration is shown, schematically, in figs. 13 and 14. The balls were contained in a polymer cup and were free to roll. In this way, the contact conditions characteristic for a ball bearing could be modelled, since the cup represents the outer race, the three balls represent the bearing's rolling elements and the upper test piece can be regarded as the inner race. Bearings of the design as described above are now commercially available [12]. The load on the assembly was applied through a piston and a lever arm by means of a deadweight. The upper test piece was assembled to a spindle via a collet. The spindle was driven by a shunt-type electric motor of which the speed could be continuously changed. A flat belt drive was used to secure the smoothness of operation. The test time and rotational velocity of the spindle were monitored by a timer and a tachometer, respectively. The apparatus could be set to halt at a pre-set number of revolutions or, by using a vibration sensor, when the level of vibration exceeds a pre-set value due to failure within the contact.

4.2. Geometry, load and stress cycle calculations

The geometry of the test assembly is shown in fig. 15. The angle denoted α is a function of the dimensions characterizing the configuration, and is given by

$$\alpha = \arcsin\left(\frac{R_{ce} - R_c}{2R_p + R_a - R_c}\right),$$

where R_c is the radius of the cup race, R_{ce} is the external radius of the cup, R_a is

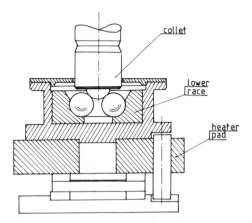

Fig. 13. Schematic representation of the cup assembly in the four-ball apparatus.

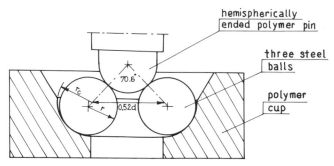

Fig. 14. Dimensions of the ball assembly.

the radius of curvature of the upper test piece and R_p is the radius of the lower
ball (planet). The angle denoted β is defined by the following expression:

$$\beta = \arctan\left(\frac{\sin(90-\alpha)}{\left[R_p/(R_a+R_p)\sin\alpha\right]\cos(90-\alpha)}\right).$$

If the radius of curvature of the upper test piece is equal to the radius of the lower
ball then, by symmetry, angle α will equal angle β. The third characteristic angle
of the configuration is given by,

$$\gamma = 90 - (\alpha + \beta)$$

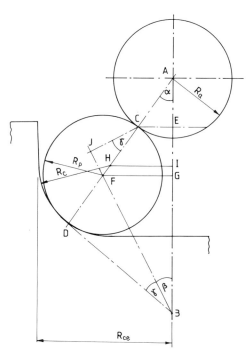

Fig. 15. Geometrical configuration of the contact between the lower and upper test pieces.

The angular velocity of the lower ball (planet) is (see fig. 15)

$$\omega_p = \frac{\omega_a R_a}{2(R_a + R_p)},$$

and its linear velocity is

$$V_p = \frac{\omega_a R_a^2 \sin \alpha}{2(R_a + R_p)}.$$

The lower-ball spin is given by (see fig. 15),

$$V_s = V - V_p,$$

and

$$V_s = \omega_a R_a \sin(\alpha) \left[1 - \frac{R_a}{2(R_a + R_p)} \right],$$

thus,

$$\omega_s = \frac{\omega_a R_a \sin \alpha}{R_p \cos \gamma} \left[1 - \frac{R_a}{2(R_a + R_p)} \right].$$

The contact region between the lower ball and the upper test piece is characterized by a degree of slip resulting from the complex motion of the lower ball. This is shown, schematically, in fig. 15 and, depending on the location within the contact region, given by the following expressions:

   (i) slip at point C is equal to zero (centre of contact),
   (ii) slip at point C′:

$$\frac{V_{C'} - (V_{p'} + V_{s'})}{V_{C'}},$$

   (iii) slip at point C″:

$$\frac{V_{C''} - (V_{p''} + V_{s''})}{V_{C''}}.$$

The magnitude of the slip within the contact area depends, among other things, on the elastic properties of the contacting materials. The slip contributes to such phenomena as heating, softening of the tested polymer and eventually to sliding wear, which must be distinguished from the wear resulting from surface fatigue produced by the rolling motion. The number of load cycles experienced by the upper test piece can be estimated from the relationship

$$L = z \left(\frac{R_a + 2R_p}{2R_a + R_p} \right),$$

where z denotes the number of lower balls in the assembly.

The normal load applied to the contact between the lower ball and the upper test piece is given by

$$N = \frac{196.2 M}{3 \cos \theta_1},$$

where M is the mass applied to the lever arm and θ_1 is the contact angle between the upper test piece and the lower ball.

Knowing the normal load on the contact it is possible to estimate the contact stresses:

(i) The peak compressive pressure (normal to the contact area):

$$p_0 = \left(\frac{6 N E^2}{\pi^3 R^4} \right)^{1/3}.$$

(ii) The principal surface plane stresses:

$$\sigma_x = \sigma_y = -\frac{p_0(1 + 2\nu)}{2}.$$

The contact-region dimensions are

$$a = b = 0.88 \sqrt[3]{\frac{NR}{E}}.$$

The maximum shear stress (at a depth $0.638a$) is

$$\tau_{\max} = \tfrac{1}{3} p_0.$$

The maximum tensile stress (at radius a) is

$$\sigma_{\max} = \tfrac{1}{3}(1 - 2\nu) p_0.$$

4.3. Results of experimental testing

The main objective of the studies [29,30] from which selected results are presented here, was the determination of the polymeric-material response to rolling contact fatigue conditions and the assessment of the suitability of some engineering polymers for rolling contact bearing components. The polymers chosen for testing were:

(i)   nylon 6.6,
(ii) acetal (Hostaform C2521),
(iii) modified acetal (Hostaform C9021K),
(iv) poly(etheretherketone).

Nylon 6.6 is a popular polymer and has already been used successfully in gears, cams, pipe fittings and a number of domestic appliances. Both acetals were kindly supplied by Hoechst Plastics. They have a better dimensional stability than nylon 6.6, mainly because of their low moisture absorption under varying humidity conditions and are very versatile polymers which have found widespread applica-

TABLE 2
Mechanical properties of nylon 6.6.

Tensile strength (MPa)	82.7
Modulus (MPa)	2450
Compressive strength (MPa)	87.4
Modulus (MPa)	1750
Flexural strength (MPa)	91.2
Modulus (MPa)	2450

TABLE 3
Mechanical properties of acetals.

	C2521	C9021K
Yield stress (MPa)	62	62
Flexural stress at 3.5% strain (MPa)	69	68
Flexural strength (MPa)	94	90
Elastic modulus (MPa)	2750	2950
Brinell hardness, 30 s value (MPa)	144	145

tion in the automotive industry. The modified acetal (C9021K) used in the study had improved sliding properties, resulting from an additive in the form of specially prepared chalk.

Poly(etheretherketone) or in short PEEK is a high-temperature engineering polymer. It is chemically identified as an aromatic linear poly(aryletherketone). This thermoplastic polymer is semicrystalline, which, together with its stable, inert chemical structure, accounts for its high resistance to such aggressive agents as organic solvents, acids and bases. It is characterized by a number of notable properties such as a high service temperature (250°C), high mechanical strength and good abrasion resistance.

Some of the mechanical properties of polymers tested are summarized in tables 2, 3 and 4.

Before embarking on the main series of tests it was necessary to decide upon the test conditions and parameters as well as the method of assessing the fatigue damage of a polymer test piece. Therefore, a series of preliminary tests was carried out involving the following:

(i) Applying a load to the stationary polymer pin in contact with the three steel balls in order to determine the onset of gross plastic deformation.

TABLE 4
Mechanical properties of PEEK.

Tensile strength (MPa)	92–114
Compressive strength (MPa)	~120
Melting temperature (°C)	334
Service temperature (°C)	250

(ii) Running short tests under the load determined in the way described in (i) to find out the speed at which there were no signs of polymer melting due to excessive heat generation within the contact. These two preliminary tests were carried out for all polymers tested to determine the magnitude of the load and speed at which long-duration fatigue tests could be run.

A speciment to be tested was placed into the apparatus and subjected to varying loads on the assembly, starting at 38 N. The specimen was then removed and inspected in order to determine whether permanent deformation had taken place. It was found that at a load on the assembly of 154 N only the nylon 6.6 specimen appeared to have an only slight indentation on its surface. Therefore, the acceptable working range of loads was 38–154 N. In order to determine the maximum permissible rotational speed the following procedure was adopted. The specimen was cleaned with a solvent, Genklene, and weighed on a precision balance. The steel balls and raceway were also thoroughly cleaned to ensure that there was no dirt or grit particles, which might lead to abrasive wear, or any traces of oil or grease, which could act as a lubricant. Then the specimen was tested under a load of 38 N at 400 rpm for half an hour under dry conditions and at room temperature. At the end of the test, the specimen was carefully inspected for signs of melting or gross plastic deformation and then weighed again to find out whether any material had been removed from the surface. Since, after this test, there had been no signs of melting or plastic flow as well as no significant weight loss from any of the polymer specimens tested, it was decided to repeat the test with an increased load on the assembly of 77 N. Exactly the same procedure was followed and, after half an hour a negligibly small loss of material was detected. Therefore, it was decided to use a load on the assembly of 77 N as test load.

The main test procedure was as follows:
(i)   The polymer specimen, steel balls and raceway were cleaned with Genklene and the polymer specimen was weighed prior to testing.
(ii) Each test was carried out under a load of 77 N (corresponding Hertzian stress is 34 MPa) and a speed of 400 rpm.
(iii) The tests were carried out under dry and lubricated conditions with Shell Vitrea Oil 100 as a lubricant.

In addition to weighing, the surface of the polymer specimen was inspected under an optical microscope and a scanning electron microscope and photographs were taken in order to identify the prevailing mode of wear.

The first test, under dry conditions, lasting 3 h, showed that the nylon 6.6 specimen underwent excessive wear and it was decided to test this polymer in 2 h intervals. Lubricated tests were carried out as planned, with the only exception that the acetal was tested under a load on the assembly of 154 N (Hertzian stress 68 MPa) because at the lower load no wear was observed.

The test results obtained under dry conditions are shown in fig. 16, where the weight loss of material was plotted against the number of load cycles. First, a significant weight loss incurred by nylon 6.6 should be noted. Damage in the form of deep pits and cracks could be seen on the surface of the wear track. Photographs taken with a scanning electron microscope show the damage more clearly

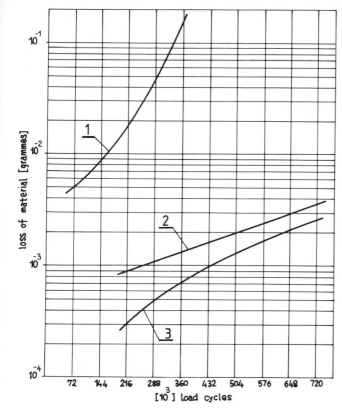

1-nylon 6.6; 2-acetal; 3-modified acetal

Fig. 16. Material loss in gramme as a function of the number of load cycles. Test conditions: speed of the upper test piece is 400 rpm; load at the end of the loading lever is 10 N; dry conditions.

(fig. 17). It can be seen from these micrographs that cracking and flaking of the surface material has occurred. These features point to the fatigue type of wear, since the load on the contact was of cyclic nature.

The weight losses of the acetals were considerably less than that of the nylon 6.6. On inspection, the surface of the contact track had a shiny appearance and no damage could be detected. The photographs taken under a microscope give a more detailed picture. Figure 18a shows the contact track, which is smooth and no serious damage can be seen. Figure 18b shows, at a greater magnification, the same area. Cracks of a characteristic shape are clearly seen, and could be the result of the spin of the three steel balls in contact with the upper polymer test piece. The rate of weight loss at the beginning of the test is slightly slower than towards the end. This is especially true in the case of the modified acetal, which weight loss was equal to that of acetal. Additional tests with nylon 6.6 were carried out in order to establish the time required to produce any visible surface damage. It was found that, under the test conditions adopted, damage with fatigue characteristics probably occurs after 1.5–2 h.

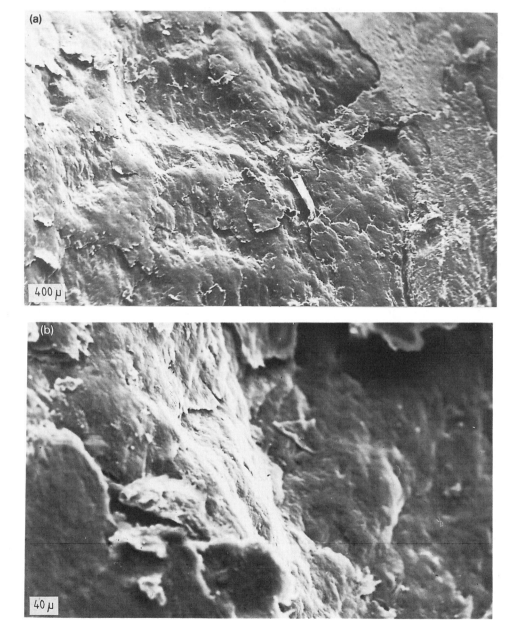

Fig. 17. Scanning electron micrograph of the contact zone on the surface of a nylon upper specimen after 3 h of testing at 400 rpm and a load of 10 N applied at the end of the loading lever. Dry conditions.

PEEK represents a special case and, therefore, deserves a separate discussion. Because of the strength of this polymer the standard test involved applying a load of 125 N to each of the contacts between the hemispherical pin and the three

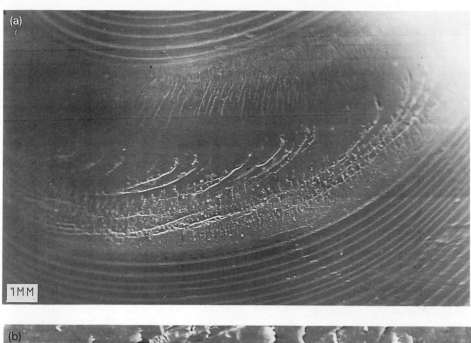

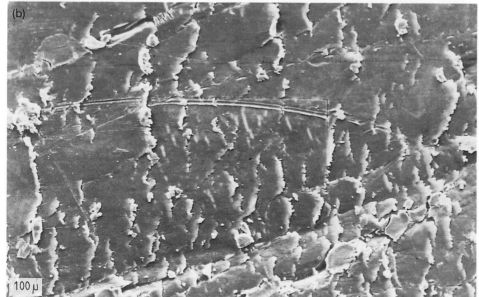

Fig. 18. Scanning electron micrograph of the contact zone on the surface of an acetal specimen after 9 h of testing under conditions as specified in fig. 17.

lower steel balls at 200 rpm of the upper test piece. The duration of the test was typically 20 h, run without lubrication.

Under the conditions of testing employed, no damage resulting from surface

Fig. 19. Appearance of a PEEK upper specimen after 5×10^5 load cycles and a normal load on the contact of 125 N.

fatigue was noted. A typical form of damage was that inflicted by plastic deformation and flow of the material. This is clearly seen in fig. 19, showing the appearance of the PEEK pin after 20 h test, which corresponds to 5×10^5 load cycles. Figure 20 shows the PEEK raceway after a test carried out under similar conditions. Again, no measurable loss of material was recorded. It seems that the combination of contact stress and rolling velocity constitute a clear limitation for the use of PEEK as a material for rolling contact. This is supported by the tests carried out under lubricated conditions. For example, the extent of plastic flow of material was much less when PEEK was tested with a water emulsion as lubricant. The reason for the improvement could be better heat removal from the contact, resulting in less material softening. The results emphasize the potential of PEEK as a material for rolling contacts, provided that the combination of load and speed is properly selected.

The overall performance of the other polymers tested improved considerably under lubricated conditions (fig. 21). In the case of the acetals, no weight losses were recorded after 3 h of testing. Therefore, subsequent tests were carried out at a load on the assembly of 154 N. Under these conditions there was some very small weight loss, but its magnitude was of the order of the weighing accuracy. On inspection, the surface of the contact track had the same shiny appearance as that in the case of the dry tests. This very substantial improvement in performance of the acetals could be attributed to elastohydrodynamic effects within the contact

Fig. 20. Appearance of a PEEK raceway after 5×10^5 load cycles at a normal load on the contact of 125 N.

area, which is quite plausible if the test conditions and mechanical properties of the polymers are taken into account.

Nylon 6.6 also showed improved performance when lubricated (fig. 21). However, after 3 h, damage of a fatigue nature had occurred. When tested for periods of 1 and 2 h, there appeared to be no damage after 1 h but after 2 h some damage, although less developed than that occuring after 3 h, was detected. This is shown in fig. 22 in the form of a scanning electron micrograph. Well-developed pits are clearly seen, pointing to the fatigue type of damage.

5. Concluding discussion

An attempt has been made to demonstrate that the fatigue contribution to the loss of material in fibre reinforced polymer composites depends on the orientation and properties of the fibre and the mechanical behaviour of the matrix. In a brittle matrix, such as epoxy resin, cracks generated in the matrix propagate right through the fibre when the bonding between the fibre and the matrix is strong. In a highly ductile matrix, such as, for example, polyurethane, the cracks cannot propagate through the matrix and the fibre. Therefore, the fibre bends with the matrix under the influence of contact load, and the rate of material removal of this kind of composite is controlled by the wear of the fibre.

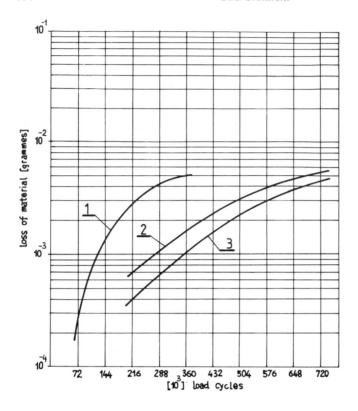

1-nylon 6.6;   2-acetal;   3-modified acetal

Fig. 21. Material loss in gramme as a function of the number of load cycles. Test conditions: speed is 400 rpm; load at the end of loading lever is 10 N (for nylon) and 20 N (for acetals). Lubricated conditions.

Experimental results [31] show that the ends of worn fibres of a graphite–poly-urethane composite are elliptical rather than circular, although the fibres were perpendicular to the surface. Moreover, it was found that the ends of these fibres were very well polished. All that points to the fact that the fibres were worn gradually in very small increments when the fibres are bent under the influence of slip, which is present to some degree in almost every rolling contact.

Test results obtained in a ball-on-polymer disc experiment [32] indicate that for such polymers as polycarbonate, rigid polyvinyl chloride and ultra-high molecular weight polyethylene, the predominant mode of wear was fatigue. However, under appropriate loads, the polymer did not begin to wear until it has been subjected to a certain number of stress cycles. An attempt to correlate the fatigue strength in this type of contact with other mechanical properties of the polymers tested has not been very successful. This is, to some extent, caused by the peculiar nature of polymer composites and the manner in which damage accumulates in them. It is

Fig. 22. Scanning electron micrograph of the contact zone on the surface of a nylon upper specimen after 3 h of testing at 400 rpm and a load of 10 N applied at the end of the loading lever. Lubricated conditions.

important to emphasize that no general conclusions, relating to polymer composites as a class of engineering materials, can be drawn from any set of results obtained for one type of composite.

The limited experimental results on rolling contact fatigue of polymers clearly show that typical surface fatigue damage can occur in certain polymers. This is especially true for nylon 6.6. On the other hand, acetal does not show a similar type of damage, when tested under nominally the same conditions. The mode of failure of PEEK tested under relatively high loads, seems to be plastic flow rather than fatigue. The limit of rolling contact performance of PEEK is probably defined by its ability to dissipate the energy generated at a particular combination of contact pressure and rolling velocity in order to avoid excessive softening. The same observation is probably true for other unfilled thermoplastic polymers.

It is interesting to note that, in general, lubrication of a rolling contact improves the performance of all polymers tested. In the case of softer polymers, such as nylon or acetal, this improvement can be attributed to elastohydrodynamic film formation. Mechanically stronger polymers, such as PEEK, could perform better under lubricated conditions because of improved heat removal from the contact area.

It may finally be concluded that some engineering polymers and their composites have a potential for use in rolling contact applications, provided that the load and speed are carefully selected. Polymers such as acetal and PEEK can certainly be included in this group.

List of symbols

A	area of contact (m^2)
E	Young's modulus (MPa)
G	shear modulus (MPa)
K	bulk modulus (MPa)
k_m	stiffness (N/m)
N	load normal to contact (N)
P	load (N)
p	contact pressure (MPa)
R	radius (m)
T_g	glass transition temperature (°C)
T_m	melting point temperature (°C)
U	energy of a system (N m)
V	linear velocity (m/s)
ε	strain
Δ	damage factor
ω	angular velocity (1/s)
μ_r	coefficient of rolling friction
σ	normal stress (MPa)
τ	shear stress (MPa)

References

[1] R.W. Hertzberg and J.A. Manson, Fatigue of Engineering Plastics (Academic Press, New York, 1988).
[2] E.H. Andrews, J. Mater. Sci. 9 (1974) 887.
[3] J.G. Williams, Adv. Polym. Sci. 27 (1978) 67.
[4] E.H. Andrews and W. Brown, eds, Fatigue in Polymers – Testing of Polymers, Vol. 4 (Wiley, New York, 1969).
[5] J.A. Manson and R.W. Hertzberg, CRC Crit. Rev. Macromol. Sci. (1973).
[6] R.W. Hertzberg, Deformation and Fracture Mechanics of Engineering Materials (Wiley, New York, 1976).
[7] J.D. Ferry, Viscoelastic Properties of Polymers (Wiley, New York, 1970).
[8] F. Bueche, Physical Properties of Polymers (Wiley, New York, 1962).
[9] P. Meares, Polymers: Structure and Bulk Properties (Van Nostrand, New York, 1965).
[10] A.V. Tobolsky, Properties and Structure of Polymers (Wiley, New York, 1960).
[11] H. Leaderman, in: Rheology, Vol. 2, ed. F.R. Eirich (Academic Press, New York, 1958) ch. 8.
[12] P. Eliades, private communication (BNL Limited, Knaresborough, 1988).
[13] O. Reynolds, Philos. Trans. R. Soc. 106 (1875) 155.
[14] R.Z. Stribeck, Ver. Dtsch Ing. 45 (1901) 73.
[15] J. Goodman, Min. Proc. Instn. Civ. Eng. 188 (1912) 82.
[16] A. Palmgren, Ball and Roller Bearing Engineering (SKF Industries, 1945).
[17] G. Tomlinson, Philos. Mag. 7 (1929) 905.
[18] D. Tabor, Proc. R. Soc. A 229 (1955) 199.
[19] K.L. Johnson, Contact Mechanics (Cambridge University Press, Cambridge, 1985).
[20] T.A. Stolarski, Int. J. Eng. Sci. 27 (1989) 169.
[21] K.H. Boller, Mod. Plast. 41 (1964) 145.
[22] M.J. Owen and P.T. Bishop, J. Phys. D 5 (1972) 1621.
[23] R.J. Howe and M.J. Owen, in: Proc. 8th Int. Reinforced Plastics Congress (BPF, London, 1972) pp. 137–148.
[24] J.T. Fong, in: Damage in Composite Materials, ed. K.L. Reifsnider, ASTM STP 775 (ASTM, Philadelphia, PA, 1982) pp. 234–266.
[25] C.J. Jones, R.F. Dickson, T. Adam, H.R. Teiter and B. Harris, Proc. R. Soc. A 396 (1984) 315–338.
[26] M.J. Owen, in: Glass Reinforced Plastics, ed. B. Parkyn (Cliffe, London, 1970) pp. 251–267.
[27] B.J. Briscoe, T.A. Stolarski and G.J. Davies, Tribol. Int. 17 (1984) 129–137.
[28] R.G. Shaver and E.F. Abrams, in: Proc. Annual Technical Conference SPE, Vol. 17 (1971) p. 378.
[29] C.C. Lawrence, Friction and Wear of Polymers in Rolling Contact, report submitted as part of B.Sc. degree, Department of Mechanical Engineering (Brunel University, 1988).
[30] M.P. Stock, Suitability of Thermoplastics for Rolling Contact Bearings, report submitted as part of B.Sc. degree, Department of Mechanical Engineering (Brunel University, 1989).
[31] N.P. Suh and S.M. Burgess, Wear 100 (1984) 81–90.
[32] N.S. Eiss Jr and J.R. Potter, Fatigue wear of polymers, in: Polymer Wear and Its Control, ACS Symposium Series, Vol. 287 (1985) p. 59.

Advances in Composite Tribology
edited by K. Friedrich

Chapter 18

Fretting and Fretting Fatigue of Advanced Composite Laminates

K. SCHULTE

DLR, Köln. Now with: Polymer and Composites Section, Technical University Hamburg-Harburg, W-2100 Hamburg 90, Germany

K. FRIEDRICH

Institute of Composite Materials, University of Kaiserslautern, W-6750 Kaiserslautern, Germany

O. JACOBS

Deutsche Airbus GmbH, 2800 Bremen, Germany

Contents

Abstract

Polymer matrix based composite materials offer substantial improvements over metals for structural applications. Their light weight, high strength and high stiffness make them candidate materials for primary components. Since an increasing number of technical parts are now frequently made of fibre reinforced polymer composites, fretting wear and fretting fatigue problems more and more arise. Both these problems will be discussed in this chapter.

In the fretting wear studies carbon-, glass- and aramid-fibre reinforced epoxy composites were exposed to oscillatory sliding against various metallic counterparts (steel, aluminium, brass and titanium) in a flat-on-flat arrangement. The peak-to-peak oscillation width amounted to 700 μm, the contact pressure was 15 MPa. The formation of an interface layer of wear debris, whose composition was determined by the reinforcing fibre, was found to govern the friction and wear mechanisms. The particulate debris of the glass and carbon fibres were more effective in separating the counterparts than that of the soft polymeric aramid fibres. Furthermore, the carbon fibre, and even more the glass fibre, particles acted as an abrasive interlayer, which increased wear. The abrasive effect of the fibre debris was more detrimental for the softer (aluminium, brass) than for the harder counterparts (steel, titanium). In the case of carbon fibres, the abrasive action was superimposed by the dry lubricating effect of the graphite debris emerging from the mechanical decomposition of the carbon fibres.

In the fretting fatigue studies continuous carbon-fibre reinforced epoxy resin laminates were exposed to tension–tension fatigue loading (R-ratio of $R = 0.1$). Simultaneously, cylindrical metallic pins with flat, polished front surfaces were pressed with an apparent contact pressure between 10 and 22 MPa onto two opposite sides of the specimen. The mechanisms of damage development of carbon fibre-reinforced laminates in such a fretting fatigue situation were studied and a quantitative measure for the degree of fretting fatigue damage was introduced. The fatigue life of the composite could drastically be reduced by an additional fretting component, if load bearing 0°-layers were damaged by fretting. Fretting fatigue damage was found to penetrate proportionally with time into the bulk material; the action of notch effects could be denied. However, cracking and delamination of fibre layers predamaged by fretting caused synergistic interaction

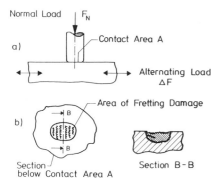

Fig. 1. Principle of the fretting fatigue combined load (a) and possible damage of the specimen surface due to fretting (b) [3].

between fretting and fatigue loading. The influence of the loading conditions was investigated.

1. Introduction

Favourable specific mechanical properties of polymer matrix composites reinforced with continuous fibres have made them an attractive material for application in many engineering structures. The reason for using these materials are mainly weight reduction, better fatigue performance, corrosion resistance, good vibration damping and/or design and manufacturing advantages.

Relative cyclic motion of two surfaces in intimate contact constitutes the condition for a wear process known as fretting. This wear process can often result in a significant reduction in the fatigue resistance of a particular material [1]. A fretting fatigue situation occurs when repeated loading of a structural part causes oscillatory sliding movement at material interfaces in the design. This movement, in turn, must induce at or near the points of contact between the faces stresses of sufficient intensity to cause surface damage, for example in form of cracking [2]. A schematic of the principle of the process and the resulting mechanisms is given in fig. 1 [3].

It is known from numerous engineering applications that these fretting fatigue requirements are often encountered, and that they have often led to failure in an unpredictable manner. The most common examples are flanges and bolted joints [4]. Other design details that can give fretting fatigue problems are multi-layer leaf springs, collars and bushes [5]. In the past, most of these structural components have been made of metallic materials. Therefore, it is not surprising that most of the scientific investigations which concentrated on the problem of fretting fatigue, were performed with metals.

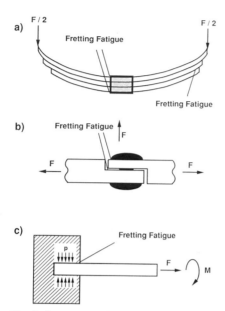

Fig. 2. Some examples of components made of polymer matrix composites where fretting fatigue can occur. (a) Multilayer leaf spring. (b) Bolted joint. (c) Interference or shrink fit.

In recent years it has been found that many of the design examples mentioned above are now frequently made of polymer composites, consisting of a certain type of fibre reinforcement in a polymeric matrix. The use of these materials often requires some changes in the design principles relative to those applied for manufacturing these components of metallic materials. But even in spite of differences in the design, it cannot be neglected that similar fretting fatigue load collectives may act on structural composite components as well. Figure 2 gives some technical components nowadays made of polymer composites. The analogies between metals and composites in applications are obvious, and it can be expected that these composite structures may also experience simultaneous fretting and fatigue-loading situations during their practical use.

First, the pure fretting wear performance of several fibre/matrix composites was studied. On this basis the fretting fatigue behaviour of continuous fibre laminates was compared with the fatigue performance. A data-reduction method for the quantitative analysis of the fretting fatigue damage was developed. Figure 3 gives an overview of the scheme used in this study [6]. Surface damage via fretting wear influenced the fatigue behaviour and vice versa the fatigue behaviour influ-

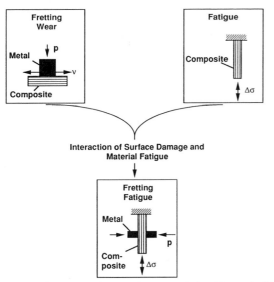

Fig. 3. Overview of the scheme used in this study [6]; p = contact pressure, $\Delta\sigma$ = fatigue stress amplitude, ν = frequency.

enced the wear process in the contact area. Fretting fatigue was the result of fretting wear and fatigue interaction.

2. Experimental

2.1. Material investigated

2.1.1. Materials used in the fretting wear tests

The samples consisted of a polymer matrix reinforced by unidirectionally oriented continuous fibres. The toughened epoxy resin R 5212 (EP) from BASF was employed as matrix system. Carbon (CF), aramid (AF), and glass fibres (GF), respectively, were chosen as reinforcements. The fibre volume content amounted to about 60%. Table 1 lists all laminates used in the fretting wear tests.

TABLE 1

Materials employed in fretting wear tests. EP = epoxy resin R 5212, CF = carbon fibres (T 300), AF = Kevlar 49, GF = E-glass fibres.

Designation	Fibre volume content V_F (%)	Density ρ (g/cm^3)	Manufacturer
EP	–	1.17	BASF
GF/EP	61	2.06	BASF
CF/EP	59	1.62	BASF
AF/EP	62	1.36	BASF

TABLE 2
Materials tested under fretting fatigue load (HC = hybrid composites).

Code	Matrix	Fibre	Stacking sequence	Fracture stress (MPa)	Young's modulus (GPa)
CF1	BASF R 5212	T 300	$[0^\circ_2, 90^\circ_2]_{2s}$	850	72
CF2	Ciba Geigy 914 C	T 300	$([0^\circ_2, \pm45^\circ, 0^\circ_2, \pm45^\circ, 90^\circ]_s)_3$	799	70
AF	BASF R 5212	Kevlar 49	$[0^\circ_2, 90^\circ_2]_{2s}$	648	36
HC	BASF R 5212	K 49, T 300	$[0^\circ(AF)_2, 90^\circ(CF)_2, 0^\circ(CF)_2, 90^\circ(CF)_2]s$	629	59
CF3	Ciba Geigy 914 C	T 300	$[\pm45, 0, \pm45_3, 90, \pm45]_{2s}$	296	27
CF4	Ciba Geigy LY 556	HTA	$[0_2, 90_2, 0_2, 90_2]_s$	996	77

Cylindrical metal pins (diameter of 5 mm) with flat front surfaces served as fretting pins. The front surfaces were ground and polished. The final surface roughness amounted to about 0.12 μm. The edges of the front surfaces were rounded off in order to avoid cutting of the laminate by sharp edges. Prior to starting the tests, the pins were cleaned with acetone. A low-carbon NiCr-steel (Vickers hardness of HV = 296), a titanium alloy (TiAl6V4, HV = 305), an aluminium alloy (HV = 135) and brass (HV = 160) were chosen as pin materials.

2.1.2. Materials used in the fatigue and fretting fatigue tests

The samples consisted of an epoxy matrix reinforced with aligned continuous carbon fibres and aramid fibres. The fibre volume content amounted to about 60%. Table 2 lists the various laminates tested. With one exception, the covering layers had 0°-orientation. In addition, a hybrid laminate with aramid fibres in the surface plies was investigated. Figure 4 is an example of the microstructure of a cross-ply $[0_2, 90_2, 0_2, 90_2]_s$ laminate.

Most of the laminates were manufactured using a peel ply. This is a layer of an open-weave material which separates during curing the laminate from the bleeder cloth, which picks up excessive matrix resin. After curing the peel ply is removed, but it leaves a resin-rich layer, with the impression of the fabric on the surface of the sample.

For the fatigue and fretting fatigue tests, the laminates were cut with a diamond saw to rectangular beam-shaped test pieces (free length of $l = 280$ mm, thickness between 2 and 6.3 mm, width between 6.3 and 8.3 mm), which were clamped to the servohydraulic test-machine by grips. Aluminium end tabs enabled easy load transfer and protected the specimen surface against the action of the grips.

Since fretting fatigue is a surface phenomenon, cylindrical metal pins (the same as those used in the fretting wear tests) with flat front surfaces were pressed against two opposite surfaces of the specimen, thus serving as fretting pins. In

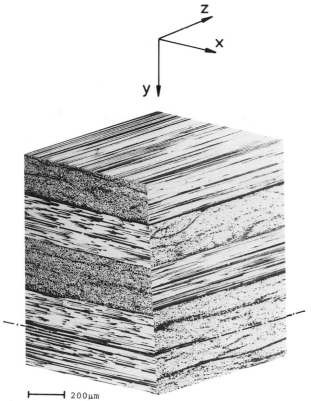

Fig. 4. Cubic section of the laminates investigated, with abraded covering layer: laminate CF4, half lay-up, stacking sequence $[0_2, 90_2, 0_2, 90_2]_s$.

addition, also a carbon-fibre reinforced laminate (CFRP) was used in some fretting fatigue tests as pin material. The stacking sequence of the laminate from which the CFRP pins were produced was: $[0_4, +45, 90, -45, 0_4, +45, 0_4, +45, 90, -45, 0_4]_s$. They were pressed with the ends of the 0°-fibres against the laminates to be investigated. Figure 5 shows a light micrograph of a section of the fretting surface of a CFRP pin before utilisation.

In contrast to metallic materials, polymer composite laminates can have three-dimensional anisotropy, with the properties parallel to the length axis of the samples quite different from those perpendicular to their flat surfaces or to their edges. Since fretting is a surface phenomenon, a fretting component can, there-fore, act quite differently when applied either to the flat surfaces of a specimen with a square or rectangular cross section, or to its free surfaces at the edges. Therefore, the application of the fretting attack was performed in both directions, i.e. either perpendicular to the flat surfaces (position 1, fig. 6) or perpendicular to the free edges (position 2, fig. 6).

├──┤   200µm

Fig. 5. Light micrograph of a section of the fretting surface of a CFRP pin before utilisation.

2.2. Fretting wear testing

Fretting wear studies were performed using a fretting test device commercially available from Optimol Instruments (SRV System). A schematic of the system is given in fig. 7. Specimens of the composite were fixed on a quartz force measurement unit. The fretting pins were pressed onto the specimen surface and slid oscillatorily with the sliding direction parallel to the fibre orientation in the specimen, of which fibres and matrix form the surface. Friction causes shearing of the quartz element, whose piezoelectric response was used as a measure for the

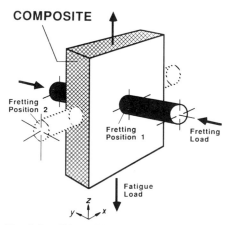

Fig. 6. Possible positions of the fretting pins on a flat test coupon.

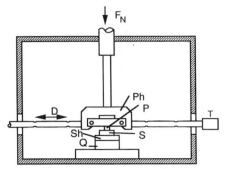

Fig. 7. Schematic illustration of the SRV fretting test system of Optimol Instruments. F_N = normal load, P = fretting pin, Ph = pin holder, S = specimen, Sh = specimen holder, Q = quartz element, D = inductive drive, T = inductive displacement transducer.

tangential force acting on the sample. The pin displacement could be measured by an inductive displacement transducer. With this setup, a graph of the coefficient of friction for each load cycle and peak-to-peak averaged as a function of the testing time could be recorded.

The total displacement of the fretting pin includes elastic deformation due to the compliance of the whole apparatus as well as real sliding. The continuous recording of the friction force versus pin displacement allowed separation of the real sliding phase from the elastic deformation and enabled the control and adjustment of the sliding amplitude. The standard testing conditions were: normal load $F_N = 300$ N (contact pressure $p = 15.3$ MPa), frequency $\nu = 20$ Hz, full sliding width from turning point to turning point of the motion $A = 700$ μm. Loading parameters deviating from these standard conditions but applied in some of the studies, are specially mentioned in the text or figure captions.

The mass loss was measured with a Sartorius balance (accuracy of 0.01 mg), after removal of loose wear particles from the fretted region with the aid of an air blowing device. Reference specimens were simultaneously balanced in order to separate the weight changes due to absorption and desorption of moisture. From the net mass loss, Δm (Δm = mass loss − mass changes due to sorption), the specific wear rate, $\dot{w}_s$, was calculated according to eq. (1). The inverse of the specific wear rate is called wear resistance.

$$\dot{w}_s = \frac{\Delta m}{\rho L F_N}, \tag{1}$$

where ρ is the density of the sample material, $L = 2A\nu t$ the total sliding distance, ν the frequency, A the full oscillation width and t the testing time.

2.3. Fretting fatigue testing

Fretting fatigue studies are usually performed in order to explore how an additional fretting load may affect the fatigue performance of a material under

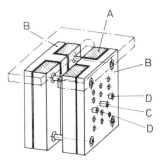

Fig. 8. Positioning device for fretting pins: A = steel block, B = plates, C = fretting pin, D = pins for holding the fretting load frame in position.

certain fatigue-loading conditions. Fretting fatigue testing should be carried out in the same servohydraulic test-machine in which also the plain fatigue experiments were performed.

Derived from a fretting fatigue test device used by Gaul and Duquette [7], for this investigation a system was developed, which can be attached to the grips of the servohydraulic test-machine [8]. The system consists of a positioning device (fig. 8) for cylindrical fretting pins and a fretting-load frame (fig. 9a). The positioning device was fixed to the upper grips. This device consists mainly of a steel block (A) into which two mutually perpendicular slids were machined, to which two plates (B) can be fixed opposite to each other. The plates can be positioned with their central holes either parallel or perpendicular to the length axis of the gripping system. Each plate contains five central holes (C), through which two fretting pins can symmetrically be pressed against two opposite sides of the specimen. Thus, by changing the gauge length, five different slip amplitudes can be selected. Besides the holes for the fretting pins, the plates contain two parallel rows of holes for positioning the fretting-load frame (D).

The fretting load frame allows a symmetrical loading, exact adjustment and control of the fretting load. The heart of this system is a small load cell on one side and a screw-adjustable coil compression spring on the other side. Depending on the stiffness of the coil spring, the fretting load can self-reduce more or less quickly with material removal at the fretting contact points. Since the load is permanently controlled by the load cell, readjustments can be made from time to time in order to maintain a constant contact pressure during the test.

Figure 9b is a photograph of the whole system. The upper grips are connected to the load cell of the servohydraulic test-machine, while the lower grips are attached to the actuator applying the fatigue load. The fretting pins (G) are pressed against the specimen (F). The cyclic straining of the laminate causes a relative motion between the fretting pins and the specimen.

For this investigation, tension–tension fatigue load was applied with a frequency of 10 s^{-1} at an R-ratio of $R = \sigma_l/\sigma_u = 0.1$ (σ_u is the upper and σ_l the minimum

a)

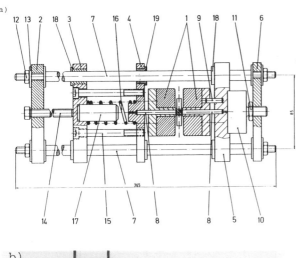

b)

Fig. 9. (a) Assembly of the fretting load frame: 1 = cross section through the positioning device, 2–6 = tension plates, 7 = guide rod, 8 = fretting pins, 9 = positioning pins (D in fig. 8), 10 = load cell, 16 = coil spring, 17 = stabilisation bolt, 18 and 19 = ball bearings. (b) Assembly of the complete fretting fatigue device: A = upper and lower grips, P = positioning device, C = fretting fatigue load frame with coil spring (D) and load cell (E), F = test sample, G = fretting pins.

fatigue stress in each consecutive load cycle). The relative slip, A, between the pins and the laminate can be estimated according to

$$A = \frac{\Delta\sigma}{\sigma_T}\varepsilon_T l_1,$$ (2)

with $\Delta\sigma$ the stress amplitude, σ_T the ultimate strength, ε_T the strain to failure (1.2%) and l_1 the distance of the fretting pins from the upper, fixed clamping grip.

For a stress amplitude of 700 MPa on laminate CF1 (upper fatigue stress $\sigma_u = 780$ MPa), eq. (2) gives a slip of 650 μm. A measurement with a mechanical

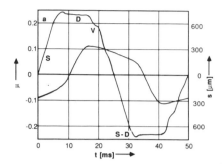
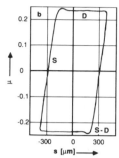

Fig. 10. Course of tangential force during one single load cycle for fretting wear of CF/EP against steel. (a) Friction coefficient, μ, and pin displacement, s, versus time, t. (b) Friction coefficient as a function of the pin displacement. S = phase of static friction; S–D = transition from static to dynamic friction; D = real sliding; V = elastic vibrations.

displacement transducer resulted in a value of about 700 μm, which is slightly higher than the calculated value.

X-ray radiography visualised the development of cracks and delaminations inside the laminate. For this purpose, the edges of the specimens were spread with a zinc iodide solution as a contrast agent which easily penetrated into cracks and delaminations.

3. Fretting wear performance

3.1. The load cycle

Figure 10 illustrates the course of the tangential force during one load cycle in a fretting wear test, which is typical for the materials listed in table 1. The driving force obeyed a sinusoidal function. At the beginning of motion there was a phase of static friction (S), during which the specimen, the counterpart and their fixtures elastically deformed. Above a particular value for the oscillation width the elastic deformation force exceeds the static-friction force and the fretting pin starts to slide across the specimen surface (S–D). Subsequently, the system changes over to real sliding (D), while the tangential force diminishes to the value for dynamic friction. At the turning point of motion the pin adheres again to the sample and the whole system performs characteristic elastic vibrations (V), which can only be observed in the μ–t diagram because in the μ–s diagram they would show up as movements on the straight line of elastic deformation.

It should be mentioned that the friction record presented in fig. 10 is characteristic for hard-elastic materials. Thermoplastics with a softening point (glass transition temperature T_g) in the region of or below the test temperature produce friction curves as shown in fig. 11. The static value of the coefficient of friction is lower than the dynamic value, which again is dependent on the sliding speed [9].

Only the phase of real sliding (D) can contribute to the actual wear process. The elastic deformations (S) may cause at the most some subsurface fatigue. Figure 12

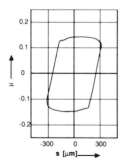

Fig. 11. Graph of the tangential force during one single load cycle for fretting wear of neat PA 6.6 ($T_g \approx 50°C$) against steel [9].

shows an EP specimen which was subjected for 30 h to fretting with an amplitude slightly below the transition to sliding. The main part of the fretting contact exhibits a smooth and undamaged appearance, even after long-term loading. Only around some impurities included in the interface and along the rim of the impression, where the contact pressure was reduced, small scratches indicate that real sliding with a small amplitude took place. Here, the small slip produced a formation and extrusion of polymeric rolls and scratches, but the mass loss was too small to be measured gravimetrically. However, the surface cracks may act as notches when the material is simultaneously subjected to a tensile load.

3.2. Effect of fibre reinforcement

The surface conditions of the two rubbing bodies and the interface composition are a function of time. In general, the tribological process can be subdivided into a

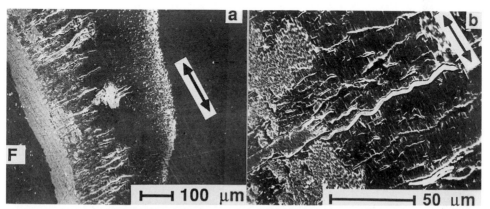

Fig. 12. Scanning electron micrographs of a neat EP specimen subjected to fretting with an amplitude slightly below transition to real sliding. The arrow indicates the direction of motion, the letter F the fretting contact region. (b) is a magnification of a detail of (a).

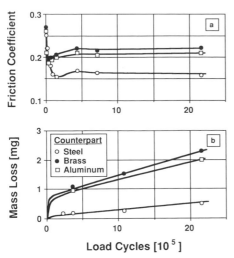

Fig. 13. Fretting of CF/EP. (a) Coefficient of friction averaged from peak to peak. (b) Mass loss of composite versus cycle number for different counterparts.

running-in period and a steady-state period, but the particular physical processes contributing to the several phases of the wear process vary significantly with the materials considered.

3.2.1. CF composites

Figure 13a presents the peak-to-peak averaged friction coefficient of CF/EP fretted versus steel as a function of the number of load cycles. the initial value of the friction coefficient was on a rather low level, near 0.1 (not resolved in fig. 13a because of the large time scale). The friction graph had a nearly quadrangular shape at the beginning of the fretting process with no peaks at the transition points from static to dynamic friction (fig. 14a). This referred to a low contribution of adhesive mechanisms to the friction process. The reason is that initially the real contact area of the two rubbing bodies was very small due to the micro-rough surface structures [10].

During the first 2000 load cycles, the asperities in the tribocontact were evened up and the real contact area grew. Therefore, not only the friction coefficient drastically increased, to a value near 0.25 (fig. 13a), but, simultaneously, a pronounced adhesion peak developed in the friction graph at the transition from static to dynamic friction (fig. 14b). The latter indicates an increasing contribution of adhesive friction mechanisms. This mutual adaption of the rubbing surfaces to each other was accompanied by fast material removal (fig. 13b).

Once the wear process has commenced, the wear debris was entrapped in the contact region [11] and this led to a gradual separation of the rubbing bodies [12]. Adhesive friction mechanisms became less effective. The friction coefficient decreased (fig. 13a) and the adhesion peak in the course of the tangential force during one load cycle gradually disappeared after some hundred thousand cycles

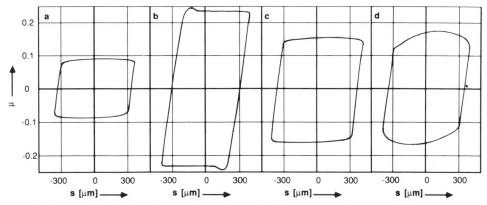

Fig. 14. Fretting of CF/EP against steel. Development of the course of the tangential force during one single load cycle with time, (a) immediately after test start, (b) after 2 min (2400 load cycles), (c) after 2 h (150 000 load cycles) and (d) after 25 h (1.8 million load cycles).

(figs. 14c and 14d). Instead, the formation of a "third body" in the interfacial region caused a transition from solid-body friction to a kind of dry lubricated friction. Accordingly, the tangential force did no longer remain constant during the sliding period of the load cycle but varied with sliding speed (fig. 14d).

The rate of material removal became constant when the coefficient of friction reached steady-state conditions (fig. 13b). The mass loss proceeded faster during fretting against the soft aluminium counterpart than for the harder steel pins. However, the difference in mass loss was smaller for steady-state conditions than for the running-in periods (fig. 15). The partial separation of the rubbing bodies by entrapped wear debris reduced the real contact area between pin and specimen and, thus, the influence of the counterpart material.

3.2.2. GF composites

Figure 16a shows the variation of the friction coefficient for GF/EP versus steel due to the number of load cycles. Distinct differences with the behaviour of

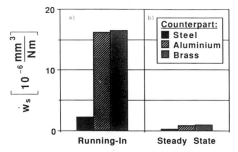

Fig. 15. Fretting wear of CF/EP against different counterparts. Specific wear rate, $\dot{w}_s$, averaged over the first hour of the test, $\dot{w}_{s,i}$ (a), and at steady-state conditions, $\dot{w}_{s,s}$ (b).

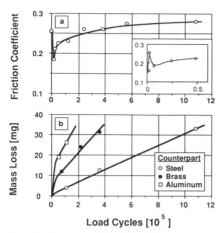

Fig. 16. Fretting of GF/EP. (a) Coefficient of friction averaged from peak to peak (counterpart steel). (b) Mass loss of composite versus time for different pin materials.

CF/EP against steel (fig. 13a) can be seen. The maximum of the friction coefficient (indicated in fig. 16a) at the beginning of the fretting wear test was equal to about $\mu = 0.25$; however, the maximum was very small. The friction coefficient dropped to a minimum and then increased again to the relatively high value of $\mu = 0.28$, where it remained for the further test.

The variation of the friction coefficient during different load cycles is shown in fig. 17: while the initial friction coefficient is relatively low (fig. 17a), after a few cycles the first maximum is reached (fig. 17b). At this time, the adhesion peak at the transition from static to sliding friction is not as pronounced as in the case of CF/EP against steel (fig. 16b). The probable reason is that the brittle glass fibres

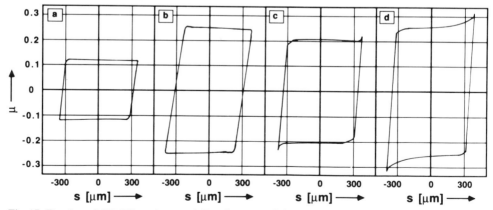

Fig. 17. Fretting of GF/EP against steel. Development of the course of the tangential force during one single load cycle with time, (a) immediately after test start, (b) after 2 min (2400 load cycles), (c) after 2 h (150 000 load cycles) and (d) after 25 h (1.8 million load cycles).

quite rapidly fracture, thus leading to fast formation of a debris granulate separating the sliding surfaces. This assumption is also supported by the high initial rate of material removal (fig. 16b).

A new feature of the course of the tangential force during one load cycle arose after the running-in process: a sharp peak at the turning point of motion (fig. 17d). This peak originated in bumping of the fretting pin against the wall of the hole produced by fretting wear. The appearance of this peak required the existence of a sufficiently deep impression of the fretting pin. This was even more pronounced for fretting wear of AF and GF composites, which were worn rather fast. For CF/EP this peak occurred only under conditions which caused high wear rates (very long test period, aluminium or brass pin, high contact load, high frequency).

The steady-state coefficient of friction was about twice the value as measured for CF composites. This points to a smaller lubricating effect of the GF debris in comparison to the CF debris, while the GF particles cause an additional high contribution of abrasive mechanisms to the friction process.

3.2.3. AF composites

Aramid fibres consist of a highly oriented liquid crystal polymer. In contrast to glass and carbon fibres, the AF fibres are rather soft and flexible. Under wear loadings, AF tends to fibrillate and smeer-off, whereas no brittle fracture occurs. The wear debris did not build up as granulates but formed flat thin flakes [13]. This type of interlayer, of course, is less mobile than the powder of GF and CF and may strongly adhere to one or both of the contacting surfaces. The debris patches separate the rubbing bodies (forming an interface between the composite and the fretting pin), but cannot accommodate the relative motion. The tribological process, thus, remains that of solid-body friction during the full testing time.

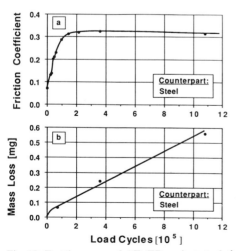

Fig. 18. Fretting wear of AF/EP against steel. (a) Coefficient of friction averaged from peak to peak. (b) Mass loss of composite versus time.

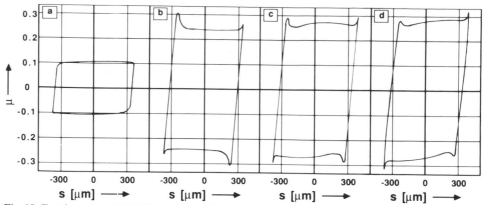

Fig. 19. Fretting wear of AF/EP against steel. Development of the course of the tangential force during one single load cycle with time, (a) immediately after test start, (b) after 0.5 h (36 000 load cycles), (c) after 2 h (150 000 load cycles) and (d) after 15 h (1.08 million load cycles).

Figure 18a shows the variation of the friction coefficient due to the number of load cycles. The curve does not show an initial peak, but gradually increases to the high, stable value of $\mu = 0.32$, which is even higher than the peak values for CF/EP and GF/EP.

Figure 19 shows how the course of the friction force during a single load cycle develops with time. The initial increase of the friction coefficient, which is related to the growth of the real contact area (as indicated by the comparison of fig. 19a and 19b), is slower than in the case of CF and GF composites. Afterwards, the shape of the friction curve changed only little (figs. 19c and 19d) and the averaged friction coefficient remained on an almost constant level at about $\mu \approx 0.32$. This points to a dominance of adhesive friction mechanisms and supports the denial of a lubricating effect of the interface layer built up from the AF wear debris. Accordingly, the adhesion peak in the friction graph (figs. 19c, d) is strongly pronounced and remains so over the total testing time. The small changes in the friction mechanisms during fretting of AF composites were responsible for the relatively constant rate of material removal (fig. 18b).

3.2.4. Comprehensive comments

Friction and wear (even if unlubricated) are no two-body but rather a "three body" problem [14]. This is especially important for fretting conditions, where small slip amplitudes favour the entrapment of debris and where an interphase is formed between the fretting wear partner [15]. In fretting under steady-state conditions the bulk material properties become less important, because the wear process is governed by the interphase formed from the wear debris [16]. In the previous section it was shown that different fibre reinforcements produce different interphases, leading to different interface dynamics:

– CF and GF composites lead to the formation of debris in granulate form. The relative motion between the rubbing surfaces is accommodated by a kind of "shear flow" within the powder bed.

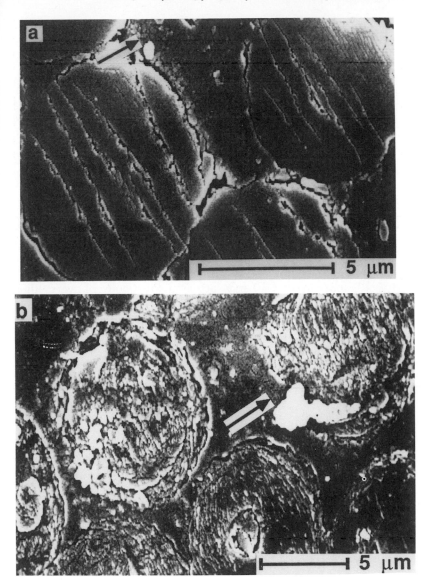

Fig. 20. Scanning electron micrographs of CF/EP subjected to sliding wear against smooth steel. Fibres oriented normal to the sliding plane. The arrow indicates the sliding direction. (a) Development of cracks in the fibres transverse to the sliding direction; advanced state of damage development. (b) The single graphite crystals are more or less separated. The crystallites are arranged tangentially at the surface and randomly in the interior, as typical for PAN-based carbon fibres.

– CF debris exhibits some lubricating effect [17]. For GF composites the lubricating effect is superposed by the additional action of strong abrasive friction mechanisms. The micrographs in fig. 20 show that the carbon fibres can be decomposed into graphite crystals, which are frequently used as a solid lubricant

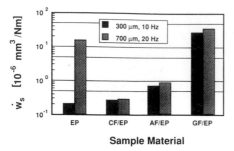

Fig. 21. Specific steady-state wear rates of the materials fretted against steel at two different loading conditions.

[18]. This is possible because the chemical bonds transverse to the fibre axis are weak van der Waals forces. In contrast, the glass fibres consist of a three-dimensional network of covalent bonds, which leads to hard and sharp-edged particles after fracture.
- The fibre particles (especially of GF) may act as a third-body abrasive, thus superposing the beneficial lubrication effect.
- AF composites build up an interface layer of relatively large, flat flakes, which are situated at the surfaces of the specimen and the counterpart instead of forming a lubricating interphase.

3.3. The wear rates

3.3.1. Effect of fretting wear velocity

The histogram in fig. 21 compares the specific steady-state wear rates of the several EP composites exposed to fretting against steel. At mild loading conditions (frequency $\nu = 10$ Hz, oscillation width $A = 250$ μm; low fretting wear velocity) the lowest wear rates are found for the neat matrix resin. Any type of fibre reinforcement accelerates material removal. However, the specific wear rates of the CF composite were only slightly above those for neat EP. The highest wear rates were found for GF composites, and some intermediate wear rates for AF composites.

The SEM micrograph in fig. 22a shows that neat EP predominantly was worn by surface fatigue and adhesive mechanisms. When carbon or glass fibres are incorporated, they tend to crack under wear loading. The fibre particles may act as a third-body abrasive. Accordingly, the micrographs of fretted CF and GF composites show much more scratches than are seen for neat EP specimens (fig. 22b). The carbon fibres are less abrasive than glass fibres, because they easily can be decomposed to small graphite crystals (fig. 20). Aramid fibres are flexible and soft due to their polymeric nature and are, therefore, not very abrasive. The micrograph in fig. 22c shows that the worn surfaces of AF composites are very smooth, but, on the other hand, the poor fibre–matrix bonding, which is typical for AF composites [19], causes extended fibrillation of the aramid fibres [20].

Fig. 22. (a) Optical micrograph of a plain EP specimen after fretting against steel (amplitude $A = 250\ \mu m$, frequency $\nu = 10$ Hz). The arrow indicates the sliding direction. Scratches are visible only along the rim of the impression. The main part of the contact zone shows a dense system of cracks. (b) Light micrograph of a CF/EP specimen after fretting against steel. The arrow indicates the sliding direction. $A = 250\ \mu m$, $\nu = 10$ Hz. (c) Scanning electron micrograph of an AF/EP specimen after fretting against steel. The arrow indicates the sliding direction. The fibrillation predominantly along the rim of the fretting contact region is typical for AF composites.

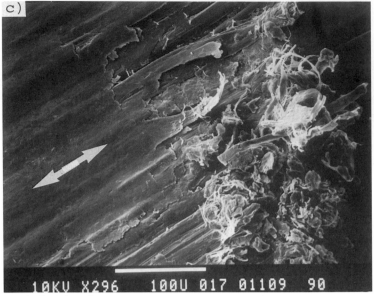

Fig. 22 (continued).

The difference between running-in and steady-state wear rates (table 3) was very small for neat EP and AF/EP and significant for GF and even more for CF composites, which produce debris in the form of loose granulate. The high lubricating efficiency of the CF composites and their low abrasiveness cause a strong reduction in the wear rate during the transition from running-in to steady state. The strongly abrasive effect of the GF particles was responsible for a rather high level of the wear rate even at steady state.

At more severe loading conditions ($\nu = 20$ Hz, $A = 700$ µm; high fretting wear velocity) neat polymer was worn very rapidly (fig. 21). Now a fibre reinforcement acted beneficially. The specific wear rates of the CF and AF composites are slightly higher, but almost the same as under mild loading conditions. Only the composites reinforced with the strongly abrasive GF still exhibit higher wear rates than the pure EP.

TABLE 3

Fretting wear against steel. Specific wear rates as averaged over the first hour of the test, $\dot{w}_{s,i}$, and during steady state, $\dot{w}_{s,s}$, for all materials tested. The last column contains the ratio $\dot{w}_{s,i} / \dot{w}_{s,s}$.

Material	$\dot{w}_{s,i}$ (10^{-6} mm³/N m)	$\dot{w}_{s,s}$ (10^{-6} mm³/N m)	$\dot{w}_{s,i} / \dot{w}_{s,s}$
EP	12.3	11.0	1.1
GF/EP	53.9	34.4	1.6
CF/EP	2.23	0.28	8.0
AF/EP	1.11	0.85	1.3

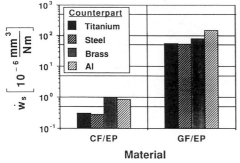

Fig. 23. Specific wear rates of CF and GF composites for fretting against several counterparts. Standard loading conditions.

The intense brown colouring of the EP surface after fretting pointed to the occurrence of thermal decomposition. Additionally, the friction heat promotes creeping of the polymer. The incorporation of fibre reinforcement improves the thermal conductivity of the polymeric sample, thus reducing the heat concentration in the fretting contact, and enhances the load-carrying capacity and resistance to creep [21]. This explains why the composites exhibit a higher wear resistance than neat EP at severe loading conditions.

3.3.2. Effect of counterpart material

The specific wear rates of EP and AF/EP did not significantly depend on the counterpart material. This is because the metallic counterparts, anyway, are much harder than the samples.

The fretting wear of CF/EP and GF/EP was strongly affected by the counterpart. As fig. 23 shows, steel and titanium pins, which have almost the same hardness, lead to equal specific wear rates. The soft aluminium pins produce a far more rapid material removal; the specific wear rate of CF/EP against brass (with an intermediate hardness) is nearly the same as for aluminium but slightly higher. Figure 15 depicts that this influence of the counterpart is smaller at steady state than during running-in. A similar trend was found for GF composites, with the exception that the aluminium pins lead to a higher wear rate than the brass pins. Furthermore, the absolute values of the wear rates were higher for GF/EP than for CF/EP and the differences between the various counterpart materials were more pronounced. Several reasons are responsible for these trends:

– The hard carbon and glass fibres are able to scratch through the counterpart surfaces, thus producing metallic debris. This metallic debris together with the fibre particles are entrapped in the interface. The softer pin materials are more susceptible to abrasion and lead to a higher concentration of metallic particles in the interface than the harder counterpart materials. However, when the hardness of the metallic particles is small compared to the hardness of the fibres, their abrasive efficiency drops.

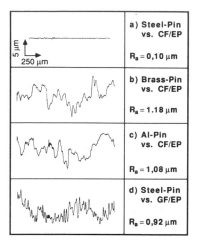

Fig. 24. Surface profiles of steel, brass and aluminium pins after fretting against CF/EP and GF/EP, respectively. Standard loading conditions. Roughness before testing $R_a = 0.12$ µm.

- Fibre debris and metallic particles entrapped in the interface region roughen the counterpart surface, thus increasing the abrasiveness of the counterpart. The harder materials are more resistant against this roughening by a third body (fig. 24). On the other hand, the abrasiveness of the produced asperities increases with hardness.

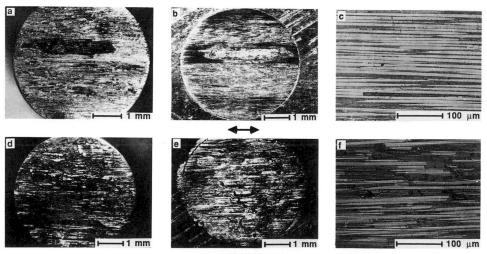

Fig. 25. Optical micrographs of fretting pins and CF/EP specimens after fretting. (a) Steel pin after 30 h fretting against CF/EP at standard loading conditions, (b) shows the corresponding CF/EP specimen, (c) detail of (b). (d) Aluminium pin after 5 h fretting against CF/EP at standard loading conditions, (e) the corresponding CF/EP specimen, (f) detail of (e).

– Protruding edges of broken, but still embedded, fibres are pressed into the counterpart surfaces and this is easier the softer the pins are. Subsequent relative motion causes cracking and removal of these fibres.

As conclusion, one can say: the softer the pin material, the more it can be attacked by the fibres, while, simultaneously, the abrasive efficiency of the roughened counterpart surfaces increases with increasing counterpart hardness. This may explain why the specific wear rate of CF/EP exhibited a maximum for the brass pin with intermediate hardness. Glass fibres can be softened by frictional heat [22]. The maximum of the specific wear rate may be shifted to lower counterpart hardness than for CF/EP, so that GF/EP is more rapidly worn by aluminium than by brass.

Figure 25 shows some optical micrographs of both the steel and aluminium pins fretted against CF/EP, and the corresponding composite specimens. It can clearly be seen that the aluminium pin was much more attacked by the wear process, while the surface of the steel pin still was rather smooth and reflective. The specimen which was worn against steel showed some scratches and some cracking of single fibres. The specimen fretted against aluminium contains much more broken fibres and extended areas from which material was removed.

3.3.3. *Effect of contact pressure*

For studying the influence of the contact pressure, fretting pins with a diameter of only $d = 4$ mm were used, because otherwise contact pressures above 22 MPa led to frictional forces that exceeded the capacity of the apparatus' inductive drive. Figure 26 presents results of this measurement series: up to a contact pressure of about 20 MPa the specific wear rate remains nearly independent of the contact pressure on a low level; beyond this region the specific wear rate increases by a factor of about six and reaches a maximum at 40 MPa. The levelling out at about 40 MPa is possibly an artefact, because at these high normal loads the inductive drive reaches the limits of its capacity.

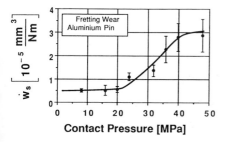

Fig. 26. Specific wear rate as a function of the contact pressure. Fretting wear of CF/EP against Al. $A = 700$ μm, $\nu = 10$ Hz.

4. Fretting fatigue performance

4.1. Mechanical tests

4.1.1. Influence of contact pressure, pin material and laminate stacking sequence

Figures 27a, b and 28 present plots of the upper fatigue load applied versus the resulting lifetime for fatigue and fretting fatigue conditions. On a semilogarithmic scale the plain fatigue curves can be approached by a linear function [23]:

$$\sigma_u = \sigma_T(1 - m \log N), \tag{3}$$

with σ_T the ultimate tensile strength and m experimental parameter.

In the present case, $m \approx 0.02$ for laminate CF1 and 0.06 for CF2, respectively. Obviously, CF2 reacts much more sensitive to a change of the fatigue load than CF1.

Application of an additional fretting load leads to a deviation of the σ–log N curve from the simple logarithmic rule in eq. (3). This deviation may strongly depend on the particular loading conditions. The hard steel pins as counterparts did not produce any significant fretting fatigue effect up to a contact pressure of

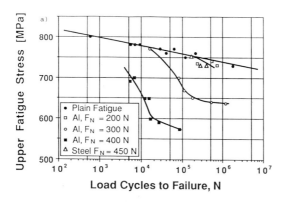

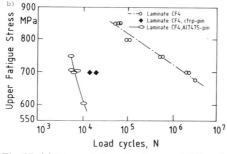

Fig. 27. (a) Upper fatigue load versus lifetime for plain fatigue and fretting fatigue (against aluminium pins) of laminate CF1. Variation of the contact pressure and pin material. (b) Fretting fatigue results for laminate CF4 at $F_N = 450$ N. Variation of pin material. The Al pin shows machining grooves on the contact surface.

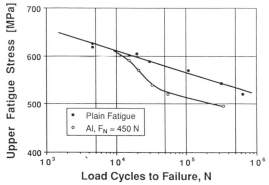

Fig. 28. Upper fatigue load versus lifetime for plain fatigue and fretting fatigue (against aluminium pins) of laminate CF2.

23 MPa ($F_N = 450$ N). When the fretting pins consisted of aluminium, the fretting fatigue life of the laminate was strongly dependent on the contact pressure (fig. 27a). This was also found for brass pins.

Figure 27b additionally shows an example of laminate CF4, where the pin material was systematically varied while the contact pressure remained the same ($F_N = 450$ N) [24]. Using the Al pin, a strong reduction can be observed due to fretting fatigue. In the case of a CFRP pin – in this case only one investigation was carried out at the upper fatigue stress level, $\sigma_u = 700$ MPa – a pronounced reduction in the fatigue life was observed, but the reduction was distinctly smaller than with the metal pins. This might be due to the fact that carbon particles from the CFRP pin have a lubricating effect, thus their damaging effect on the laminate is, besides their low Vickers hardness, less pronounced than from metallic pins.

4.1.2. Influence of laminate plane on fretting fatigue

Because of the anisotropy of the composite laminates fretting fatigue tests were applied in the *xy*- and *xz*-plane, respectively (cf. fig. 6). Figure 29 summarises the results on laminate CF3, with a normal load on the fretting pins of $F_N = 450$ N [24]. The fretting fatigue in the *xy*-plane (position 1 in fig. 5) is denoted as CF3.1 and in the *xz*-plane (position 2 in fig. 5), i.e. the edge surface of the specimen, is denoted as CF3.2.

The pure fatigue tests are marked by the closed circles and the straight line. Only a slight deviation of fretting fatigue relative to the behaviour for pure fatigue was observed. Fretting – and this applies equally to TiAl6V4 and Al 7475 fretting pins – results in a somewhat longer lifetime. This was more marked at higher maximum loading in a fatigue cycle than at lower maximum loading. Using lower normal loads, i.e. $F_N = 300$ or 150 N, gave the same results.

In fig. 29b the results of fretting fatigue against the edge surface of the specimen are plotted (laminate CF3.2). Results similar to those for laminate CF3.1 were found. Again, the fretting fatigue values attain higher numbers of load cycles if σ_u is high, and tend to be in the range of the pure fatigue test at low σ_u-values.

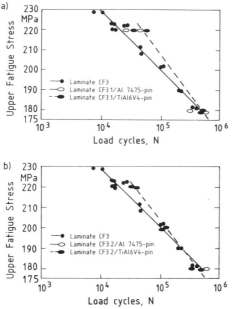

Fig. 29. Fretting fatigue results for laminate CF3.1 (a) and laminate CF3.2 (b) at $F_N = 450$ N.

4.1.3. Fretting fatigue of hybrid composites

Table 2 shows the static performance of the laminates tested. The elastic modulus and the tensile strength of the AF laminate are smaller than those of the CF laminates. The modulus of the hybrid is in accordance with the simple rule-of-mixtures. Its strength, in contrast, is not between the data for the monolithic composites but clearly lower, and can be calculated in terms of the constant-strain criteria [25].

Figures 30a and b show the Wöhler diagrams (plot of the upper fatigue stress applied versus the resulting lifetime on a logarithmic scale) for plain fatigue and fretting fatigue of the CF1, AF and HC laminates tested [26]. Some features are worth to be mentioned:

(a) The fatigue curves of the AF laminate (fig. 30a) and the hybrid (fig. 30b) showed a rather odd shape, with a transition from a very fatigue-resistant to a pronounced fatigue-sensitive behaviour [27].

(b) The fatigue curve of the hybrid (fig. 30b) is also composed of a more flat and a steep branch. The relative slope of the flat part was almost the same as for the neat CF/EP laminate, whereas in the steep part the curve runs parallel to the curve obtained for the AF composite.

The fatigue life of the CF laminate is drastically reduced (arrow 1 in fig. 30b) when aluminium pins were pressed onto the specimen surfaces with a contact load of $F_N = 450$ N during fatigue loading. The reason for this drastic effect of an additional fretting loading is the low wear resistance of CF/EP during the running-in period. Steady-state wear cannot be achieved under fretting fatigue

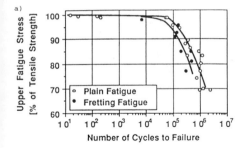

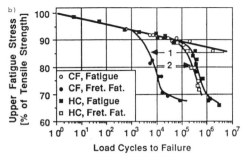

Fig. 30. (a) Fatigue and fretting fatigue of AF/EP. Counterpart aluminium, $F_N = 450$ N. (b) Fatigue and fretting fatigue of CF/EP and an AF/CF hybrid (HC). Counterpart aluminium, $F_N = 450$ N.

conditions, because cracking and peeling of covering fibre layers prevent the formation of equilibrium contact conditions [28].

The effect of an additional fretting component applied using aluminium pins, on the fatigue performance of AF/EP was as small as expected: the Wöhler curves for fretting fatigue and plain fretting were nearly indistinguishable (fig. 30a).

A hybrid was designed with CF-layers in the core and AF-layers at the surfaces. The intention was to combine the fatigue performance of CF/EP with the resistance to fretting fatigue against aluminium of AF/EP. In fact, the hybrid showed a similar fretting fatigue resistance as the AF laminate: the Wöhler-curves for plain fatigue and for fretting fatigue were close together (fig. 30b). The stresses plotted in fig. 30b, however, are normalized fatigue stress levels (related to the static tensile strength); the absolute stresses at which the CF laminates fail after a certain number of load cycles are much higher (see table 2) than for the hybrid, and the advantages of the AF covering layers disappear. Nevertheless, for a thick CF laminate with a thin AF surface layer, the fatigue Wöhler curve of the hybrid will approach the Wöhler curve for the monolithic CF composite. In this case, the high fretting wear resistance of the AF surface layers may improve the fretting fatigue performance of the laminate (arrow 2 in fig. 30b) also in absolute values.

4.2. Failure mechanisms

The development of fretting fatigue damage in homogeneous, isotropic materials can be described by means of fracture-mechanics concepts [29,30]. In the area of the fretting contact, local stress peaks due to friction and normal forces initiate cracks. These cracks act as sharp notches and lead to a high concentration of the fatigue stress at their tips. As a result, the cracks propagate into the bulk and, thus, cause premature failure.

In continuous fibre reinforced composites, the situation is considerably different. The simple application of fracture mechanics is not possible because cracks do not always propagate perpendicular to the main loading direction, but advance preferentially parallel to the fibres [31]. Neither the initiation and accelerated growth of single cracks, as in homogeneous, isotropic materials, nor morphological changes characterise the failure mechanisms in laminated composites under fatigue loading. Instead, multiple matrix cracking along the fibres causes a reduced load-carrying capacity of the off-axis plies and accordingly enhanced stresses in the 0°-layers. Starting from intersections of matrix cracks and from edges, delamination between the differently oriented plies develops [32,33]. Subsequently, the stress in the off-axis plies is reduced while the 0°-plies have to carry an increasing part of the applied load. Final failure occurs when the stress in the 0°-plies locally exceeds their strength [31], which may additionally be reduced by random cracking of 0°-fibres [34]. Figure 31 schematically illustrates the stress situation in a 0°-layer of a cross-ply laminate containing transverse cracks in the 90°-plies.

Laminate CF2 contains a large number of 45°-plies, which have a higher load-carrying capacity than the 90°-layers in CF1 and CF4. Therefore, it is reasonable that the residual strength after a given number of load cycles is more reduced in laminate CF2 than in CF1 and CF4. The effect of an additional fretting component on the fatigue damage mechanisms of CF/EP laminates will be discussed later in more detail with the examples laminate CF3 and CF4.

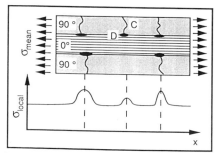

Fig. 31. Schematic illustration of the stress distribution in a 0°-layer after cracking (C) and local delamination (D) of the neighbouring 90°-plies.

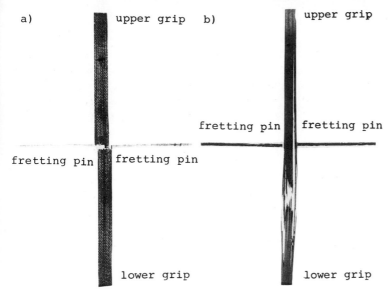

Fig. 32. (a) Typical fracture pattern for laminate CF3.2; (b) typical fracture pattern for laminate CF4.

4.2.1. General failure mechanisms

Fretting fatigue of laminate CF3, both in the CF3.1 and CF3.2 configuration, did not lead to a reduction in the fatigue life, when compared to plain fatigue. On the contrary, a slight increase in the fatigue life was observed (cf. fig. 29). This astonishing behaviour that fretting fatigue, in spite of damaging the surface layers of the test samples at the contact area of the fretting pins, does not reduce the fatigue life, can be attributed to the location of specimen failure, which was never at the fretting position but always between the lower grips and the pins [24]. This is illustrated in fig. 32a by the example of failure of laminate CF3.2. In fact, this result was repeatedly found in all the tests carried out with laminate CF3. An appropriate explanation may be as follows:

(1) In the case of laminate CF3.1, the fretting pins were pressed against the $\pm 45°$ surface layers of the laminate. The damage induced was limited to the surface plies. They carry only a small amount of the total load and protect the underlying load-bearing $0°$-plies.

(2) In the region between the upper grip and the fretting pin a load F_I (cf. fig. 33) was acting which was lower than the load F_{II} acting on the specimen between the fretting pins and the lower grip (connected to the servohydraulic actuator). The actual difference was dependent on the frictional force, F_R, generated by the two fretting pins, via:

$$F_R = \mu F_N, \tag{4}$$

where μ is the "effective coefficient of friction" between the fretting pins and the composite test sample. The frictional force was transferred to the upper grip through the fretting fatigue test device, which was attached to it [8].

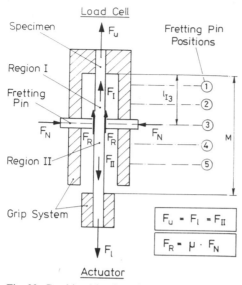

Fig. 33. Combined loads acting on the test sample during fretting fatigue.

(3) In the case of laminate CF3.2, only a small amount of 0°-fibres could be harmed by fretting against the side surface. The load transfer through the test device, as described above, is superimposed on this effect, so that failure could again occur between the fretting pins and the lower grips.

Failure of the test coupons, therefore, even in the case of fretting fatigue, was a pure fatigue failure, and the deviation to longer life in the fretting fatigue situation can either be attributed simply to scatter or to the smaller sample dimensions, which means that the length of the test coupon is reduced by the distance between the upper grip and the fretting pins.

The fretting fatigue of laminate CF4 showed a reduction in the fatigue life of up to three orders of magnitude. For this laminate, but also for CF1 and CF2, the fretting pins were pressed against the surface layers, which were the load bearing 0°-plies. The test samples always failed at the fretting pin position, independent of the pin material used (cf. fig. 32b).

Fretting against the load-bearing 0°-surface layers led to a significant reduction in the fatigue life. Because of fretting, early damage of the fibre surfaces occurred, the fracture stress of each of the damaged fibres is drastically reduced and fibre failure initiates early, well before the nominal fracture stress of the fibres is reached. Since the 0°-fibres carry most of the load, a transfer of stresses to the neighbouring fibres increases their overall stress and initiates early failure of the whole laminate. The mechanisms described above will be discussed in more detail in the following sections.

4.2.2. Failure mechanisms in laminate CF3

As shown above, fretting fatigue of laminate CF3 did not affect the fatigue life, despite the fact that essential damage occurred. Figures 34a–c show in reflected-

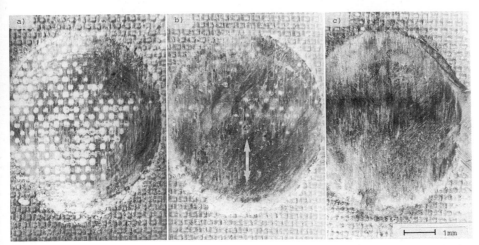

Fig. 34. Light micrographs of laminate CF3.1 ($\sigma_u = 220$ MPa, TiAl6V4 pin): overall view of the fretting area. (a) $F_N = 150$ N, (b) $F_N = 300$ N, (c) $F_N = 450$ N.

light micrographs an overall view of the fretting area of laminate CF3.1 after fretting against the front surfaces. TiAl6V4 pins were used. Varying the contact load, F_N, from 150 (fig. 34a) to 300 (fig. 34b) and 450 N (fig. 34c), led to a significant increase in damage in the fretting area. At the lowest load level ($F_N = 150$ N), the thin epoxy layer on the surface (as reflected by the pattern of the covering fabric removed prior to the experiments) is not quite destroyed even after about 40 000 load cycles. Increasing the contact load to 300 N gives a stronger fretting effect, which results in almost total removal of the epoxy surface coating after about 33 000 cycles. For a similar time period with the higher contact load of 450 N even deeper damage is caused, completely fretting away the epoxy surface layer and some protrusions of fibre layers near the edge of the fretting area. However, this significant surface damage did not affect the internal-damage development in laminate CF3.1. Rupture occurred in all cases between the fretting pins and the lower grips. This is further illustrated in fig. 35, which is an X-ray radiograph of laminate CF3.1. The specimen failed after 589 480 load cycles. The main part of the damaged area, with extensive development of transverse intra-ply cracks in the $+45°$-, $-45°$- and 90°-layers of the laminates, is between the fretting pins and the lower grips. In the fretting area under the pins, which is highly damaged (cf. fig. 34), far fewer cracks can be observed. Obviously, pressing the pins against the laminate delays the crack formation. Only in the direct vicinity of the fretting area more damage, namely in the $+45°$-surface plies, can be observed. This area, i.e. a segment of the fretting region, is shown in a SEM micrograph in fig. 36a. At higher magnification details of abraded particles (matrix coagulates and fragments of fibres) can be distinguished (fig. 36b). In the interior of the fretting area, sliding wear of carbon fibres, fibre fractures, fibre/matrix debonding, cavities and debris of fibres and matrix can all be found simultaneously (fig. 36c).

upper grip

position of
fretting pin

5 mm

lower grip

Fig. 35. X-ray radiograph of laminate CF3.1 (σ_u = 180 MPa, F_N = 450 N, Al 7475 pin). Damage pattern after specimen failure.

Fig. 36. SEM micrographs of laminate CF3.1 (σ_u = 180 MPa, F_N = 450 N, TiAl6V4 pin), fretting area. (a) Segment of a fretting area, overall view – debris in the boundary area; (b) debris and fibre fragments in the boundary area – detail of (a); (c) fracture of 45°–fibre, debonding, cavities and debris – detail of the centre of the fretting area.

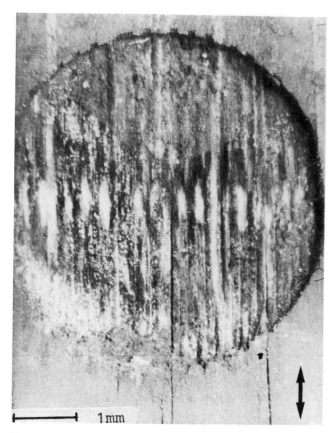

1 mm

Fig. 37. Light micrographs of laminate CF3.2 ($\sigma_u = 180$ MPa, $F_N = 450$ N, TiAl6V4 pin). Fretting area on the side surface of the testpiece.

With fretting fatigue against the edge surface in laminate CF3.2 significant surface damage was observed again. Figure 37a is a light micrograph of the fretting area. The specimen had failed between the pins and the lower grip. At the instant of failure the specimen split lengthwise by delamination as a result of a sudden release of the high level of stored energy.

The extensive damage produced by fretting fatigue against the edge surface is further demonstrated in fig. 38a, which is a SEM micrograph taken from laminate CF3.2 before final failure took place at a position away from this point. The pin deeply marked its location of fretting in the specimen. A groove nearly 1 mm deep was produced. At the right-hand edge of the fretting area, which corresponds to the part next to the upper grip, a triangular-shaped area is visible (arrow). Within this area and at the opposite left-hand edge of the fretting area in fig. 38, debris of fibre and matrix particles are mainly found (fig. 38b). In the centre of the fretting area the main wear processes due to fretting fatigue can be observed. Figure 38c shows a detail of the centre of fig. 38a with worn and broken 0°-fibres. Figure 38d

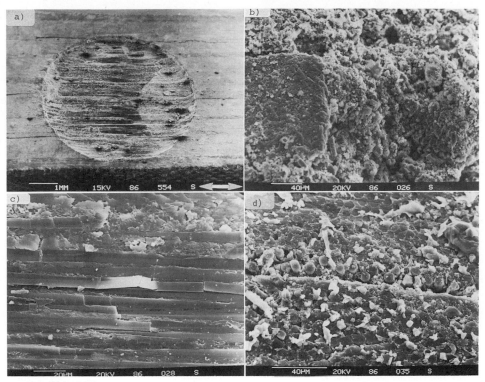

Fig. 38. SEM micrographs of laminate CF3.2 (σ_u = 180 MPa, F_N = 450 N, Al 7475 pin), fretting area. (a) Overall view; (b) detail, lower edge of the fretting area – debris of fibre and matrix particles; (c) detail, centre of the fretting area – worn and broken 0°-fibres; (d) detail, centre of the fretting area – boundary area between (top) 90°- and (bottom) 45°-plies.

is again a detailed view of the central fretting area, showing the boundary area between a 90°- and 45°-ply. The fibre ends in the 45°-ply are oval-shaped, while those in the 90°-ply are circular. The 90°-fibres next to the 90°/45°-ply boundary are sticking out. In the boundary itself the onset of delamination is visible.

Damage is not only associated with the composite, but also the fretting pins can show appreciable wear. Figure 39 shows that sliding grooves have formed. The deepest groove, in the centre of the figure, was formed by the 90°-ply in laminate CF3.2.

4.2.3. Failure mechanisms in laminate CF4

Fretting fatigue of laminate CF4 with fretting against the surface 0°-ply has been shown to have a significant influence on the fatigue life.

The imprint from the fretting pins, at a comparable upper load, σ_u = 700 MPa (approximately 80% of the fracture stress), and identical contact loads, F_N, is therefore more clearly marked than in the case of laminate CF3.1. This is shown in figs. 40a (F_N = 150 N), 40b (F_N = 300 N) and 40c (F_N = 450 N) for laminate CF4.

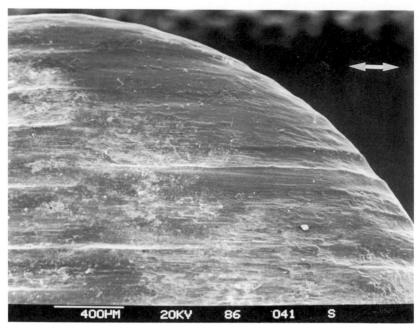

Fig. 39. Al 7475 pin – grooves formed during fretting fatigue. The different depths of the grooves can be related to the different abrasive wear behaviour of the various out-of-axis plies of the counterpart (laminate CF3.2).

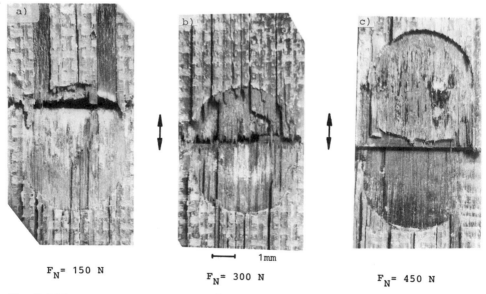

$F_N = 150$ N $F_N = 300$ N $F_N = 450$ N

Fig. 40. Light micrographs of laminate CF3 ($\sigma_u = 700$ MPa, TiAl6V4 pin), overall view of fretting area. (a) $F_N = 150$ N, (b) $F_N = 300$ N, (c) $F_N = 450$ N.

upper grip

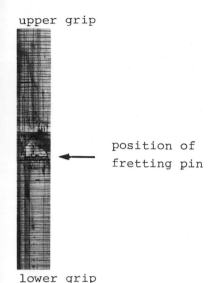

position of
fretting pin

lower grip

Fig. 41. X-ray radiograph of laminate CF4 ($\sigma_u = 700$ MPa, $F_N = 450$ N, AL 7475 pin), damage pattern. Sample has exceeded 80% (4800 cycles) of its expected fatigue life.

The penetration depth of the fretting pins into the first layer of laminate increased with increasing F_N. Specimen failure of laminate CF4 was observed to be always under the fretting pins. It can be seen that fracture occurred in the lower area of the fretting imprint at high F_N (450 N), in the middle area at $F_N = 300$ N and in the upper part of the fretting surface at $F_N = 150$ N. This is probably due to the different magnitudes of the shear, tensile and compressive stress fields as a function of F_N, resulting in different positions of the maximum of the equivalent stress, according to the deformation-energy hypothesis [35].

Figure 41 is an X-ray radiograph of a test piece that has exceeded about 80% of its expected fatigue life. The damage pattern typical for a cross-ply laminate under fatigue loading can be recognised: transverse and longitudinal interply cracks, edge delamination and interior delaminations at cross sections of longitudinal and transverse cracks [31]. In addition, at the centre of the test sample, further damaged areas deriving from the fretting pin can be recognised. The start of a fracture front, progressing through the centre of the pin imprint across the longitudinal axis, can clearly be seen. There are numerous fibre failures. Additionally, delaminations between the 0°-plies and the underlying 90°-plies can be observed, often terminated by longitudinal cracks. In contrast to the case of fretting fatigue in laminate CF3, the test pieces of laminate CF4 broke under the action of the fretting pins.

Figure 42 shows SEM micrographs taken from the fretting area of a test sample of laminate CF4. The sample failed near the centre of the fretting area, in which extensive damage can be recognised (fig. 42a). The fretting pin particularly affected the 0°-fibres, since it was only pressed against the outer 0°-layer. This

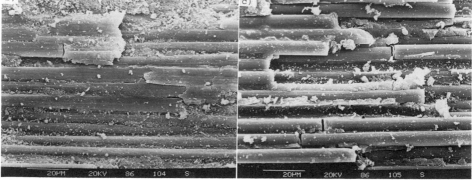

Fig. 42. SEM micrographs of laminate CF4 ($\sigma_u = 700$ MPa, $F_N = 450$ N, Al 7475 pin). (a) Fretting fatigue fracture, overall view; (b) worn and broken 0°-fibres plus wear debris; (c) broken 0°-fibres.

resulted in fibre wear. Fibre and matrix debris are distributed all over the fretting area, to half of the original sample thickness (fig. 42b). At other locations, already broken 0°-fibres, which have not yet been worn but have been completely displaced by fretting, can be observed (fig. 42 c).

Figure 43 shows the fretting area in a 0°-ply from the centre of the test piece. it was protected from the fretting pin pressing against the outer 0°-ply by another ply (90_2), which was approximately 0.25 mm thick. Nonetheless, the effect of fretting fatigue extends even to lower laminate plies of the test piece, leading to local delamination. The figure clearly indicates pressure marks from the fretting pin.

4.2.4. Fretting fatigue failure mechanisms. A comprehensive comment

The fretting fatigue failure mechanisms [36] of the different laminates investigated will not be discussed separately in the following, since they all show some general similarities. The only difference lies in the direction of the relative motion of the fretting pins, due to the orientation of the fibres in the individual plies of the laminate.

4.2.4.1. Fretting fatigue parallel to the fibre direction.

Figure 44a schematically summarizes the different mechanisms observed as a result of fretting fatigue

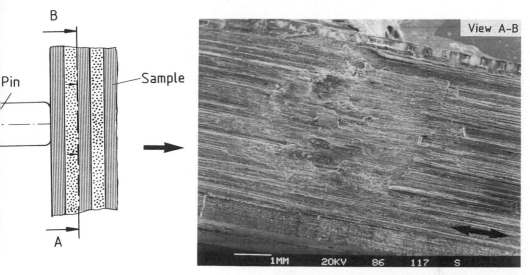

Fig. 43. SEM micrograph of laminate CF4 ($\sigma_u = 700$ MPa, $F_N = 450$ N, TiAl6V4 pin). Surface of a central 0°-ply far from direct contact with the fretting pin.

parallel to the fibre direction. When fretting fatigue occurs at an angle of 45° to the parallel fibre direction, very similar damage mechanisms can be observed. The following summary of the damage mechanisms, is therefore, also admissible for that case:

  (a) Matrix wear due to plastic deformation, ploughing and cutting, favoured by the local heating-up of the matrix. Formation of matrix debris (figs. 36, 38c, 42b).
  (b) Wear thinning of the fibre reinforcement, with fibre pulverisation (figs. 36c, 38c).
  (c) Fibre cracking (figs. 36c, 38c, 42c).
  (d) Fibre/matrix interface separation (figs. 36c, 38c, 42b) and delamination wear (fig. 43).
  (e) Peeling-off of the fibres from the matrix (figs. 38c, 42c).
  (f) Matrix cracking, mainly observed during fretting fatigue under 45° to the parallel direction (fig. 36c).

4.2.4.2. Fretting fatigue normal to the fibre direction. In fig. 44b the different damage mechanisms due to fretting fatigue normal to the fibre direction, as well as under 45° to the normal direction, are schematically summarized:

  (a) Fibre and matrix sliding wear with the formation of traces by the fretting pin (fig. 45).
  (b) Fragmental fracture at fibre ends (fig. 38d).
  (c) Fibres sticking out of the matrix with fibre/matrix debonding (fig. 38d).
  (d) Delamination growth due to the relative fibre motion of the 90°- and 45°-fibres (fig. 38d).

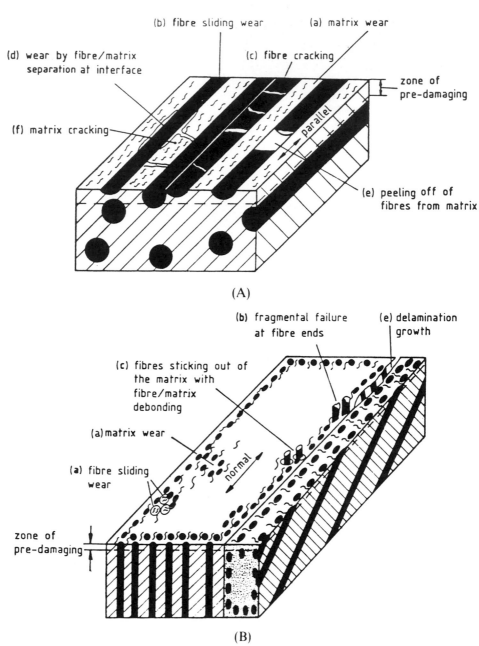

Fig. 44. Model of different fretting fatigue mechanisms on the surface of continuous fibre reinforced polymers. (a) Parallel to the fibre direction; (b) normal to the fibre direction.

Fig. 45. SEM micrograph of laminate CF3.2 ($\sigma_u = 180$ MPa, $F_N = 450$ N). Formation of traces by the fretting pin, detail of fig. 38d.

4.3. Mathematical modelling of fretting fatigue

4.3.1. The mathematical model

The observations on the fretting fatigue failure mechanisms suggest:

(1) The absence of notch effects causes a relatively uniform distribution of tensile stresses across any cross section of the undamaged 0°-plies. Therefore, the change of the load-carrying capacity, dF, should be proportional to the reduction of the cross section, dQ, of the load-bearing 0°-layers in the fretting contact region, between N and $N + dN$ load cycles:

$$dF = \sigma_u \, dQ, \tag{5}$$

with σ_u the upper fatigue load at which the specimen fails after N load cycles.

(2) In first-order approximation the damage development can be considered to proceed proportionally with time:

$$\frac{dQ}{dN} = \text{const.} \tag{6}$$

The reduction of the cross section of the 0°-layers in the fretted region leads to an enhanced fatigue stress in the remaining 0°-fibres. The specimen finally fails when this stress exceeds the strength of the 0°-plies. However, the actual cross section is not continuously monitored and, thus, the true stress in the laminate

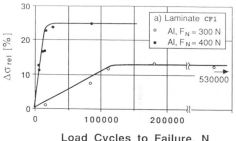

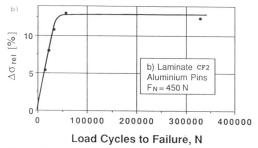

Fig. 46. Relative fatigue strength reduction ($\Delta\sigma_{rel}$) as a function of the fretting fatigue life for laminate CF1 (a) and CF2 (b) at several contact loads.

between the fretting pins cannot be determined. But the following consideration can be made: the apparent upper fatigue stress, σ_u, at which the specimen fails after N load cycles, diminishes proportionally with the cross section of the load bearing 0°-layers. Accordingly, σ_u can be calculated by changing eq. (3) as follows:

$$\sigma_u = \sigma_T \left(1 + \frac{dQ\,N}{Q_0\,dN} \right) - (1 - m\,\log\,N), \tag{7}$$

with Q_0 the initial total cross section of the 0°-plies.

Inserting eq. (7) into eq. (5) and integrating the resulting differential equation [28] leads to:

$$\Delta\sigma_{rel} = \frac{\sigma_F - \sigma_{FF}}{\sigma_F} \approx \frac{\Delta Q}{Q_0} - 0.5 \frac{\Delta Q^2}{Q_0}, \tag{8}$$

with $\Delta\sigma_{rel}$ the relative fatigue strength reduction, σ_F the upper fatigue load at which the specimen fails after N load cycles and σ_{FF} the upper fretting fatigue load at which the specimen fails after N cycles.

Presuming the validity of eq. (6), eq. (8) suggests that, in first-order approximation, $\Delta\sigma_{rel}$ increases proportionally with time. To check this assumption, fig. 46 presents a plot of $\Delta\sigma_{rel}$ versus the fretting fatigue life, N, for two different laminates. Initially, all curves increase proportionally with time but reach a constant value after a certain number of load cycles. The maximum fretting fatigue damage is expected when the covering 0°-layers are totally penetrated by the

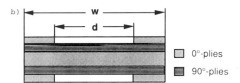

0°-plies

90°-plies

Fig. 47. (a) Optical micrograph of a specimen of laminate CF2 subjected to fretting fatigue against an aluminium pin. $F_N = 450$ N, failure after 345 900 load cycles. (b) Cross section of the specimen in the region of fretting contact, schematically.

fretting pins. Further fretting only damages off-axis plies and influences the fatigue performance of the laminate insignificantly [24]. Figure 47 presents an optical micrograph of a specimen of laminate CF2, which failed after 345 900 load cycles, i.e. within the plateau region of the $\Delta\sigma_{rel}$–N curve. The covering 0°-layers were totally penetrated and fretting already took place on the 45°-plies. Based on these considerations, the maximum value for $\Delta\sigma_{rel}$ can be calculated from the maximum reduction of the cross section, ΔQ_{max}, of the 0°-layers (see fig. 47b):

$$\Delta Q_{max} = \frac{d}{w} \frac{i_{F,max}}{i_{tot}}, \qquad (9)$$

where d is the diameter of the pin, w the width of the specimen (6.3 mm for CF2 and 8.4 mm for CF1), $i_{F,max}$ the maximum number of fretted 0°-plies ($= 4$) and i_{tot} the total number of 0°-plies. Accordingly, the maximum fretting fatigue damage can be calculated using eq. (10). It is a simple geometric quantity depending only on the stacking sequence of the laminate and on the contact geometry.

$$\Delta\sigma_{rel,max} = \frac{\Delta Q_{max}}{Q_0} - 0.5\left(\frac{\Delta Q_{max}}{Q_0}\right)^2, \tag{10}$$

which gives 0.25 for CF1 and 0.125 for CF2.

These values are in good agreement with the experimental results when considering aluminium pins at a contact pressure of $p = 23$ MPa ($F_N = 400$ N, fig. 46). However, at lower contact pressures, the maximum fretting fatigue damage is considerably smaller than calculated. This effect will be presented and discussed later in this chapter.

From the slope of the linear part of the $\Delta\sigma_{rel}-N$ curve, the rate of damage development, dQ/dN, can be derived. For reasons of comparability with the fretting wear tests, this quantity will be transformed to a "specific pseudo-wear rate", $\dot{w}_s^*$, in analogy to the concept of the specific wear rate, $\dot{w}_s$:

$$\dot{w}_s^* = \frac{\Delta V}{2F_N L} = \frac{\pi}{16}\frac{dQ_0}{F_N AN}, \tag{11}$$

where ΔV is the volume of removed or cracked $0°$-plies, L the total sliding distance ($= 2AN$) and F_N the normal load. The factor $\frac{1}{2}$ results from the fact that two pins simultaneously rub against the laminate and only one half of the accumulated damage of the laminate can be related to one pin. Resolving eq. (10) to ΔQ and inserting this result into eq. (11) gives:

$$\dot{w}_s^* = \frac{\pi}{16}\frac{dQ_0}{F_N AN}\left(1 - \sqrt{1 - 2\Delta\sigma_{rel}}\right). \tag{12}$$

The quantity $\dot{w}_s^*$ measures how deep the fretting fatigue damage penetrates the laminate, and includes material removal due to pure fretting wear as well as cracking of predamaged fibres and fibre bundles. Therefore, $\dot{w}_s^*$ should be greater than $\dot{w}_s$.

4.3.2. Experimental proof of the model

In order to avoid confusion, it should be emphasized that the measured points along the $\Delta\sigma_{rel}-N$ curve in fig. 46 represent different specimens which were subjected to different fatigue stress amplitudes. The fact that these points lie on a straight line primarily means that the fretting fatigue damage accumulated until final failure is proportional to the fretting fatigue life, regardless the fatigue stress level. On the other hand, the presented model states that during the test of one particular specimen the effective cross section of the load bearing $0°$-layers, when exposed to fretting, decreases proportional with time. Therefore, an additional test was necessary to check this assumption of the model.

Several specimens of laminate CF1 were exposed to a fretting fatigue loading equivalent to one half of the expected fretting fatigue life. The contact load was set to $F_N = 300$ N, the upper fatigue stress was $\sigma_u = 700$ MPa. From fig. 27a an

TABLE 4
Residual fatigue life and relative fatigue strength reduction of CF/EP laminates ($\sigma_u = 740$ MPa) predamaged by 45 000 load cycles under fretting fatigue ($F_N = 300$ N, $\sigma_u = 700$ MPa) conditions.

No.	Number of fatigue cycles to failure	$\Delta\sigma_{rel}$ (%)	$\Delta\sigma_{rel,mean}$ (%)
1	7320	5.1	
2	29 970	3.8	4.5
3	15 630	4.5	

expected lifetime of 79 000 load cycles can be derived under these conditions. After 45 000 load cycles, the test was interrupted, which was in accordance with a theoretical fatigue strength reduction of 4.5% (fig. 46). Subsequently, the tests were continued under plain fatigue conditions ($\sigma_u = 740$ MPa) until final failure. Table 4 lists the resulting lives and the according relative deviation from the σ versus log N curve for plain fatigue of the undamaged laminate ($\Delta\sigma_{rel}$). The measured mean value of $\Delta\sigma_{rel}$ amounts to 4.5%, which is exactly the theoretical value calculated under the assumption that the fretting fatigue damage proceeds proportional with time during one single test.

4.3.2.1. Effect of contact pressure and counterpart material. Figure 48 shows the specific pseudo-wear rate, $\dot{w}_s^*$, calculated from the linear part of the $\Delta\sigma_{rel}$ versus N curve of laminate CF1, as a function of the contact pressure p. Obviously, there exists a critical value where the propagation of damage development jumps from an insignificant to a high level (boundary value). The specific fretting wear rate, $\dot{w}_s$, follows a similar behaviour (fig. 26).

For aluminium pins, $\dot{w}_s^*$ is smaller than 8×10^{-6} mm^3/N m at a contact load of 10 MPa ($F_N = 200$ N) and, thus, has a similar magnitude as the specific fretting wear rate (6×10^{-6} mm^2/N m). At 15 MPa, still below the boundary value of the contact pressure, $\dot{w}_s^*$ has already increased considerably to a value of about 3×10^{-5} mm^3/N m, while the specific fretting wear rate, $\dot{w}_s$, still remained constant below its boundary value.

As conclusion one can say that fretting fatigue damage proceeds as fast as material removal due to plain fretting wear below a critical contact pressure.

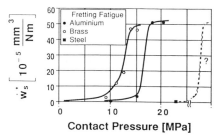

Fig. 48. Specific pseudo-wear rate, $\dot{w}_s^*$, of laminate CF1 as a function of the contact pressure for different counterpart materials.

Above this critical contact pressure the fretting wear and fretting fatigue both increase accelerated. But under fretting fatigue conditions, this boundary pressure lies at considerably lower values ($p_{crit} \approx 17$ MPa) than under plain fretting ($p_{crit} \approx 30$ MPa for aluminium pins). Above the boundary value of the contact pressure, fretting fatigue damage proceeds much faster than the pure material removal due to plain fretting wear:

$$\dot{w}_s^*(p \approx 20 \text{ MPa}) = 5 \times 10^{-4} \text{ mm}^2/\text{N m},$$

$$\dot{w}_s(p \approx 45 \text{ MPa}) = 2.8 \times 10^{-5} \text{ mm}^3/\text{N m}.$$

A similar behaviour of the specific pseudo-wear rate as a function of contact pressure can be found for the brass pins. However, the boundary value of the pressure is shifted to lower values, compared to aluminium. This correlates with the fretting wear of CF/EP, which was found to be higher for brass than for aluminium counterparts [9]. When steel pins were used as counterparts, no stepwise increase in the specific pseudo-wear rate was observed up to a contact pressure of 23 MPa ($F_N = 450$ N), which is in accordance with the low fretting wear rate of CF/EP against steel pins.

A probable reason for the small influence of the hard steel pins on the fatigue life of laminate CF1 is their high resistance against abrasion by fibre debris. The surface of the steel pins remains rather smooth during fretting and, in turn, acts only little abrasively on the sample material. Aluminium pins, however, can easily be roughened by the fibre debris, but the arising asperities are rather soft and, thus, less abrasive than in the case of brass. A significant influence of electrical contact corrosion seems to be unambiguous, because a similar trend in the effect of the counterpart material was observed for glass fibre (GF/EP) composites. Table 5 compares the specific fretting wear rates of CF/EP and GF/EP worn against steel and aluminium, respectively. Fretted against steel, CF/EP and GF/EP exhibit similar wear rates. Changing to aluminium pins, the wear of GF/EP is much more accelerated than that of CF/EP. This cannot be explained by contact corrosion, which is not effective in the case of glass fibres. But the glass-fibre particles have a more severe effect and, thus, are more effective in roughening the counterpart.

Another effect of the contact load can be seen in fig. 46a. At a contact pressure of $p = 15$ MPa ($F_N = 300$ N), the relative fatigue strength reduction, $\Delta\sigma_{rel}$, be-

TABLE 5
Relative specific fretting wear rates for several sample/counterpart combinations normalised to that of CF/EP versus steel. Matrix: R 5212, CF = T 300, GF = E-glass fibres. The fibres were aligned parallel to the sliding direction. $F_N = 300$ N, the peak-to-peak oscillation width is 500 μm, the number of load cycles is 72 000.

	steel	aluminium
CF/EP	1	1.8
GF/EP	1.3	26.7

comes constant after about 150 000 load cycles at a value of $\Delta\sigma_{rel} = 14\%$ and does not reach the theoretical value of 25%, as was found for $p = 23$ MPa ($F_N = 450$ N).

Several authors [37–39] reported that also in metallic materials the influence of an additional fretting component on the material fatigue is effective only during crack initiation and the early stage of crack growth. After a certain number of load cycles the crack propagation rate under fretting fatigue approaches that under plain fatigue [40]. This can be explained as follows. The initial value for the stress concentration at a crack tip is enhanced by normal and shear stresses caused by the fretting pin. While the crack propagates into the interior, this stress concentration diminishes to that value which is also active under plain fatigue conditions [41]. This crack-propagation behaviour would be in agreement with the fact that $\Delta\sigma_{rel}$ remains constant after a certain number of load cycles.

However, the absence of notch effects in continuous fibre reinforced layers with 0°-orientation opposes the application of this model to the CF/EP laminates investigated here. In fact, it was observed that specimens exposed to fatigue stresses above 640 MPa (contact load of $F_N = 300$ N) showed a more pronounced surface damage after a given number of cycles than specimens which were fatigued at a lower stress level. Especially, no peeling-off of fibre bundles occurred at fatigue stress levels below 640 MPa even after some hundred thousand load cycles (fig. 49a). These observations suggest that exceeding a particular number of load cycles is not responsible for the lowering of the propagation rate of fretting fatigue damage, but rather the reduction of the fatigue stress level below a critical value ($\sigma_{u,crit} = 640$ MPa). In this case, probably, the tensile stresses in the fretting region and the shear stresses between cracked and undamaged layers are not high enough to cause amplification of the surface damage due to fretting. In this context, it should be remembered that the various points along the $\Delta\sigma_{rel}$-curve (fig. 46) represent differently loaded specimens. The damage progress in the specimens which failed in the plateau region, may not have proceeded along the drawn line but along any other course with a smaller slope which ends at the measured points.

According to the above consideration on damage development, it can be assumed that the $\Delta\sigma_{rel}$ versus N curve for each individual specimen follows a straight line until final failure. Table 6 represents an experimental check of this concept. Some specimens were exposed to fretting fatigue at an upper fatigue stress of 640 MPa and a contact pressure of $p = 15$ MPa ($F_N = 300$ N) for 200 000 load cycles. According to fig. 29, the expected fretting fatigue life of these specimens amounts about 500 000 load cycles. Figure 49a shows that these specimens did not exhibit any peeling-off after 200 000 load cycles, in contrast to specimens subjected to higher fatigue loads. Presuming a linear increase of $\Delta\sigma_{rel}$ until final failure after 500 000 load cycles at $\Delta\sigma_{rel} = 14\%$, a relative fatigue strength reduction after 200 000 load cycles of about 5.5% would be expected (see fig. 46). Subsequent to fretting fatigue loading, the specimens were run under plain fatigue at $\sigma_u = 750$ MPa until final failure occurred. During this time, the layers predamaged by fretting rapidly cracked and delaminated (fig. 49b). Table 6 lists the number of plain fatigue cycles to failure and the resulting deviation from the σ_u versus log N curve for the undamaged laminate. The mean value for this

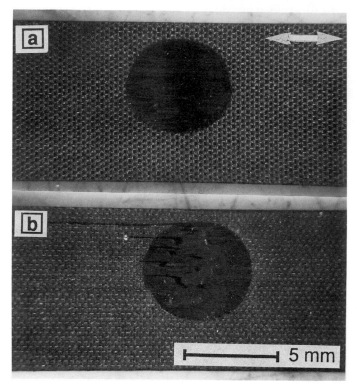

Fig. 49. Optical micrographs of laminate CF1 after exposure to fretting fatigue. (a) Fretting fatigue against aluminium (F_N = 300 N) at an upper fatigue load of σ_u = 640 MPa for 200 000 load cycles. (b) Without fretting (plain fatigue), σ_u = 750 MPa.

difference is $\Delta\sigma_{rel}$ = 6%. The coincidence of this result with the expected value confirms the assumptions that
- fretting fatigue damage proceeds proportional with time in continuous fibre reinforced laminates and
- at low contact loads ($F_N \leqslant$ 300 N), the rate of fretting damage development is controlled by the upper fatigue stress, σ_u.

TABLE 6
Relative fatigue strength reduction due to fretting fatigue predamage. The specimens (laminate CF1) were first subjected to 200 000 fretting fatigue load cycles (F_N = 300 N, σ_u = 640 MPa) and subsequently exposed to plain fatigue (σ_u = 750 MPa) until failure.

No.	Number of cycles to failure	$\Delta\sigma_{rel}$ (%)	$\Delta\sigma_{rel,mean}$ (%)
1	390	6.6	
2	1380	5.4	6.0
3	780	6.0	

However, for high contact loads ($F_N \geqslant 400$ N) the development rate is independent of the fatigue stress level (see above). This can be explained by assuming the equivalent stress amplitude in the subsurface region instead of the fatigue stress level to be responsible for peeling off of predamaged $0°$-layers, and thus for the amplification of fretting-induced surface damage.

5. Conclusion

When relative movement can occur between mating parts and one of them is subjected to cyclic stresses, then the fatigue life can be reduced. This phenomenon is referred to as fretting fatigue. To understand fretting fatigue, insight in the wear and fretting wear behaviour is presumed.

Wear, generally, cannot be considered as a material property, but rather as depending on the whole tribological system under consideration. The tribological system itself may change with time. Therefore, it is not easy to predict the wear behaviour of a sample material from simple material constants. Only some qualitative trends can be formulated:

- The fretting wear performance of a fibre reinforced composite is strongly influenced by the interface conditions, which reduce the role of the material and counterpart. On the other hand, the sample and counterpart materials determine the composition of the third body.
- At mild loading conditions, any type of fibre reinforcement accelerates wear, AF fibres are poorly bonded to the matrix, so that AF composites tend to fibrillation. CF, and more pronounced GF, builds up an abrasive interphase layer, which attacks the counterpart, thus producing a higher counterpart roughness and metallic debris, which again act as a third-body abrasive.
- The wear-promoting effect of CF and GF is less strong for hard counterpart materials.
- Under severe loading conditions a reinforcement by AF and CF may significantly improve the wear performance of the polymer matrix material.

Under fretting fatigue conditions the following conclusions can be drawn:
- An additional fretting component may drastically reduce the fatigue life of a CF/EP laminate if the fibres subjected to fretting have a $0°$-orientation.
- The occurrence and magnitude of the influence of fretting on the fatigue performance sensitively depends on the particular loading conditions (counterpart material, contact pressure). The use of hard steel counterparts prevents a fretting fatigue effect, at least up to a contact pressure of 23 MPa, under conditions given in the previously described experiments. For softer counterparts (aluminium, brass), a fretting fatigue effect occurred above a critical contact pressure.
- As a quantitative measure for the degree of fretting fatigue damage, the relative fatigue strength reduction

$$\Delta\sigma_{rel} = \frac{\sigma_F - \sigma_{FF}}{\sigma_F}$$

was proposed (eq. (8)). No notch effects were observed.

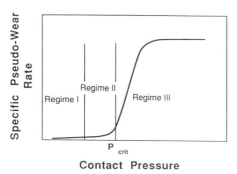

Fig. 50. Schematic presentation of the source of the specific pseudo-wear rate as a function of the contact pressure.

– The mechanisms of interaction between surface damage due to fretting and fatigue are different for different loading conditions. Figure 50 schematically distinguishes three regimes of fretting fatigue. At low contact pressures (*regime I*), there exists no synergism between fretting and fatigue. The rate of damage development is small and is determined by the resistance against plain fretting wear. The specimens live nearly as long as under plain fatigue. At higher pressures (*regime II*), the fretting fatigue damage proceeds already considerably faster than the pure material removal due to plain fretting wear. Fibre bundles exposed to fretting tend to crack and delaminate. However, if the fatigue stress descends below a certain value, the mutual amplification of fretting and fatigue damage becomes less effective, because cracking and peeling-off of fibre bundles decelerates. Above a critical contact pressure (*regime III*), fretting fatigue damage proceeds about 15 times faster than the pure material removal due to fretting. Fibre bundles which are predamaged by fretting rapidly crack and peel off. The fretting fatigue damage proceeds until the fretting pins reach the off-axis plies, which carry only a small part of the applied fatigue stress. The particular value of the critical contact pressure (p_{crit}) in fig. 50 depends on the hardness of the counterpart material.

Acknowledgements

This investigation was financed by the Ministry for Research and Technology of the FRG (BMFT 03 M 1022). Thanks are due to BASF and MBB for supplying the sample materials.

List of symbols

A	full oscillation width
d	diameter
F_N	normal load
F_R	frictional force

F_I	load on specimen between upper grip and fretting pin
F_{II}	load on specimen between lower grip (actuator) and fretting pin
D_F	change of the load-bearing capacity
HV	Vickers hardness
$i_{F,max}$	maximum number of fretted 0°-plies
i_{tot}	total number of 0°-plies
l	length
l_1	distance of fretting pins from the upper fixed grip
L	total sliding distance
m	experimental parameter
Δm	mass loss
N	number of load cycles
p	contact pressure
p_{crit}	critical boundary pressure
Q_0	initial total cross section of 0°-ply
dQ	reduction of cross section
ΔQ_{max}	maximum reduction of cross section
s	pin displacement
t	time
T_g	glass transition temperature
ΔV	volume of removed cracked 0°-plies
w	specimen width
$\dot{w}_s$	specific wear rate
$\dot{w}_s^*$	specific pseudo-wear rate
ε_T	strain to failure
μ	friction coefficient
ν	frequency
ρ	density
σ_l	minimum fatigue stress
σ_u	maximum fatigue stress
σ_F	upper fatigue load at which specimen fails after N cycles
σ_{FF}	upper fretting fatigue load at which specimen fails after N cycles
σ_T	ultimate strength
$\Delta\sigma$	stress amplitude
$\Delta\sigma_{rel}$	relative fatigue strength reduction
$\Delta\sigma_{rel,max}$	maximum relative fatigue strength reduction.

References

[1] R.B. Waterhouse, in: Fretting Fatigue, ed. R.B. Waterhouse (Applied Science Publishers, London, 1981) pp. 159–175.
[2] T.C. Chieven and S.C. Gordelier, Wear 96 (1984) 153–157.
[3] D.W. Hoeppner, in: Fretting Fatigue, ed. R.B. Waterhouse (Applied Science Publishers, London, 1981) pp. 143–158.

[4] P.J.E. Forsyth, in: Fretting Fatigue, ed. R.B. Waterhouse (Applied Science Publishers, London, 1981) pp. 99–125.

[5] Y. Mutoh, K. Tanaka and K. Takeda, I Mech E, C 251/86 (1986) 203–209.

[6] O. Jacobs, Dr.-Ing. Dissertation (TU Hamburg, Hamburg, 1991).

[7] D.J. Gaul and D.J. Duquette, Metall. Trans. A 11 (1980) 1555.

[8] K. Friedrich, S. Kutter and K. Schulte, Compos. Sci. & Technol. 30 (1987) 19.

[9] O. Jacobs, K. Friedrich and K. Schulte, in: Wear Testing of Advanced Materials, ASTM-STP 1167 (1992) 87–96.

[10] F.T. Barwell, in: Treatise on Materials Science and Technology, Vol. 13, ed. D. Scott (Academic Press, 1979) p. 1.

[11] R.B. Waterhouse, in: Treatise on Materials Science and Technology, Vol. 13, ed. D. Scott (Academic Press, 1979) p. 259.

[12] M. Godet, Wear 100 (1984) 437.

[13] K. Friedrich and O. Jacobs, Compos. Sci. & Technol. 43 (1992) 77–84.

[14] Y. Berthier, L. Vincent and M. Godet, Tribol. Int. 22 (1989) 235.

[15] R.E. Pendlebury, Wear 125 (1988) 3.

[16] Y. Berthier, C. Colombie, G. Lofficial, L. Vincent and M. Godet, in: Mechanisms and Surface Distress, eds D. Dawson, C.M. Taylor, M. Godet and D. Berthe (Butterworths, London, 1986) p. 81.

[17] T. Tsukizoe and N. Ohmae, Industr. Lubr. & Tribol. Jan./Feb. (1976) 19.

[18] O. Jacobs, J. Mater. Sci. Lett. 10 (1991) 838–839.

[19] R.W. Lang, H. Stutz, M. Heym and D. Nissen, Angew. Makromol. Chem. 145/146 (1986) 267.

[20] K. Friedrich, O. Jacobs, M. Cirino and G. Marom, in: Proc. Int. Conf. on Tribology of Composite Materials, Oak Ridge, TN, May 1–3, 1990, eds P.K. Rohatgi, C.S. Yust and P.J Blau (ASM International, 1990).

[21] Anonymous, LNP-Bulletin 254–988 (LNP-Plastics Nederland, 4940 AA Raamsdonksveer, Netherlands, 1990).

[22] K. Tanaka, in: Proc. Int. Solid Lubrication Symp., Tokyo (1975).

[23] K.L. Reifsnider, K. Schulte and J.C. Duke, in: Long Term Behavior of Composites, STP 813, ed. T.K. O'Brien (ASTM, Philadelphia, PA, 1983) p. 136.

[24] K. Schulte, K. Friedrich and S. Kutter, Compos. Sci. & Technol. 30 (1987) 203.

[25] T.-W. Chou and A. Kelly, Ann. Rev. Mater. Sci. 10 (1980) 229.

[26] O. Jacobs, K. Friedrich and K. Schulte, in: Proc. ICCM-VIII, Honolulu, Hawaii (1991) pp. 38-6-1–38-6-7.

[27] G. Fernando, R.F. Dickson, T. Adam, H. Reiter and B. Harris, J. Mater. Sci. 23 (1988) 3732.

[28] O. Jacobs, K. Friedrich and K. Schulte, Wear 145 (1991) 167–188.

[29] P.R. Edwards, in: Fretting Fatigue, ed. R.B. Waterhouse (Applied Science Publishers, Barking, 1981).

[30] D.A. Hills, D. Nowell and J.J. O'Connor, Wear 125 (1988) 129.

[31] K. Schulte, in: Damage Development and Failure Processes in Composite Materials, eds I. Verpoest and M. Wevers (Leuven, Belgium, 1987) p. 39.

[32] R.D. Jamison, K. Schulte, K.L. Reifsnider and W.W. Stinchcomb, ASTM-STP 836 (1984) 21.

[33] T.K. O'Brien, M. Rigamonti and C. Zanotti, Int. J. Fatigue 11 (1989) 379.

[34] M. Bader, in: Damage Development and Failure Processes in Composite Materials, eds I. Verpoest and M. Wevers (Leuven, Belgium, 1987) p. 8.

[35] E. Broszeit, K.H. Kloos and B.O. Schweighöfer, Z. Werkstofftech. 16 (1985) 187–193.

[36] K. Schulte, K. Friedrich and S. Kutter, Compos. Sci. & Technol. 33 (1988) 155.

[37] K. Endo and G. Goto, Wear 38 (1976) 311.

[38] J.A. Alic and A.L. Hawley, Wear 56 (1979) 377.

[39] K. Sato, Wear 125 (1988) 163.

[40] K. Sato, H. Fujii and S. Kodama, Wear 107 (1986) 245.

[41] D.W. Hoeppner, Wear 43 (1977) 267.

Advances in Composite Tribology
edited by K. Friedrich

Chapter 19

Tribological Facets of Polymer Composites Fabrication

DARYL R. WILLIAMS

Department of Chemical Engineering and Chemical Technology, Imperial College of Science, Technology and Medicine, Kensington, London SW7 2BY, UK

Contents

Abstract

In the wide range of manufacturing processes which are currently used to form polymeric composite materials, the rheological and tribological behaviour of the precursor materials and the final product is critically important. Whilst for polymeric materials processing the importance of rheological behaviour has been very well established in the literature, the importance of tribological phenomena, or the boundary conditions, is less clearly established.

The objective of this review is to provide an overview of the materials processing literature in which tribological, and related, phenomena play an important role. The fabrication processes which are explicitly discussed in this article include more recent fabrication techniques such as pultrusion and powder based fabrication techniques, as well as traditional consolidation processes such as autoclaving and compression moulding. The importance of tribological phenomena at both a gross macroscopic level, as in the primary processing operations, as well as at the microscopic level, as within the microstructure of fibre reinforced composites, will be examined through selected examples. It is hoped that this review will stimulate other workers to consider the role played by tribological or boundary phenomena at both the macroscopic and the microscopic level in the fabrication and manufacture of polymer composite materials. A better understanding of tribological phenomena in these contexts will surely assist in allowing optimised material manufacture and material performance to be attained.

1. Introduction

Whilst certain facets of the science and engineering of composite materials manufacturing processes are relatively well understood, in other areas our understanding is immature and is continuing to evolve. Consequently, in these latter areas, manufacturing and fabrication practice and methodology are still strongly determined by experience and by other empirical indicators. Major advances are continuing to be made in improving our fundamental understanding of materials processing via innovative experimental studies and are underpinned by computer modelling, which may now mimic the complex phenomena which occur within typical processing environments. However, it is not always practical to anticipate or predict all of the various phenomena which might determine an important attribute of a multiphase polymeric material which is evolving via complex manufacturing processes. Therefore, before a more sophisticated understanding of these fabrication processes can be achieved, it is necessary first to know what the fundamental processes are that are occurring during fabrication. It is the thesis of this chapter that tribological processes are one important, though poorly appreciated, facet of polymer composite materials processing.

In the wide range of manufacturing processes which are currently used to form polymeric composite materials, the rheological and tribological characteristics of the materials, as a function of the processing variables, such as pressure, temperature and time, are critical. An understanding of these material characteristics, and their interplay, is the key to gaining an important insight into the production process, which would allow optimal material processing and optimum material performance to be attained. Although the rheological and tribological properties are intrinsically linked, this chapter will specifically consider the role played by tribological processes in certain facets of polymer composite fabrication.

Although our understanding of materials processing has advanced considerably, it has not attained a level of sophistication at which the role played by tribological phenomena in the processing operation may be fully described and understood. As an aside in this context, the tribological phenomena may be regarded as a manifestation of an interfacial rheological response. Indeed, the process engineer concerned with the manufacturing process is likely to view the interfacial facets of the materials' processing operation in terms of interfacial rheology or boundary conditions, and may be totally unfamiliar with the tribological descriptions adopted in this chapter.

A large number of the manufacturing routes for polymer composite fabrication are derived from established polymer processing methods. Processes such as screw extrusion, calendering, injection moulding and compression moulding are well established routes for the manufacture of polymeric materials and many of these have been adapted for the fabrication of composite materials. The fundamentals of polymer process engineering of these materials has been described in a number of excellent books, including those by Pearson [1], Fenner [2] and Tadmor and Gogos [3]. It is beyond the scope of the current review to consider in detail the basics of polymer processing and the readers are referred to the above references. It is however pertinent to consider those general areas of tribological significance for the processing of unfilled polymers, some of which will be discussed in greater detail later in this chapter in the context of polymer composite materials.

The conveyancing and compaction of powders and agglomerates is the important early phase of many polymer processing operations, since most raw materials are in a particulate or pellet form. The tractions produced within the particulate assemblies, as well as those between these materials and the confining surfaces, such as in the feed zones or within the processing engines such as extruders, controls the processing behaviour of these products. The solid–solid contacts, which are an important feature of the material locomotion, result in the tribological description of these processing operations. For example, in a single-screw extruder (fig. 1) the friction between the material and the barrel must be high relative to the friction between the material and the screw for optimal conveying efficiency. Fenner is more explicit in this matter and notes that the surface of the screw should be "smooth" and the barrel "rough" to achieve optimal throughput [2]. In engineering parlance "smooth" and "rough" surfaces are considered to provide low and high friction, respectively. Similarly, frictional behaviour is important during the introduction of feedstock materials into the manufacturing engine or through the first stage of extrusion, where the compaction of agglomerates or powders must occur prior to melting/consolidation.

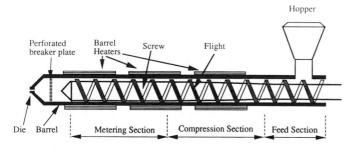

Fig. 1. Diagram of a single-screw plasticating extruder (after Fenner [2]).

The transport of the material within the processing equipment, once it has become a consolidated melt, is generally ascribed as being well modelled by drag flow in which stick, as opposed to slip, is assumed to occur at the walls of the processing engine. In this case the problem becomes one of melt rheology, of which much has been written. However, recent work has highlighted a range of materials systems where this wall-stick assumption is not well founded, including polyethylene, polyvinyl chloride and other materials [4]. Meijer and Verbraak [4] have highlighted the importance of both microscopic and macroscopic slip during the flow of certain types of materials, and have modelled the effects of various boundary conditions on material flow during extrusion.

The tribological facets of particulate materials have been considered in an extensive monograph by Adams and Briscoe, which includes chapters on particulate flow, compaction, attrition, agglomeration as well as slurry behaviour [5].

Frictional behaviour is also important in many other more routine macroscopic facets of polymer processing. However, in many cases the importance of tribological phenomena may be not practically apparent due to the extensive use of processing aids. These additives are typically lubricants and other types of tribological modifying agents which become an integral part of the polymeric material system. However, in many cases, the absence of these processing aids would have a catastrophic effect on material manufacture, either in terms of the productivity of the process, or in terms of product quality. For example, the rapid and damage-free transport of fine polymeric fibres over rollers and guides during their manufacture depends very much on the tribology of the fibre-guide contacts, which is governed by the presence of propriety processing agents on the fibre surfaces. During extrusion for example, material die exit phenomena, including problems of material adhesion to the die walls as well as die wall wear, depend on the tribology of the contacting materials.

In general, tribological phenomena would be expected to be crucial during both the initial and the final stages of the polymer processing cycle, at which stages the raw ingredients and the final product are in a solid state. During the other stages of the process the melting/softening of the material results in dominant importance of the rheological behaviour. However, during the processing of polymer composite materials the situation is necessarily more complex. The presence of a solid particulate phase, in the form of fibres or other reinforcing inclusions, results in the presence of a solid phase in all stages of the processing cycle. The presence of this solid phase has a number of profound implications for the manufacturing process, a number of which are tribological in nature and which will be discussed in this chapter.

All the manufacturing processes which are in current usage, depend to some extent on the rheological and tribological behaviour of the polymer/reinforcement system. Although these material characteristics are intrinsically linked, this chapter will focus on the role played by tribological processes in certain facets of polymer composite fabrication. The first three sections of this review will provide preliminary information relevant to the discussion of the manufacturing and fabrication techniques detailed in the remaining sections. The first section provides an introductory overview of polymeric frictional phenomena and is followed by a general summary of the consolidation behaviour of both powder and fibrous assemblies, in which frictional behaviour plays an important role. The final three sections explicitly discuss frictional facets of three important polymer composite manufacturing processes.

2. Fundamentals of polymer tribology

Tribology is the study of the science and engineering associated with the interaction of contacting solid bodies in relative motion. In its broader definitions it encompasses adhesion, lubrication, wear and friction, though these topics are strongly interrelated and should not be viewed as wholly independent phenomena. In the case of polymer-based materials, it is friction and lubrication which are the most relevant to understanding the tribology associated with materials processing.

2.1. Friction

The frictional interactions between solid bodies are generally the result of two main mechanisms for energy dissipation. The first mechanism involves the dissipation of energy in the interfacial layers between the two phases, and results from interfacial shearing of the bodies at the points of adhesive contact. The second mechanism involves the energy dissipation in relatively large subsurface volumes within the interacting bodies. Energy dissipation in this case is the result of the surface asperities of the harder solid deforming, or ploughing into, the surface of the softer solid phase. This localised deformation results in energy losses due to viscous, viscoelastic or plastic processes. Bowden and Tabor combined the interfacial and bulk dissipation processes into a two-term independent model; the adhesion model of friction [6]. This seminal model has been successfully applied to many materials problems and is reviewed in detail by Briscoe [7] and by Briscoe and Tabor [8]. An abbreviated description of this model follows below.

The adhesion and bulk energy dissipation mechanisms result in the adhesion and the bulk-deformation components of the total frictional work. Both dissipation mechanisms are treated as independent and energetically additive processes [6], though for some situations these terms may take on an interactive character. These two differing frictional mechanisms may also be distinguished by the differing sizes of the energy dissipation regions. In the case of adhesive friction, the zone for energy dissipation will correspond to the distances through which surface stresses propagate and will thus be generally less than 20 nm in thickness. Whilst for subsurface deformations, the energy dissipation zone would be greater in size and would be comparable in scale to the size of the asperities or the contact areas between the bodies. The diagram, fig. 2, shows the two primary zones for energy dissipation for a rigid body contacting a soft polymer substrate and for a soft polymer contacting a more rigid substrate.

The relative importance of these two mechanisms is dependent on the chemical and physical properties of the materials, the contact geometry, as well as the

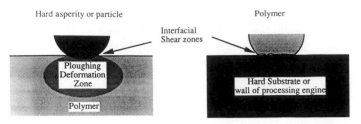

Fig. 2. A schematic diagram showing the primary zones of energy dissipation for a hard body contacting a soft polymer substrate and for a soft polymer contacting a more rigid substrate. Energy may be dissipated in the shear interfacial zone and/or the bulk deformation subsurface zone.

surface roughness and the type of relative motion of the bodies. For relatively smooth surfaces, such as high-modulus fibres, in contact with a glassy matrix phase, the adhesive component of the friction would be expected to be the dominant term. Whilst, for an elastomer containing a high-modulus reinforcement phase, the bulk or ploughing component of friction is likely to be dominant.

The primary quantity which characterises the frictional properties of a sliding contact, is the interfacial shear strength, τ_i. The interfacial shear strength may be described as the interfacial force per unit area of contact, or alternately as the energy dissipated per unit area of contact per unit sliding length. The total adhesive contribution to the interfacial friction is given by

$$F = A\tau_i,\tag{1}$$

where F is the frictional force per unit area, τ_i the interfacial or adhesive shear stress and A the real area of contact. The quantity τ_i is sensitive to the surface chemical and morphological character of the materials in contact. The contact area, A, is defined by the surface topography as well as the gross contact geometry, the normal load, W, and the bulk mechanical properties of the materials. The most important variable affecting τ_i is the mean contact pressure, P, which is defined as the normal load per unit of real contact area, W/A. A number of studies of τ_i as a function of P for polymers have been reported [9–12], and it has been demonstrated that this dependency is well modelled by the expression:

$$\tau_i = \tau_0 + \alpha P,\tag{2}$$

where τ_0 is the interfacial shear strength at zero contact pressure and α is the pressure coefficient of the shear strength. Values for τ_0 and α have been reviewed in detail by Briscoe and Smith [13]. More recently the use of eq. (2) led to an alternate expression for the coefficient of friction:

$$\mu = \tau_0/P + \alpha.\tag{3}$$

For large contact pressures, this equation reduces to:

$$\mu \approx \alpha.\tag{4}$$

Studies of the relationship between polymer morphology/chemistry and the values of τ_i have been undertaken by Amuzu et al. [10,11]. They have reported on the variations in τ_i for a homologous series of poly n-alkyl methacrylates which was rationalised in terms of the molecular engagements of protruding segments of the different size pendant alkyl groups. Pooley and Tabor [12] have interpreted the interfacial shear properties in terms of the molecular profile or "roughness" of the polymer molecules. PTFE has a smooth helical structure, which can be perturbed by the introduction of perfluoromethyl groups in place of the fluorine atoms. The frictional forces of PTFE were observed to increase significantly with the introduction of these large substituents. Briscoe and Smith [13] have reported on the effects of polymer conformation and morphology on the interfacial shear behaviour of a wide range of polymeric films.

A critical facet of the frictional interactions between contacting bodies is the real contact area, A, between bodies. A number of theoretical models based on contact mechanics have been proposed which allow the variation of A with the normal load, W, to be predicted. Since it is difficult, if not impossible, to measure A directly, these models have attained a critical importance. For contacting bodies, the Hertzian mechanics model of an elastic contact predicts that the area of contact between a flat plate and a sphere of radius r_S is given by

$$A = \pi [DWr_S]^{2/3}, \tag{5}$$

$$D = \frac{3}{4}\left[\frac{1 - \nu_1^2}{E_1} + \frac{1 - \nu_2^2}{E_2}\right], \tag{6}$$

where eq. (6) allows D to be defined in terms of Young's modulus E and Poisson's ratio ν for contacting materials 1 and 2 [14]. These same equations also apply for a pair of elastic cylinders in an orthogonal contacting geometry in which case the sphere radius, r_S, would be replaced by the cylinder radius, r_C.

Using eqs. (2) and (5) an important result can be derived for the frictional force between a pair of contacting bodies:

$$F = \tau_0 \pi [DWr_S]^{2/3} + \alpha W, \tag{7}$$

which leads to the conclusion that the frictional force will scale with the normal load to a power of between 2/3 and 1. Experimentally this load dependency is observed, the lower-order load dependency corresponding to simple Hertzian elastic frictional contacts, whilst the higher-order linear load dependency is attributed to rough surfaces (multi-asperity contacts) or plastic deformations. A number of investigations on the frictional behaviour of monofilament systems have confirmed the usefulness of the adhesion model of friction and the other principles described above [15–17]. However, empirically eq. (8) gives a good fit to experimental data, where n is known as the load index, k is the friction constant and may be theoretically derived from eq. (7):

$$F = kW^n. \tag{8}$$

This section has so far concentrated on the frictional processes associated with the surface regions of the material, and we will now describe the bulk subsurface deformations introduced earlier in this section. The bulk or ploughing component of the friction, as illustrated in fig. 2, refers to the dissipation of frictional energy in the bulk of the material, as might occur for example during the sliding of a hard asperity across a softer substrate. The work dissipated per unit of sliding distance will be a function of the modulus, or flow stress, as well as the contact conditions and mode of deformation. The strain associated with these deformations is defined by the contact geometry. The bulk strain will scale with tan θ, for a substrate in contact with a conical asperity with a total included angle of $(180 - 2\theta)°$. Briscoe et al. [18] have reported on the sliding frictional behaviour of PMMA in contact with various conical indentors. For blunt indentors, with small θ, energy was dissipated via reversible viscoelastic processes. As θ increases, plastic flow accompanies the viscoelastic processes, which culminates in the onset of brittle tearing and cracking.

These types of dissipation mechanisms can result in the formation of debris associated with wear in the contact zone. This leads into a particularly important and complex facet of tribology, which involves the so-called "third-body" effects, which involve the formation and entrainment of debris within the contact zone, the detailed discussion of which is beyond the scope of this review.

2.2. Lubrication

Lubricants play an important role in many materials processing operations by reducing the friction and wear, which would otherwise occur during solid–solid interactions. The role of a lubricant is to interpose itself between the two contacting bodies, and in doing so, reduces the extent of solid–solid contact via the formation of a weak interfacial layer. Frictional energy dissipation therefore becomes localised in this lubricant layer, which, by virtue of its chemical and physical structure, is able to accommodate large strains and strain rates without mechanical or chemical failure. Although the lubrication process may be described in a conceptually simple manner, the intrinsic theory and practical application of the concepts of lubrication to real problems can be very complex.

There are three major regimes of lubrication [19]:
(i) hydrodynamic lubrication,
(ii) elasto-hydrodynamic lubrication and
(iii) boundary lubrication.

The presence of a liquid film constrained between two ideal, rigid surfaces in sliding contact may result in hydrodynamic lubrication (see fig. 3). Because of the fluids viscosity, η, a hydrodynamic pressure builds up between the contacting surfaces, which allows the surfaces to maintain an equilibrium separation. Optimal hydrodynamic lubrication will occur when the equilibrium film thickness, δ, is greater than the typical asperity height, h, on the contacting surfaces. The shearing of the contacting bodies thus becomes dependent on the shear of the interposed lubricant layer and is specifically determined by the viscosity of the liquid lubricant layer. The lubrication process is commonly described by the dimensionless parameter $\eta N/P$ where η is the liquid viscosity, N is the relative velocity of the contact and P is the normal pressure. Since the only relevant material property is η, it becomes a matter of some importance to understand what variables will effect η.

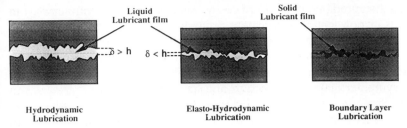

Fig. 3. The three major regimes of lubrication.

The three most significant factors are the shear rate, since this will control the adiabatic heating of the liquid film with consequent changes in η, the possibility of shear thinning for a non-Newtonian fluid and the effects of pressure.

In many lubricated contacts, the assumption of non-deforming contacts, implicit within the models of hydrodynamic lubrication, are unrealistic. In many real bodies elastic deformations between the contacting bodies, and most specifically the asperities present on their surfaces, are the rule rather than the exception. In the cases in which elastic contacts occur between the sliding bodies in the presence of a liquid lubricant, then elasto-hydrodynamic lubrication may occur. For these types of lubricated contacts, the viscosity of the fluid as a function of pressure is important and is normally described by Barus's equation, which is a common feature of elasto-hydrodynamic models:

$$\eta = \eta_0 \exp(\alpha P), \tag{9}$$

where P is the pressure, α is a constant and η_0 is the viscosity for zero pressure. Therefore, a significant increase in viscosity will accompany a modest increase in the pressure.

The final type of lubrication to be considered is boundary lubrication, which refers to lubrication by a solid film. Boundary lubrication is typically provided by weak solid layers which are typically less than 200 nm in thickness. These layers may be chemically reactive and reactively bond to the underlying substrate to provide the lubricant action, or more commonly, they may be applied via simple solution deposition techniques. Common boundary lubricants include certain organic polymers, surfactant species and metallic dichalcogenides such as MoS_2. Unlike the preceding forms of lubrication, asperity contact will occur during sliding, since the lubricant film thickness will be typically less than the surface roughness. The frictional tractions which develop in boundary lubricants are similar to those exhibited by highly compressed liquid lubricant films. Other facts of the behaviour of boundary lubricated materials may be taken from section 2.1 on friction.

3. Powder and particulate processing

The frictional properties of polymeric materials, such as powders or fibres, are important in many material manufacturing and fabrication processes. The previous section reviewed the fundamentals of tribology, and this section will survey the tribological facets of powder and particulate processing.

There is an abundance of literature in this area and the reader will be introduced to the key topics via recent work in this area. The information considered will be of particular relevance to the fabrication of polymer-based composites involving particulate compaction and/or particulate flow.

In a majority of polymer composite processing operations the raw products are initially in a fibrous, powder, granular or pellet form. Two of the earliest stages of processing are the consolidation of this particulate material and the transport of the feedstock into the processing engine, or indeed transport of the product over the surfaces of the processing engine. All of these operations will both involve particle–wall or particle–particle friction.

The friction associated with particulate materials flowing over a wall can be a significant factor in a number of processing or material conveying operations. Adams et al. [20] have reviewed this topic and have highlighted the success in modelling wall friction via the use of a semi-continuum model based on the mechanics of a single particle–wall frictional contact. This model predicts that the wall friction will be independent of particle size for smooth materials sliding against smooth walls. This behaviour is the case since the increase in the particle size, and thus the decrease in the number of wall contacts per unit area, is directly offset by the increased Hertzian contact area of these contacts. However, the analysis is far more complex for rough particles in contact with either smooth or rough walls. Generally, plug flow will normally occur for particulate flow involving smooth walls, whereas for rough walls the particle flow is accommodated via a shear zone within the bulk particulate materials. Estimates of the position of the zone vary between 6 and 12 particle diameters from the wall and the material flow is accompanied by a dilation necessary to accommodate this particle rearrangement.

The behaviour of powders in hoppers and other bulk storage systems has been considered by Tuzun [21], who has reviewed the frictional properties of powders from this bulk handling standpoint. He clearly identifies the coefficient of internal friction, μ_i, of the powder and the coefficient of wall friction, μ_w, as being the primary variables. μ_i determines the stress distribution within the deforming powder bed, whilst μ_w will determine the magnitude of the stresses between the materials and the walls of the container. This author has detailed the yield and failure criteria for the flow of both cohesive and cohesiveless powders, and has highlighted the successful use of discrete particle approaches to the modelling of powder flow, compared to stochastic and continuum approaches, especially once powder discharge has commenced.

Few polymeric materials exist which have not at some stage been processed through an extruder. As was previously introduced, the wall behaviour of materials within extruder systems is complex and may not be well described by the rheological models of wall drag in common usage. As was noted earlier, Meijer and Verbraak [4] have highlighted the existence of slip boundary conditions during the processing of certain polymers. Other workers have observed slip boundary be-

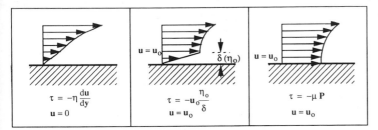

Fig. 4. Differing boundary conditions for wall traction in an capillary. Left to right: viscous drag, low viscosity slip and Coulombic friction.

haviour for multiphase materials, including food, bulk moulding compounds as well as sheet moulding compounds [22] and unfilled/filled elastomers [23]. For the modelling of extrusion with slip boundary conditions, Meijer and Verbraak [4] have considered the isothermal flow of a Newtonian fluid in a capillary with three different boundary conditions. These boundary conditions are listed below and shown in fig. 4:

(i) Boundary with viscous stick or drag. The material velocity at the boundary is zero.

(ii) Hydrodynamic lubrication. Boundary with velocity-dependent wall shear stress for materials in which a very thin, low-viscosity layer is sheared between the wall and the extrudate.

(iii) A pressure-dependent Coulombic frictional boundary condition, similar to that used for the plug flow modelling of processing within the solids-conveying zone of extruders.

The possible operation of three distinct boundary conditions for the processing of filled polymer materials has a number of implications. Complex transitions between differing wall boundary conditions during processing could be expected to occur, for example, from boundary slip to wall boundary stick due to pressure-induced melting at the boundaries [24,25], which would inevitably affect the final product as well as its processing. The situation within a real extruder, as distinct to a simple capillary, is even more complex due to the existence of two independent surfaces: the barrel and the wall. This results in four, or more, different permutations of slip/stick boundary condition behaviour [4]. These workers concluded that the nature of the boundary condition behaviour will have a dramatic effect on the pumping efficiency of extruder systems as determined by the material velocity profiles, residence times and the extruder efficiency.

In extruder processing of polymer-based materials, a range of solid-state properties is important in the processing operation. Although the chemical engineer will be concerned with the thermal conductivity and specific heat of the materials since these properties control the ease with which melting may occur in the processing engine, the tribologist would be equally concerned about both the internal friction between the solid polymer granules or powders as well as the external friction between the particles and the constraining walls in the processing system. Of course, once consolidation has occurred within the extruder the primary concern tribologically will be the friction of the material at the screw and barrel surfaces.

The consolidation of particulate materials to facilitate processing, or directly as a manufacturing process, involves complex particle–particle interactions with associated elastic, plastic and viscoelastic deformations. The compaction behaviour of particulate materials is commonly expressed in terms of an applied pressure–bulk volume relationship. These expressions may be derived on an empirical or a semi-empirical basis and are somewhat analogous to the stress–strain curve that would be obtained for a continuous solid body, although, unfortunately, they do not convey the same level of fundamental information as given by a stress–strain curve. Although both wall and bulk frictional processes play a critical role in the compaction process, a number of these models are based on continuum approaches which do not consider friction explicitly, whilst some other models, which consider the discrete nature of the particulate interactions, include frictional phenomena. The continuum models are typified by the early approaches of Janssen [26] and the later Walker [27] models, whilst the Spencer [28] model typifies one of the earlier particulate approaches. More recently, a number of workers have modified these approaches to take account of wall friction [29] or to separate wall friction from the bulk compaction phenomena [7,8]. Amongst workers studying metallurgical and pharmaceutical powder compression, a number of simple expressions have become widespread in their usage. Assuming that compaction could be treated like a first-order chemical reaction, Heckel has derived a simple expression for the pressure–volume behaviour of a powder [29,30]. This equation is shown below, where D is the relative density, D_0 is the relative apparent density, P is the pressure and K is an empirical constant:

$$\ln\left(\frac{1}{1-D}\right) = KP + \left(\frac{1}{1-D_0}\right). \tag{10}$$

An alternate model is the Kawakita–Ludde equation, which has been found to offer more widespread applicability [30,31]:

$$\frac{P}{C} = \frac{1}{ab} + \frac{P}{a}, \tag{11}$$

$$C = \frac{V_0 - V_p}{V_0}, \tag{12}$$

where C is the volume reduction on compression, P is the pressure, a is a constant equal to the initial bed porosity and b is an empirical constant related to the ease of powder compression. On the assumption that compression proceeds via a particle rearrangement mechanism, Cooper and Eaton [34] have proposed an expression that allows the pressures required at which densification occurs to be estimated via extrapolation:

$$\frac{E_0 - E}{E_0(1-E)} = a_1 \exp\left(\frac{-k_1}{P}\right) + a_2 \exp\left(\frac{-k_2}{P}\right), \tag{13}$$

where E_0 is the initial bed porosity, E is the porosity of the compact and P is the pressure. a_1, a_2, k_1 and k_2 are empirical constants which pertain to the rearrangement {1} and compression at infinite pressure {2}. Pressure–volume compaction relationships have been reviewed by Kawakita and co-workers [33,34], whilst more recently various workers have provided significant bibliographical detail on the wall frictional facets of compaction phenomena [35,36].

Unfortunately however, the current generation of pressure–volume relationships does not provide insight into the fundamental mechanisms or micromechanics of the compression/compaction process, due to their strong empirical bias. They, therefore, can only provide a limited insight in compression phenomena and do not allow the frictional facets of the process to be explicitly deduced. They do, however, underline the critical importance of frictional processes.

4. Mechanics of fibrous assemblies

4.1. General mechanics

The mechanical properties of fabrics and fibrous assemblies are complex and do not exhibit the linear behaviour typified by simple continuous bodies such as monofilaments. In general, fabrics exhibit large levels of non-linear deformation and significant levels of hysteresis. The mechanics of flexible fibrous assemblies has been reviewed in important monographs by Hearle et al. [37,38]. Grosberg and Swani [39] have noted that significant levels of strain energy can be "locked in" by the friction at contact points within a fibre array. In reviewing the role of friction in the mechanical performance of fabrics, Grosberg [40] has commented on the three major regions of fabric deformation and response.

(i) Initial non-linear deformations dominated by frictional effects.

(ii) A linear deformational region where the material performance is determined by the elastic behaviour of the component fibres.

(iii) A non-linear region at large deformations, dominated essentially by plasticity and creep behaviour.

The primary ramification of the frictional behaviour of fibrous assemblies is the significant levels of hysteresis observed during the cyclic deformation of the materials. The relative motion of the fibres and yarns within the assemblies, the contact forces between the elements and the coefficient of friction will define the frictional energy dissipated during deformation.

The original seminal work on the modelling of fabric properties as a function of the geometry of fabric weave was reported by Pierce [41]. The basic nature of this model remains unchanged, although it has since been modified by a number of other workers. It forms the basis of most analyses of specific-fabric mechanical properties. The mechanical properties of fabrics are mathematically described in terms of the yarn spacing, the weave angle, the yarn diameters, the crimp heights and the thread spacings. Research on the tensile and bending properties of fabrics has been carried out by a relatively small number of workers; Hearle, Grosberg, Leaf and Abbot. The tensile characteristics of typical fabrics have been analysed by

Hearle et al. [38] as well as by Leaf [42] and Grosberg [43]. Bending properties of fabrics have been studied by Grosberg and co-workers [44] as well as by Gibson and Postle [45]. In the cases of both tensile and bending deformations, these researchers have concluded that inter-fibre and inter-yarn friction is a very significant factor in the mechanical response of the fabrics. The ability of fibres and yarns to migrate, and thus accommodate the imposed stresses, directly depends on the frictional forces associated with these fibre interactions. However, none of these models directly describes the effects of fibre friction, but rather include this behaviour in a general hysteresis parameter.

Fibre migration has been considered by a number of workers in the context of twisted yarn systems. These models have been reviewed by Motamedi [46], who points out that none of these models accounts directly for the inter-fibre frictional forces. Recent work has shown that the frictional properties of aramid fabrics with or without lubricants markedly affect the static modulus and also the ballistic energy dissipation capacity [47]. Briscoe and co-workers have also demonstrated the importance of frictional effects in determining the mechanical behaviour of "Kevlar" fabrics under indentation loading [48,49]. The effects of friction were successfully modelled on the basis of changes in the effective tensile modulus of the fibre assembly. Low friction at the filaments junctions reduces the tensile modulus of the yarns and also the effective modulus of the fabric.

4.2. Compressional behaviour

Many processes for the manufacture of polymer composite materials involve the compaction or compression of fibrous assemblies. Developments in our current understanding of compaction/compression of fibrous assemblies have been dominated by the early work of van Wyk [50]. This important theoretical work predicted the relationship between the applied pressure and the volume of a random fibrous assembly by considering the fibre assembly to be a system of bending units. This theory combines the elastic beam theory bending of component fibres with a mean fibre–fibre contact length for the random fibrous assembly. The model takes no account of fibre twisting, extension or frictional slippage at the fibre–fibre junction. The derived relationship is

$$p = A\left[\frac{1}{V^3} - \frac{1}{V_0^3}\right], \tag{14}$$

where p is the pressure, A is a constant, while V and V_0 represent the assembly volume for pressures p and the pressure $p = 0$.

Carnaby [51] has reviewed this topic and has highlighted the limitations of the van Wyk model. Dunlop [52] has reported only modest agreement between eq. (14) and the experimental behaviour of random wool fibre assemblies, noting particularly the poor fit for those fibre assemblies in which inter-fibre frictional slippage occurred. Nevertheless, the van Wyk model is generally found to be a useful relationship for many fibrous systems.

Few explicit models have been reported for the compressional properties of planar fabrics. Although it is suggested that these models are likely to produce relationships which are similar in form to eq. (14) [12], this topic has rarely been studied. One of the more recent equations was proposed by Gutowski and co-workers [53,54]. Their equation is shown below and is similar in form to that proposed by van Wyk:

$$p = A\left[\frac{V_f}{V_0} - 1\right]\Bigg/\left[\frac{1}{V_f} - \frac{1}{V_a}\right]^4, \tag{15}$$

where A is a "spring constant", V_f is the original fibre volume, V_a is the available fibre volume and V_0 is the fibre volume for zero pressure p. The authors report good agreement between this expression and experimental results for fibre bundles. Batch and co-workers have developed a series of models for the compaction of random and aligned fibrous materials [55,56].

The importance of fibre compaction has been noted in the manufacture by compression moulding of short-fibre reinforced polymers. Silva-Nieto et al. [57] have reported that for sheet moulding compound with 25% glass fibres, the force response for squeezing flow between a pair of parallel disk plates does not exhibit a simple exponential monotonic behaviour with the gap height, as would be predicted on the basis of the Scott equation for a viscous fluid [58]. The possible formation of an elastic network within the fibres, due to the frictional stick between the fibre–fibre contacts, was directly supported by the work of Zentner [59], who measured the stiffness of the fibre polymer network by compressing the SMC material between a pair of parallel disk plates. Via sinusoidal mechanical loading he confirmed the change from a viscous, non-elastic behaviour to one with an elastic, non-viscous character of the material with decreasing plate separation.

Predicting the moulding forces during the manufacture of sheet moulding compounds (SMC) is necessary for optimised material fabrication. Castro and Tomlinson have proposed a model which predicts the compressional force necessary for SMC manufacture under non-isothermal conditions [60]. The model requires two materials parameters, namely the material resistance to extension and the frictional coefficient. The model includes slip at the boundary walls and assumes that the resistance to material flow is thus constrained by resistance of the SMC to extension as well as friction at the surface. Lubricating the walls was found to affect the material velocity profile during processing.

The permeability and compressibility of aligned and cross-ply carbon fibre beds during composites processing has been recently investigated by Lam and Kardos [61]. The compressibility of wet (water or silicon liquid) impregnated fibre beds was found to be independent of both the liquid surface tension and liquid viscosity, and was therefore related to the elastic properties of the fibrous assembly.

Briscoe and Williams [62] have investigated the relationship between fibre surface properties and the compressive behaviour of "Kevlar" fabrics. Fibres with a low interfacial shear strength, and thus low inter-fibre friction, were created by coating the fibres with poly(dimethyl siloxane), a well known lubricant. Fibres with

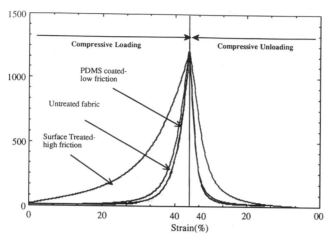

Fig. 5. Compressive loading and unloading cycle for a range of surface-modified "Kevlar" fabrics.

an increased interfacial shear strength, and thus high friction, relative to the "as-received" "Kevlar" fibres were prepared by modifying of the fibre surface. Eight layers of these different "Kevlar" fabrics were horizontally stacked together and compressed against a flat platen and the results obtained are discussed below.

Figure 5 shows the compressive loading and unloading cycle for three different surface-modified "Kevlar" fabrics. The PDMS lubricated fabric and the untreated fabrics both exhibit similar compressive behaviour, both during loading and unloading. This behaviour was anticipated, since it has been previously shown that commercial "Kevlar" materials contain significant amounts of organic processing aids which are lubricants by nature [63]. The higher-friction fabric shows dramatically different behaviour, including a significant level of hysteresis as would be anticipated for a high-friction material and in contrast to the lower level of hysteresis exhibited by the lubricated fabrics. The macroscopic mechanical behaviour of fabrics is directly related to the inter-fibre friction and fig. 5 shows the clear and substantial differences evident. The shape of the loading curve indicates that the high-frictional fabric is a significantly more compliant system than the other two fabrics. These other two fabrics show a steeper gradient in the load versus strain curve, indicating that these materials are effectively stiffer systems mechanically. Using the compliance of the fibre assembly as a qualitative measure of the inter-fibre friction, a ranking order of the frictional coefficients may be deduced:

Surface treated ≫ "as-received" > PDMS coated.

The compressive deformation of aramid fabric, shown above, is a particularly appropriate method to sense fibre–fibre friction, since the fabrics undergo a similar deformation during the fabrication of laminate specimens in the autoclave.

5. Pultrusion

Pultrusion is an important technique by which polymer composites containing continuous fibres, with either thermoplastic or thermoset matrices, may be manufactured. It is conceptually similar to extrusion, except that the product is pulled through the die rather than being pushed through the die. Due to the continuous nature of the manufacturing process and the quality of the final product, including the high-quality surface finish, this process is becoming more extensively used. During the past five years a number of models have been reported for the pultrusion process. However, in common with all composite manufacturing processes, these models are necessarily complex due to the simultaneous changes in the thermal, mechanical and chemical properties during the manufacturing process.

Figure 6 shows a schematic diagram of a generic pultrusion process. The pultrusion process involves the pulling of a series of prepreg tapes of a continuous fibre, or prepregged fabrics, through a heated die or channel. The matrix phase may be either thermoplastic or thermoset and will be melt flowed or cured into the desired shape during the passage through the heated die. The die is normally a long thin channel which will taper down to a final exit dimension.

Modelling of the pultrusion process for thermoset polymeric materials has been undertaken by Han et al. [64], Walsh and Charmchi [65], Lo et al. [66] and Batch [67]. More recently work has focussed on the pultrusion of thermoplastic materials; Bibbi and Gutowski [68], Larock et al. [69] and Wilson et al. [70]. This review will consider the very recent modelling of the pulling force necessary to extract the final material from the pultrusion machine.

Lee et al. [71] have produced a model of thermoplastic pultrusion for fabric-based composites. The model consists of three components: a thermochemical

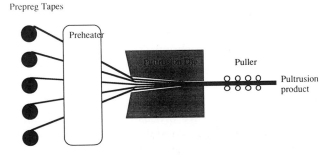

Fig. 6. Schematic diagram of pultrusion process.

submodel, a consolidation submodel and a pulling-force submodel. It is the pulling-force model which will be of interest to us, since frictional processes are crucial element of this model. There are two major frictional components of the pulling force. These are:

(i) Coulombic friction between solid composite and the die wall.

(ii) Hydrodynamic friction resulting from the shearing of a thin "fluid" resin layer constrained between the composite and the die wall.

These workers do not, however, produce any experimental data for confirming the validity or accuracy of this model. Anstrom and Pipes [72,73] have also modelled the pultrusion of continuous thermoplastic composites. Their model of the pulling force considers the frictional, compactional and viscous forces which resist the pultrusion process. The authors specifically propose that the forces resisting pultrusion consist of an interfacial viscous force, F_v, a frictional force, F_f, and a force that resists compaction of the material through the die, F_c.

$$F_v = \eta_0 \frac{U}{\delta} \cos \beta \left[1 + \left(\frac{\lambda U}{\delta} \right)^2 \right]^{(n-1)/2}, \tag{16}$$

$$F_c = (P_m + P_f \cos \beta) \sin \beta, \tag{17}$$

$$F_f = \mu P_f \cos^2 \beta, \tag{18}$$

where η_0 is the zero shear rate viscosity, β is the angle of taper in the die, λ is a time constant relating to shear thinning, μ is the coefficient of friction, n is a dimensionless constant, U is the velocity of the material and δ is the thickness of the boundary layer as given by

$$\delta = r_f \left[\sqrt{\frac{2\pi}{\sqrt{3} \, v_f}} - 2 \right], \tag{19}$$

where v_f is the fibre volume packing and r_f is the fibre radius. This approach uses the Carreau model for a shear-thinning non-Newtonian fluid [74], and these authors present experimental data which confirm the general correctness of this approach. Indeed, the practical problems associated with friction during this type of processing operation have led to the obvious expedient of the introduction of pultrusion dies with inner surfaces coated with solid lubricants (Kanzaki [75]).

6. Powder fabrication processes

One important trend in the design of many new polymer composite materials is the need for new materials with improved high-temperature performance. However, the development of these new thermoplastic and high-temperature thermoset matrix phases is limited by the lack of cost effective processing routes. Indeed, a number of workers have evaluated the use of powder-based fabrication techniques, including the impregnation of fibrous assemblies with powdered matrix phases as well as the formation of thermoplastic materials via powder compaction and sintering. In both cases, these approaches for composite fabrication are at an early stage in development, though of course the use of powder-based manufacturing methods for the formation of ceramic and metallic materials is quite advanced. Although it is not possible yet to conclude what the final importance of tribological phenomena will be for these polymer powder fabrication methods, a preliminary insight is available by surveying the literature.

6.1. Powder prepregging

A number of workers have reported on the direct use of thermoplastic polymeric powders for the formation of composite prepreg materials [76–81]. The basic strategy involves passing the fibres or roving through a bed of polymer powder, either loosely packed or within a fully fluidised powder bed. Particles are electrostatically attracted to the fibre surface and the entire material is passed through a heater and finally a die, resulting in a thermoplastic prepreg material. The impregnation chamber which is used to transfer the powder to the fibres, is a critical item in the fabrication process and must deliver a controlled concentration of powder to the fibres during the coating process. The earliest reported work was that of Price, who reported the successful prepregging of glass fibres using a powdered matrix phase [76]. The fluidisation of fine particles (< 10 μm) is complicated by the presence of intermolecular forces which can result in the formation of agglomerates.

A significant tribological facet of the processing, observed by Iyer et al. [80], is the tribo-electrification of the fibres during processing. Fibres or particles tend to acquire charge during the frictional contacts with other bodies. This phenomenon is known as contact or tribo-electrification. Polymer particles tend to acquire negative charges very easily and due to their low dielectric constants, charge dissipation will be slow. Iyer et al. [80] observed that the tribo-electrification resulted in strong adhesion being generated between the graphite fibres and a range of thermoplastic powder particles, and this adhesion was sufficient to allow the final prepreg material to be formed by passing it through a heated oven.

6.2. Polymer powder compaction

Various high-temperature polymers, such as PEEK, are difficult and expensive to process using traditional melt processing routes. Recent work has considered the scope for the use of compaction and sintering processes, as are commonly used for the fabrication of metals and ceramic materials, for the formation of polymeric products [81–88]. Reilly and Kamel [81] have specifically evaluated the cold compaction properties of PEEK powders. A range of particle sizes were evaluated, and it was found that the particles with the smaller size had the roughest surface and the highest inter-particle friction.

Bigg [88] has reported on the compaction and sintering of both thermoset and thermoplastic materials, although the emphasis is on the sintering rather than the compaction processes. Umeya and Hara have examined the rheology of compacting polystyrene powders in some detail [86–87]. From stress-relaxation measurements the authors concluded that relaxation at shorter times was due to sliding between the particles. Addition of a lubricant, magnesium stearate, had a significant effect in reducing the interparticle friction, and thus facilitating the compaction process.

7. Autoclaving of laminates

Autoclaved fabrication of composites is one of the most popular routes for the manufacture of larger components. Composite laminates manufactured from continuous thermoset prepreg materials are fabricated by arranging the prepreg material into the necessary shape, followed by curing of the thermoset material. The curing process involves the application of elevated temperatures so that polymerisation of the prepolymer may proceed. Simultaneously, pressure is applied to the laminate, so as to consolidate the plies, as well as allowing the removal of excess matrix phase and the compression of void bubbles as the thermoset polymer cures. The overall manufacturing process becomes further complicated since neither the pressure or the temperature are kept constant during processing.

It is currently common industrial practice to prepare test specimens for different operating conditions and to determine the optimal conditions by testing the performance of the test materials. This approach is clearly inefficient and implicitly assumes that one of the chosen test conditions is near the optimal processing conditions. To help overcome this problem, a number of models have been developed for simulating the autoclaving process during the last ten years.

In this section it is the compaction properties of the prepreg layers during curing which have our principal interest. It will be argued that the fibres or fabric within a laminate system, during cure constitute a non-elastic material in which the compressional deformational characteristics are directly related to the inter-fibre friction. The compression of fabrics was explicitly considered in a previous section.

7.1. Modelling of the compaction during autoclaving

During the last ten years a number of models of autoclave processing have been reported. Earlier models, such as those by Springer and Loos [89,90], have emphasised the resin-flow aspects of the composite fabrication process but have not explicitly considered the deformation of the fibre network. Similarly, Halpin et al. [91] and Lindt [92] have modelled the complex temperature- and pressure-dependent resin flow through a porous fibrous network along similar lines.

One of the earlier models that considered elastic deformation of the fibre network was of Bartlett [93], whilst the more recent work of Springer and co-workers [94] also recognises the importance of these types of deformations. Gutowski et al. [95,96] have recently considered in some detail this phenomena and have pointed out that during autoclaving the resin viscosity falls to very low values, resulting in the applied load, the processing pressure, being carried almost entirely by the fibre network. They went on to conclude, therefore, that the final volume fraction of the laminate would thus be solely dependent on the deformational characteristics of the fibre assembly, and therefore, in general, independent of the resin-flow processes. Although the importance of multiple fibre–fibre contact was recognised, especially for high volume fractions, the authors did not comment on the inherent frictional character of the deforming assembly.

Yoon et al. [97] have investigated the compaction properties of unidirectional graphite/epoxy laminates during cure. Their study involved the in situ measurement of the laminate thickness as a function of time during autoclaving of a unidirectional laminate. They were able to clearly show that the extent of compaction was a function of the cure pressure as well as the cure temperature. The final compacted thickness showed small variations with temperature and pressure, though the authors have not considered the intrinsic mechanisms associated with the compaction processes.

Recent work by Batch and Cumiskey [98] has considered the compaction facets of fibres in composites processing in some detail. These workers consider both fibre compaction and fibre flow, and attempt to deal with these two complex processes during composites processing. The fibre compaction process is modelled as a system of stiffening non-Hookean springs at higher compaction pressures and as a series of Hookean springs at lower pressures.

7.2. Microscopic influences of tribology

Gutowski [95] has commented that the fibre volume of an autoclaved laminate is dependent entirely on the deformation behaviour of the fibrous network, which makes up the laminate, during fabrication. In this section we consider the next important implication of processing. Specifically, an interesting relationship between the fibre microstructure, processing, fracture toughness and fabric friction will be presented as an example of the complex interrelationships which can exist between materials processing, tribology and composite morphology.

Briscoe and Williams [99] have investigated the fracture performance of aramid–epoxy laminates in which the laminates were fabricated using fabrics with three differing surface treatments. The compressive properties of these fabrics were discussed earlier in this chapter. The mode I fracture properties of laminates fabricated from these materials, were measured and the average results obtained are shown below. The fracture performance of the PDMS and the "As-received" or untreated fabrics gave an "R curve", consistent with the presence of significant levels of fibre bridging, which was also visually confirmed. However, the high-friction surface treatment resulted in fracture energies which were independent of crack length, and thus were associated with an absence of fibre bridging. The authors concluded that the surface modification of the fibres directly influences the migration characteristics of the filaments within the weave during compression, and that this process was controlled by the fibre–fibre friction. As a result fibre bridges were created in the wake of the inter-laminar crack, which toughened the composite by effectively increasing the volumetric viscoelastic work associated with the fracture process. A detailed analysis of the fracture data has allowed the size

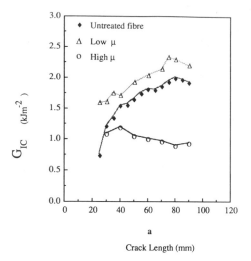

a

Crack Length (mm)

Fig. 7. G_{IC} versus crack length, a, for laminates made from various surface-treated "Kevlar" fabrics.

of the fibre bridging zone, a_b, to be estimated. This estimation was accomplished by plotting the reduced bending force as a function of the crack length, followed by a subsequent extrapolation to zero load, and is detailed elsewhere (see fig. 7) [99]. The results obtained using this analysis are summarised in table 1 and clearly indicate that a_b varies significantly amongst the various materials.

A simple pragmatic argument may be advanced on the basis of the fabric compression experiments. Fabrics used in the formation of laminates are typically exposed to uniaxial and hydrostatic pressures of between 0.5 and 1.0 MPa during autoclaved fabrication. The fabric compression results shown in fig. 5 indicate that for this pressure range the normal compressive strain of the various surface-treated fabrics varied significantly. The lubricated fabrics underwent a more significant compressive strain deformation than the surface-treated materials for a specific compressive loading. A direct consequence of this observation is that the lubricated material will have accommodated significant fibre migration, across the fabric inter-ply zones, during compression. It is argued that this same process occurs during manufacture and that the resulting differing fibre distributions thus account for the variations in the extent of fibre bridging observed during fracture.

TABLE 1
The fibre bridging zone size, a_b, for various "Kevlar"–epoxy laminates [99].

	a_b (mm)
Untreated fabric	19
Poly(dimethyl siloxane) coated (low μ)	23
Surface treated (high μ)	5

Thus, certain important facets of this fibre microstructure which were responsible for the occurrence of fibre bridging, were deduced as being controlled by the fibre–fibre friction experienced during manufacture.

Hunston et al. have observed that the fracture performance of composite materials depends strongly on the manufacturing process and the fabrication methodology [100]. These unexpected dependencies are commonly described as being due to local fibre nesting within the materials. One possible tribological explanation for why these "fibre nests" are not disrupted during manufacture is the possible existence of strong friction junctions within the fibre assemblies, although this thesis has not been explicitly investigated.

More recently this author has considered the fibre bridging phenomena within aramid–epoxy laminates, so as to further clarify the importance of local fibre microstructure [101]. Aramid fabric surfaces were abraded with an emery paper so as to create a felt-like fabric surface, with a high surface density of fibre ends. It was anticipated that this fabric modification would result in a different fibre microstructure within the laminate, since the free fibre ends would have a mobility during processing analogous to that exhibited by the lubricated fibres reported in the preceding study. The mode I fracture results showed a dramatic increase in fracture toughness for the laminates manufactured from surface abraded fabrics. Polished cross sections taken from these laminates reveal a subtle but discernible difference in the local fibre distribution. This observation supports the hypothesis that the local fibre structure within laminates was highly sensitive to the ability of fibres to migrate during manufacture, which in turn could be influenced by inter-fibre friction.

8. Conclusions

Selected facets of the tribology associated with polymer composites fabrication and manufacture have been reviewed. Although tribological phenomena are significant in various manufacturing processes, little cohesive research has been reported on this facet of materials process engineering. However, two areas have been clearly identified which typify the way in which tribological phenomena can play an important role in material manufacture at both macroscopic and microscopic levels. The first concerns interactions between the walls or constraining surfaces of processing machines and the composite material, such as in extruder barrels or pultrusion dies. The second area concerns the consolidation or compaction behaviour of materials, in which frictional effects can play an important role in the evolution of the microstructure. It is hoped that this review will encourage other workers to investigate the role played by tribology in their manufacturing or fabrication environment.

List of symbols

a	constant equal to the initial bed porosity
a	crack length (mm)

a_b	length of fibre bridging zone (mm)
a_1, a_2	empirical constants
A	real area of contact (m^2)
A	constant (Pa m^3)
A	stiffness (N/m)
b	empirical constant related to the ease of powder compression
C	volume reduction on compression
D	relative density
D_0	relative apparent density
E	Young's modulus (GPa)
E	porosity of the compact
E_0	initial bed porosity
F	frictional force (N)
F_c	force of resistance (N)
F_f	frictional force (N)
F_v	interfacial viscous force (N)
G_{IC}	interlaminar mode I fracture toughness (kJ/m^2)
h	asperity height (m)
k	power law constant
k_1, k_2	empirical constants
K	empirical constant
n	power law index
n	dimensionless constant
N	relative velocity of the contact (m/s)
P	mean contact pressure (MPa)
P	normal pressure (Pa)
r_f	fibre radius (μm)
r_S	sphere of radius (m)
U	velocity of the material (m/s)
v_f	fibre volume packing
V	assembly volume for pressures p (m^3)
V_0	assembly volume for zero pressures (m^3)
V_0	fibre volume for zero pressure
V_a	available fibre volume
V_f	original fibre volume
W	the normal load (N)
α	pressure coefficient of the shear strength
α	constant
β	angle of taper in the die (°)
δ	thickness of the boundary layer (m)
δ	equilibrium film thickness (m)
η	fluids viscosity (Pa s)
η_0	viscosity for zero pressure (Pa s)
η_0	zero shear rate viscosity (Pa s)
θ	angle of asperity (°)

λ	time constant (s)
μ	coefficient of friction
μ_i	coefficient of internal friction
μ_w	coefficient of wall friction
ν	Poisson's ratio
τ_0	interfacial shear strength at zero contact pressure (MPa)
τ_i	interfacial shear strength (MPa)

Subscripts

1	material phase 1
2	material phase 2

References

[1] J.R.A. Pearson, Mechanics of Polymer Processing (Elsevier, New York, 1985).

[2] R.T. Fenner, Principles of Polymer Processing (Macmillan, London, 1979).

[3] Z. Tadmor and C.G. Gogos, Principles of Polymer Processing (Wiley, New York, 1979).

[4] H.E.H. Meijer and C.P.J.M. Verbraak, Polym. Sci. Eng. 28 (1988) 758–771.

[5] M.J. Adams and B.J. Briscoe, eds, Tribology in Particulate Technology (Adam Hilger, Bristol, 1987). G.A. Carnaby, in: Mechanics of Flexible Fibre Assemblies, NATO ASIS Series E No. 38, eds J.W.S. Hearle, J.J. Thwaites and J. Amirbayat (Sijthoff and Noordhoff, Netherlands, 1979) pp. 99–112.

[6] F.P. Bowden and D. Tabor, The Friction and Lubrication of Solids, Part I (Clarendon Press, Oxford, 1950).

[7] B.J. Briscoe, Interfacial friction of polymer composites: General fundamental principles, in: Friction and Wear of Composite Materials, ed. K. Friedrich (Elsevier, Amsterdam, 1986).

[8] B.J. Briscoe and D. Tabor, General outlines of the mechanisms of polymer friction and wear, in: Polymer Surfaces, eds D. Clark and W.J. Feast (Wiley, New York, 1978).

[9] B.J. Briscoe, B. Scruton and F.R. Willis, Proc. R. Soc. A 333 (1973) 99.

[10] J.K.A. Amuzu, B.J. Briscoe and D. Tabor, Trans. ASLE 20 (1977) 152.

[11] J.K.A. Amuzu, B.J. Briscoe and D. Tabor, Trans. ASLE 20 (1977) 354.

[12] C.M. Pooley and D. Tabor, Proc. R. Soc. A 329 (1972) 251.

[13] B.J. Briscoe and A.C. Smith, J. Appl. Polym. Sci. 28 (1983) 3827.

[14] S.P. Timoshenko and J.N. Goodier, Theory of Elasticity (McGraw-Hill, New York, 1970).

[15] I.C. Roselman and D. Tabor, The friction of carbon fibres, J. Phys. D 9 (1976) 2517.

[16] M.J. Adams, B.J. Briscoe and S.L. Kremnitzer, in: Physicochemical Aspects of Polymer Surface, ed. K. Mittal (Plenum, New York, 1983) pp. 425–450.

[17] C.M. Pooley and D. Tabor, Proc. R. Soc. A 329 (1972) 251.

[18] B.J. Briscoe, P.D. Evans and J.K. Lancaster, in: Proc. 12th Leeds–Lyon Tribology Symp., Leeds (1985) Paper II (ii), pp. 39–43.

[19] B.J. Briscoe and D. Tabor, in: Interfacial Phenomena in Apolar Media, eds H. Friedrich-Eickhe and G.D. Parfitt (Marcel Dekker, New York, 1987).

[20] M.J. Adams, B.J. Briscoe and L. Pope, in: Tribology in Particulate Technology, eds M.J. Adams and B.J. Briscoe (Adam Hilger, Bristol, 1987).

[21] U. Tuzun, in: Tribology in Particulate Technology, eds M.J. Adams and B.J. Briscoe (Adam Hilger, Bristol, 1987).

[22] J.L. den Otter, Rheol. Acta 14 (1975) 329.

[23] F. Buschhaus, Internal Report (IKV, Aachen, Germany, 1983).

[24] Z. Tadmor, in: Technical Papers, 29th Ann. Technical Conf., Washington, SPE (1971) p. 136.

[25] Z. Tadmor and E. Broyer, Polym. Eng. Sci. 12 (1972) 378.

[26] Z. Janssen, Ver. Dtsch. Ing. 39 (1985) 1045–1049.

[27] D.M. Walker, Chem. Eng. Sci. 21 (1966) 975–997.
[28] R.S. Spencer, G.D. Gilmore and R.M. Wiley, J. Appl. Phys. 21 (1950) 527–531.
[29] B.J. Briscoe, M.S.D. Fernando and A.C. Smith, in: Tribology in Particulate Technology, eds M.J. Adams and B.J. Briscoe (Adam Hilger, Bristol, 1987).
[30] D.P. Isherwood, in: Tribology in Particulate Technology, eds M.J. Adams and B.J. Briscoe (Adam Hilger, Bristol, 1987).
[31] D.P. Isherwood, Powder Technol. 48 (1987) 253–262.
[32] R.W. Heckel, Trans. Metal. Soc. AIME 221 (1961) 671.
[33] R.W. Heckel, Trans. Metal. Soc. AIME 221 (1961) 1001.
[34] A.R. Cooper and L.E. Eaton, J. Am. Ceram. Soc. 45 (1962) 97.
[35] K. Kawakita and Y. Tsutsumi, Jpn. J. Appl. Phys. 4 (1965) 56.
[36] K. Kawakita and K.H. Ludde, Powder Technol. 4 (1970) 61.
[37] J.W.S. Hearle, P. Grosberg and S. Backer, Structural Mechanics of Fibres, Yams and Fabrics, Vol. 1 (Wiley Interscience, New York, 1969).
[38] J.W.S. Hearle, J.J. Thwaites and J. Amirbayat, eds, Mechanics of Flexible Fibre Assemblies, NATO ASIS Series E No. 38 (Sijthoff and Noordhoff, Netherlands, 1979).
[39] P. Grosberg and N.M. Swani, Text. Res. J. 36(4) (1966) 332.
[40] P. Grosberg, in: Surface Characteristics of Fibers and Textiles, Part II, ed. M.J. Schick (Marcel Dekker, New York, 1977).
[41] F.T. Pierce, J. Text. Inst. 28 (1937) T45.
[42] G.A.V. Leaf, in: Mechanics of Flexible Fibre Assemblies, NATO ASIS Series E No. 38, eds J.W.S. Hearle, J.J. Thwaites and J. Amirbayat (Sijthoff and Noordhoff, Netherlands, 1979).
[43] P. Grosberg, J. Text Inst. 54 (1963) T223–T233.
[44] C.M. Abbott, P. Grosberg and G.A.V. Leaf, Text. Res. J. 41 (1971) 345.
[45] V.L. Gibson and R. Postle, Text. Res. J. 48 (1978) 14.
[46] F. Motamedi, Interfacial Friction in Fabric Mechanics, Ph.D. Dissertation (Imperial College, London, 1989).
[47] B.J. Briscoe, F. Motamedi and D.R. Williams, A static model of the transverse ballistic capture efficiency of modified aramid fabrics, in: Proc. Conf. on Technology and Tomorrow Fibres (Textile Institute, Manchester, UK, 1988).
[48] A.I. Bailey, B.J. Briscoe, F. Motamedi and D. Tabor, Text. Res. J. 59 (1989) 160–172.
[49] S.A.R.D. Sebastian, A.I. Bailey, B.J. Briscoe and D. Tabor, Text. Res. J. 56 (1986) 604–611.
[50] C.M. van Wyk, J. Text. Inst. 37 (1946) T285.
[52] J.I. Dunlop, J. Text. Inst. 65 (1974) 532–536.
[53] T.G. Gutowski, J. Kingery and D. Boucher, SPE Ann. Tech. Conf. Proc.. (1986).
[54] T.G. Gutowski, SAMPE Quart. 16 (1985).
[55] G.L. Batch and C.W. Macosko, American Institute of Chemical Engineering, Fall Meeting (Minneapolis, 1988).
[56] G.L. Batch and S. Cumiskey, SPI 45th Annual Conference 9A (1990).
[57] R.J. Silva-Nieto, B.C. Fisher and A.W. Birley, Polym. Eng. Sci. 21 (1981) 499.
[58] C.L. Tucker, in: Injection and Compression Moulding Fundamentals, ed. A.I. Isayv (Marcel Dekker, New York, 1987).
[59] M. Zentner, M.Sc. Thesis (Department of Mechanical and Industrial Engineering, University of Illinios at Urbana-Champaign, 1984).
[60] J.M. Castro and G. Tomlinson, Polym. Eng. Sci. 30 (1990) 1568.
[61] R.C. Lam and J.L. Kardos, Polym. Eng. Sci. 31 (1991) 1064.
[62] B.J. Briscoe and D.R. Williams, J. Adhesion Sci. & Technol. 5 (1991) 23–38.
[63] P.J.C. Chappell and D.R. Williams, in: Proc. 6th Int. Conf. on Composite Materials, Vol. 3 (Elsevier, Amsterdam, 1987) pp. 346–355.
[64] C.D. Han, D.S. Lee and H.B. Chin, Polym. Sci. Eng. 26 (1986) 393–404.
[65] S.M. Walsh and M. Charmchi, in: Proc. 1988 Heat Transfer Conf. (ASME, 1988) pp. 23–28.
[66] Y.C. Lo, D.W. Lamb and A.G. Gibson, Flow and Cure of Polymers – Measurement and Control (Rapra Technologies, 1990).
[67] G.L. Batch, Crosslinking Free Radical Kinetics and the Pultrusion Processing of Composites, Ph.D. Thesis (University of Minnesota, Minneapolis, 1989).

[68] M.A. Bibbi and T.G. Gutowski, SPE Technical papers, ANTEC'86 (1986) pp. 1430–1432.

[69] J.A. Larock, H.T. Hahn and D.J. Evans, J. Thermoplast. Compos. Mater. 2 (1989) 216–228.

[70] M.L. Wilson, J.D. Buckley, G.E. Dickerson, G.S. Johnson, E.C. Taylor and E.W. Covington, J. Thermoplast. Compos. Mater. 2 (1989) 230–246.

[71] W.I. Lee, G.S. Springer and F.N. Smith, J. Compos. Mater. 25 (1991) 1632–1652.

[72] B.T. Anstrom and R.B. Pipes, SPI Composites Annual Conference 46, Vol. 4A (1991).

[73] B.T. Anstrom and R.B. Pipes, SPI Composites Annual Conference 47, Vol. 13A (1992).

[74] P.J. Carreau, Trans. Soc. Rheol. 16 (1972) 99.

[75] T. Kanzaki, US Patent 4 861 621 (1989).

[76] R.V. Price, US Patent 3 993 726 (1976).

[77] R. Ganza, French Patent FR 2 548 084-A1 (1983).

[78] J.D. Muzzy, in: Proc. ASME Symp. on Manufacturing Science of Composites (ASM International, Metals Park, OH, 1988).

[79] L.E. Allen, D.D. Edie, G.C. Lickfield and J.R. McCollum, J. Thermoplast. Compos. Mater. 1 (1988) 371–379.

[80] S.R. Iyer, M. Ball and L.T. Drzal, SPI Composites Annual Conference 46, Vol. 10A (1991).

[81] J. Reilly and I. Kamel, SPE ANTEC Conference (1988) pp. 555–558.

[82] R.J. Crawford and D.W. Paul, J. Mater. Sci. 17 (1982) 2267.

[83] G.W. Halldin and I.L. Karnel, Polym. Eng. Sci. 17 (1977) 21.

[84] R.W. Truss, K.S. Han, J.F. Wallace and P.H. Geil, Polym. Eng. Sci. 20 (1980) 747.

[85] G.S. Jayaraman, J.F. Wallace, P.H. Geil and E. Baer, Polym. Eng. Sci. 16 (1976) 529.

[86] K. Umeya and R. Hara, Polym. Eng. Sci. 18 (1978) 366.

[87] K. Umeya and R. Hara, Polym. Eng. Sci. 20 (1980) 778.

[88] D.M. Bigg, Polym. Eng. Sci. 17 (1977) 691.

[89] A.C. Loos and G.S. Springer, J. Compos. Mater. 17 (1983) 135–169.

[90] G.S. Springer, J. Compos. Mater. 16 (1982) 400–410.

[91] J.C. Halpin, J.L. Kardos and M.P. Dudukovic, Pure and Applied Chem. 55 (1983) 893–906.

[92] J.T. Lindt, SAMPE Quart. 13 (1982).

[93] C.J. Bartlett, J. Elastomers & Plast. 10 (1978) 369.

[94] J.M. Tang, W.I. Lee and G.S. Springer, J. Compos. Mater. 21 (1987) 421–440.

[95] T.G. Gutowski, SAMPE Quart. 16 (1985).

[96] T.G. Gutowski, T. Morigaki and Z. Cai, J. Compos. Mater. 21 (1987) 172–188.

[97] K.J. Yoon, T.W. Kim, W.I. Lee and E.J. Jun, in: Proc. 6th Int. Conf. on Composite Materials, Vol. 1 (Elsevier, Amsterdam, 1987) pp. 81–86.

[98] G.L. Batch and S. Cumiskey, SPI 45th Annual Conference, Vol. 9A (1990).

[99] B.J. Briscoe and D.R. Williams, Compos. Sci. & Technol. 46 (1993) 277–286.

[100] D.L. Hunston, R.J. Moulton, N.J. Johnston and W.D. Bascom, in: Toughened Composites, ASTM STP 937, ed. N.J. Johnston (ASTM, Philadelphia, PA, 1987) pp. 74–94.

[101] B.J. Briscoe, R.S. Court and D.R. Williams, Compos. Sci. & Technol. 47 (1993) 261–270.

AUTHOR INDEX

Roman numbers refer to pages on which the author (or his work) is mentioned. Italic numbers refer to reference pages. Numbers between brackets are the reference numbers. No distinction is made between first author and co-author(s).

SUBJECT INDEX